FOUNDATIONS OF SPACE DYNAMICS

Aerospace Series – Recently Published Titles

Sustainable Aviation Technology and Operations: Research and Innovation Perspectives	Sabatini	Dec 2020
Essentials of Supersonic Commercial Aircraft Conceptual Design	Torenbeek	July 2020
Design and Development of Aircraft Systems, 3rd Edition	Seabridge	April 2020
Design of Unmanned Aerial Systems	Sadraey	April 2020
Flight Dynamics and Control of Aero and Space Vehicles	Yedavalli	February 2020
Future Propulsion Systems and Energy Sources in Sustainable Aviation	Farokhi	January 2020
Conceptual Aircraft Design: An Industrial Approach	Kundu	April 2019
Helicopter Flight Dynamics: Including a Treatment of Tiltrotor Aircraft, 3rd Edition	Padfield	November 2018
Space Flight Dynamics, 2nd Edition	Kluever	May 2018
Performance of the Jet Transport Airplane: Analysis Methods, Flight Operations, and Regulations	Young	October 2017
Differential Game Theory with Applications to Missiles and Autonomous Systems Guidance	Faruqi	May 2017
Advanced UAV Aerodynamics, Flight Stability and Control: Novel Concepts, Theory and Applications	Marques and Da Ronch	April 2017
Introduction to Nonlinear Aeroelasticity	Dimitriadis	April 2017
Introduction to Aerospace Engineering with a Flight Test Perspective	Corda	March 2017
Aircraft Control Allocation	Durham, Bordignon and Beck	January 2017
Remotely Piloted Aircraft Systems: A Human Systems Integration Perspective	Cooke, Rowe, Bennett. Jr. and Joralmon	October 2016
Adaptive Aeroservoelastic Control	Tewari	March 2016
Theory and Practice of Aircraft Performance	Kundu, Price and Riordan	November 2015
The Global Airline Industry, Second Edition	Belobaba, Odoni and Barnhart	July 2015
Modeling the Effect of Damage in Composite Structures: Simplified Approaches	Kassapoglou	March 2015
Introduction to Aircraft Aeroelasticity and Loads, 2nd Edition	Wright and Cooper	December 2014
Aircraft Aerodynamic Design: Geometry and Optimization	Sóbester and Forrester	October 2014
Theoretical and Computational Aerodynamics	Sengupta	September 2014
Aerospace Propulsion	Lee	October 2013
Aircraft Flight Dynamics and Control	Durham	August 2013
Civil Avionics Systems, 2nd Edition	Moir, Seabridge and Jukes	August 2013
Modelling and Managing Airport Performance	Zografos, Andreatta and Odoni	July 2013
Advanced Aircraft Design: Conceptual Design, Analysis and Optimization of Subsonic Civil Airplanes	Torenbeek	June 2013
Design and Analysis of Composite Structures: With Applications to Aerospace Structures, 2nd Edition	Kassapoglou	April 2013
Aircraft Systems Integration of Air-Launched Weapons	Rigby	April 2013

See more Aerospace Series titles at www.wiley.com

FOUNDATIONS OF SPACE DYNAMICS

First Edition

Ashish Tewari

Indian Institute of Technology Kanpur

WILEY

This edition first published 2021

Registered Offices
John Wiley & Sons, Inc., 111 River Street, Hoboken, NJ 07030, USA
John Wiley & Sons Ltd, The Atrium, Southern Gate, Chichester,West Sussex, PO19 8SQ, UK

Editorial Office
The Atrium, Southern Gate, Chichester,West Sussex, PO19 8SQ, UK

For details of our global editorial offices, customer services, and more information about Wiley products visit us at www.wiley.com.

Wiley also publishes its books in a variety of electronic formats and by print-on-demand. Some content that appears in standard print versions of this book may not be available in other formats.

Library of Congress Cataloging-in-Publication Data

Names: Tewari, Ashish, author.
Title: Foundations of space dynamics / Ashish Tewari.
Description: First edition. | Hoboken, NJ : Wiley, [2020] | Series: Aerospace series | Includes bibliographical references and index.
Identifiers: LCCN 2020033397 (print) | LCCN 2020033398 (ebook) | ISBN 9781119455349 (paperback) | ISBN 9781119455332 (adobe pdf) | ISBN 9781119455325 (epub) | ISBN 9781119455301 (obook)
Subjects: LCSH: Aerospace engineering. | Astrodynamics. | Orbital mechanics.
Classification: LCC TL545 .T385 2020 (print) | LCC TL545 (ebook) | DDC 629.4/11–dc23
LC record available at https://lccn.loc.gov/2020033397
LC ebook record available at https://lccn.loc.gov/2020033398

Cover Design: Wiley
Cover Image: © Philip Wallick/Getty Images

Typeset in 10/12pt TimesLTStd by SPi Global, Chennai, India

SKYDC21106D-2BA2-4116-B9A9-50D6392B7125_112420

To the loving memory of my daughter, Manya (24.1.2000 - 9.7.2019)

Contents

Preface

Foundations of Space Dynamics is written as a textbook for students, as well as a ready reference covering the essential concepts for practicing engineers and researchers. It introduces a reader to the basic aspects of both orbital mechanics and attitude dynamics. While many good textbooks are available on orbital mechanics and attitude dynamics, there is a need for a direct, concise, yet rigorous treatment of both the topics in a single textbook. Important derivations from basic principles are highlighted, while offering insights into the physical principles which can often be hidden by mathematical details. While the emphasis is on analytical derivations, the essential computational tools are presented wherever required, such as the iterative root-finding methods and the numerical integration of ordinary differential equations.

The objective of this book is to provide a physically insightful presentation of space dynamics. The usage of simple ideas and numerical tools to illustrate advanced concepts is inspired by the work of the original masters (Newton, Liebnitz, Laplace, Gauss, etc.), and is combined with the application and terminology of modern space dynamics.

A student of space dynamics in the past generally possessed a strong background in analytical mechanics, often reinforced by such classical treatises as those by Whittaker, Lanczos, Truesdell, and Mach. Today, the exposure to analytical dynamics is often based upon a single undergraduate course. This book therefore includes a basic introduction to analytical mechanics by both Newtonian and Lagrangian approaches.

The contents of the textbook are arranged such that they may be covered in two successive courses: *Space Dynamics I* could focus on Chaps. 1–7 and 11, while the following course, *Space Dynamics II*, could cover Chaps. 8–10 and 12, supplemented by a semester project exploring a specific research topic. However, the arrangement of the chapters in the book offers sufficient flexibility for them to be covered in a single comprehensive course, if so required. There are a multitude of exercises at the end of the chapters which can serve as homework assignments and quiz problems. Solutions to selected exercises is also provided.

I would like to thank the editorial and production staff of Wiley, Chichester, for their constructive suggestions and valuable insights during the preparation of the manuscript.

Ashish Tewari
May 2020

1

Introduction

This chapter gives an introduction to the basic features of space flight, which is predominated by the quiet space environment and gravity. The essential differences with atmospheric flight are discussed, and the important time scales and frames of reference for space flight are described. Topics in space dynamics are classified as the translational motion (orbital mechanics) and rotational motion (attitude dynamics) of a rigid spacecraft. Classification of the various practical spacecraft is given according to their missions.

1.1 Space Flight

Space flight refers to motion outside the confines of a planetary atmosphere. It is different from atmospheric flight in that no assistance can be derived from the atmospheric forces to support a vehicle, and no benefit of planetary oxygen can be utilized for propulsion. Apart from these major disadvantages, space flight has the advantage of experiencing no (or little) drag due to the resistance of the atmosphere; hence a spacecraft can achieve a much higher flight velocity than an aircraft. Since atmospheric lift is absent to sustain space flight, a spacecraft requires such high velocities to balance the force of gravity by a centrifugal force in order to remain in flight. The trajectories of spacecraft (called *orbits*) – being governed solely by gravity – are thus much better defined than those of aircraft. Since gravity is a conservative force, space flight involves a conservation of the sum of kinetic and potential energies, as well as that of the angular momentum about a fixed point. Therefore, space flight is much easier to analyze mathematically when compared to atmospheric flight.

1.1.1 Atmosphere as Perturbing Environment

When can the effects of the atmosphere be considered negligible so that space flight can come into existence? The atmosphere of a planetary body – being bound by gravity – becomes less dense as the distance from the planetary surface (called *altitude*) increases, owing to the inverse-square diminishing of the acceleration due to gravity from the planetary centre. For an atmosphere completely at rest, this relationship between the atmospheric density,

Foundations of Space Dynamics, First Edition. Ashish Tewari.

ρ, and the altitude, z, can be derived from the following differential equation of *aerostatic equilibrium* (Tewari, 2006):

$$dp = -\rho g dz \tag{1.1}$$

where p refers to the atmospheric pressure, and g the *acceleration due to gravity* prevailing at a given altitude. For a spherical body of radius r_0, the gravity obeys the inverse-square law discovered by Newton, given by

$$g = g_0 \frac{r_0^2}{(r_0 + z)^2} \tag{1.2}$$

where g_0 is the acceleration due to gravity at the surface of the body (i.e., at $z = 0$). When Eq. (1.2) is substituted into Eq. (1.1), and the thermodynamic properties of the atmospheric gases are taken into account, the differential equation, Eq. (1.1), can be integrated to yield an algebraic relationship between the *atmospheric density*, ρ, and the altitude, z, called an atmospheric model. For Earth's atmosphere, one such model is the *U.S. Standard Atmosphere 1976* (Tewari, 2006), whose predicted density variation with the altitude in the range $0 \leq z \leq 250$ km is listed in Table 1.1. It is evident from Table 1.1 that the atmospheric density, ρ, can be considered to be negligible for a flight for $z \geq 120$ km around Earth. A similar (albeit smaller) value of ρ is obtained on Mars at $z = 120$ km. Hence, for both Earth and Mars, $z = 120$ km can be taken to be the boundary above which the *space* begins.

The flight of a spacecraft around a large spherical body of radius r_0 is assumed to take place outside the atmosphere, (such as $z > 120$ km for Earth and Mars), and is governed by the gravity of the body, with acceleration given by Eq. (1.2). Space-flight trajectories are well defined *orbits*

Table 1.1 Variation of density with altitude in Earth's atmosphere

Altitude, z (km)	Density, ρ kg/m^3
0	1.2252
1	1.1119
5	0.7366
10	0.4136
20	0.0891
30	0.0185
40	0.0041
50	0.0011
60	3.24×10^{-4}
70	8.65×10^{-5}
80	2.04×10^{-5}
90	3.90×10^{-6}
100	6.94×10^{-7}
110	1.37×10^{-7}
120	3.40×10^{-8}
150	2.57×10^{-9}
200	4.66×10^{-10}
250	1.41×10^{-10}

due to the simple nature of Eq. (1.2). However, since the atmospheric density in a very low orbit (e.g., $120 < z < 250$ km on Earth), albeit quite small, is not exactly zero, the flight of a spacecraft can be gradually affected, to cause significant deviations over a long period of time from the orbits predicted by Eq. (1.2). This is due to the fact that the atmospheric forces and moments are directly proportional to the flight dynamic pressure, $1/2\rho v^2$, where v is the flight speed. The high orbital speed, v, required for space flight makes the dynamic pressure appreciable, even though the density, ρ, is by itself negligible. The atmospheric *drag* (the force resisting the motion) causes a slow but steady decline in the flight speed, until the latter falls below the magnitude where an orbital motion can be sustained. Thus atmospheric drag can cause a low-orbiting satellite to slightly decay in altitude after every orbit, and to ultimately enter the lower (dense) portions of the atmosphere, where the mechanical stress created by the ever increasing dynamic pressure, as well as the heat generated by atmospheric friction, lead to its destruction. Therefore, for predicting the life of a satellite in a low orbit, the atmospheric effects must be properly taken into account. Figure 1.1 shows an example of the decay in the orbit of a spacecraft initially placed into a circular orbit of $z = 200$ km around Earth. In this simulation obtained by a *Runge-Kutta method* (Appendix A), the spacecraft is assumed to be a sphere of 1 m diameter, with a constant free-molecular drag coefficient of 2.0 (Tewari, 2006). As seen in the figure, the altitude decays quite rapidly as the number of orbits, N, increases. The initial average rate of altitude loss seen in Fig. 1.1 – 1 km per 4 orbits – is likely to increase as the spacecraft descends lower, thereby encountering a higher density. When the spacecraft is placed in a circular orbit of $z = 180$ km, its altitude decays very rapidly, and it re-enters the

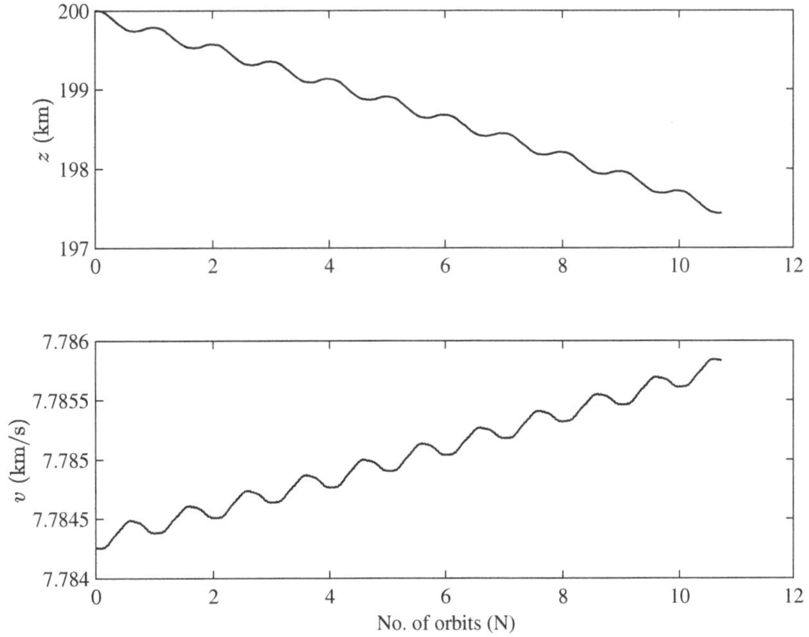

Figure 1.1 Decay in the orbit due to atmospheric drag for a spacecraft initially placed in a circular orbit of $z = 200$ km around Earth.

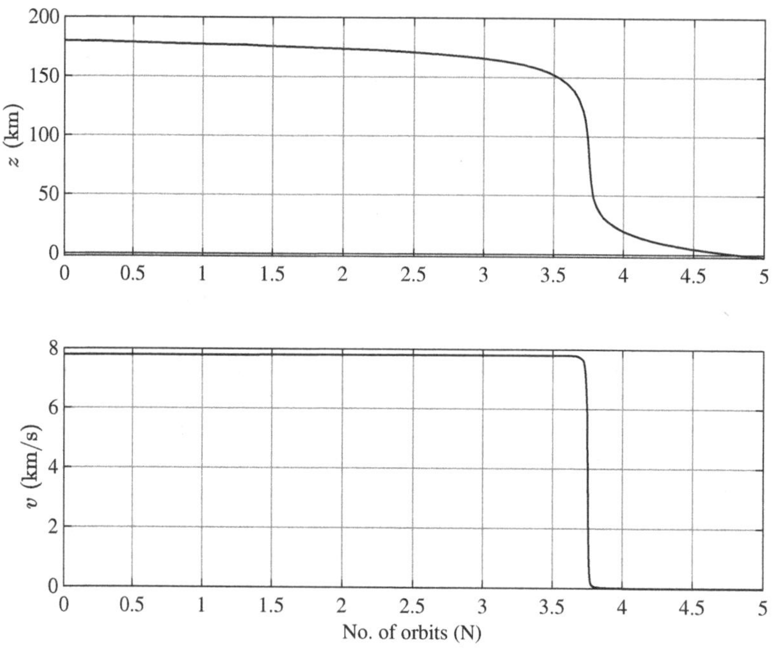

Figure 1.2 Decay in the orbit due to atmospheric drag for a spacecraft initially placed in a circular orbit of $z = 180$ km around Earth.

atmosphere after only 3.5 orbits (Fig. 1.2). Hence, the life of the spacecraft is only about 3.5 revolutions in a circular orbit of altitude 180 km above Earth. As Figs. 1.1 and 1.2 indicate, a stable orbit around Earth for this spacecraft should have $z > 200$ km at all times.

Apart from the atmospheric effects, there are other environmental perturbations to a spacecraft's flight around a central body, which is assumed to be spherical as required by Eq. (1.2). These are the gravity of the actual (non-spherical) shape of the central body, as well as the gravity of other remote large bodies, and the solar radiation pressure. However, such effects are typically small enough to be considered small perturbations when compared to the spherical gravity field of the central body given by Eq. (1.2). Such effects can be regarded as small perturbations applied to the orbit governed by Eq. (1.2), and should be carefully modelled in order to predict the actual motion of the spacecraft.

1.1.2 Gravity as the Governing Force

Space flight is primarily governed by gravity. "Governing" implies dictating the path a given body describes in a three-dimensional space. Aircraft and rocket flights are *not* primarily governed by gravity, because there are other forces acting on the body, such as the lift and the thrust, which are of comparable magnitudes to that of gravity and therefore determine the flight path. Discovered and properly analyzed for the first time by Newton in the late 17th century, gravity can be expressed simply, but has profound consequences. For example, by applying Newton's law of gravitation, it could have been inferred that the universe cannot be static, because gravity would cause all the objects to collapse towards a single point. However, this simple fact escaped the notice of all physicists ranging from Newton himself to Einstein, until it was observed by Hubble in 1924 that the universe is expanding at a rate which increases with the distance between any two objects. A reader may be cautioned against the complacency which often arises by treating the motion governed by gravity as simple (even trivial) to understand. There are many surprising and interesting consequences of gravity being the governing force in flight, such as Kepler's third law of planetary motion, which implies that the time period of an orbiting body depends only upon the mean radius, and is independent of the shape of the orbit. A larger part of a course on space dynamics involves understanding gravity and its effects on the motion of a body in space.

1.1.3 Topics in Space Dynamics

Space dynamics consists of two parts: (a) *orbital mechanics*, which describes the translation in space of the centre of mass of a rigid body primarily under the influence of gravity, and (b) *attitude dynamics*, which is the description of the rotation of the rigid body about its own centre of mass. While these two topics are largely studied separately, in some cases orbital mechanics and attitude dynamics are intrinsically coupled, such as when the rigid body experiences an appreciable gravity-gradient torque during its orbit. Furthermore, when designing an attitude control system for a spacecraft, it is necessary to account for its orbital motion. Therefore, while elements of orbital mechanics and attitude dynamics can be grasped separately, their practical application involves a combined approach.

1.2 Reference Frames and Time Scales

Space flight requires a definite background of objects to measure distances, as well as to orient the spacecraft in specific directions. Since fixed objects are hard to come by in practice, navigation and attitude determination are non-trivial problems in space flight. Such a problem does not exist for the motion taking place on, or very close to, a solid surface, where ground-fixed objects can serve as useful references for both navigation and orientation of the vehicles.

1.2.1 Sidereal Frame

Three mutually perpendicular straight lines joining distant objects constitute a *reference frame*. Generally, distant objects in the universe are moving with respect to one another; hence the straight lines joining them would rotate, as well as either stretch out or contract with time. Suppose one can find two objects which are fixed relative to each other. Then a straight line joining them would be fixed in length, and a vector pointing from one object to the other would always have a constant direction. A reference frame consisting of axes which have fixed directions is said to be a *sidereal frame*. There are certain directions which can be used to orient a sidereal frame. For example, the orbital plane of Earth around the sun, called the *ecliptic*, intersects Earth's equatorial plane along a straight line called the *line of nodes*. The *nodes* are the two specific points where this line intersects Earth's orbit, as shown in Fig. 1.3. One of the two

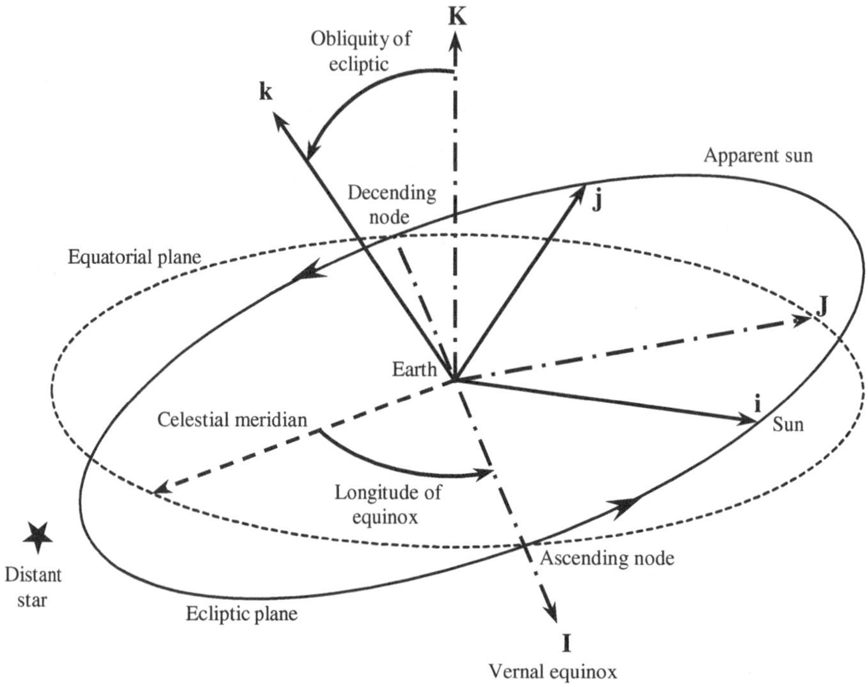

Figure 1.3 The equinoctial sidereal frame (**I**, **J**, **K**), the ecliptic synodic frame (**i**, **j**, **k**), and Earth centred celestial meridian.

nodes is an *ascending node*, where the apparent motion of the sun as seen from Earth (called the *apparent Sun*) occurs from the south to the north of the equator. This happens at the *vernal equinox*, occurring every year around March 21. The *descending node* of the apparent sun is at the *autumnal equinox*, which takes place around September 22. Since the vernal equinox points in a specific direction from the centre of Earth, it can be used to orient one of the axes of the sidereal frame, as the axis **I** in Fig. 1.3. Another axis of the sidereal frame can be taken to be normal to either the ecliptic or the equatorial plane (axis **K** in Fig. 1.3), and the third axis can be chosen to be perpendicular to the first two (axis **J** in Fig. 1.3).

The rate of rotation of Earth on its own axis (normal to the equatorial plane) is from the west to the east, and can be measured in a sidereal reference frame oriented with the vernal equinox direction. This rate is called the *sidereal rotation rate*, and would be the true rotation rate of Earth if the vernal equinox were a constant direction. A *sidereal day* is the period of rotation of Earth measured from the vernal equinox. If the sun is used for timing the rotational rate of Earth, the period from noon to noon is a *mean solar day* (m.s.d.) of 24-hour duration. However, the mean solar day is not the true rotational rate of Earth because of Earth's orbit around the sun, which also takes place from the west to the east. To calculate the sidereal day from the mean solar day, a correction must be applied by adding the average rate at which Earth orbits the sun. The *tropical year* is the period of Earth's orbit around the sun measured from one vernal equinox to the next, and equals 365.242 mean solar days. This implies that the mean apparent sun is slightly less than one degree per day ($360^\circ/365.242$). Such a correction gives the sidereal day as the following:

$$\frac{1}{\frac{1}{24\times3600}+\frac{1}{365.242\times24\times3600}} = 86164.0904\ \text{s}\,, \tag{1.3}$$

or 23 hr., 56 min., 4.0904 s.

Unfortunately, the vernal equinox is not a constant direction because of the slow *precession* of Earth's axis (thus the equatorial plane) caused by the gravitational influence of the sun and the moon (called the *luni-solar attraction*). When a spinning rigid body, such as Earth, is acted upon by an external torque, such as due to the gravity of the sun and the moon, its spin axis undergoes a complex rotation called "precession" and "nutation", which will be explained in detail in Chapter 11. This rotation of the equatorial plane causes the two equinoxes to shift towards the west, and is thus called the *precession of the equinoxes*. The period of the precession is about 25772 yr., which implies that the sidereal day differs only slightly from the true rotational period of Earth. It also means that an equinoctial sidereal reference frame, such as the frame $(\mathbf{I},\mathbf{J},\mathbf{K})$ in Fig. 1.3, rotates very slowly against a background of distant stars. Hence the vernal equinox (and the equinoctial sidereal reference frame) can be approximated to be the fixed references for most space flight applications. However, for a long flight time of several years' duration, the calculations must be brought to a common reference at a specific time (called an *epoch*[1]) by applying the necessary corrections, which take into account the slow movement of the vernal equinox towards the west. The equinox is given for various epochs by the *International Earth Rotation and Reference Systems Service* (IERS) in terms of the longitude of the equinox measured from a *celestial meridian* (see Fig. 1.3). The inclination of Earth's spin axis from the normal to the ecliptic is called the *obliquity of the ecliptic* (Fig. 1.3), and also varies with time due to the *nutation* caused by the luni-solar attraction. (The precession and nutation, discussed in detail in Chapter 11, cause Earth's spin axis to rotate with time due to the luni-solar attraction.) The value of the obliquity of the ecliptic in the current epoch is measured by IERS to be about 23°26'21". The period of nutation of Earth's spin axis is about 41000 yr., which is considerably longer than the period of its precession. The precession and nutation are explained in Chapter 11 when considering the rotation of a rigid body (such as Earth).

Apart from the precession and the nutation of Earth's spin axis, there is also a precession of the ecliptic caused by the gravitational attraction of the other planets. This is a much smaller variation in the equinoxes (about 100 times smaller than that caused by luni-solar attraction).

Since the vernal equinox moves slightly westward every year, the tropical year is not the true period of revolution of Earth in its orbit around the sun. The true period of revolution is the *sidereal year*, which is measured by timing the passage of Earth against the background of distant stars, and equals 365.25636 mean solar days. Thus a tropical year is shorter than the actual year by 20 hr., 40 min., and 42.24 s.

[1] An epoch is a moment in time used as a reference point for a time-varying astronomical quantity, such as the orbital elements specifying the shape and the plane of an orbit, the direction of the spin axis of a body, the coordinates of important celestial objects, etc.

1.2.2 Celestial Frame

For a motion taking place inside the solar system, any two stars (except the sun) appear to be fixed for the duration of the motion. Hence, a reference frame constructed out of three mutually perpendicular axes, each of which are pointing towards different distant stars, would appear to be fixed in space, and can serve as a sidereal reference frame. A reference frame fixed relative to distant stars is termed a *celestial reference frame*. For example, the rate of rotation of Earth about its own axis can be measured by an observer standing astride the North Pole by timing the rate at which a straight line joining Earth to a distant star, called a celestial meridian (see Fig. 1.3), appears to rotate. This rate gives the true rotational time period of Earth, called the *stellar day*, which is measured by IERS to be 23 hr., 56 min., 4.0989 s. Hence, the sidereal day is shorter than the stellar day by about 8.5×10^{-3} s.

1.2.3 Synodic Frame

When two objects orbit one another at nearly constant rates on a fixed plane, a reference frame can be defined by two of its axes on the plane of rotation and rotating at the constant rate, and the third axis normal to the plane. Such a rotating reference frame is called a *synodic frame*. An example of a synodic frame is the *ecliptic frame*, which is a reference frame constructed out of the ecliptic plane, such as the frame $(\mathbf{i}, \mathbf{j}, \mathbf{k})$ in Fig. 1.3. The motion of an object measured relative to a synodic frame must be corrected by a vector subtraction of the motion of the frame itself, as exemplified by the calculation of the sidereal day from the observed rotation in the ecliptic frame. The ecliptic frame has been used as a reference since the earliest days of astronomical observations. The division of the circle into 360° arose out of the apparent motion of the sun per day, which subtends an arc of one diameter every 12 hours when seen from Earth. Since the moon's apparent diameter from Earth is roughly the same as that of the sun, the eclipses of the sun and the moon are observed in the ecliptic (thus the name). However, since the moon's orbital plane around Earth is tilted $\pm 5.1°$ relative to the ecliptic, the eclipses happen only along the intersection (i.e., the line of nodes) of the two planes.

The Earth-moon line provides another synodic reference frame for space flight. The Earth and the moon describe coplanar circles about the common centre of mass (called the *barycentre*) every 27.32 mean solar days relative to the vernal equinox (called a *sidereal month*). This rotational period appears in the synodic frame to be 29.53 mean solar days (a *synodic month*) from one new moon to the next, which is obtained from the sidereal month by subtracting the rate of revolution of Earth-moon system around the sun.

1.2.4 Julian Date

Instead of the *calendar year* of 365 mean solar days, the tropical year of 365.242 mean solar days, and the sidereal year of 365.25636 mean solar days, it is much more convenient to use a *Julian year* of 365.25 mean solar days, which avoids the addition of leap years in carrying out astronomical calculations. A *Julian day number* (*JDN*) is defined to be the continuous count of the number of mean solar days elapsed since 12:00 noon *universal time* (UT) on January 1, 4713 BC. Universal time refers to the time taken as 12:00 noon when the sun is directly over the Greenwich meridian (which is defined to be zero longitude). The Julian day number 0 is assigned to the day starting at that time on the Julian proleptic calendar. The *Julian date* of a general time instant is expressed as the JDN plus the fraction of the 24-hour day elapsed since the preceding noon UT. Julian dates are thus expressed as a Julian day number plus a decimal fraction. For example, the Julian date for 10:00 a.m. UT on April 21, 2020, is given by J2458960.91667, and the JDN is 2458960. Epochs are listed in ephemeris charts and nautical almanacs according to their Julian dates. Hence a Julian date serves as a common time measure for astronautical calculations involving two events separated in time.

Computation of the Julian date (JD) from a Gregorian calendar date is complicated due to the three calendar cycles used to produce the Julian calendar, namely the solar, the lunar, and the indiction cycles of 28, 19, and 15 year periods, respectively (Seidelmann, 1992). A product of these gives the *Julian period* of 7980 years. The Julian period begins from 4713 BC, which is chosen to be the first year of solar, lunar, and indiction cycles beginning together. The next epoch when the three cycles begin together will happen at noon UT on January 1, 3268. The following conversion formula for the JDN, truncated to the last integer, uses the numbering of the months from January to December as $M = 1, 2, \dots, 12$; the Gregorian calendar years are numbered such that the year 1 BC is the year zero, $Y = 0$, (i.e., 2 BC is $Y = -1$, 4713 BC is $Y = -4712$, etc.); and the day number, D, is the last completed day of the month up to noon UT:

$$\begin{aligned} JDN &= D + \frac{1461}{4}\left(Y + 4800 + \frac{M-14}{12}\right) \\ &\quad - \frac{3}{400}\left(Y + 4900 + \frac{M-14}{12}\right) - 31708 \,. \end{aligned} \tag{1.4}$$

This formula calculates the JDN for 09:25 a.m. UT on June 25, 1975, by taking $Y = 1975$, $M = 6$, $D = 24$, and yields the last truncated integer value as $JDN = 2442589$. Then the time elapsed from noon UT on June 24 to 09:25 a.m. UT on June 25 is added as a fraction to give the following Julian date:

$$JD = JDN + \frac{12 + 9 + 25/60}{24} = 2442589.892361 \,.$$

An epoch in the Julian date is designated with the prefix J, and the suffix being the closest Gregorian calendar date. For example, $J2000$ refers to 12:00 noon UT on January 1, 2000, and has the Julian date of 2451545. Similarly, the epoch $J1900$, which occurs exactly 100 Julian years *before* 12:00 noon UT on January 1, 2000, must refer to 12 noon UT on January 0, 1900; hence its date in the Gregorian calendar is December 31, 1899, and its Julian date is 2415020. The difference in the epochs $J2000$ and $J1900$ is therefore $2451545 - 2415020 = 36525$ mean solar days (which is exactly 100 Julian years).

Since Julian day numbers with the epoch $-J4712$ can become very large, it is often convenient to use a later epoch for computing JD. Epochs can be chosen with simpler JDN figures, such as 12:00 hr. UT on November 16, 1858, which has $JDN = 2400000$. Then Julian dates can be converted to this epoch by replacing JD with $JD - 2400000$. For example, the Julian date for 09:25 a.m. UT, June 25, 1975, converted to the epoch of Nov. 16, 1858, is $JD = 42589.892361$. For the consistency of data, all modern astronomical calculations are reduced to the epoch, $J2000$, by international agreement. This means that all the Julian dates must be converted to this epoch by replacing JD with $JD - 2451545$.

1.3 Classification of Space Missions

Spacecraft are classified according to their missions. A large majority of spacecraft orbit Earth as artificial satellites for observation, mapping, thermal and radio imaging, navigation, scientific experimentation, and telecommunications purposes. These satellites are classified according to the shapes and sizes of their orbits. A spacecraft orbiting a central body at altitudes smaller than the mean radius, r_0, of the body, $z < r_0$, is termed a *low-orbiting spacecraft*. Examples of such spacecraft for Earth ($r_0 = 6378.14$ km) are the *low-Earth orbit* (LEO) satellites, which orbit the planet in nearly circular orbits of $200 \leq z \leq 2000$ km. Orbital periods of LEO satellites range from 90 to 127 min., and are mainly used for Earth observation, photo reconnaissance, resource mapping, and special sensing and scientific missions. The *International Space Station* is a manned LEO spacecraft with a nearly circular orbit of mean altitude, $z = 400$ km. There are hundreds of active LEO satellites in orbit at any given time, launched by various nations for civil and military applications.

A *medium-Earth orbit* (MEO) satellite has a period of about 12 hours. Examples of such spacecraft are the *Global Positioning System* (GPS) navigational satellites in circular orbits of altitudes about 20,000 km, and *Molniya* telecommunications satellites of Russia in highly eccentric elliptical orbits inclined at 63.435° relative to Earth's equatorial plane.

The highest altitude of Earth satellites is for those in the *geosynchronous equatorial orbit* (GEO), which is a circular orbit in the equatorial plane of a period exactly matching a sidereal day, i.e., 23 hr., 56 min., 4.0904 s. This translates into an altitude of $z = 35786.03$ km. Since the orbital frequency of a GEO satellite equals the rate of rotation of Earth on its axis, such a satellite returns to the same point above the equator after each sidereal day, thereby appearing to be stationary to an observer on the ground. Hence, a GEO satellite is used as a telecommunications relay platform for signals between any two ground stations directly in the line of sight of the satellite. Due to the high altitude of the GEO satellite, a broad coverage of signals is provided to the receiving stations on the ground, and is the basis of modern television broadcasts and mobile telephone communications.

A small number of spacecraft are put into highly specialized lunar, interplanetary, and asteroid/cometary intercept orbits for the exploration of the solar system. Due to the typically large distances involved in their missions, which might include the time spent beyond the line-of-sight of Earth, such spacecraft must be fully autonomous in terms of their basic operations. The spacecraft which are sent to explore the outer planets (such as NASA's *Voyager 1* and *Voyager 2*, *Cassini*, *Galileo*, and *New Horizons*) must also have an onboard electrical power source for charging their batteries, due to the unavailability of effective solar power (the sun is too dim at such large distances).

Exercises

1. Using the following exponential atmosphere model for Earth with the scale height, $H = 6.7$ km, and base density, $\rho_0 = 1.752\ \mathrm{kg/m^3}$, calculate the atmospheric density at the altitude, $z = 150$ km:

$$\rho = \rho_0 e^{-z/H}$$

Compare the result with that given in Table 1.1.

2. Calculate the Julian date for 3:30 p.m. UT on October 15, 2007, referring to the $J2000$ epoch.
3. What is the exact time difference between two events happening at 11:05 a.m. on July 28, 1993, and 8:31 p.m. on November 3, 2005, respectively?

References

Tewari A 2006. *Atmospheric and Space Flight Dynamics*. Birkhäuser, Boston.
Seidelmann KP (ed.) 1992. *Explanatory Supplement to the Astronomical Almanac*. University Science Books, Sausalito, CA.

2

Dynamics

Dynamics is the study of an object in motion, and pertains to a change in the position and orientation of the object as a function of time. This chapter introduces the basic principles of dynamics, which are later applied to the motion of a vehicle in the space.

2.1 Notation and Basics

The vectors and matrices are denoted throughout this book in boldface, whereas scalar quantities are indicated in normal font. The elements of each vector are arranged in a column. The *Euclidean norm* (or *magnitude*) of a three-dimensional vector, $\mathbf{A} = (A_x, A_y, A_z)^T$, is denoted as follows:

$$|\mathbf{A}| = A = \sqrt{A_x^2 + A_y^2 + A_z^2}. \tag{2.1}$$

All the variables representing the motion of a spacecraft are changing with *time*, t. The overdots represent the time derivatives, e.g., $\mathrm{d}x/\mathrm{d}t = \dot{x}$, $\dot{A} = \mathrm{d}A/\mathrm{d}t$, $\ddot{A} = \mathrm{d}^2A/\mathrm{d}t^2$. The time derivative of a vector $\mathbf{A}$, which is changing both in its magnitude and its direction, requires an explanation.

The time derivative of a vector, $\mathbf{A}$, which is changing both in magnitude and direction can be resolved in two mutually perpendicular directions – one along the original direction of $\mathbf{A}$, and the other normal to it on the plane of the rotation of $\mathbf{A}$. The instantaneous *angular velocity*, $\boldsymbol{\omega}$, of $\mathbf{A}$ denotes the vector rate of change in the direction, whereas $\dot{A}$ is the rate of change in its magnitude. By definition, $\boldsymbol{\omega} \times \mathbf{A}$ is normal to the direction of the unit vector, $\mathbf{A}/A$, and lies in the instantaneous plane of rotation normal to $\boldsymbol{\omega}$. The rotation of $\mathbf{A}$ is indicated by the *right-hand rule*, where the thumb points along $\boldsymbol{\omega}$, and the curled fingers show the instantaneous direction

Foundations of Space Dynamics, First Edition. Ashish Tewari.

of rotation,[1] $\boldsymbol{\omega} \times \mathbf{A}$. The time derivative of $\mathbf{A}$ is therefore expressed as follows:

$$\dot{\mathbf{A}} = \frac{\mathrm{d}\mathbf{A}}{\mathrm{d}t} = \dot{A}\frac{\mathbf{A}}{A} + \boldsymbol{\omega} \times \mathbf{A}\,, \tag{2.2}$$

where the term $\mathbf{A}/A$ represents a unit vector in the original direction of $\mathbf{A}$, and $\boldsymbol{\omega} \times \mathbf{A}$ is the change normal to $\mathbf{A}$ caused by its rotation. Equation (2.2) will be referred to as the *chain rule* of vector differentiation in this book.

Similarly, the second time derivative of $\mathbf{A}$ is given by the application of the chain rule to differentiate $\dot{\mathbf{A}}$ as follows:

$$\begin{aligned}\ddot{\mathbf{A}} &= \frac{\mathrm{d}^2\mathbf{A}}{\mathrm{d}t^2} = \frac{\mathrm{d}\dot{\mathbf{A}}}{\mathrm{d}t} = \frac{\mathrm{d}(\dot{A}/A)}{\mathrm{d}t}\mathbf{A} + \left(\frac{\dot{A}}{A}\right)\dot{\mathbf{A}} + \dot{\boldsymbol{\omega}} \times \mathbf{A} + \boldsymbol{\omega} \times \dot{\mathbf{A}} \\ &= \left(\frac{A\ddot{A} - \dot{A}^2}{A^2}\right)\mathbf{A} + \left(\frac{\dot{A}}{A}\right)\left(\dot{A}\frac{\mathbf{A}}{A} + \boldsymbol{\omega} \times \mathbf{A}\right) + \dot{\boldsymbol{\omega}} \times \mathbf{A} + \boldsymbol{\omega} \times \left(\dot{A}\frac{\mathbf{A}}{A} + \boldsymbol{\omega} \times \mathbf{A}\right) \\ &= \ddot{A}\left(\frac{\mathbf{A}}{A}\right) + \boldsymbol{\omega} \times (\boldsymbol{\omega} \times \mathbf{A}) + 2\boldsymbol{\omega} \times \left(\frac{\dot{A}}{A}\right)\mathbf{A} + \dot{\boldsymbol{\omega}} \times \mathbf{A}\,. \end{aligned} \tag{2.3}$$

Applying Eq. (2.1) to the time derivative of the angular velocity, $\boldsymbol{\omega}$, we have the following expression for the *angular acceleration* of $\mathbf{A}$:

$$\dot{\boldsymbol{\omega}} = \dot{\omega}\left(\frac{\boldsymbol{\omega}}{\omega}\right) + \boldsymbol{\Omega} \times \boldsymbol{\omega}\,, \tag{2.4}$$

where $\boldsymbol{\Omega}$ is the instantaneous angular velocity at which the vector $\boldsymbol{\omega}$ is changing its direction. Hence, the second time derivative of $\mathbf{A}$ is expressed as follows:

$$\ddot{\mathbf{A}} = \left[\ddot{A}\left(\frac{\mathbf{A}}{A}\right) + \boldsymbol{\omega} \times (\boldsymbol{\omega} \times \mathbf{A})\right] + \left(\frac{2\dot{A}}{A} + \frac{\dot{\omega}}{\omega}\right)\boldsymbol{\omega} \times \mathbf{A} + (\boldsymbol{\Omega} \times \boldsymbol{\omega}) \times \mathbf{A}\,. \tag{2.5}$$

The bracketed term on the right-hand side of Eq. (2.5) is parallel to $\mathbf{A}$, while the second term on the right-hand side is perpendicular to both $\mathbf{A}$ and $\boldsymbol{\omega}$. The last term on the right-hand side of Eq. (2.5) denotes the effect of a time-varying axis of rotation of $\mathbf{A}$.

2.2 Plane Kinematics

As a special case, consider the motion of a point, P, in a fixed plane described by the *radius* vector, $\mathbf{r}$, which is changing in time. The vector $\mathbf{r}$ is drawn from a fixed point, o, on the plane, to the moving point, P, and hence denotes the instantaneous radius of the moving point from o. The instantaneous rotation of the vector $\mathbf{r}$ is described by the angular velocity, $\boldsymbol{\omega} = \omega\mathbf{k}$, which is fixed in the direction given by the unit vector $\mathbf{k}$, normal to the plane of motion. Thus we have the following in Eq. (2.4):

$$\boldsymbol{\Omega} = 0\,.$$

[1] A reference frame, $(\mathbf{i}, \mathbf{j}, \mathbf{k})$, consisting of three mutually perpendicular axes, $\mathbf{i}$, $\mathbf{j}$, and $\mathbf{k}$, is termed a *right-handed frame* if it satisfies the right-hand rule of vector multiplication of the first two axes in the proper sequence, to produce the third axis:

$$\mathbf{i} \times \mathbf{j} = \mathbf{k}\,.$$

The net *velocity* of the point, P, is defined to be the time derivative of the radius vector, $\mathbf{r}$, which is expressed as follows according to the chain rule of vector differentiation:

$$\mathbf{v} = \dot{\mathbf{r}} = \dot{r}\left(\frac{\mathbf{r}}{r}\right) + \boldsymbol{\omega} \times \mathbf{r} \tag{2.6}$$

and consists of the *radial velocity* component, $\dot{r}/r$, and the *circumferential velocity* component, $|\,\boldsymbol{\omega} \times \mathbf{r}\,|$. Similarly, when the chain rule is applied to the velocity, $\mathbf{v}$, the result is the net *acceleration* of the moving point, P, which is defined to be the time derivative of $\mathbf{v}$, or the second time derivative of $\mathbf{r}$. In this special case of the radius vector, $\mathbf{r}$, always lying on a fixed plane, its angular velocity vector, $\boldsymbol{\omega}$, is always perpendicular to the given plane (hence the direction $\mathbf{k} = \boldsymbol{\omega}/\omega$ is constant), but can have a time-varying magnitude, ω. Hence, Eq. (2.4) yields the following expression for the time derivative of $\boldsymbol{\omega}$:

$$\dot{\boldsymbol{\omega}} = \dot{\omega}\left(\frac{\boldsymbol{\omega}}{\omega}\right) . \tag{2.7}$$

When these results are substituted into Eq. (2.3), the following expression for the acceleration of the point, P, is obtained:

$$\mathbf{a} = \dot{\mathbf{v}} = \ddot{\mathbf{r}} = \left(\frac{\ddot{r}}{r}\right)\mathbf{r} + \boldsymbol{\omega} \times (\boldsymbol{\omega} \times \mathbf{r}) + \left(\frac{2\dot{r}}{r} + \frac{\dot{\omega}}{\omega}\right)\boldsymbol{\omega} \times \mathbf{r} . \tag{2.8}$$

The net acceleration of the point, P, parallel to the instantaneous radius vector, $\mathbf{r}$, is identified from Eq. (2.8) to be the following:

$$\left(\frac{\ddot{r}}{r}\right)\mathbf{r} + \boldsymbol{\omega} \times (\boldsymbol{\omega} \times \mathbf{r}) .$$

The direction of the term $\boldsymbol{\omega} \times (\boldsymbol{\omega} \times \mathbf{r}) = -\omega^2\mathbf{r}$ is always towards the instantaneous centre of rotation (i.e., along $-\mathbf{r}/r$). The other radial acceleration term, $\ddot{r}\mathbf{r}/r$, is caused by the instantaneous change in the radius, r, and is positive in the direction of the increasing radius (i.e., away from the instantaneous centre of rotation).

The component of acceleration along the vector $\boldsymbol{\omega} \times \mathbf{r}$ in Eq. (2.8) is perpendicular to both $\mathbf{r}$ and $\boldsymbol{\omega}$, and is given by

$$\left(\frac{2\dot{r}}{r} + \frac{\dot{\omega}}{\omega}\right)\boldsymbol{\omega} \times \mathbf{r} = 2\dot{r}\left(\boldsymbol{\omega} \times \frac{\mathbf{r}}{r}\right) + \dot{\omega}(\mathbf{k} \times \mathbf{r}) .$$

In terms of the *polar coordinates*, (r, θ), we have $\boldsymbol{\omega} = \dot{\theta}\mathbf{k}$; hence the motion is resolved in two mutually perpendicular directions, $(\mathbf{r}/r, \mathbf{i}_\theta)$, where $\mathbf{i}_\theta$ is a unit vector along the direction of increasing θ (called the *circumferential direction*), defined by

$$\mathbf{i}_\theta = \mathbf{k} \times \frac{\mathbf{r}}{r} . \tag{2.9}$$

Thus the rotating frame, $(\mathbf{r}/r, \mathbf{i}_\theta, \mathbf{k})$, constitutes a right-handed triad. In this rotating coordinate frame, the motion of the point, P, is represented as follows:

$$\dot{\mathbf{r}} = \dot{r}\frac{\mathbf{r}}{r} + r\dot{\theta}\mathbf{i}_\theta \tag{2.10}$$

$$\frac{\mathrm{d}(\mathbf{r}/r)}{\mathrm{d}t} = \boldsymbol{\omega} \times \frac{\mathbf{r}}{r} = \dot{\theta}\mathbf{k} \times \frac{\mathbf{r}}{r} = \dot{\theta}\mathbf{i}_\theta \tag{2.11}$$

$$\frac{\mathrm{d}\mathbf{i}_\theta}{\mathrm{d}t} = \boldsymbol{\omega} \times \mathbf{i}_\theta = \dot{\theta}\mathbf{k} \times \mathbf{i}_\theta = -\dot{\theta}\frac{\mathbf{r}}{r} \tag{2.12}$$

$$\ddot{\mathbf{r}} = (\ddot{r} - r\dot{\theta}^2)\frac{\mathbf{r}}{r} + (2\dot{r}\dot{\theta} + r\ddot{\theta})\mathbf{i}_\theta \,. \tag{2.13}$$

It is clear from Eq. (2.13) that in the rotating coordinate system, $(\mathbf{r}/r, \mathbf{i}_\theta, \mathbf{k})$, the acceleration along the instantaneous radius vector, $\mathbf{r}$, is given by

$$(\ddot{r} - r\dot{\theta}^2)\frac{\mathbf{r}}{r}$$

and consists of the acceleration towards the instantaneous centre of rotation, $-r\dot{\theta}^2\mathbf{r}/r$, as well as that away from the instantaneous centre, $\ddot{r}\mathbf{r}/r$. Of the acceleration normal to the instantaneous radius vector $\mathbf{r}$, the term $2\dot{r}\dot{\theta}\mathbf{i}_\theta$ is caused by a change of the radius in the rotating coordinate frame, $(\mathbf{r}/r, \mathbf{i}_\theta, \mathbf{k})$, whereas the other term, $r\ddot{\theta}\mathbf{i}_\theta$, is due to the variation of the angular velocity of rotation, $\dot{\omega} = \ddot{\theta}$, in the same rotating frame.

An alternative representation of the motion of the point P is via *Cartesian coordinates*, (x, y), measured in a reference frame whose axes are fixed in space. Let us consider $(\mathbf{i}, \mathbf{j}, \mathbf{k})$ as such a fixed, right-handed coordinate system with $\mathbf{i} \times \mathbf{j} = \mathbf{k}$, and $(\mathbf{i}, \mathbf{j})$ being the constant plane of rotation. The radius vector and its time derivatives in the fixed frame are then given by

$$\mathbf{r} = x\mathbf{i} + y\mathbf{j} \tag{2.14}$$

$$\dot{\mathbf{r}} = \dot{x}\mathbf{i} + \dot{y}\mathbf{j} \tag{2.15}$$

$$\ddot{\mathbf{r}} = \ddot{x}\mathbf{i} + \ddot{y}\mathbf{j} \,. \tag{2.16}$$

In general, a time variation of the radius vector, $\mathbf{r}$, gives rise to a radial acceleration, $\ddot{\mathbf{r}}$, which is resolved in a fixed coordinate frame, $(\mathbf{i}, \mathbf{j}, \mathbf{k})$, without resorting to any rotational acceleration terms. Such a coordinate frame whose axes are fixed in space is termed an *inertial reference frame*, and the acceleration measured by such a frame is termed the inertial (or "true") acceleration. The inertial acceleration, $\ddot{\mathbf{r}}$, can be thought of as being directed towards (or away from) an instantaneous centre of rotation, which itself could be a moving point. For example, a point moving along an arc of a constant radius, $r = \sqrt{x^2 + y^2}$, at a constant angular rate, ω, has its acceleration directed towards the arc's centre, $\ddot{\mathbf{r}} = -\omega^2\mathbf{r}$.

2.3 Newton's Laws

In 1687 Newton gave his three famous laws of motion, which are valid for the motion of all objects (unless they are moving at speeds comparable to the speed of light). Stated briefly, they are the following:

(i) An object continues to move in a straight line at a constant *velocity*, unless acted upon by a *force* applied to it by another object.

(ii) The time rate of change of the velocity (called the *acceleration*) of an object is directly proportional to the force applied to the object. The constant of proportionality is a property of the object, called the *mass*.

(iii) If an object, A, applies a force on another object, B, then B applies a force on A of the same magnitude, but opposite in direction to that applied by A.

The consequences of these laws are profound, as they govern the motion of all objects (which are moving quite slowly when compared to the speed of light[2]). The first law implies that there is no *absolute position* in space, because two observers moving in parallel straight lines at a constant speed, c, relative to one another find two events separated in time to take place at different positions. Such observers can be regarded to be located at the origins of two different *reference frames*, $oxyz$ and $o'x'y'z'$, such that the relative motion takes place along the parallel axes, ox, and $o'x'$. The distances travelled by a moving object during the time, t, measured in the two frames are different, as given by the following *Galilean transformation*:

$$\begin{aligned} x' &= x + ct \\ y' &= y \\ z' &= z \end{aligned} \tag{2.17}$$

Newton's laws applied to a moving object are equally valid in the two reference frames, $oxyz$ and $o'x'y'z'$; hence they are both referred to as *inertial reference frames*. Another consequence of the first law is that it postulates an absolute quantity called the *time*, which is the same in all reference frames, and is therefore unaffected by the motion. The second law assigns a property called the *mass* to all material objects, which can be determined by measuring the force applied to the object and the corresponding acceleration experienced by it, while the third law defines the force applied by two isolated objects upon each other.

2.4 Particle Dynamics

A *particle* is defined to have a finite mass but infinitesimal dimensions, and is therefore regarded to be a point mass. Since a particle has negligible dimensions, its position in space is completely determined by the *radius* vector, $\mathbf{r}$, measured from a fixed point, o, at any instant of *time*, t. The components of $\mathbf{r}$ are resolved in a right-handed *reference frame* with origin at o, having three mutually perpendicular axes denoted by the unit vectors $\mathbf{i}, \mathbf{j}$, and $\mathbf{k}$, and are defined to be the Cartesian position coordinates, (x, y, z), of the particle along the respective axes, $(\mathbf{i}, \mathbf{j}, \mathbf{k})$, as shown in Fig. 2.1. The *velocity*, $\mathbf{v}$, of the particle is defined to be the time derivative of the radius vector, and given by

$$\dot{\mathbf{r}} = \frac{\mathrm{d}\mathbf{r}}{\mathrm{d}t} = \mathbf{v}\,. \tag{2.18}$$

If the reference frame, $(\mathbf{i}, \mathbf{j}, \mathbf{k})$, used to measure the velocity of the particle is at rest, then the components of the velocity, $\mathbf{v}$, resolved along the axes of the frame, $\mathbf{i}, \mathbf{j}$, and $\mathbf{k}$, are simply the time derivatives of the position coordinates, $\dot{x}, \dot{y}$, and $\dot{z}$, respectively. However, if the origin, o, of the reference frame itself is moving with a velocity, $\mathbf{v}_o$, and the frame, $(\mathbf{i}, \mathbf{j}, \mathbf{k})$, is rotating with an

[2] This caveat must be added because, as postulated by Einstein in 1905, Newton's laws cease to apply to objects travelling close to the speed of light, which is the same in all reference frames.

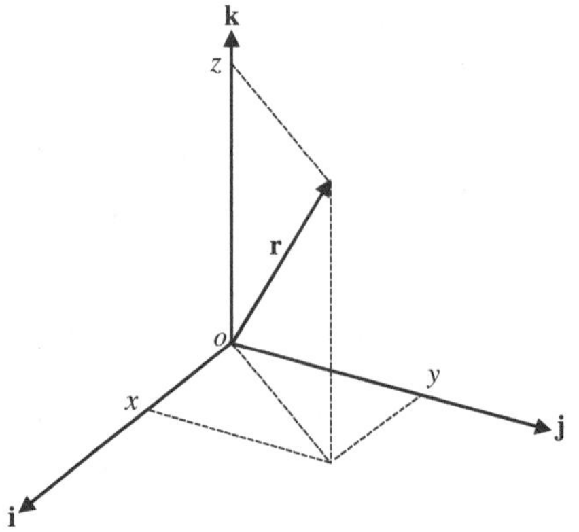

Figure 2.1 The position vector, $\mathbf{r}$, of a particle resolved in an inertial reference frame using Cartesian coordinates, (x, y, z).

angular velocity, $\boldsymbol{\omega}$, with respect to an *inertial frame*[3], then the velocity of the moving reference frame must be vectorially added to that of the particle in order to derive the net velocity of the particle in the stationary frame as follows:

$$\mathbf{v} = \dot{\mathbf{r}} = \frac{\mathrm{d}\mathbf{r}}{\mathrm{d}t} = \mathbf{v}_o + \dot{r}\frac{\mathbf{r}}{r} + \boldsymbol{\omega} \times \mathbf{r} \, . \tag{2.19}$$

The last term on the right-hand side of Eq. (2.19) is the change caused by rotating axes, $(\mathbf{i}, \mathbf{j}, \mathbf{k})$, each of which have the same angular velocity, $\boldsymbol{\omega}$. The relationship between the position and velocity described by a vector differential equation, either Eq. (2.18) or Eq. (2.19), is termed the *kinematics* of the particle.

Since the velocity of the particle could be varying with time, the *acceleration*, $\mathbf{a}$, of the particle is defined to be the time derivative of the velocity vector, and is given by

$$\mathbf{a} = \dot{\mathbf{v}} = \frac{\mathrm{d}\mathbf{v}}{\mathrm{d}t} = \frac{\mathrm{d}^2\mathbf{r}}{\mathrm{d}t^2} \, , \tag{2.20}$$

with the understanding that the derivatives are taken with respect to a stationary reference frame. If the reference frame in which the position and velocity of the particle are resolved is itself moving such that its origin, o, has an instantaneous velocity, $\mathbf{v}_o$ and an instantaneous acceleration,

[3] Rotation of the frame, $(\mathbf{i}, \mathbf{j}, \mathbf{k})$, refers to the rotation of its axes. Since $\mathbf{i}, \mathbf{j}$, and $\mathbf{k}$ are unit vectors, they cannot change in magnitude. Hence the time derivative of each axis is just due to its rotation by an angular velocity, $\boldsymbol{\omega}$. This angular velocity is common to all axes because they must always be mutually perpendicular and right handed. Therefore, we have

$$\frac{\mathrm{d}\mathbf{i}}{\mathrm{d}t} = \boldsymbol{\omega} \times \mathbf{i}$$

$$\frac{\mathrm{d}\mathbf{j}}{\mathrm{d}t} = \boldsymbol{\omega} \times \mathbf{j}$$

$$\frac{\mathrm{d}\mathbf{k}}{\mathrm{d}t} = \boldsymbol{\omega} \times \mathbf{k} \, .$$

$\mathbf{a}_0 = \mathrm{d}\mathbf{v}_0/\mathrm{d}t$, and its axes are rotating with an instantaneous angular velocity, $\boldsymbol{\omega}$, all measured in a stationary frame, then the net acceleration of the particle is given by

$$\mathbf{a} = \frac{\mathrm{d}\mathbf{v}}{\mathrm{d}t} = \mathbf{a}_0 + \boldsymbol{\omega} \times \mathbf{v}_0 + \ddot{r}\frac{\mathbf{r}}{r} + \dot{r}\left(\boldsymbol{\omega} \times \frac{\mathbf{r}}{r}\right) + \dot{\boldsymbol{\omega}} \times \mathbf{r} + \boldsymbol{\omega} \times \dot{\mathbf{r}}$$
$$= \mathbf{a}_0 + 2\boldsymbol{\omega} \times \mathbf{v}_0 + \ddot{r}\frac{\mathbf{r}}{r} + 2\dot{r}\left(\boldsymbol{\omega} \times \frac{\mathbf{r}}{r}\right) + \dot{\boldsymbol{\omega}} \times \mathbf{r} + \boldsymbol{\omega} \times (\boldsymbol{\omega} \times \mathbf{r})\,. \tag{2.21}$$

Equation (2.21) is an alternative kinematical description of the particle's motion, and can be regarded as being equivalent to that given by Eq. (2.19), which has been differentiated in time according to the chain rule. Equation (2.21) is useful in finding the acceleration of the particle from the position and velocity measured in a moving reference frame. The first two terms on the right-hand side of Eq. (2.21) represent the net acceleration due to the origin of the moving frame. The term $2\dot{r}(\boldsymbol{\omega} \times \mathbf{r}/r)$ is the *Coriolis acceleration*, and $\boldsymbol{\omega} \times (\boldsymbol{\omega} \times \mathbf{r})$ the *centripetal acceleration* of the particle in the moving reference frame. The term $\dot{\boldsymbol{\omega}} \times \mathbf{r}$ is the effect of the *angular acceleration* of the reference frame, whereas $\ddot{r}(\mathbf{r}/r)$ is the acceleration due to a changing magnitude of $\mathbf{r}$, and would be the only acceleration had the reference frame been stationary.

The application of Newton's second law to the motion of a particle of a fixed mass, m, and acted upon by a *force*, $\mathbf{f}$, gives the following important relationship – called the *kinetics* – for the determination of the particle's acceleration:

$$\mathbf{f} = m\mathbf{a} = m\frac{\mathrm{d}\mathbf{v}}{\mathrm{d}t}\,. \tag{2.22}$$

The *linear momentum*, $\mathbf{p}$, of the particle is defined as the product of its mass, m, and velocity, $\mathbf{v}$:

$$\mathbf{p} = m\mathbf{v}\,. \tag{2.23}$$

Since the particle's mass is constant, the second law of motion given by Eq. (2.22) can alternatively be expressed as follows:

$$\mathbf{f} = \frac{\mathrm{d}\mathbf{p}}{\mathrm{d}t} = \dot{\mathbf{p}}\,, \tag{2.24}$$

which gives rise to the principle of linear momentum conservation if no force is applied to the particle.

The *angular momentum*, $\mathbf{H}$, of the particle about a point, o, is defined to be the vector product of the radius vector, $\mathbf{r}$, of the particle from o and its linear momentum, $\mathbf{p}$:

$$\mathbf{H} = \mathbf{r} \times \mathbf{p} = m\mathbf{r} \times \mathbf{v}\,. \tag{2.25}$$

By virtue of Eq. (2.24), it is evident that the angular momentum of the particle about o can vary with time, if and only if a *torque*, defined by $\mathbf{M} = \mathbf{r} \times \mathbf{f}$, acts on the particle about o:

$$\mathbf{M} = \mathbf{r} \times \mathbf{f} = \mathbf{r} \times \frac{\mathrm{d}\mathbf{p}}{\mathrm{d}t}$$
$$= \frac{\mathrm{d}\mathbf{H}}{\mathrm{d}t} - \dot{\mathbf{r}} \times \mathbf{p} = \dot{\mathbf{H}} - m\mathbf{v} \times \mathbf{v} \tag{2.26}$$
$$= \dot{\mathbf{H}}\,.$$

This results in the principle of angular momentum conservation if no torque acts on the particle about o.

The *work* done on a particle by a force while moving from point A to point B is defined by the following integral of the scalar product of the force, $\mathbf{f}$, and the particle's displacement, $\mathrm{d}\mathbf{r}$:

$$W_{AB} = \int_A^B \mathbf{f} \cdot \mathrm{d}\mathbf{r} \,. \tag{2.27}$$

The application of Newton's second law for the constant mass particle, Eq. (2.22), results in the following expression for the work done:

$$W_{AB} = m \int_A^B \frac{\mathrm{d}\mathbf{v}}{\mathrm{d}t} \cdot \mathbf{v}\mathrm{d}t = m \int_A^B v\mathrm{d}v = \frac{1}{2}m(v_B^2 - v_A^2)\,, \tag{2.28}$$

where v_A and v_B are the speeds of the particle at the points A and B, respectively. Thus the net work done on a particle equals the net change in its *kinetic energy*, $1/2mv^2$.

2.5 The *n*-Body Problem

Gravity, being the predominant force in space flight, must be understood before constructing any model for space flight dynamics. Consider two particles of masses, m_1 and m_2, whose instantaneous positions in an inertial frame, *OXYZ*, are denoted by the vectors, $\mathbf{R}_1(t)$ and $\mathbf{R}_2(t)$, respectively. The relative position of mass, m_2, with respect to the mass, m_1, is given by the vector $\mathbf{r}_{21} = \mathbf{R}_2 - \mathbf{R}_1$. By Newton's law of gravitation, the two particles apply an equal and opposite attractive force on each other, which is directly proportional to the product of the two masses, and inversely proportional to the distance, $r_{21} =\mid \mathbf{R}_2 - \mathbf{R}_1 \mid$, between them. The equations of motion of the two particles are expressed as follows by Newton's second law of motion:

$$m_1 \frac{\mathrm{d}^2\mathbf{R}_1}{\mathrm{d}t^2} = G\frac{m_1 m_2}{r_{21}^3}\mathbf{r}_{21} \tag{2.29}$$

$$m_2 \frac{\mathrm{d}^2\mathbf{R}_2}{\mathrm{d}t^2} = -G\frac{m_1 m_2}{r_{21}^3}\mathbf{r}_{21}\,, \tag{2.30}$$

G being the *universal gravitational constant.* Adding the two equations of motion yields the important result that the *centre of mass* of the two particles is non-accelerating:

$$\frac{\mathrm{d}^2\mathbf{R}_c}{\mathrm{d}t^2} = \mathbf{0}\,, \tag{2.31}$$

where

$$\mathbf{R}_c = \frac{m_1\mathbf{R}_1 + m_2\mathbf{R}_2}{m_1 + m_2} \tag{2.32}$$

is the position of the centre of mass. This approach can be extended to a system of n particles, where the ith particle has the following equation of motion:

$$m_i \frac{\mathrm{d}^2\mathbf{R}_i}{\mathrm{d}t^2} = G\sum_{j=1}^{n} \frac{m_i m_j}{r_{ji}^3}\mathbf{r}_{ji}\,; \qquad (i = 1, 2, \ldots, n, j \neq i)\,, \tag{2.33}$$

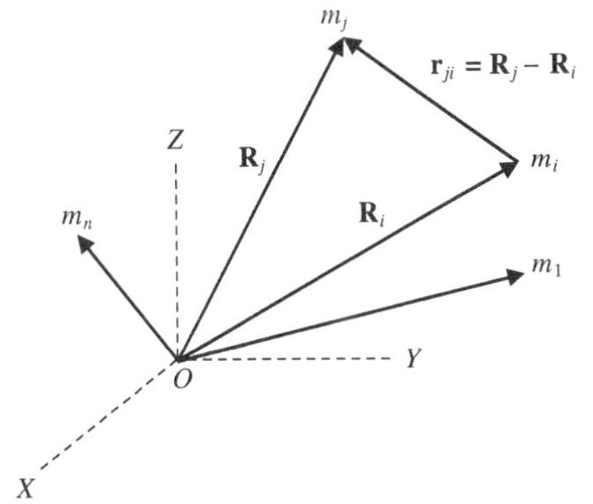

Figure 2.2 A system of n particles in an inertial reference frame *OXYZ*.

where $\mathbf{r}_{ji} = \mathbf{R}_j - \mathbf{R}_i$ locates the jth particle from the ith particle (Fig. 2.2). A summation of Eq. (2.33) for all the particles produces the following result:

$$\sum_{i=1}^{n} m_i \frac{\mathrm{d}^2\mathbf{R}_i}{\mathrm{d}t^2} = G \sum_{i=1}^{n} \sum_{j=1}^{n} \frac{m_i m_j}{r_{ji}^3} \mathbf{r}_{ji} = \mathbf{0} \,; \qquad (j \neq i) \tag{2.34}$$

because $\mathbf{r}_{ij} = -\mathbf{r}_{ji}$. Thus Eq. (2.31) is seen to be valid for the n-particles problem, where the centre of mass is located by

$$\mathbf{R}_c = \frac{\sum_{i=1}^{n} m_i \mathbf{R}_i}{\sum_{i=1}^{n} m_i} \,. \tag{2.35}$$

The non-accelerating centre of mass is the result of the law of conservation of linear momentum in the absence of a net external force on the system of particles. Integrated twice with time, Eq. (2.31) shows that the centre of mass moves in a straight line at a constant velocity:

$$\mathbf{R}_c(t) = \mathbf{v}_{c0} t + \mathbf{R}_{c0} \,, \tag{2.36}$$

where $\mathbf{R}_{c0}, \mathbf{v}_{c0}$ are constants (the initial position and the constant velocity, respectively, of the centre of mass).

For the convenience of notation, consider the overdot to denote the time derivative relative to the inertial reference frame. Taking a scalar product of Eq. (2.33) with $\dot{\mathbf{R}}_i$, and summing over all particles, we have

$$\sum_{i=1}^{n} m_i \ddot{\mathbf{R}}_i \cdot \dot{\mathbf{R}}_i = G \sum_{i=1}^{n} \sum_{j=1}^{n} \frac{m_i m_j}{r_{ji}^3} \mathbf{r}_{ji} \cdot \dot{\mathbf{R}}_i \,; \qquad (j \neq i) \,. \tag{2.37}$$

The term on the left-hand side of Eq. (2.37) is identified to be the time derivative of the net *kinetic energy* of the system given by

$$T = \frac{1}{2} \sum_{i=1}^{n} m_i \dot{\mathbf{R}}_i \cdot \dot{\mathbf{R}}_i \,. \tag{2.38}$$

Before proceeding further, it is necessary to consider that gravity is a *conservative* force, because, as will be seen later, it has no influence on the net energy, $T + V$, of a system. A force which depends only upon the relative position of the masses (as gravity does) is a conservative force, and can be expressed as the *gradient* of a scalar function, called the *gravitational potential*. The gradient of a scalar, U, with respect to a position vector, $\mathbf{r} = (x, y, z)^T$, is defined to be the following derivative of the scalar with respect to the given vector,

$$\frac{\partial U}{\partial \mathbf{r}} = \left(\frac{\partial U}{\partial x}, \frac{\partial U}{\partial y}, \frac{\partial U}{\partial z} \right) . \tag{2.39}$$

Thus, the gradient of a scalar with respect to a column vector is a row vector.

Consider, for example, an isolated pair of masses, m_1, m_2. The gravitational attraction on m_1 due to m_2 is given by the force, $\mathbf{f}_{21}$. By Newton's law of gravitation, we have

$$\mathbf{f}_{21} = G\frac{m_1 m_2}{r_{21}^3}\mathbf{r}_{21} , \tag{2.40}$$

where the relative position of mass m_2 from the mass m_1 is given by the vector $\mathbf{r}_{21} = \mathbf{R}_2 - \mathbf{R}_1$. Let U be the gravitational potential at the location of the particle, m_1, defined by

$$U = \frac{Gm_2}{r_{21}} . \tag{2.41}$$

The gradient of U with respect to $\mathbf{R}_1$ is the following:

$$\begin{aligned} \frac{\partial U}{\partial \mathbf{R}_1} &= -\frac{Gm_2}{r_{21}^2}\frac{\partial r_{21}}{\partial \mathbf{R}_1} \\ &= \frac{Gm_2}{r_{21}^2}\frac{\partial r_{21}}{\partial \mathbf{r}_{21}} = \frac{Gm_2}{r_{21}^3}\mathbf{r}_{21}^T . \end{aligned} \tag{2.42}$$

On comparing Eqs. (2.40) and (2.42), we have

$$\mathbf{f}_{21}^T = m_1 \frac{\partial U}{\partial \mathbf{R}_1} . \tag{2.43}$$

The acceleration of the mass m_1 due to the gravitational field created by the mass m_2 is therefore given by

$$\mathbf{g}_1 = \frac{\mathbf{f}_{21}}{m_1} = \left(\frac{\partial U}{\partial \mathbf{R}_1} \right)^T , \tag{2.44}$$

and is independent of the test mass, m_1.

The concept of gravitational potential between a pair of isolated masses, m_1, m_2, can be extended to a system of n point masses, where the net gravitational acceleration caused by $n - 1$ point masses, $m_j, j = 1, \ldots, n - 1$, on the nth particle of mass m_i is the vector sum of all the individual gravitational accelerations given by

$$\mathbf{g}_i = G\sum_{j=1}^{n} \frac{m_j}{r_{ji}^3}\mathbf{r}_{ji} ; \qquad (i = 1, 2, \ldots, n, j \neq i) , \tag{2.45}$$

or

$$\mathbf{g}_i^T = \frac{\partial U_i}{\partial \mathbf{R}_i}, \tag{2.46}$$

where

$$U_i = G\sum_{j=1}^{n} \frac{m_j}{r_{ji}}; \qquad (j \neq i) \tag{2.47}$$

is the net *gravitational potential* experienced by the ith particle due to the gravity of all the other $n-1$ particles.

The *potential energy*, V, of the n-particle system can be defined by

$$V = -\frac{1}{2}\sum_{i=1}^{n} m_i U_i \tag{2.48}$$

to be the net work done by the gravitational forces to assemble all the particles, beginning from an infinite separation, $r_{ij} \to \infty, i = 1, \ldots, n, j = 1, \ldots, n, (j \neq i)$, where $V = 0$. Thus a finite separation of the particles results in a negative potential energy (a potential well), escaping from which requires a positive energy expenditure.

The gradient of V with respect to $\mathbf{R}_i$ gives the negative of the gravitational force, $-m_i\mathbf{g}_i$, on the ith particle as follows:

$$\begin{aligned}\frac{\partial V}{\partial \mathbf{R}_i} &= -\frac{\partial V}{\partial \mathbf{r}_{ji}} + \frac{\partial V}{\partial \mathbf{r}_{ij}} = -2\frac{\partial V}{\partial \mathbf{r}_{ji}} \\ &= -G\sum_{j=1}^{n} \frac{m_i m_j}{r_{ji}^2}\frac{\partial r_{ji}}{\partial \mathbf{r}_{ji}} \\ &= -G\sum_{j=1}^{n} \frac{m_i m_j}{r_{ji}^3}\mathbf{r}_{ji}; \qquad (j \neq i)\end{aligned} \tag{2.49}$$

Hence the right-hand side of Eq. (2.37) is expressed as follows:

$$\begin{aligned}G\sum_{i=1}^{n}\sum_{j=1}^{n} \frac{m_i m_j}{r_{ji}^3}\mathbf{r}_{ji} \cdot \dot{\mathbf{R}}_i &= -\sum_{i=1}^{n} \frac{\partial V}{\partial \mathbf{R}_i} \cdot \dot{\mathbf{R}}_i \\ &= -\frac{\mathrm{d}V}{\mathrm{d}t}.\end{aligned} \tag{2.50}$$

A substitution of Eqs. (2.38) and (2.50) into Eq. (2.37) yields the important result that the *total energy* of the system is conserved:

$$\frac{\mathrm{d}}{\mathrm{d}t}(T + V) = 0 \tag{2.51}$$

or $T + V = c_1 = \text{const}$. This is true for any system solely governed by gravity.

To demonstrate another constant of the n-particle system, consider the vector product of Eq. (2.33) with $\mathbf{R}_i$, followed by summing over all particles:

$$\sum_{i=1}^{n} m_i \ddot{\mathbf{R}}_i \times \mathbf{R}_i = G\sum_{i=1}^{n}\sum_{j=1}^{n} \frac{m_i m_j}{r_{ji}^3}\mathbf{r}_{ji} \times \mathbf{R}_i; \qquad (j \neq i) \tag{2.52}$$

Because $\mathbf{r}_{ji} \times \mathbf{R}_i = -\mathbf{r}_{ij} \times \mathbf{R}_i$, all the terms on the right-hand side of Eq. (2.52) vanish, resulting in the following:

$$\sum_{i=1}^{n} m_i \ddot{\mathbf{R}}_i \times \mathbf{R}_i = \frac{\mathrm{d}}{\mathrm{d}t} \sum_{i=1}^{n} m_i \dot{\mathbf{R}}_i \times \mathbf{R}_i = \mathbf{0} \tag{2.53}$$

or

$$\sum_{i=1}^{n} m_i \dot{\mathbf{R}}_i \times \mathbf{R}_i = \mathbf{H} = \text{const.} \tag{2.54}$$

This implies that the n-particle motion takes place in a constant (or *invariant*) plane containing the centre of mass. The constant vector $\mathbf{H}$ is normal to the invariant plane, and is termed the net *angular momentum* of the system about the origin O. This is the law of conservation of angular momentum in the absence of a net external torque about O.

The conservation of linear and angular momentum, as well as the total energy of the n-particle system, is valid for any system ruled only by gravitational forces. The conservation principles are also valid for n-bodies of arbitrary shapes, as no restrictions have been applied in deriving those principles for the n-particle system. A *body* is defined to be a collection of a large number of particles. Thus the particles can be grouped into several bodies, each translating and rotating with respect to a common reference frame. However, solving for the motion variables (linear and angular positions and velocities) of a system of n bodies (referred to as the *n-body problem*) requires a numerical determination of the individual gravity fields of the bodies, as well as an integration of the $6n$ first-order, ordinary differential equations governing their motion. The next section discusses how such differential equations are derived for a body. The solar system is an example of the n-body system. Numerical approximations and simplifying assumptions are invariably employed in the solution of the n-body problem. For example, when the separations between the centres of mass of the respective bodies given by r_{ij}, $(i \neq j)$, are always large, the problem is approximated as that of n-bodies of spherical shape with radially symmetrical mass distributions.

2.6 Dynamics of a Body

The motion of a body is described by the motion of the particles constituting the body. A pure *translation* of a body is a motion in which all the particles constituting the body are moving in parallel straight lines with the same velocity. If the body is rigid, then the distance between any two of its particles is fixed; hence it is possible for the body to have a pure *rotation*, defined as the motion in which all the particles describe concentric circles about a fixed axis, and thus have velocities that are proportional to their respective distances from the axis of rotation. A rigid body in a combined translation and rotation has its constituent particles travelling in curved paths of different shapes relative to a stationary reference frame. A non-rigid body can have structural deformation as it translates and rotates, wherein the relative distances of the particles varies with time. The general motion of a body therefore consists of a combination of translation, rotation, and structural deformation, whose complete description requires a determination of the spatial trajectories of the particles constituting the body.

Consider a body with the centre of mass, o, with a particle of elemental mass, $\mathrm{d}M$, located at $\mathbf{r}$ relative to o (see Fig. 2.3). Also consider an inertial reference frame, $OXYZ$, with origin at O, and unit vectors, $\mathbf{I}, \mathbf{J}, \mathbf{K}$, along OX, OY, and OZ, respectively. The positions of o and the elemental

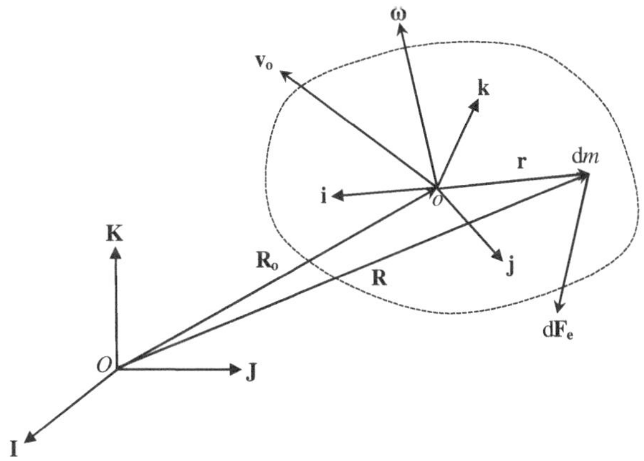

Figure 2.3 A body as a collection of large number of particles of elemental mass, dm, with centre of mass o.

mass relative to the origin, O, are given by $\mathbf{R}_0$ and $\mathbf{R}$, respectively, whose time derivatives in the inertial frame are the respective velocities, $\mathbf{v}_0$ and $\mathbf{v}$. If the net force experienced by the elemental mass is $\mathrm{d}\mathbf{F}$, then its equation of motion by Newton's second law is expressed as follows:

$$\mathrm{d}m\frac{\mathrm{d}\mathbf{v}}{\mathrm{d}t} = \mathrm{d}\mathbf{F}\,, \tag{2.55}$$

where the net force, $\mathrm{d}\mathbf{F}$, is a sum of all internal ($\mathrm{d}\mathbf{F}_\mathrm{i}$) and external ($\mathrm{d}\mathbf{F}_\mathrm{e}$) forces applied to the elemental mass, $\mathrm{d}\mathbf{F} = \mathrm{d}\mathbf{F}_\mathrm{i} + \mathrm{d}\mathbf{F}_\mathrm{e}$. The velocities, $\mathbf{v}_0$ and $\mathbf{v}$, are related by the following kinematic equation:

$$\mathbf{v} = \mathbf{v}_0 + \frac{\mathrm{d}\mathbf{r}}{\mathrm{d}t}\,, \tag{2.56}$$

where the time derivative is taken relative to the inertial frame, ($OXYZ$). Integrated over all the mass particles constituting the body, Eq. (2.55) yields the following result:

$$\int \frac{\mathrm{d}\mathbf{v}}{\mathrm{d}t}\mathrm{d}m = \int \mathrm{d}\mathbf{F} = \int \mathrm{d}\mathbf{F}_\mathrm{e} = \mathbf{F}_\mathrm{e}\,, \tag{2.57}$$

where all the internal forces (consisting of equal and opposite pairs) cancel out by Newton's third law, and $\mathbf{F}_\mathrm{e}$ is the net external force acting on the body.

The term on the left-hand side of Eq. (2.57) can be simplified by moving the time derivative outside the integral, because the total mass, $m = \int \mathrm{d}m$, of the body is a constant. This fact, along with Eq. (2.56), leads to the following:

$$\begin{aligned}\int \frac{\mathrm{d}\mathbf{v}}{\mathrm{d}t}\mathrm{d}m &= \frac{\mathrm{d}}{\mathrm{d}t}\int \mathbf{v}\mathrm{d}m \\ &= \frac{\mathrm{d}}{\mathrm{d}t}\int \mathbf{v}_0\mathrm{d}m + \frac{\mathrm{d}^2}{\mathrm{d}t^2}\int \mathbf{r}\mathrm{d}m \\ &= m\frac{\mathrm{d}\mathbf{v}_0}{\mathrm{d}t}\,,\end{aligned} \tag{2.58}$$

where we have $\int \mathbf{r}\mathrm{d}m = 0$ by the virtue of o being the centre of mass of the body. Thus Eqs. (2.57) and (2.58) result in the following equation of the body's *translation*:

$$\mathbf{F_e} = m\frac{\mathrm{d}\mathbf{v_o}}{\mathrm{d}t}. \tag{2.59}$$

Thus the translational motion of the body is described by the motion of its centre of mass, as if all the mass were concentrated at that point.

The rotational *kinetics* of the body are described by taking moments of Eq. (2.55) about the centre of mass, o, and integrating over the body as follows:

$$\int \mathbf{r}\times \mathrm{d}m\frac{\mathrm{d}\mathbf{v}}{\mathrm{d}t} = \int \mathbf{r}\times \mathrm{d}\mathbf{F}, \tag{2.60}$$

where all the internal torques cancel out (being equal and opposite pairs), resulting in only the net external torque, $\mathbf{M_e} = \int \mathbf{r}\times \mathrm{d}\mathbf{F_e}$, appearing on the right-hand side. The left-hand side of Eq. (2.60) is derived as follows, once again by the virtue of Eq. (2.56) and the fact that o is the centre of mass:

$$\begin{aligned}
\frac{\mathrm{d}}{\mathrm{d}t}\int \mathbf{r}\times \mathbf{v}\mathrm{d}m &= \int \mathbf{r}\times \frac{\mathrm{d}\mathbf{v}}{\mathrm{d}t}\mathrm{d}m + \int \frac{\mathrm{d}\mathbf{r}}{\mathrm{d}t}\times \mathbf{v}\mathrm{d}m \\
&= \int \mathbf{r}\times \frac{\mathrm{d}\mathbf{v}}{\mathrm{d}t}\mathrm{d}m + \int (\mathbf{v}-\mathbf{v_o})\times \mathbf{v}\mathrm{d}m \\
&= \int \mathbf{r}\times \frac{\mathrm{d}\mathbf{v}}{\mathrm{d}t}\mathrm{d}m - \mathbf{v_o}\times \int \mathbf{v}\mathrm{d}m \\
&= \int \mathbf{r}\times \frac{\mathrm{d}\mathbf{v}}{\mathrm{d}t}\mathrm{d}m - \mathbf{v_o}\times \left(\int \mathbf{v_o}\mathrm{d}m + \frac{\mathrm{d}}{\mathrm{d}t}\int \mathbf{r}\mathrm{d}m\right) \\
&= \int \mathbf{r}\times \frac{\mathrm{d}\mathbf{v}}{\mathrm{d}t}\mathrm{d}m.
\end{aligned} \tag{2.61}$$

A substitution of Eq. (2.61) into Eq. (2.60) yields the following equation of *rotational kinetics* of the body:

$$\mathbf{M_e} = \frac{\mathrm{d}\mathbf{H}}{\mathrm{d}t}, \tag{2.62}$$

where

$$\mathbf{H} = \int \mathbf{r}\times \mathbf{v}\mathrm{d}m \tag{2.63}$$

is the *angular momentum* of the body about its centre of mass, o. Thus a net external torque about the centre of mass of a body equals the time derivative of its angular momentum about the centre of mass.

If the body is *rigid*, then the distance between any two of its particles is a constant. Hence, the velocity of the elemental mass relative to o is given by

$$\frac{\mathrm{d}\mathbf{r}}{\mathrm{d}t} = \left(\frac{\mathrm{d}r}{\mathrm{d}t}\right)\frac{\mathbf{r}}{r} + \boldsymbol{\omega}\times \mathbf{r} = \boldsymbol{\omega}\times \mathbf{r}, \tag{2.64}$$

where $\dot{r} = \mathrm{d}r/\mathrm{d}t = 0$ because $r = |\mathbf{r}| = $ const. Here $\boldsymbol{\omega}$ is the angular velocity of a local reference frame, *oxyz*, rigidly attached to the body at o, with unit vectors $\mathbf{i}, \mathbf{j}, \mathbf{k}$ along *ox*, *oy*, and *oz*, respectively (Fig. 2.3), and is measured relative to the inertial frame, (*OXYZ*). Such a reference frame, *oxyz*, is termed a *body-fixed frame*.

2.7 Gravity Field of a Body

The principles derived for the n-body problem earlier in this chapter can be extended to the determination of the gravitational acceleration caused by a body whose mass, M, is distributed over a large number of elemental masses, m_i. In the following discussion, it is assumed that the other $n-1$ point masses are densely clustered together to form a body, away from the test mass, m_1; hence the test mass has the following equation of motion:

$$m_1 \frac{d^2\mathbf{R}_1}{dt^2} = -\frac{m_1 + M}{M} \left(\frac{\partial V}{\partial \mathbf{R}_1} \right), \tag{2.65}$$

where

$$V = -\frac{1}{2} \sum_{i=1}^{n} \sum_{j=1}^{n} \frac{Gm_i m_j}{s_{ji}}; \qquad (j \neq i) \tag{2.66}$$

is the total potential energy of the system,

$$M = \sum_{i=2}^{n} m_i \tag{2.67}$$

is the mass of the body consisting of the last $n-1$ particles,

$$\mathbf{r} = \mathbf{R}_1 - \mathbf{R}_o; \qquad \mathbf{s}_{1i} = \mathbf{R}_1 - \mathbf{R}_i, \tag{2.68}$$

with $\mathbf{R}_1$ being the location of the *test mass*, m_1, in an inertial reference frame, $(OXYZ)$, and $\mathbf{R}_o$ being the location of the centre of mass of the attracting body consisting of the remaining $n-1$ particles, which are located at $\mathbf{R}_i$, $i = 2, \ldots, n$. If it is further assumed that the test mass is negligible in comparison with the combined mass of the remaining $n-1$ particles constituting the body, that is, $m_1 \ll M$, then the test mass, m_1, causes a negligible acceleration on the body. Consequently, the body can be assumed to be at rest, and the origin of the inertial reference frame, $OXYZ$, is moved to the centre of mass of the body, i.e., $\mathbf{r} = \mathbf{R}_1$, $\mathbf{R}_o = \mathbf{0}$, and $\mathbf{s}_{1i} = \mathbf{r} - \mathbf{R}_i$. Hence, the equation of motion of the test mass becomes the following:

$$m_1 \frac{d^2\mathbf{r}}{dt^2} = -\left(\frac{\partial V}{\partial \mathbf{r}} \right) \tag{2.69}$$

or, since the partial derivative on the right-hand side yields only the terms for which either i or j equals 1, we have

$$m_1 \frac{d^2\mathbf{r}^T}{dt^2} = \frac{\partial}{\partial \mathbf{r}} \sum_{i=2}^{n} \frac{Gm_i}{s_{1i}}. \tag{2.70}$$

In terms of the gravitational potential of the body at the location of the test mass, which is given by

$$U = \sum_{i=2}^{n} \frac{Gm_i}{s_{1i}}, \tag{2.71}$$

the acceleration of the test mass is expressed by dividing the right-hand side of Eq. (2.70) by m_1 as follows:

$$\mathbf{g}^T = \frac{\partial U}{\partial \mathbf{r}} = \sum_{i=2}^{n} \frac{\partial(Gm_i/s_{1i})}{\partial \mathbf{s}_{1i}} = -\sum_{i=2}^{n} \frac{Gm_i}{s_{1i}^3} \mathbf{s}_{1i}. \tag{2.72}$$

For all the $n - 1$ particles constituting the mass, M, of the attracting body, let the limit of an infinitesimal *elemental mass*, $m_i \to \mathrm{d}M$, be taken as $n \to \infty$, whereby the summation in Eq. (2.71) is replaced by the following integral:

$$U = \int \frac{G}{s} \mathrm{d}M \,, \tag{2.73}$$

which results in the following expression for the acceleration of the test mass:

$$\mathbf{g}^T = \frac{\partial}{\partial \mathbf{r}} \int \frac{G}{s} \mathrm{d}M \,, \tag{2.74}$$

where s is the distance of the test mass, m_i, from the elemental mass, dM, as depicted in Fig. 2.4, and can be expressed as follows:

$$s = \sqrt{r^2 + \rho^2 - 2r\rho\cos\gamma} \,, \tag{2.75}$$

with $\mathbf{r}$ and $\boldsymbol{\rho}$ being the position vectors of the test mass, m_i, and the elemental mass, dM, respectively, from the centre of mass of the attracting body, and γ, being the angle between $\mathbf{r}$, and $\boldsymbol{\rho}$ as shown in Fig. 2.4.

From Fig. 2.4 it follows that

$$\mathbf{r} = \mathbf{s} + \boldsymbol{\rho} \,, \tag{2.76}$$

and $\rho = \mid \boldsymbol{\rho} \mid$ is a constant, because the attracting body is assumed to be a rigid body. When the position vectors $\mathbf{r}$ and $\boldsymbol{\rho}$ are resolved in the Cartesian coordinates, we have

$$\mathbf{r} = \begin{Bmatrix} X \\ Y \\ Z \end{Bmatrix} ; \quad \boldsymbol{\rho} = \begin{Bmatrix} \xi \\ \eta \\ \zeta \end{Bmatrix} ; \quad \mathbf{s} = \begin{Bmatrix} X - \xi \\ Y - \eta \\ Z - \zeta \end{Bmatrix} , \tag{2.77}$$

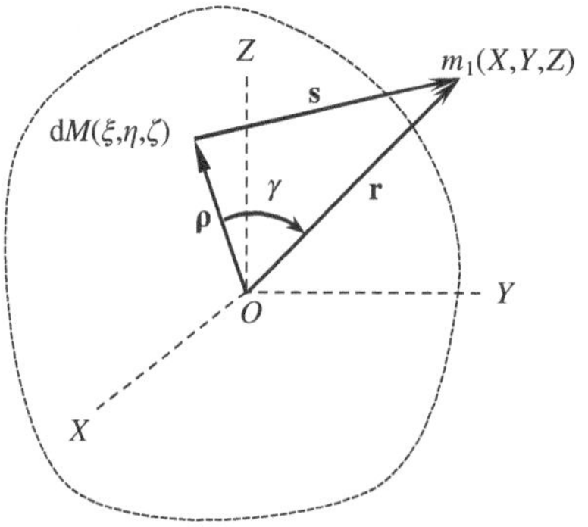

Figure 2.4 An elemental mass, dM, of a body with centre of mass O, and a test mass, m_1, located away from the body.

the gravitational potential of the mass distribution is given by

$$U = \int \frac{GdM}{\sqrt{(X-\xi)^2 + (Y-\eta)^2 + (Z-\zeta)^2}} = \int \frac{GdM}{\sqrt{r^2 + \rho^2 - 2r\rho\cos\gamma}}, \tag{2.78}$$

and the gravitational acceleration at $\mathbf{r}$ from the centre of mass of the attracting body is the following:

$$\mathbf{g}^T = \frac{\partial U}{\partial \mathbf{r}}. \tag{2.79}$$

2.7.1 Legendre Polynomials

To carry out the integration in Eq. (2.78), it is assumed that the body is entirely contained within the radius r measured from its centre of mass; that is, $r > \rho$ for all points on the body. It is then convenient to expand the integrand in the following series:

$$\frac{1}{\sqrt{r^2 + \rho^2 - 2r\rho\cos\gamma}} = \frac{1}{r}\left[1 + \frac{\rho}{r}\cos\gamma + \frac{1}{2}\left(\frac{\rho}{r}\right)^2(3\cos^2\gamma - 1) + \frac{1}{2}\left(\frac{\rho}{r}\right)^3(5\cos^2\gamma - 3)\cos\gamma + \cdots\right]. \tag{2.80}$$

Equation (2.80) is an infinite series expansion in polynomials of $\cos\gamma$, and is commonly expressed as follows:

$$\frac{1}{\sqrt{r^2 + \rho^2 - 2r\rho\cos\gamma}} = \frac{1}{r}\left[P_0(\cos\gamma) + \frac{\rho}{r}P_1(\cos\gamma) + \left(\frac{\rho}{r}\right)^2 P_2(\cos\gamma) + \cdots + \left(\frac{\rho}{r}\right)^k P_k(\cos\gamma) + \ldots\right], \tag{2.81}$$

where $P_k(\nu)$ is the *Legendre polynomial* of degree k, defined by

$$P_k(\nu) = \sum_{i=0}^{[k/2]} \frac{(-1)^i(2k-2i)!}{2^k i!(k-i)!(k-2i)!}\nu^{k-2i}, \tag{2.82}$$

with $[k/2]$ denoting the largest integer value of $k/2$ given by

$$[k/2] = \begin{cases} \dfrac{k}{2} & (k \text{ even}) \\[2ex] \dfrac{k-1}{2} & (k \text{ odd}) \end{cases}. \tag{2.83}$$

The first few Legendre polynomials are the following:

$$\begin{aligned} P_0(v) &= 1 \\ P_1(v) &= v \\ P_2(v) &= \frac{1}{2}(3v^2 - 1) \\ P_3(v) &= \frac{1}{2}(5v^3 - 3v) \\ P_4(v) &= \frac{1}{8}(35v^4 - 30v^2 + 3) \\ P_5(v) &= \frac{1}{8}(63v^5 - 70v^3 + 15v) \end{aligned} \tag{2.84}$$

Clearly the Legendre polynomials satisfy the condition $| P_k(\cos\gamma) | \leq 1$, which implies that the series in Eq. (2.81) is convergent. Therefore, one can approximate the integrand of Eq. (2.78) by retaining only a finite number of terms in the series.

By writing $x = \rho/r$ and $v = \cos\gamma$, the general expression for the Legendre polynomials is given in terms of the following *generating function*, $\mathcal{L}(x, v)$:

$$\mathcal{L}(x, v) = \frac{1}{\sqrt{1 + x^2 - 2vx}} = \sum_{k=0}^{\infty} P_k(v)x^k . \tag{2.85}$$

The generating function can be used to establish some of the basic properties of the Legendre polynomials, such as the following:

$$\begin{aligned} P_k(1) &= 1 \\ P_k(-1) &= (-1)^k \\ P_k(-v) &= (-1)^k P_k(v) \\ P'_{k+1}(v) &= (k + 1)P_k(v) + vP'_k(v) , \end{aligned} \tag{2.86}$$

where the prime stands for the derivative with respect to the argument, $\mathrm{d}/\mathrm{d}v$. The generating function, $\mathcal{L}(x, v)$, is also used to generate a Legendre polynomial from those of lower degrees with the help of *recurrence formulae*, such as

$$P_k(v) = \frac{(2k - 1)vP_{k-1}(v) - (k - 1)P_{k-2}(v)}{k} . \tag{2.87}$$

The reciprocal of the generating function, $\sqrt{1 + x^2 - 2vx}$, can be regarded as the radical portion, $\sqrt{1 - 4ac}$, of the real root, y, to the following quadratic equation:

$$ay^2 + y + c = 0 , \tag{2.88}$$

where the positive sign is taken to correspond to the smaller of the two roots. The choice $a = -x/2$ and $c = (x - 2v)/2$ yields

$$\frac{\partial y}{\partial v} = (1 + x^2 - 2vx)^{-1/2} = \mathcal{L}(x, v) , \tag{2.89}$$

or

$$y = v + x\left(\frac{y^2 - 1}{2}\right). \tag{2.90}$$

Since $x < 1$, the following series expansion (called *Lagrange's expansion theorem*) can be applied to Eq. (2.90) (Abramowitz and Stegun 1974):

$$y = v + \sum_{k=1}^{\infty} \frac{x^k}{k!} \frac{\mathrm{d}^{k-1}}{\mathrm{d}v^{k-1}} \left(\frac{v^2 - 1}{2}\right)^k. \tag{2.91}$$

The differentiation of Eq. (2.91) with v results in

$$\begin{aligned}\mathcal{L}(x, v) = \frac{\partial y}{\partial v} &= 1 + \sum_{k=1}^{\infty} \frac{x^k}{k!} \frac{\mathrm{d}^k}{\mathrm{d}v^k} \left(\frac{v^2 - 1}{2}\right)^k \\ &= \sum_{k=0}^{\infty} \frac{x^k}{k!} \frac{\mathrm{d}^k}{\mathrm{d}v^k} \left(\frac{v^2 - 1}{2}\right)^k \\ &= \sum_{k=0}^{\infty} P_k(v) x^k,\end{aligned} \tag{2.92}$$

where $x = 0$ corresponds to the root $y = v$. Equation (2.92) yields the following expression for the Legendre polynomials, called *Rodrigues' formula*:

$$P_k(v) = \frac{1}{2^k k!} \frac{\mathrm{d}^k}{\mathrm{d}v^k} (v^2 - 1)^k. \tag{2.93}$$

The gravitational potential is expressed as follows in terms of the Legendre polynomials by substituting Eq. (2.81) into Eq. (2.78):

$$U = \frac{G}{r} \sum_{k=0}^{\infty} \int \left(\frac{\rho}{r}\right)^k P_k(\cos\gamma) \mathrm{d}M. \tag{2.94}$$

It is possible to further simplify the gravitational potential before carrying out the complete integration. The integral arising out of $P_0(\cos\gamma)$ in Eq. (2.94) yields the mass, M, of the planet, thereby resulting in

$$\frac{G}{r} \int P_0(\cos\gamma) \mathrm{d}M = \frac{GM}{r}. \tag{2.95}$$

2.7.2 Spherical Coordinates

To evaluate the gravitational potential given by Eq. (2.94), it is necessary to introduce the spherical coordinates for the mass distribution of the body, as well as the location of the test mass. Let the right-handed triad, **I**, **J**, and **K**, represent the axes, *OX*, *OY*, and *OZ*, respectively, of the inertial frame, (*OXYZ*), with the origin, *O*, at the centre of the body. The location of the test mass, **r**, is resolved in the spherical coordinates, (r, ϕ, θ), as follows (Fig. 2.5):

$$\mathbf{r} = r(\sin\phi\cos\theta\mathbf{I} + \sin\phi\sin\theta\mathbf{J} + \cos\phi\mathbf{K}), \tag{2.96}$$

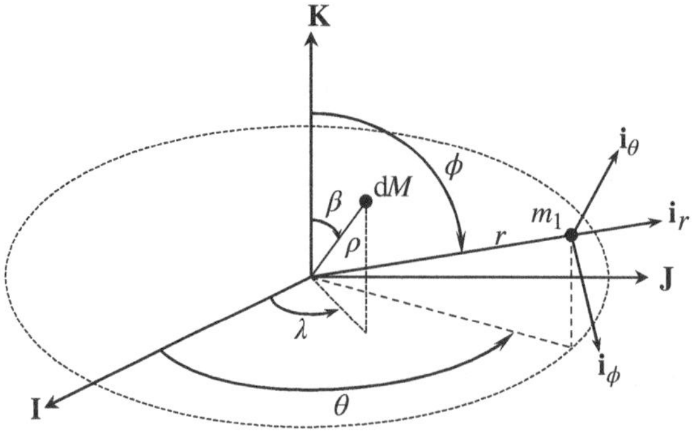

Figure 2.5 Spherical coordinates for the gravitational potential of a body.

where ϕ is the *co-latitude* and θ is the *longitude*. Similarly, let an elemental mass, dM, on the body be located by $\boldsymbol{\rho}$ using the spherical coordinates (ρ, β, λ) as follows (Fig. 2.5):

$$\boldsymbol{\rho} = \rho(\sin\beta\cos\lambda\mathbf{I} + \sin\beta\sin\lambda\mathbf{J} + \cos\beta\mathbf{K}), \tag{2.97}$$

where β and λ are the co-latitude and longitude, respectively, of the elemental mass.

The coordinate transformation between the spherical and Cartesian coordinates for the elemental mass is the following:

$$\begin{Bmatrix} \xi \\ \eta \\ \zeta \end{Bmatrix} = \rho \begin{Bmatrix} \sin\beta\cos\lambda \\ \sin\beta\sin\lambda \\ \cos\beta \end{Bmatrix}, \tag{2.98}$$

differentiating which produces

$$\begin{Bmatrix} \mathrm{d}\xi \\ \mathrm{d}\eta \\ \mathrm{d}\zeta \end{Bmatrix} = \begin{Bmatrix} \sin\beta\cos\lambda \\ \sin\beta\sin\lambda \\ \cos\beta \end{Bmatrix} \mathrm{d}\rho + \rho \begin{Bmatrix} \cos\beta\cos\lambda\mathrm{d}\beta - \sin\beta\sin\lambda\mathrm{d}\lambda \\ \cos\beta\sin\lambda\mathrm{d}\beta + \sin\beta\cos\lambda\mathrm{d}\lambda \\ -\sin\beta\mathrm{d}\beta \end{Bmatrix}, \tag{2.99}$$

or the following in the matrix form:

$$\begin{Bmatrix} \mathrm{d}\xi \\ \mathrm{d}\eta \\ \mathrm{d}\zeta \end{Bmatrix} = \begin{pmatrix} \sin\beta\cos\lambda & \cos\beta\cos\lambda & -\sin\beta\sin\lambda \\ \sin\beta\sin\lambda & \cos\beta\sin\lambda & \sin\beta\cos\lambda \\ \cos\beta & -\sin\beta & 0 \end{pmatrix} \begin{Bmatrix} \mathrm{d}\rho \\ \rho\mathrm{d}\beta \\ \rho\mathrm{d}\lambda \end{Bmatrix}. \tag{2.100}$$

An inversion of the square matrix on the right-hand side (called the *Jacobian* of the coordinate transformation) yields the following result:

$$\begin{Bmatrix} \mathrm{d}\rho \\ \rho\mathrm{d}\beta \\ \rho\mathrm{d}\lambda \end{Bmatrix} = \frac{1}{\sin\beta} \begin{pmatrix} \sin^2\beta\cos\lambda & \sin^2\beta\sin\lambda & \sin\beta\cos\beta \\ \sin\beta\cos\beta\cos\lambda & \sin\beta\cos\beta\sin\lambda & -\sin^2\beta \\ -\sin\lambda & \cos\lambda & 0 \end{pmatrix} \begin{Bmatrix} \mathrm{d}\xi \\ \mathrm{d}\eta \\ \mathrm{d}\zeta \end{Bmatrix}. \tag{2.101}$$

Since the determinant of the matrix on the right-hand side of Eq. (2.100) equals $\sin\beta$, and that of the inverse matrix in Eq. (2.101) is $1/\sin\beta$, the two sets of coordinates are related by the following expression for the elemental volume at (ρ, β, λ):

$$\begin{aligned} d\xi d\eta d\zeta &= \sin\beta d\rho(\rho d\beta)(\rho d\lambda) \\ &= \rho^2 \sin\beta d\rho d\beta d\lambda \,. \end{aligned} \quad (2.102)$$

If the mass density at the location of the elemental mass is given by $D(\rho, \beta, \lambda)$, then the elemental mass is the following:

$$\begin{aligned} dM &= D(\xi, \eta, \zeta) d\xi d\eta d\zeta \\ &= D(\rho, \beta, \lambda)\rho^2 \sin\beta d\rho d\beta d\lambda \,. \end{aligned} \quad (2.103)$$

The angle γ between $\mathbf{r}$, and $\boldsymbol{\rho}$ (Fig. 2.4) is related to the spherical coordinates by the following *cosine law* of the scalar product of two vectors:

$$\begin{aligned} \cos\gamma = \mathbf{r}\cdot\boldsymbol{\rho} &= \sin\phi\cos\theta\sin\beta\cos\lambda + \sin\phi\sin\theta\sin\beta\sin\lambda + \cos\phi\cos\beta \\ &= \sin\phi\sin\beta\cos(\theta-\lambda) + \cos\phi\cos\beta \,. \end{aligned}$$

To derive the gravitational potential given by Eq. (2.94) in spherical coordinates, it is necessary to expand the cosine law (Eq. (2.104)) in terms of the Legendre polynomials. To do so, consider the following *associated Legendre functions* of the first kind, degree k and order j (Abramowitz and Stegun 1974):

$$P_k^{(j)}(\nu) = (1-\nu^2)^{j/2} \frac{d^j}{d\nu^j} P_k(\nu) \,, \quad (2.104)$$

where $P_k(\nu)$ is the Legendre polynomial of degree k. Some of the commonly used associated Legendre functions are

$$P_1^{(1)}(\cos\alpha) = \sin\alpha \,; \quad P_2^{(1)}(\cos\alpha) = 3\sin\alpha\cos\alpha \,; \quad P_2^{(2)}(\cos\alpha) = 3\sin^2\alpha \,.$$

In terms of the associated Legendre functions and the Legendre polynomials of the first degree, Eq. (2.104) becomes

$$P_1(\cos\gamma) = P_1(\cos\phi)P_1(\cos\beta) + P_1^{(1)}(\cos\phi)P_1^{(1)}(\cos\beta)\cos(\theta-\lambda) \,, \quad (2.105)$$

which is referred to as the *addition theorem* for the Legendre polynomials of the first degree, $P_1(\nu)$. In terms of the Legendre polynomials of the second degree, $P_2(\nu)$, we have

$$\begin{aligned} P_2(\cos\gamma) = P_2(\cos\phi)P_2(\cos\beta) &+ \frac{1}{3}P_2^{(1)}(\cos\phi)P_2^{(1)}(\cos\beta)\cos(\theta-\lambda) \\ &+ \frac{1}{12}P_2^{(2)}(\cos\phi)P_2^{(2)}(\cos\beta)\cos 2(\theta-\lambda) \,, \end{aligned} \quad (2.106)$$

which is the addition theorem for the Legendre polynomials of the second degree, $P_2(\nu)$. Extending this procedure leads to the following addition theorem for the Legendre polynomials of degree k, $P_k(\nu)$:

$$P_k(\cos\gamma) = P_k(\cos\phi)P_k(\cos\beta) + 2\sum_j^k \frac{(k-j)!}{(k+j)!} P_k^{(j)}(\cos\phi)P_k^{(j)}(\cos\beta)\cos j(\theta-\lambda)\,. \tag{2.107}$$

The substitution of the addition theorem into Eq. (2.94) results in the following expansion of the gravitational potential:

$$U(r,\phi,\theta) = \frac{GM}{r} + \sum_{k=1}^{\infty} \frac{A_k}{r^{k+1}} P_k(\cos\phi) + \sum_{k=1}^{\infty}\sum_{j=1}^{k} \frac{B_k^{(j)}}{r^{k+1}} P_k^{(j)}(\cos\phi)\cos j\theta + \sum_{k=1}^{\infty}\sum_{j=1}^{k} \frac{C_k^{(j)}}{r^{k+1}} P_k^{(j)}(\cos\phi)\sin j\theta\,, \tag{2.108}$$

where

$$A_k = G\int_0^{2\pi}\int_0^{2\pi}\int_0^{r_m} \rho^{k+2} D(\rho,\beta,\lambda)P_k(\cos\beta)\sin\beta \mathrm{d}\rho\mathrm{d}\beta\mathrm{d}\lambda\,, \tag{2.109}$$

$$B_k^{(j)} = 2G\frac{(k-j)!}{(k+j)!}\int_0^{2\pi}\int_0^{2\pi}\int_0^{r_m} \rho^{k+2} D(\rho,\beta,\lambda)P_k^{(j)}(\cos\beta)\sin\beta\cos j\lambda \mathrm{d}\rho\mathrm{d}\beta\mathrm{d}\lambda\,, \tag{2.110}$$

$$C_k^{(j)} = 2G\frac{(k-j)!}{(k+j)!}\int_0^{2\pi}\int_0^{2\pi}\int_0^{r_m} \rho^{k+2} D(\rho,\beta,\lambda)P_k^{(j)}(\cos\beta)\sin\beta\sin j\lambda \mathrm{d}\rho\mathrm{d}\beta\mathrm{d}\lambda\,, \text{ and} \tag{2.111}$$

with r_m denoting the maximum radial extent of the body. The mass of the body is evaluated by

$$M = \int_0^{2\pi}\int_0^{2\pi}\int_0^{r_m} \rho^2 D(\rho,\beta,\lambda)\sin\beta \mathrm{d}\rho\mathrm{d}\beta\mathrm{d}\lambda\,. \tag{2.112}$$

Equation (2.108) is a general expansion of the gravitational potential which can be applied to a body of an arbitrary shape and an arbitrary mass distribution. However, the evaluation of the series coefficients by Eqs. (2.109)–(2.111) is often a difficult exercise for a body of a complicated shape, and requires experimental determination (such as the acceleration measurements by a low-orbiting satellite).

2.7.3 Axisymmetric Body

A body whose mass is symmetrically distributed about the polar axis, OZ, has its density varying only with the radius and the latitude; that is, $D = D(\rho,\beta)$. Upon neglecting the longitudinal (λ) variations in the mass distribution, the additional theorem for the Legendre polynomial of degree k, Eq. (2.107), becomes the following:

$$P_k(\cos\gamma) = P_k(\cos\beta)P_k(\cos\phi)\,, \tag{2.113}$$

whose substitution into the triple integrals in Eqs. (2.110) and (2.111) leads to the integration in the longitude, λ, being carried out independently of ρ and β as follows:

$$\int_0^{2\pi} \cos j\lambda \mathrm{d}\lambda = \int_0^{2\pi} \sin j\lambda \mathrm{d}\lambda = 0 \qquad (j = 1, 2, \ldots ,) . \tag{2.114}$$

This implies that $B_k^{(j)} = C_k^{(j)} = 0$ and

$$A_1 = G \int_0^{2\pi} \int_0^{2\pi} \int_0^{r_0} \rho^3 D(\rho, \beta) P_1(\cos\beta) \sin\beta \mathrm{d}\rho \mathrm{d}\beta \mathrm{d}\lambda = G \int \rho \cos\beta \mathrm{d}M = 0 . \tag{2.115}$$

These simplifications allow the gravitational potential of an axisymmetric body to be expressed as follows:

$$U(r, \phi) = \frac{GM}{r} + \sum_{k=2}^{\infty} \frac{A_k}{r^{k+1}} P_k(\cos\phi) , \tag{2.116}$$

where

$$A_k = G \int_0^{2\pi} \int_0^{\pi} D(\rho, \beta) \rho^{k+2} P_k(\cos\beta) \sin\beta \mathrm{d}\rho \mathrm{d}\beta \mathrm{d}\lambda . \tag{2.117}$$

A more useful expression for the gravitational potential can be obtained as follows in terms of the non-dimensional distance, $\frac{r}{r_0}$, where r_0 is the equatorial radius of the axisymmetric body:

$$U(r, \phi) = \frac{\mu}{r} \left[1 - \sum_{k=2}^{\infty} \left(\frac{r_0}{r} \right)^k J_k P_k(\cos\phi) \right] , \tag{2.118}$$

where $\mu = GM$ and

$$J_k = -\frac{A_k}{\mu r_0^k} \tag{2.119}$$

are called *Jeffery's constants*, and are unique for a body of a given mass distribution. Jeffery's constants represent the spherical harmonics of the mass distribution, and diminish in magnitude as the order, k, increases. The largest of these constants, J_2, denotes a non-dimensional difference between the moments of inertia about the polar axis, OZ, and an axis in the equatorial plane (**I** or **J** in Fig. 2.5), and is a measure of the *ellipticity* (or *oblateness*) of the body. The higher order term, J_3 indicates the pear-shaped or triangular harmonic, whereas J_4 and J_5 are the measures of square and pentagonal shaped harmonics, respectively. For a reasonably large body, it is seldom necessary to include more than the first four Jeffery's constants. For example, Earth's spherical harmonics are given by $J_2 = 0.00108263$, $J_3 = -0.00000254$, and $J_4 = -0.00000161$.

The acceleration due to gravity of an axisymmetric body is obtained by taking the gradient of the gravitational potential, Eq. (2.118), with respect to the position vector, **r**, and can be resolved in the radial, $\mathbf{i_r} = \mathbf{r}/r$, and the north polar, $\mathbf{K}$, directions (Fig. 2.5) as follows:

$$
\begin{aligned}
\mathbf{g} &= \left(\frac{\partial U}{\partial \mathbf{r}}\right)^T \\
&= -\frac{\mu}{r^2}\mathbf{i_r} + \frac{\mu}{r^2}\sum_{k=2}^{\infty}(k+1)J_k\left(\frac{r_0}{r}\right)^k P_k(\cos\phi)\mathbf{i_r} - \frac{\mu}{r}\sum_{k=2}^{\infty}J_k\left(\frac{r_0}{r}\right)^k P_k'(\cos\phi)\left(\frac{\partial\cos\phi}{\partial\mathbf{r}}\right)^T \\
&= -\frac{\mu}{r^2}\left\{1 - \sum_{k=2}^{\infty}J_k\left(\frac{r_0}{r}\right)^k[(k+1)P_k(\cos\phi) + P_k'(\cos\phi)\cos\phi]\right\}\mathbf{i_r} \\
&\quad -\frac{\mu}{r^2}\sum_{k=2}^{\infty}J_k\left(\frac{r_0}{r}\right)^k P_k'(\cos\phi)\mathbf{K} \qquad (2.120)\\
&= -\frac{\mu}{r^2}\left\{\mathbf{i_r} - \sum_{k=2}^{\infty}J_k\left(\frac{r_0}{r}\right)^k[P_{k+1}'(\cos\phi)\mathbf{i_r} - P_k'(\cos\phi)\mathbf{K}]\right\},
\end{aligned}
$$

where the following identities have been employed:

$$
\begin{aligned}
\frac{\partial\cos\phi}{\partial\mathbf{r}} &= \frac{\partial}{\partial\mathbf{r}}\frac{\mathbf{r}^T\mathbf{K}}{r} = -\frac{\mathbf{r}^T\mathbf{K}}{r^2}\left(\frac{\mathbf{r}^T}{r}\right) + \frac{1}{r}\mathbf{K}^T, \\
P_{k+1}'(\nu) &= (k+1)P_k(\nu) + \nu P_k'(\nu).
\end{aligned}
$$

The acceleration can be alternatively resolved in two mutually perpendicular directions, $\mathbf{i_r}$ and $\mathbf{i}_\phi$ (see Fig. 2.5). The unit vectors $\mathbf{i_r}$ and $\mathbf{i}_\phi$ denote the radial and southward directions, as shown in Fig. 2.5, while the unit vector $\mathbf{i}_\theta = \mathbf{i_r} \times \mathbf{i}_\phi$ signifies the direction of the increasing longitude; that is, the eastward direction. These unit vectors constitute a moving coordinate frame, $(\mathbf{i_r}, \mathbf{i}_\phi, \mathbf{i}_\theta)$, attached to the test mass as shown in Fig. 2.5. Such a frame is called a *local-horizon frame* (see Chapter 5). The acceleration given by Eq. (2.120) is resolved in $\mathbf{i_r}$ and $\mathbf{i}_\phi$ by substituting $\mathbf{K} = \mathbf{i_r}\cos\phi - \mathbf{i}_\phi\sin\phi$, resulting in

$$
\mathbf{g} = g_r\mathbf{i_r} + g_\phi\mathbf{i}_\phi, \qquad (2.121)
$$

where

$$
\begin{aligned}
g_r &= -\frac{\mu}{r^2}\left[1 - \sum_{k=2}^{\infty}(k+1)J_k\left(\frac{r_0}{r}\right)^k P_k(\cos\phi)\right] \\
&= -\frac{\mu}{r^2}\left[1 - 3J_2\left(\frac{r_0}{r}\right)^2 P_2(\cos\phi) - 4J_3\left(\frac{r_0}{r}\right)^3 P_3(\cos\phi)\right. \\
&\quad \left. - 5J_4\left(\frac{r_0}{r}\right)^4 P_4(\cos\phi) - \cdots\right], \qquad (2.122)
\end{aligned}
$$

and

$$
g_\phi = \frac{\mu}{r^2}\sum_{k=2}^{\infty}J_k\left(\frac{r_0}{r}\right)^k P_k'(\cos\phi)\sin\phi
$$

$$= \frac{\mu}{r^2} \sum_{k=2}^{\infty} J_k \left(\frac{r_0}{r}\right)^k P_k^{(1)}(\cos\phi)$$

$$= \frac{3\mu}{r^2}\left(\frac{r_0}{r}\right)^2 \sin\phi\cos\phi \left[J_2 + \frac{1}{2}J_3\left(\frac{r_0}{r}\right)\sec\phi(5\cos^2\phi - 1) + \frac{5}{6}J_4\left(\frac{r_0}{r}\right)^2(7\cos^2\phi - 3) + \cdots\right]. \quad (2.123)$$

Due to a non-zero transverse gravity component, g_ϕ, the direction of $\mathbf{g}$ differs from the radial direction, while its radial component, g_r, is smaller in magnitude when compared to that predicted by a spherical gravity model; that is, the model derived by assuming a perfectly spherical mass distribution of a body of radius r_0.

2.7.4 *Spherical Body with Radially Symmetric Mass Distribution*

A spherical body of constant radius r_0 with a radially symmetrical mass distribution has the density $D = D(\rho)$, varying only with the radius, ρ. A body consisting of concentric spherical shells, each of which has a different (either constant or radially varying) density, has a radially symmetrical mass distribution. Such a distribution results in the vanishing of all the integrals of β and λ in Eqs. (2.109)–(2.111), which produces $A_k = B_k^{(j)} = C_k^{(j)} = 0$, thereby implying a potential which is independent of the co-latitude and longitude. Hence the gravitational potential of a body with a spherically symmetric mass distribution of radius, r_0, is given by

$$U(r) = \frac{GM}{r} \qquad (r \geq r_0). \quad (2.124)$$

Clearly, a spherical body with a radially symmetric mass distribution behaves exactly as if all its mass were concentrated at its centre. The acceleration due to gravity is thus given by

$$\mathbf{g} = \left(\frac{\partial U}{\partial \mathbf{r}}\right)^T = -\frac{GM}{r^2}\mathbf{i_r} \qquad (r \geq r_0). \quad (2.125)$$

Exercises

1. A particle, P, is launched vertically upward with velocity, v, from the surface of a spherical planet of radius R, rotating about the polar axis, OA, from west to east at the rate Ω. The launch takes place at the latitude, δ, and longitude, λ, measured from a fixed meridian, OB, as shown in Fig. 2.6. Derive the expressions for the net velocity and acceleration of the particle, and resolve the vector components in a reference frame, $(OABC)$, with its origin, O, at the centre of the planet, and with one axis along the polar axis, OA, another along the meridian, OB, and the third axis, OC, normal to both OA and OB.

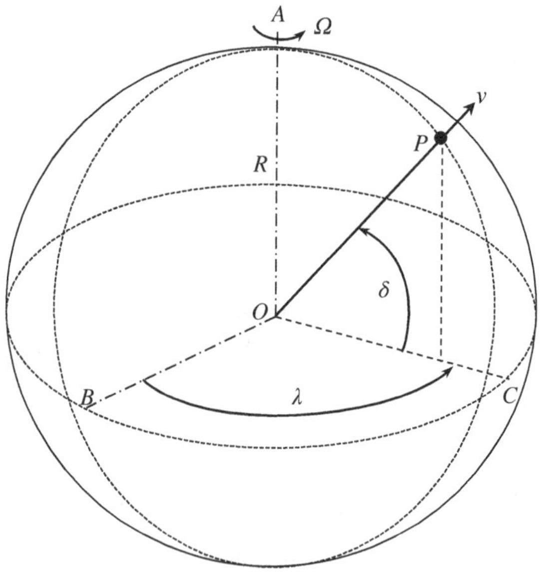

Figure 2.6 Geometry for Exercise 1.

2. A point, P, moves on the horizontal plane, OAB, describing a circle of radius R around the centroid, O, of a cylinder of length L and radius r, which is rotating about the axis, AB, at a constant angular speed, ω, as shown in Fig. 2.7. The angle made by the line OP with the axis AB is θ, which varies with time at a constant rate, $\dot{\theta} = n$. Another point, Q, is on the edge of the cylinder at its base, BCD, located at the semi-length, $L/2$, from the median section, OEF. The radial line, BQ, makes an angle, ϕ, with the vertical axis, CD, passing through B (Fig. 2.7). Derive the expressions for the position, velocity, and acceleration vectors of Q relative to P as functions of time, t, assuming that $\theta = \phi = 0$ when $t = 0$, and resolve the components in the reference frame, $(OEPG)$, with the mutually perpendicular axes, OE, OP, and OG, shown in Fig. 2.7.

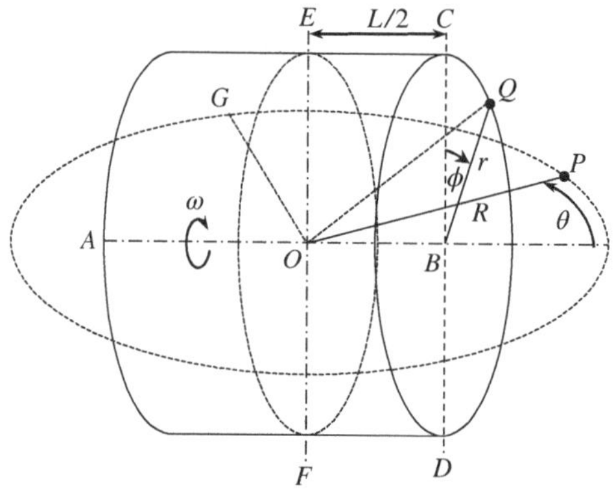

Figure 2.7 Geometry for Exercise 2.

3. Transform the equations of motion of the system of n particles under mutual gravitational attraction to the following *generalized coordinates*, $\mathbf{q} = (q_1, q_2, \dots, q_{3n})^T$, of Lagrange:

$$\mathbf{R}_i = \mathbf{R}_i[t, \mathbf{q}(t)] \,; \qquad \mathbf{v}_i = \dot{\mathbf{R}}_i = \frac{\partial \mathbf{R}_i}{\partial t} + \frac{\partial \mathbf{R}_i}{\partial \mathbf{q}}\dot{\mathbf{q}} = \mathbf{v}_i[t, \dot{\mathbf{q}}(t)] \,,$$

and show that they can be expressed as follows:

$$\frac{\mathrm{d}}{\mathrm{d}t}\frac{\partial L}{\partial \dot{\mathbf{q}}} = \frac{\partial L}{\partial \mathbf{q}} \,,$$

where $L = T - V$ is the *Lagrangian* of the system, with T and V being the net kinetic and potential energies.

4. Transform the system of equations derived in Exercise 3 to a set of variables, $\mathbf{p} = (p_1, p_2, \dots, p_{3n})^T$, called the *generalized momenta*, defined by

$$\mathbf{p} = \left(\frac{\partial T}{\partial \dot{\mathbf{q}}}\right)^T \,,$$

and show that they can be expressed in the following *canonical form* (Whittaker, 1988):

$$\frac{\mathrm{d}\mathbf{q}^T}{\mathrm{d}t} = \frac{\partial H}{\partial \mathbf{p}} \,; \qquad \frac{\mathrm{d}\mathbf{p}^T}{\mathrm{d}t} = -\frac{\partial H}{\partial \mathbf{q}} \,,$$

where the net kinetic energy, T, is expressed as follows:

$$T = \frac{1}{2}\sum_{i=1}^{n} m_i \left[\frac{\partial \mathbf{R}_i^T}{\partial t}\frac{\partial \mathbf{R}_i}{\partial t} + 2\frac{\partial \mathbf{R}_i^T}{\partial t}\frac{\partial \mathbf{R}_i}{\partial \mathbf{q}}\dot{\mathbf{q}} + \dot{\mathbf{q}}^T\left(\frac{\partial \mathbf{R}_i}{\partial \mathbf{q}}\right)^T\frac{\partial \mathbf{R}_i}{\partial \mathbf{q}}\dot{\mathbf{q}}\right] \,,$$

and

$$H = V - \frac{1}{2}\sum_{i=1}^{n} m_i \left[\frac{\partial \mathbf{R}_i^T}{\partial t}\frac{\partial \mathbf{R}_i}{\partial t} - \dot{\mathbf{q}}^T\left(\frac{\partial \mathbf{R}_i}{\partial \mathbf{q}}\right)^T\frac{\partial \mathbf{R}_i}{\partial \mathbf{q}}\dot{\mathbf{q}}\right]$$

is the *Hamiltonian* of the system, with V being the net potential energy.

5. Referring to Exercise 4, show that if the Hamiltonian is not explicitly dependent on the time, t, i.e.,

$$H = H[\mathbf{q}(t), \dot{\mathbf{q}}(t)] \,,$$

then we have

$$\frac{\mathrm{d}H}{\mathrm{d}t} = 0 \,,$$

or $H = \text{const.}$

6. Consider four particles, each having the mass, m, such that three of them are always located at the vertices of an equilateral triangle while the fourth is at its centroid. What is the angular velocity, ω, of the triangle?

7. Derive an expression for the gravitational potential of a thin circular ring of uniform density with mass, M, radius, R, and negligible cross-sectional area, at a point located on the plane of the ring at radius r from the centre, where (a) $r \geq R$, and (b) $r < R$. (*Ans.* (a) $U(r) = -\frac{Gm}{R}\left[1 + \frac{1}{4}\left(\frac{r}{R}\right)^2 + \frac{9}{64}\left(\frac{r}{R}\right)^4 + \frac{25}{256}\left(\frac{r}{R}\right)^6 + \frac{1225}{16384}\left(\frac{r}{R}\right)^8 \cdots\right]$.)

8. Calculate the gravitational potential of a thin spherical shell of radius, R and constant density per unit surface area, σ, at a point located at radius r from the centre, where (a) $r \geq R$, and (b) $r < R$.

9. The density of a spherically symmetric mass distribution extending up to a radius R is given by

$$D(r) = \begin{cases} D_0 r^{-c} & (r \leq R) \\ 0 & (r > R) \end{cases},$$

where $c < 5/2$ is a constant. Find the gravitational potential for $r \geq R$, as well as the net potential energy, if the total mass of the distribution is M.

References

Whittaker ET 1988. *Analytical Dynamics of Particles and Rigid Bodies*. Cambridge University Press, New York.
Abramowitz M and Stegun IA 1974. *Handbook of Mathematical Functions*. Dover, New York.

3

Keplerian Motion

In the previous chapter, you saw that the motion of a rigid body can be described as the translational motion of its centre of mass and the rotational motion of the body about the centre of mass. In most applications of space dynamics, it is quite useful to approximate a spacecraft as a rigid body. For a spacecraft idealized as a rigid body, the translational dynamics of the centre of mass is termed *orbital mechanics*, while the rotational motion of the spacecraft about its centre of mass is called *attitude dynamics*. In this chapter, we shall discuss the basic problem of orbital mechanics, namely when the spacecraft is acted upon by the gravity of a much larger spherical body, such as a planet, a moon, or the Sun. The shape and the plane of a trajectory governed by the uniform gravity of a spherical central body are well defined by constant parameters. The position and velocity of the centre of mass along such a trajectory can be expressed in terms of the known constant parameters.

3.1 The Two-Body Problem

The motion of two spherical bodies under mutual gravitational attraction is the primary problem of orbital mechanics, and can be solved analytically. Since such a motion obeys the laws of planetary motion first discovered by Kepler, two-body orbital mechanics is also termed *Keplerian motion*. Consider two perfectly spherical bodies of masses m_1 and m_2 whose centres are located at $\mathbf{R}_1$ and $\mathbf{R}_2$, respectively, in an inertial reference frame (Fig. 3.1). By Newton's law of gravitation, the mass m_1 attracts the other mass by a force which acts along the straight line joining the centres of the two masses, and whose magnitude varies inversely with the distance between them. This force of gravity is expressed as follows:

$$\mathbf{f}_{12} = -G\frac{m_1 m_2}{r^3}\mathbf{r} , \tag{3.1}$$

where $\mathbf{r} = \mathbf{R}_2 - \mathbf{R}_1$, $r =| \mathbf{r} |$, and G is the universal gravitational constant. The gravitational force, $\mathbf{f}_{21}$, by which the mass m_2 attracts the mass m_1 is equal in magnitude, but opposite in direction to $\mathbf{f}_{12}$ (i.e., $\mathbf{f}_{21} = -\mathbf{f}_{12}$). By Newton's second law of motion, the acceleration experienced by m_2 is proportional to the force acting upon it, and therefore we have the following

Foundations of Space Dynamics, First Edition. Ashish Tewari.

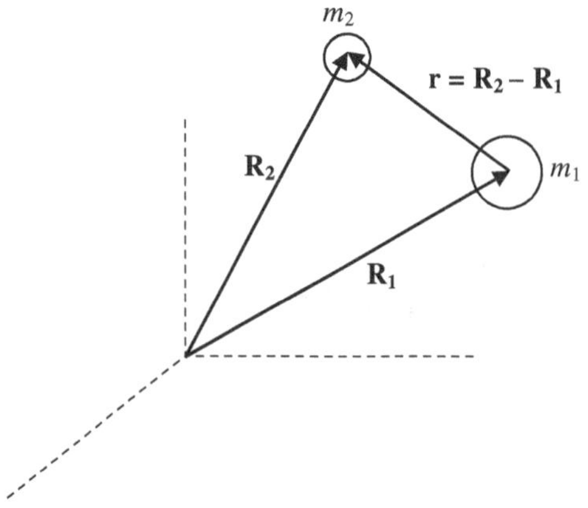

Figure 3.1 Two spherical bodies in mutual gravitational attraction.

equation of motion for the mass m_2:

$$m_2 \frac{\mathrm{d}^2\mathbf{R_2}}{\mathrm{d}t^2} = \mathbf{f_{12}} = -G\frac{m_1 m_2}{r^3}\mathbf{r} \tag{3.2}$$

Similarly, the equation of motion for the mass m_1 is given by

$$m_1 \frac{\mathrm{d}^2\mathbf{R_1}}{\mathrm{d}t^2} = \mathbf{f_{21}} = G\frac{m_1 m_2}{r^3}\mathbf{r}\,. \tag{3.3}$$

If the two equations of motion are added together, one arrives at the following equation of motion for the common centre of mass of the two-body system:

$$\frac{\mathrm{d}^2\mathbf{R_c}}{\mathrm{d}t^2} = \mathbf{0}\,, \tag{3.4}$$

where

$$\mathbf{R_c} = \frac{m_1\mathbf{R_1} + m_2\mathbf{R_2}}{m_1 + m_2} \tag{3.5}$$

represents the position of the centre of mass. Equation (3.4) implies that the centre of mass has a net zero acceleration, since the two-body system is not acted upon by a net external force (because $\mathbf{f_{21}} = -\mathbf{f_{12}}$), which agrees with Newton's third law of motion. Furthermore, Eq. (3.4) is twice integrated in time to yield

$$\mathbf{R_c}(t) = \mathbf{R_c}(0) + \mathbf{v_c}(0)t\,, \tag{3.6}$$

where $\mathbf{R_c}(0)$ and $\mathbf{v_c}(0) = \mathrm{d}\mathbf{R_c}/\mathrm{d}t(0)$ are the position and velocity, respectively, of the centre of mass when $t = 0$. Hence, Eq. (3.6) implies the centre of mass of the two-body system moves in a straight line at a constant velocity, as expected by Newton's first law of motion. These results of the two-body system are thus derived from the n-body problem discussed in Chapter 2 with $n = 2$.

While the motion of the centre of mass of the two-body system is easily derived and solved, the more interesting problem is that of the relative motion of the two masses. The multiplication of Eq. (3.2) by m_1, that of Eq. (3.3) by m_2, and the subtraction of the two resulting equations from each other leads to the following equation governing the relative motion of the mass m_2 around the mass m_1:

$$\frac{d^2\mathbf{r}}{dt^2} + \frac{\mu}{r^3}\mathbf{r} = \mathbf{0}\,, \tag{3.7}$$

where $\mu = G(m_1 + m_2)$ is the *specific gravitational constant* of the two-body system. Equation (3.7) is a non-linear, ordinary vector differential equation of sixth order, to be solved for the relative position vector, $\mathbf{r}(t)$, and velocity, $\mathbf{v}(t) = d\mathbf{r}/dt$, from the specified initial conditions, $\mathbf{r}(0)$ and $\mathbf{v}(0)$, and is referred to as the *two-body problem.* Such a problem forms the core of orbital mechanics because the flight of a spacecraft (say, the mass m_2) under the gravitational attraction of the nearest celestial body (of mass m_1) can be approximated as a two-body problem by ignoring the attraction caused by the other more distant celestial objects, as well as the actual, non-spherical shapes of the two bodies. Such an approximation is valid whenever the gravity of the other bodies, and the gravitational asymmetry resulting from non-spherical shapes, are negligible in magnitude when compared with the inverse-square gravitational attraction between the two primary bodies, m_1 and m_2. When $m_2 \ll m_1$ (which is the case when m_2 is the mass of the spacecraft), one can approximate $\mu \simeq Gm_1$, and the resulting problem is called the *restricted two-body problem.*

The solution to the two-body problem is the starting point for developing the actual trajectory of a space object primarily influenced by the gravity of a large body. Since the two differential equations governing the two-body motion – either Eqs. (3.2) and (3.3) or Eqs. (3.4) and (3.7) – are each of sixth order, a total of 12 integral constants are necessary to obtain a complete solution. The 10 known integrals of the general n-body problem were derived in Chapter 2, which for $n = 2$ leaves only two more integral constants to be found in case of the two-body problem. However, instead of deriving the remaining two integrals in terms of the translation of, and the rotation about, the centre of mass of the n-body problem with $n = 2$, it is much more intuitive to separate the two-body motion into the following two parts: (i) the motion of the centre of mass, for which six integrals of motion are already available from Eq. (3.6) as $\mathbf{R}_c(0)$ and $\mathbf{v}_c(0)$, and (ii) the relative motion of the two bodies around one another posed as Eq. (3.7), for which the relative position, $\mathbf{r}(t)$, and relative velocity, $\mathbf{v}(t)$, could be expressed in terms of another six integral constants, such as perhaps the initial quantities $\mathbf{r}(0)$ and $\mathbf{v}(0)$. This choice of studying the motion of space objects in a coordinate frame fixed at the centre of the primary body has a well established astronomical tradition, and space flight problems typically involve navigating the spacecraft around a large primary body, where $\mathbf{r}(t)$ offers a practical description of the relative position. However, being measured from a moving origin, $\mathbf{r}(t)$ is not an inertial position vector.

3.2 Orbital Angular Momentum

The non-linear differential equation, Eq. (3.7), is not expected to possess a closed-form solution. However, it can be analyzed to yield the constants of integration required for a complete solution, which is not possible for the general n-body problem with $n > 2$. The vector product of Eq. (3.7) with **r** yields the following:

$$\mathbf{r}\times\frac{d^2\mathbf{r}}{dt^2}+\mathbf{r}\times\frac{\mu}{r^3}\mathbf{r}=\mathbf{0}, \tag{3.8}$$

or, representing the time derivatives by overdots and simplifying, we have

$$\mathbf{r}\times\frac{d^2\mathbf{r}}{dt^2}=\mathbf{r}\times\ddot{\mathbf{r}}=\mathbf{0}, \tag{3.9}$$

resulting in

$$\frac{d}{dt}(\mathbf{r}\times\dot{\mathbf{r}})-\dot{\mathbf{r}}\times\dot{\mathbf{r}}=\mathbf{0}, \tag{3.10}$$

from which it follows that

$$\frac{d}{dt}(\mathbf{r}\times\mathbf{v})=\frac{d\mathbf{h}}{dt}=\mathbf{0}, \tag{3.11}$$

where

$$\mathbf{h}=\mathbf{r}\times\frac{d\mathbf{r}}{dt}=\mathbf{r}\times\mathbf{v} \tag{3.12}$$

is the *specific angular momentum* of the two-body system, which is also referred to as the *orbital angular momentum*. It must be noted that **h** is *not* the net angular momentum of the two-body system about the common centre of mass, since it is measured about a moving point (the centre of the spherical mass m_1).

Equation (3.11) implies that the orbital angular momentum of the two-body system is conserved. Because **h** is a constant vector, it represents three scalar integral constants of the two-body motion, and this fact leads to the following important consequences:

(a) The direction of $\mathbf{h}=\mathbf{r}\times\mathbf{v}$ is a constant, which implies that the vectors **r** and **v** are always in the same plane, and **h** is normal to that plane. Therefore, a coordinate frame describing the motion can be chosen with its origin at the centre of the mass, m_1, such that one of the axes is along **h**, and the orbital motion occurs in the constant plane normal to **h**. This constant plane which always contains the vectors $\mathbf{r}(t)$ and $\mathbf{v}(t)$ is called the *orbital plane*.

(b) The magnitude of **h** is constant. If the motion in the orbital plane is represented by the polar coordinates (r,θ) such that $r=|\mathbf{r}|$ and the angle θ is measured along the direction of relative motion described by the curled fingers of the right hand when the thumb is pointed along **h**, then the relative velocity can be expressed as follows:

$$\mathbf{v}=\dot{r}\frac{\mathbf{r}}{r}+\boldsymbol{\omega}\times\mathbf{r}, \tag{3.13}$$

where $\boldsymbol{\omega}=\dot{\theta}\mathbf{h}/h$ represents the angular velocity of **r** about the centre of the mass m_1. Thus the magnitude of **h** is given by

$$h=|\mathbf{r}\times\mathbf{v}|=\left|\mathbf{r}\times\left(\dot{r}\frac{\mathbf{r}}{r}+\boldsymbol{\omega}\times\mathbf{r}\right)\right|, \tag{3.14}$$

which results in the following because the vector product with **r** yields only the transverse velocity component in the angular momentum:

$$h=|\mathbf{r}\times(\boldsymbol{\omega}\times\mathbf{r})|=\left|\mathbf{r}\times\left(\dot{\theta}\frac{\mathbf{h}}{h}\times\mathbf{r}\right)\right|=r^2\dot{\theta}. \tag{3.15}$$

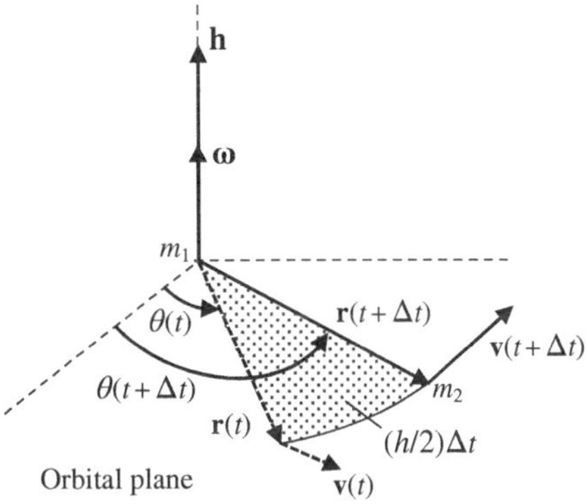

Figure 3.2 Orbital angular momentum and the constant orbital plane.

Equation (3.15) implies that the radius vector, **r**, sweeps out area at a constant rate, $h/2 = \frac{1}{2}r^2\dot{\theta}$, called the *areal velocity*, as shown in Fig. 3.2. This is the general form of *Kepler's second law of planetary motion*. The orbital angular momentum, **h**, provides an important constant parameter for classifying the two-body trajectories, wherein the orbital plane is represented by the direction of **h**, and the nature (or the shape) of the orbit is determined from the magnitude, h. For example, $h = 0$ represents the rectilinear motion along the line joining the two bodies, while $h \neq 0$ denotes the more common motion involving the rotation of the two bodies about one another (called an *orbit*), a further classification of which is derived from **h** (as explained later).

3.3 Orbital Energy Integral

Another important integral constant of the two-body motion is derived by taking the scalar product of Eq. (3.7) with $\mathbf{v} = \dot{\mathbf{r}}$ as follows:

$$\dot{\mathbf{r}} \cdot \ddot{\mathbf{r}} = -\dot{\mathbf{r}} \cdot \frac{\mu}{r^3}\mathbf{r}\,, \tag{3.16}$$

or

$$\frac{1}{2}\frac{\mathrm{d}}{\mathrm{d}t}(\dot{\mathbf{r}} \cdot \dot{\mathbf{r}}) = -\frac{\mu}{r^3}(\dot{\mathbf{r}} \cdot \mathbf{r})\,. \tag{3.17}$$

Of the two scalar products required in Eq. (3.17), the one on the left-hand side is expressed as follows:

$$\dot{\mathbf{r}} \cdot \dot{\mathbf{r}} = \mathbf{v} \cdot \mathbf{v} = v^2\,, \tag{3.18}$$

while the one the right-hand side of Eq. (3.17) is given by

$$\dot{\mathbf{r}} \cdot \mathbf{r} = \left(\dot{r}\frac{\mathbf{r}}{r} + \boldsymbol{\omega} \times \mathbf{r}\right) \cdot \mathbf{r}\,, \tag{3.19}$$

and a simplification of Eq. (3.19) yields the following result:

$$\dot{\mathbf{r}} \cdot \mathbf{r} = \dot{r}r\,. \tag{3.20}$$

The substitution of Eqs. (3.18) and (3.20) into (3.17) produces the following:

$$\frac{\mathrm{d}}{\mathrm{d}t}\left(\frac{1}{2}v^2\right) = -\frac{\mu\dot{r}}{r^2}, \tag{3.21}$$

or

$$\frac{\mathrm{d}}{\mathrm{d}t}\left(\frac{1}{2}v^2 - \frac{\mu}{r}\right) = 0, \tag{3.22}$$

which yields the following scalar integral constant of the two-body motion, called the *specific relative energy*, or the *energy integral* (or more commonly, the *orbital energy*):

$$\varepsilon = \frac{v^2}{2} - \frac{\mu}{r}. \tag{3.23}$$

The new constant, ε, thus represents the sum of the specific kinetic energy, $\frac{v^2}{2}$, and the specific potential energy, $-\frac{\mu}{r}$, of the relative motion of the two spherical masses. The constant energy, ε, is classically referred to as the *vis viva*, or *living force*, because a change in the radius r produces a corresponding change in the speed v; therefore, it is thought of as being responsible for the motion. However, remember that ε is not the total energy of the two-body system, because it does not include the kinetic energy of the mass m_1.

The energy integral is used to classify the two-body trajectories as follows. For a closed trajectory, the radius, r, is bounded, which implies that the specific potential energy, $-\frac{\mu}{r}$, must always be greater in magnitude than the specific kinetic energy, $\frac{v^2}{2}$, or $\varepsilon < 0$. For an open trajectory (called an *escape trajectory*), the specific kinetic energy must be either greater than or equal to the magnitude of specific potential energy, thereby implying $\varepsilon \geq 0$.

3.4 Orbital Eccentricity

The shape of a non-rectilinear ($h \neq 0$) trajectory is determined by another scalar integral constant of the two-body motion, which is derived by taking the vector product of Eq. (3.7) with $\mathbf{h}$:

$$\ddot{\mathbf{r}} \times \mathbf{h} + \frac{\mu}{r^3}(\mathbf{r} \times \mathbf{h}) = \mathbf{0}. \tag{3.24}$$

The first term on the left-hand side of Eq. (3.24) is expressed as follows:

$$\ddot{\mathbf{r}} \times \mathbf{h} = \frac{\mathrm{d}}{\mathrm{d}t}(\mathbf{v} \times \mathbf{h}), \tag{3.25}$$

since $\mathbf{h}$ is a constant vector. Furthermore, the identity $\mathbf{r} \cdot \mathbf{v} = r\dot{r}$, which was derived in Eq. (3.20), is employed as follows in order to evaluate the second term on the left-hand side of Eq. (3.24):

$$\begin{aligned}\frac{\mu}{r^3}(\mathbf{r} \times \mathbf{h}) &= \frac{\mu}{r^3}[\mathbf{r} \times (\mathbf{r} \times \mathbf{v})] \\ &= \frac{\mu}{r^3}[(\mathbf{r} \cdot \mathbf{v})\mathbf{r} - (\mathbf{r} \cdot \mathbf{r})\mathbf{v}] \\ &= \frac{\mu\dot{r}}{r^2}\mathbf{r} - \frac{\mu}{r}\mathbf{v}\end{aligned}$$

$$= -\mu \left(\frac{\mathbf{v}}{r} - \frac{\dot{r}}{r^2}\mathbf{r} \right)$$

$$= -\mu \frac{\mathrm{d}}{\mathrm{d}t} \left(\frac{\mathbf{r}}{r} \right) . \tag{3.26}$$

A substitution of Eqs. (3.25) and (3.26) into Eq. (3.24) yields the following result:

$$\frac{\mathrm{d}}{\mathrm{d}t} \left(\mathbf{v} \times \mathbf{h} - \frac{\mu \mathbf{r}}{r} \right) = \mathbf{0} . \tag{3.27}$$

Therefore, the following constant vector, **e**, called the *eccentricity vector*, can be deduced from Eq. (3.27):

$$\mathbf{e} = \frac{\mathbf{v} \times \mathbf{h}}{\mu} - \frac{\mathbf{r}}{r} . \tag{3.28}$$

Since **e** is a constant vector lying in the orbital plane, it is related to the orbital angular momentum by $\mathbf{e} \cdot \mathbf{h} = 0$, and its magnitude, $e = \mid \mathbf{e} \mid$, called the *orbital eccentricity*, supplies the fifth scalar constant required for solving Eq. (3.7). This imparts important information about the shape of the two-body orbit:

$$e^2 = \mathbf{e} \cdot \mathbf{e} = \frac{1}{\mu^2} (\mathbf{v} \times \mathbf{h}) \cdot (\mathbf{v} \times \mathbf{h}) - \frac{2}{\mu r} \mathbf{r} \cdot (\mathbf{v} \times \mathbf{h}) + 1 . \tag{3.29}$$

Because **v** and **h** are mutually perpendicular, it follows that $(\mathbf{v} \times \mathbf{h}) \cdot (\mathbf{v} \times \mathbf{h}) = v^2 h^2$. Furthermore, it is true that $\mathbf{r} \cdot (\mathbf{v} \times \mathbf{h}) = (\mathbf{r} \times \mathbf{v}) \cdot \mathbf{h} = h^2$. Therefore, Eq. (3.29) results in

$$1 - e^2 = \frac{v^2 h^2}{\mu^2} - \frac{2h^2}{\mu r} + 1 = \frac{h^2}{\mu} \left(\frac{2}{r} - \frac{v^2}{\mu} \right) . \tag{3.30}$$

It is useful to define a *parameter, p*, and another constant called the *semi-major axis*, *a*, both having units of length, by the

$$p = \frac{h^2}{\mu} ; \quad \frac{1}{a} = \frac{2}{r} - \frac{v^2}{\mu} , \tag{3.31}$$

which are mutually related by $p = a(1 - e^2)$, according to Eq. (3.30). The definition of a relates it to the energy integral, ε, as follows:

$$\varepsilon = \frac{v^2}{2} - \frac{\mu}{r} = -\frac{\mu}{2a} . \tag{3.32}$$

The simplest two-body, non-rectilinear trajectory is a *circular* orbit of m_2 about m_1, wherein the orbital radius, r, is constant, and Eq. (3.13) yields the following expression for the relative velocity:

$$\mathbf{v} = \boldsymbol{\omega} \times \mathbf{r} . \tag{3.33}$$

From that it follows that $\frac{\mathrm{d}^2\mathbf{r}}{\mathrm{d}t^2} = \dot{\mathbf{v}} = \boldsymbol{\omega} \times (\boldsymbol{\omega} \times \mathbf{r}) = -\omega^2 \mathbf{r}$, and substituting this result into Eq. (3.7) produces the following:

$$-\omega^2 \mathbf{r} + \frac{\mu}{r^3} \mathbf{r} = \mathbf{0} . \tag{3.34}$$

Therefore, Eq. (3.34) shows that a circular orbit is a two-body trajectory if and only if $\omega^2 = \mu / r^3$, which is *Kepler's third law* for a circular orbit, and yields the constant frequency of circular motion to be $\omega = \dot{\theta} = \sqrt{\mu/r^3}$. The orbital speed in a circular orbit is given from Eq. (3.33) by $v = r\omega = \sqrt{\mu/r}$, and substituted into the energy integral, Eq. (3.32), it leads to

$$-\frac{\mu}{2a} = \frac{v^2}{2} - \frac{\mu}{r} = \frac{\mu}{2r} - \frac{\mu}{r} = -\frac{\mu}{2r}, \tag{3.35}$$

or, $r = a$. Furthermore, since $\mathbf{h} = \mathbf{r} \times \mathbf{v} = rv\mathbf{i}_h$, it follows that $p = h^2/\mu = r^2v^2/\mu = r^4\omega^2/\mu = r$. Thus, for circular orbit, we have $r = a = p$, which implies a zero orbital eccentricity, $e = 0$.

Example 3.4.1 *Determine the orbital frequency, radius, and orbital speed of a satellite in a geosynchronous equatorial orbit (GEO). A GEO is a special circular orbit wherein the spacecraft orbits Earth in the equatorial plane such that it appears to be stationary when seen from any point on the surface of Earth. This implies that the orbital radius of a GEO satellite must be selected such that the orbital frequency, ω, matches the angular speed of rotation of Earth on its axis. Now, Earth takes 23 hours, 56 minutes, and 4.09 seconds to complete one rotation as measured against the background of distant stars. This period of inertial rotation of Earth is called its* sidereal day *(Chap. 1), and is different from the period of rotation of Earth as observed from the Sun, which is the 24-hour* mean solar day. *Thus the orbital frequency of a GEO satellite is given by*

$$\omega = \frac{2\pi}{23 \times 3600 + 56 \times 60 + 4.09} = 7.2921159 \times 10^{-5} \text{ rad./s.}$$

To carry out the calculation of the GEO orbital radius, the value of Earth's gravitational constant, μ, is necessary. Neglecting the mass of the satellite in comparison with that of Earth, this constant is $\mu \simeq Gm_1 = 398600.4 \text{ km}^3/\text{s}^2$, *which, substituted into Eq. (3.34), yields*

$$r = \left(\frac{\mu}{\omega^2}\right)^{1/3} = 42164.168 \text{ km}$$

as the radius of the GEO orbit. Finally, the orbital speed is calculated by Eq. (3.35) to be the following:

$$v = \sqrt{\frac{\mu}{r}} = 3.07466 \text{ km/s.}$$

This example deals with a special trajectory of the two-body motion, namely a circular orbit (e = 0). A more general shape of the Keplerian trajectories is derived from orbital eccentricity in Section 3.5.

Example 3.4.2 *The radius and velocity vectors of a spacecraft in an Earth-centred (geocentric), inertial reference frame,* $(\mathbf{I}, \mathbf{J}, \mathbf{K})$, *are observed to be the following:*

$$\mathbf{r} = -7000\mathbf{I} + 1000\mathbf{J} + 500\mathbf{K} \text{ km}$$

$$\mathbf{v} = 5\mathbf{I} + 7.5\mathbf{J} - 3\mathbf{K} \text{ km/s} .$$

Find the semi-major axis, orbital angular momentum, and eccentricity.

The radius and orbital speed are first computed as follows:

$$r = |\mathbf{r}| = \sqrt{7000^2 + 1000^2 + 500^2} = 7088.7234 \text{ km}$$

$$v = |\mathbf{v}| = \sqrt{5^2 + 7.5^2 + 3^2} = 9.5 \text{ km/s} .$$

Then using the value of $\mu = 398600.4\ \text{km}^3/\text{s}^2$ *for Earth, we have the following value for the orbital energy:*

$$\varepsilon = v^2/2 - \mu/r = -11.1052\ \text{km}^2/\text{s}^2\ ,$$

which yields the semi-major axis to be

$$a = \frac{-\mu}{2\varepsilon} = 17946.551\ \text{km}\ .$$

The orbital angular momentum is calculated as follows:

$$\mathbf{h} = \mathbf{r} \times \mathbf{v} = -6750\mathbf{I} - 18500\mathbf{J} - 57500\mathbf{K}\ \text{km}^2/\text{s}$$
$$h = \mid \mathbf{h} \mid = 60778.7998\ \text{km}^2/\text{s}\ .$$

Finally, the eccentricity vector is computed via Eq. (3.28) to be the following:

$$\begin{aligned}\mathbf{e} &= \frac{\mathbf{v} \times \mathbf{h}}{\mu} - \frac{\mathbf{r}}{r} \\ &= -0.233664\mathbf{I} + 0.631007\mathbf{J} - 0.1755897\mathbf{K}\ ,\end{aligned} \tag{3.36}$$

whose magnitude gives the orbital eccentricity,

$$e = \mid \mathbf{e} \mid = 0.69541417\ .$$

An alternative way of calculating e is from the magnitude of angular momentum vector, h, as follows:

$$p = h^2/\mu = 9267.5835\ \text{km}$$
$$e = \sqrt{1 - \frac{p}{a}} = 0.69541417\ .$$

3.5 Orbit Equation

Two-body (or Keplerian) trajectories have well defined shapes. Section 3.4 showed that a constant angular momentum results in a planar two-body motion, and rectilinear and circular orbits are special shapes of Keplerian trajectories. A further classification of Keplerian trajectories can be derived from the constant eccentricity vector, **e**, by taking the scalar product of $\mu\mathbf{e}$ with the position, **r**, to obtain

$$\mathbf{r} \cdot (\mathbf{v} \times \mathbf{h}) = \mathbf{r} \cdot \mu\frac{\mathbf{r}}{r} + \mu\mathbf{r} \cdot \mathbf{e}\,, \tag{3.37}$$

the left-hand side of which is simplified using a triple-product identity as follows:

$$\mathbf{r} \cdot (\mathbf{v} \times \mathbf{h}) = (\mathbf{r} \times \mathbf{v}) \cdot \mathbf{h} = \mathbf{h} \cdot \mathbf{h} = h^2\,. \tag{3.38}$$

A substitution of Eq. (3.38) into Eq. (3.37) yields

$$h^2 = \mu(r + \mathbf{r} \cdot \mathbf{e})\,. \tag{3.39}$$

If the angle between the vectors **r** and **e** measured along the direction of motion of m_2 about m_1 is defined to be the *true anomaly*, θ, then Eq. (3.39) results in the following expression relating the radius, r, with θ:

$$r = \frac{p}{1 + e\cos\theta}, \tag{3.40}$$

where $p = h^2/\mu \neq 0$ is the orbital parameter defined earlier. Equation (3.40) is called the *orbit equation* defining the shape of the trajectory in polar coordinates, r and θ. The orbit equation indicates that there is no change in the radius, r, when the sign of the true anomaly, θ, is changed, which implies that the trajectory is symmetrical about the eccentricity vector, **e**. Equation (3.40) indicates that the shape of the orbit is determined by the orbital eccentricity, e, while its size (or the span) is indicated by the parameter, p. Furthermore, the minimum separation of the two bodies, called the *periapsis*[1] radius, $r_p = p/(1+e)$, occurs when $\theta = 0$, which indicates that the constant eccentricity vector, **e**, points towards the occurrence of the minimum orbital radius (the periapsis). Equation (3.40) also dictates that the maximum separation of the two bodies, called the *apoapsis* radius, $r_a = p/(1-e)$, occurs at $\theta = \pi$ if and only if $e < 1$. For $e = 1$, the maximum orbital radius is infinite, which is reached asymptotically in the limit, $\theta \to \pi$, whereas the maximum radius does not exist for $e > 1$. For $\theta = \pi/2$ and $\theta = 3\pi/2$, the orbital radius equals the orbital parameter, $r = p$, which confirms the symmetry of the orbit about the eccentricity vector, $\theta = 0$.

An alternative yet more intuitive derivation of the orbit equation is possible using the polar coordinates. We first express the acceleration in the polar coordinates, (r, θ), with unit vectors $(\mathbf{i}_r, \mathbf{i}_\theta)$ and angular velocity, $\boldsymbol{\omega} = \dot{\theta}\mathbf{i}_h$, as follows:

$$\begin{aligned}
\mathbf{r} &= r\mathbf{i}_r \\
\dot{\mathbf{r}} &= \dot{r}\mathbf{i}_r + \boldsymbol{\omega} \times \mathbf{r} = \dot{r}\mathbf{i}_r + r\dot{\theta}\mathbf{i}_\theta \\
\ddot{\mathbf{r}} &= \ddot{r}\mathbf{i}_r + \boldsymbol{\omega} \times (\dot{r}\mathbf{i}_r) + \dot{r}\dot{\theta}\mathbf{i}_\theta + r\ddot{\theta}\mathbf{i}_\theta + \boldsymbol{\omega} \times (r\dot{\theta}\mathbf{i}_\theta) \\
&= (\ddot{r} - r\dot{\theta}^2)\mathbf{i}_r + (r\ddot{\theta} + 2\dot{r}\dot{\theta})\mathbf{i}_\theta .
\end{aligned} \tag{3.41}$$

The equation of Keplerian motion, Eq. (3.7), is resolved into its constituent radial and circumferential components as follows:

$$\begin{aligned}
\ddot{r} - r\dot{\theta}^2 + \frac{\mu}{r^2} &= \ddot{r} - \frac{h^2}{r^3} + \frac{\mu}{r^2} = 0 \\
r\ddot{\theta} + 2\dot{r}\dot{\theta} &= \frac{1}{r}\frac{\mathrm{d}}{\mathrm{d}t}(r^2\dot{\theta}) = \frac{1}{r}\frac{\mathrm{d}h}{\mathrm{d}t} = 0 .
\end{aligned} \tag{3.42}$$

The second of Eqs. (3.42) is the well known principle of conservation of orbital angular momentum, h, whereas the first of Eq. (3.42) is transformed as follows in terms of the *flight-path curvature*, $\rho = 1/r$, and the derivatives with respect to θ represented by primes, $(.)' = \mathrm{d}(.)/\mathrm{d}\theta$,

[1] The general suffix *apsis* in the noun and the corresponding *apsis* in the adjective can be replaced by a more specific *gee* when the central body, m_1, is Earth, *helion* when m_1 is the sun, *lune* when the central body is the moon, etc.

$(.)'' = \mathrm{d}^2(.)/\mathrm{d}\theta^2$, etc.:

$$\begin{aligned}\dot{r} &= r'\dot{\theta} \\ \ddot{r} &= r''\dot{\theta}^2 + r'\ddot{\theta} \\ \ddot{\theta} &= -2\frac{h}{r^3}\dot{r} = -2\frac{h^2}{r^5}r' ,\end{aligned} \tag{3.43}$$

where

$$\begin{aligned}r' &= -\frac{\rho'}{\rho^2} \\ r'' &= -\frac{\rho''}{\rho^2} + 2\frac{(\rho')^2}{\rho^3} ,\end{aligned} \tag{3.44}$$

A substitution of Eqs. (3.43) and (3.44) into the first of Eqs. (3.42) yields the following transformed equation of radial motion:

$$\rho'' + \rho = \frac{\mu}{h^2} , \tag{3.45}$$

which is a linear, ordinary differential equation in ρ, and has the following general solution:

$$\rho(\theta) = A\cos\theta + B\sin\theta + \frac{\mu}{h^2} , \tag{3.46}$$

with A, B, μ, h being constants. Since $\theta = 0$ represents the point of the minimum radius, we have $r'(0) = 0$, or $\rho'(0) = 0 = B$. Therefore the shape of the orbit is given by

$$\rho(\theta) = A\cos\theta + \frac{\mu}{h^2} \tag{3.47}$$

or, in terms of the periapsis radius, $r_p = 1/\rho(0)$, we have

$$A = \frac{1}{r_p} - \frac{\mu}{h^2} . \tag{3.48}$$

Hence, the shape of the orbit is described by

$$r(\theta) = \frac{1}{\rho(\theta)} = \frac{h^2/\mu}{1 + \left(\frac{h^2}{\mu r_p} - 1\right)\cos\theta} , \tag{3.49}$$

which in terms of the orbital parameter, $p = h^2/\mu$, and eccentricity, $e = p/r_p - 1$, becomes the orbit equation, Eq. (3.40).

When Eq. (3.40) is converted to Cartesian coordinates by the transformation, $x = r\cos\theta$ and $y = r\sin\theta$, it results in the following for $e \neq 1$ ($p \neq 0$):

$$\frac{(x + ae)^2}{a^2} + \frac{y^2}{p^2} = 1 , \tag{3.50}$$

which represents the equation of an *ellipse* for $e < 1$, and that of a *hyperbola* for $e > 1$, with one of the *foci* at the origin, $x = 0$, $y = 0$. For the case of $e = 1$ ($p \neq 0$), the given coordinate transformation leads to

$$y^2 = p(p - 2x) , \tag{3.51}$$

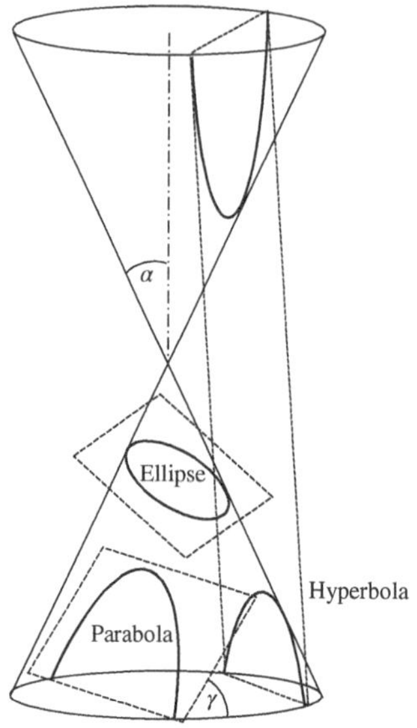

Figure 3.3 The conic section shapes of a non-rectilinear Keplerian orbit.

which is the equation of a *parabola* with the origin at one of the foci. Since an ellipse, a hyperbola, and a parabola are the shapes obtained by cutting a right-circular cone in a particular manner, as shown in Fig. 3.3, it becomes evident that the non-rectilinear ($p \neq 0$) orbit is a *conic section*. The straight lines generating the cone can be extended as shown in the figure to produce a mirror cone sharing the axis and the semi-vertex angle, α, with the original cone. The orbital plane, xy, is the sectional plane, whose orientation relative to the cone's base depends upon the orbital eccentricity, e, and produces an orbit of the corresponding shape. Let the angle, γ, between the base and the sectional plane be called the *dihedral angle*, which is measured as shown in Fig. 3.3. When the sectional plane is parallel to the cone's base (i.e., $\gamma = 0$), the conic section obtained is a circle corresponding to $e = 0$. When the section is taken at the dihedral angle, $0 < \gamma < (\pi/2 - \alpha)$, while not cutting the base, the shape produced is an ellipse ($0 < e < 1$). If the sectional plane cuts the base at the dihedral angle, $0 < \gamma < (\pi/2 - \alpha)$, the shape produced is a parabola, ($e = 1$). A section with $\gamma > (\pi/2 - \alpha)$ cuts the bases of both the cones as shown in Fig. 3.3, and produces a hyperbola ($e > 1$), which has the two branches shown in the figure.

The various conic section orbits are depicted in Figs. 3.4–3.6. The x-axis in each case is along the eccentricity vector, $\mathbf{e}$, and the y-axis is perpendicular to the eccentricity vector. The origin, $x = 0$, $y = 0$, coincides with one of the foci, O, of each conic section, which is occupied by the centre of m_1. The other focus, V, is unoccupied, and is thus termed the *vacant focus*. The *semi-major axis*, a, is seen to have a geometric significance with regard to each conic section, and determines its orbital energy, $\varepsilon = -\mu/(2a)$. The *parameter*, p, is the length of the *chord* (also called the *semi-latus rectum*) drawn from the focus normal to the x-axis. Table 3.1 summarizes the possible shapes of a two-body trajectory.

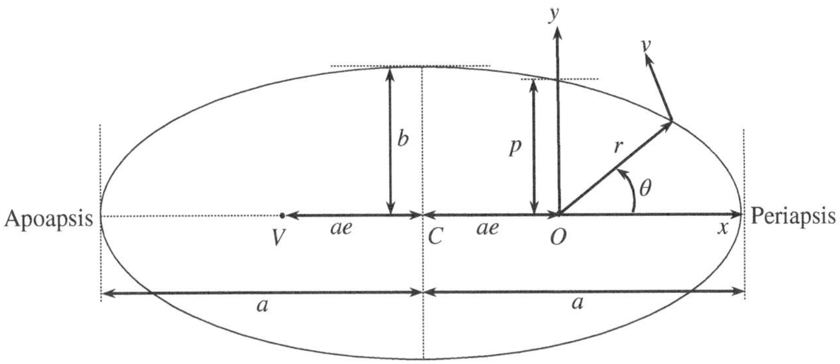

Figure 3.4 The elliptic orbit ($e < 1$).

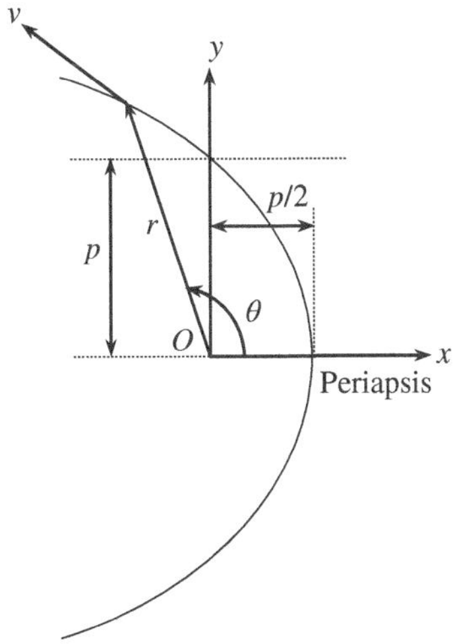

Figure 3.5 The parabolic orbit ($e = 1$).

3.5.1 *Elliptic Orbit*

The elliptic orbit has $0 \leq e < 1, a > 0, p = a(1 - e^2) > 0$, and the distance between the periapsis and the apoapsis is the *major axis*, $2a$, which is the maximum span of the orbit along the x-axis, as shown in Fig. 3.4. The *periapsis radius*, r_p, is the radius when $\theta = 0$, and can be derived from the orbit equation to be the following:

$$r_p = \frac{a(1 - e^2)}{1 + e} = a(1 - e)\,. \tag{3.52}$$

Similarly, the apoapsis radius, r_a, occurs for $\theta = \pi$, and is obtained from the orbit equation as follows:

$$r_a = \frac{a(1 - e^2)}{1 - e} = a(1 + e)\,. \tag{3.53}$$

Table 3.1 Two-body orbits.

	$(h \neq 0)$		
$0 \leq e < 1$	$a > 0$	$\varepsilon < 0$	Ellipse
$e = 1$	$a = \infty$	$\varepsilon = 0$	Parabola
$e > 1$	$a < 0$	$\varepsilon > 0$	Hyperbola
	$(h = 0)$		
$e = 1$	$a > 0$	$\varepsilon < 0$	Rectilinear ellipse
$e = 1$	$a = \infty$	$\varepsilon = 0$	Rectilinear parabola
$e = 1$	$a < 0$	$\varepsilon > 0$	Rectilinear hyperbola

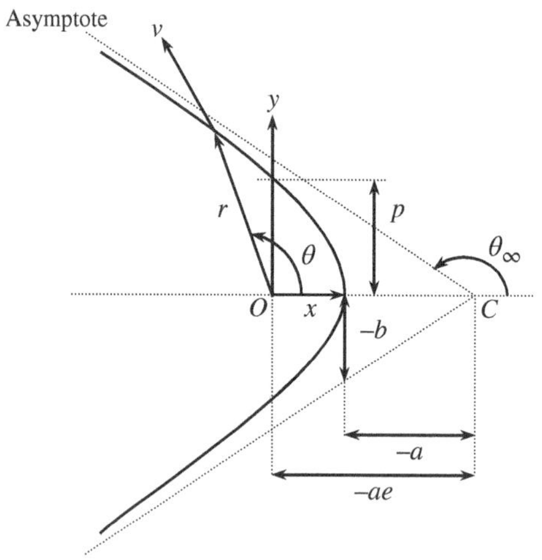

Figure 3.6 The hyperbolic orbit ($e > 1$).

The centre of the ellipse, C, is the point on the x-axis equidistant from the two focii, as well as equidistant from the periapsis and the apoapsis; hence it is located by the coordinates, $x = -ae$, $y = 0$, as shown in Fig. 3.4. The maximum distance spanned by the elliptic orbit along the y-axis is the *minor axis*, $2b$, where b is termed the *semi-minor axis*. The value of b can be derived from the orbit equation as follows:

$$b = \pm \bar{r} \sin \bar{\theta} \,, \tag{3.54}$$

where $\bar{r}$ and $\bar{\theta}$ are the polar coordinates corresponding to the position $x = -ae$, $y = \pm b$. Hence, the coordinate transformation produces $\bar{r} \cos \bar{\theta} = -ae$, which, substituted into the orbit equation, yields

$$\bar{r} = \frac{a(1 - e^2)}{1 - \frac{ae^2}{\bar{r}}} \,, \tag{3.55}$$

or $\bar{r} = a$, from which it follows that $\cos \bar{\theta} = -e$, and $\sin \bar{\theta} = \pm\sqrt{1 - e^2}$. Substitution of these into Eq. (3.54) produces

$$b = a\sqrt{1 - e^2} \,. \tag{3.56}$$

It is interesting to note that the value of the orbital radius at the midpoint of the orbit is derived from Eq. (3.55) to exactly equal the semi-major axis, $\bar{r} = a$. Therefore, the semi-major axis, a of the elliptic orbit is also called its mean radius, and is the average of the periapsis and the apoapsis radii, $a = (r_p + r_a)/2$.

Since elliptic orbit involves a periodic motion of m_2 about m_1, its period, T, can be derived from Eq. (3.15), which gives the constant areal velocity, $\frac{1}{2}r^2\dot{\theta} = h/2$, of the radius vector. The area of the ellipse is derived by using Eq. (3.50) as follows:

$$A = 2\int_{a(1-e)}^{a(1+e)} y\mathrm{d}x = 2p\int_{a(1-e)}^{a(1+e)} \sqrt{1-\frac{(x+ae)^2}{a^2}}\mathrm{d}x$$

$$= 2b\int_{-a}^{a}\sqrt{1-\frac{\xi^2}{a^2}}\mathrm{d}\xi = \frac{b}{a}\int_{-a}^{a} 2\sqrt{a^2-\xi^2}\mathrm{d}\xi \tag{3.57}$$

$$= \frac{b}{a}\pi a^2 = \pi ab\,. \tag{3.58}$$

Since $A = \pi ab$ is the area swept by the radius vector in a complete revolution, the period of the revolution, T, is obtained by dividing A by the constant areal velocity, $h/2$, as follows:

$$T = \frac{2\pi ab}{h} = 2\pi\frac{ab}{\sqrt{\mu p}} = 2\pi\frac{a^2\sqrt{1-e^2}}{\sqrt{\mu a(1-e^2)}} = \frac{2\pi}{\sqrt{\mu}}a^{3/2}\,. \tag{3.59}$$

Equation (3.59) relating the period of revolution to the semi-major axis is an important result, and verifies *Kepler's third law of planetary motion*, which states that the square of the period of revolution, T, is directly proportional to the cube of the mean radius, a. It is interesting to see that the period of the revolution is independent of the orbital eccentricity, e. The constant of proportionality indicated by Kepler's law is obtained in Eq. (3.59) to be $2\pi/\sqrt{\mu}$, which can be used to estimate the mass of the central body, m_1, by measuring T and a of an orbiting body, m_2. The average angular frequency of the orbit, called the *mean motion*, is defined by

$$n = \frac{2\pi}{T} = \sqrt{\mu/a^3}\,. \tag{3.60}$$

The circular orbit is a special ellipse with $e = 0$ and $r = a = p$, and has a constant angular frequency, $\dot{\theta} = n$.

Example 3.5.1 *Two points in an orbit have the radius and true anomaly given by* $r_1 = 6700$ *km,* $\theta_1 = 25°$*, and* $r_2 = 6900$ *km,* $\theta_2 = 120°$*, respectively. Find the semi-major axis, the orbital eccentricity, and the periapsis radius.*

Using the orbit equation for the two given positions, we write

$$\frac{r_2}{r_1} = \frac{1+e\cos\theta_1}{1+e\cos\theta_2}\,,$$

$$e = \frac{1-r_2/r_1}{\frac{r_2}{r_1}\cos\theta_2 - \cos\theta_1}$$

$$= \frac{1-69/67}{\frac{69}{67}\cos(120°)-\cos(25°)} = 0.021\,.$$

Hence, the orbital eccentricity is $e = 0.021$, *which implies an elliptic orbit. Next, the parameter is calculated as follows:*

$$p = r_1(1 + e\cos\theta_1) = 6827.5382 \text{ km} ,$$

from which the semi-major axis is obtained to be

$$a = \frac{p}{1 - e^2} = 6830.5515 \text{ km} .$$

The periapsis radius is $r_p = a(1 - e) = 6687.0866$ *km. Note that the value of the gravitational constant,* μ, *is not required in these calculations.*

3.5.2 *Parabolic Orbit*

For $e = 1$ and $h \neq 0$, the orbit is a parabola, which is an open (or escape) trajectory with $a = \infty$ and $p > 0$. The parabolic escape requires the minimum possible orbital energy, $\varepsilon = v^2/2 - \mu/r = 0$, which implies that $v = \sqrt{2\mu/r}$ and $\lim_{r\to\infty} v = 0$. The orbit equation for a parabolic orbit becomes the following:

$$r = \frac{p}{1 + \cos\theta} , \tag{3.61}$$

from which it follows that the periapsis radius is $r_p = r_{\theta=0} = p/2$ (as shown in Fig. 3.5), and the maximum possible orbital radius is $r_{\theta=\pi} = \infty$, thereby confirming the infinite semi-major axis of the parabolic orbit. It can be said that the centre, the vacant focus, and the apoapsis of a parabolic orbit are all located at an infinite distance from the focus, O, i.e., at $x = -\infty$.

Example 3.5.2 *Calculate the velocity at the true anomaly,* $\theta = 85°$, *in a parabolic orbit around Earth with a periapsis radius,* $r_p = 6800$ *km.*

The value of the radius at the given position is calculated as follows:

$$r = \frac{2r_p}{1 + \cos\theta} = \frac{2 \times 6800}{1 + \cos(85°)} = 12509.707 \text{ km} ,$$

from which the parabolic escape velocity is calculated to be the following:

$$v = \sqrt{\frac{2\mu}{r}} = 7.983 \text{ km/s} .$$

3.5.3 *Hyperbolic Orbit*

For $e > 1$, the orbit is the concave branch of a hyperbola[2] (Fig. 3.6), with $a < 0$ and $p = a(1 - e^2) > 0$, and denotes an escape trajectory with $\varepsilon > 0$, which implies that $\lim_{r\to\infty} v > 0$. The negative value of the semi-major axis, a, of the hyperbolic orbit results in the centre, C, being

[2] A hyperbola consists of two symmetric branches centred at C. However, only the concave branch of the hyperbola (i.e., the one curving towards the central body, m_1, as shown in Fig. 3.6) is relevant here because gravity is an attractive force, and hence cannot produce the convex branch as the solution to the two-body problem.

located at $x = -ae > 0$ from the occupied focus, O, as shown in Fig. 3.6. Thus it can be said that the centre has moved from $x = -\infty$ for a parabola to a location ($x = -ae > 0$) at the right of O for a hyperbola. The periapsis is located at a distance $-a$ from the centre, C, as shown in Fig. 3.6, which implies that the periapsis radius is $r_p = a(1 - e)$. This also follows from the orbit equation. The orbit equation further indicates that in the limit $r \to \infty$, the orbit approaches a straight line called an *asymptote*, passing through the centre, C, on each side of the major axis. The angle made by each asymptote with the major axis, θ_∞ (see Fig. 3.6), is derived from the orbit equation as follows:

$$\theta_\infty = \lim_{r\to\infty} \theta = \cos^{-1}\left(-\frac{1}{e}\right) . \tag{3.62}$$

The value of the semi-minor axis, b, defined to be the extent of the asymptote in the y-direction tangential to the periapsis (see Fig. 3.6), can be derived as follows:

$$b = a\tan(\pi - \theta_\infty) = -a\tan\theta_\infty = ae\sqrt{1 - \frac{1}{e^2}} = a\sqrt{e^2 - 1} . \tag{3.63}$$

The speed at infinite radius, called the *hyperbolic excess speed*, v_∞, is given by

$$v_\infty = \lim_{r\to\infty} \sqrt{\frac{2\mu}{r} - \frac{\mu}{a}} = \sqrt{-\frac{\mu}{a}} . \tag{3.64}$$

This is the amount by which the speed in a hyperbolic orbit exceeds that in a parabolic orbit when $r = \infty$.

Example 3.5.3 *Two points in an orbit have the radius and true anomaly given by* $r_1 = 6700$ *km,* $\theta_1 = 5°$, *and* $r_2 = 7200$ *km,* $\theta_2 = 30°$, *respectively. Find the semi-major axis, the orbital eccentricity, and the periapsis radius.*

The orbital eccentricity is calculated using the orbit equation for the two given positions:

$$\frac{r_2}{r_1} = \frac{1 + e\cos\theta_1}{1 + e\cos\theta_2} ,$$

$$e = \frac{1 - r_2/r_1}{\frac{r_2}{r_1}\cos\theta_2 - \cos\theta_1}$$

$$= \frac{1 - 72/67}{\frac{72}{67}\cos(30°) - \cos(5°)} = 1.13864 .$$

The value of the orbital eccentricity, $e = 1.13864$, *implies an hyperbolic orbit. The orbital parameter is calculated as follows:*

$$p = r_1(1 + e\cos\theta_1) = 14299.837 \text{ km} ,$$

from which the semi-major axis is obtained to be

$$a = \frac{p}{1 - e^2} = -48229.789 \text{ km} ,$$

which, as expected, is negative. The periapsis radius is $r_p = a(1 - e) = 6686.426$ *km.*

An additional parameter of the hyperbolic orbit is the asymptote angle, which is calculated as follows:

$$\theta_\infty = \cos^{-1}\left(-\frac{1}{e}\right) = 151.431^\circ \,.$$

The speed at infinite radius (the hyperbolic excess speed), v_∞, *is calculated by*

$$v_\infty = \sqrt{-\frac{\mu}{a}} = 2.8748 \text{ km/s}.$$

3.5.4 Rectilinear Motion

The orbit equation yields only non-rectilinear, Keplerian trajectories (i.e., those with $h \neq 0$). A rectilinear trajectory can be regarded as a conic section whose semi-minor axis vanishes ($b = 0$), which implies that $e = 1$. However, it must be distinguished from a parabolic orbit, which also has $e = 1$, but $p \neq 0$. A rectilinear trajectory has $p = h^2/\mu = a(1 - e^2) = 0$, whose energy, $\varepsilon = -\mu/(2a) = v^2/2 - \mu/r$, is determined by the semi-major axis, a. A bound trajectory (called the *rectilinear ellipse*) has a positive but finite value of a. The orbit with $p = 0$ and $a = \infty$ is the minimum-energy escape (or open) trajectory, called the *rectilinear parabola*, while that with $p = 0$ and $a < 0$ is another escape trajectory, called the *rectilinear hyperbola*.

The orbit equation becomes useless for a rectilinear trajectory, because $r\dot{\theta} = 0$. Instead, the following equation of motion is employed:

$$v = \dot{r} = \sqrt{\mu}\left(\frac{2}{r} - \frac{1}{a}\right)^{1/2} . \tag{3.65}$$

For a rectilinear ellipse ($a > 0$), Eq. (3.65) can be integrated to yield the time elapsed between any two radii, $r(t_1) = r_1$ and $r(t_2) = r_2$, as follows:

$$t_2 - t_1 = \frac{1}{\sqrt{\mu}} \int_{r_1}^{r_2} \frac{rdr}{\sqrt{2r - r^2/a}} . \tag{3.66}$$

The integral in Eq. (3.66) is evaluated by changing the integration variable to ξ as follows:

$$r = a(1 - \cos\xi) , \tag{3.67}$$

which yields

$$t_2 - t_1 = \sqrt{\frac{a^3}{\mu}} \int_{\xi_1}^{\xi_2} (1 - \cos\xi) d\xi , \tag{3.68}$$

or

$$t_2 - t_1 = \frac{1}{n}[(\xi_2 - \sin\xi_2) - (\xi_1 - \sin\xi_1)] , \tag{3.69}$$

with $r_1 = a(1 - \cos\xi_1)$, $r_2 = a(1 - \cos\xi_2)$, and $n = \sqrt{\mu/a^3}$ being the average frequency (the *mean motion*). The angle, ξ, is called an *eccentric anomaly*, which will be further explored in Chapter 4.

For a rectilinear hyperbola ($a < 0$), the integral in Eq. (3.66) is evaluated by changing the integration variable to η as follows:

$$r = a(1 - \cosh\eta)\,, \tag{3.70}$$

which yields

$$t_2 - t_1 = \sqrt{\frac{-a^3}{\mu}} \int_{\eta_1}^{\eta_2} (1 - \cosh\eta)\mathrm{d}\eta\,, \tag{3.71}$$

or

$$t_2 - t_1 = \sqrt{\frac{-a^3}{\mu}}[(\sinh\eta_2 - \eta_2) - (\sinh\eta_1 - \eta_1)]\,, \tag{3.72}$$

with $r_1 = a(1 - \cosh\eta_1)$ and $r_2 = a(1 - \cosh\eta_2)$. The variable, η, is called a *hyperbolic anomaly*, and will be discussed further in Chapter 4.

For a rectilinear parabola ($a = \infty$), we have

$$v = \dot{r} = \sqrt{\frac{2\mu}{r}}\,, \tag{3.73}$$

whose integration yields

$$t_2 - t_1 = \frac{1}{\sqrt{2\mu}} \int_{r_1}^{r_2} \sqrt{r}\mathrm{d}r = \frac{1}{3}\sqrt{\frac{2}{\mu}}(r_2^{3/2} - r_1^{3/2})\,. \tag{3.74}$$

Example 3.5.4 *A spacecraft is launched vertically upwards from the moon's surface* ($\mu = 4902.8$ km^3/s^2, $r_0 = 1737.1$ *km) with a velocity 5 km/s. What is (a) the largest radius reached by the spacecraft, and (b) the time elapsed from launch to reach the radius,* $r = 1800$ *km?*

The nature of the rectilinear trajectory is determined by its energy, calculated to be the following:

$$\varepsilon = v_0^2/2 - \mu/r_0 = 5^2/2 - 4902.8/1737.1 = 9.6776 \text{ km}^2/\text{s}^2\,,$$

and indicates a rectilinear hyperbola, which is an escape trajectory. Hence, the largest radius is infinite. The semi-major axis is $a = -\mu/(2\varepsilon) = -253.307$ *km.*

The time to reach the radius, $r = 1800$ *km, is calculated by Eq. (3.72) as follows:*

$$\eta_1 = \cosh^{-1}(1 - r_1/a) = \cosh^{-1}\left(1 + \frac{1737.1}{253.307}\right) = 2.75057$$

$$\eta_2 = \cosh^{-1}(1 - r_2/a) = \cosh^{-1}\left(1 + \frac{1800}{253.307}\right) = 2.78193$$

$$t_2 - t_1 = \sqrt{\frac{-a^3}{\mu}}[(\sinh\eta_2 - \eta_2) - (\sinh\eta_1 - \eta_1)] = 725.768 \text{ s}\,.$$

Hence, the time elapsed from the launch at the moon's surface to the radius, $r = 1800$ *km, is 725.768 s (12.096 min).*

3.6 Orbital Velocity and Flight Path Angle

The orbital velocity vector is typically resolved in a moving coordinate frame with its origin at the centre of the orbiting mass, m_2, and having mutually perpendicular axes along the radial direction, $\mathbf{i}_r = \mathbf{r}/r$, and the circumferential direction, $\mathbf{i}_\theta$. Such a coordinate system, denoted by the right-handed triad, $(\mathbf{i}_r, \mathbf{i}_\theta, \mathbf{i}_h)$, with $\mathbf{i}_r \times \mathbf{i}_\theta = \mathbf{i}_h = \mathbf{h}/h$, is called a *local-horizon frame*, and is depicted in Fig. 3.7. The third vector of the triad, $\mathbf{i}_h$, is normal to the orbital plane, and comes out of the page in Fig. 3.7. The plane, $(\mathbf{i}_\theta, \mathbf{i}_h)$, which is normal to the radius vector, is called the *local horizon*. The orbital velocity resolved in the local horizon frame is the following:

$$\mathbf{v} = \dot{r}\mathbf{i}_r + r\dot{\theta}\mathbf{i}_\theta \,. \tag{3.75}$$

To derive the radial and circumferential velocity components, the time derivative of the orbit equation, Eq. (3.40), is taken as follows:

$$\dot{r} = \frac{pe\sin\theta}{(1+e\cos\theta)^2}\dot{\theta} \,. \tag{3.76}$$

The orbital angular momentum and the orbit equation, Eq. (3.40), yield the following expression for the velocity component normal to the radius vector:

$$r\dot{\theta} = \frac{h}{r} = \frac{\sqrt{\mu p}}{r} = \sqrt{\frac{\mu}{p}}(1+e\cos\theta) \,. \tag{3.77}$$

The angular rate, $\dot{\theta} = h/r^2$, and Eq. (3.40), substituted into Eq. (3.76), provide the following expression for the radial velocity:

$$\dot{r} = \sqrt{\frac{\mu}{p}}e\sin\theta \tag{3.78}$$

Hence, the orbital velocity is expressed in the local-horizon frame as follows:

$$\mathbf{v} = \sqrt{\frac{\mu}{p}}e\sin\theta\mathbf{i}_r + \sqrt{\frac{\mu}{p}}(1+e\cos\theta)\mathbf{i}_\theta \,. \tag{3.79}$$

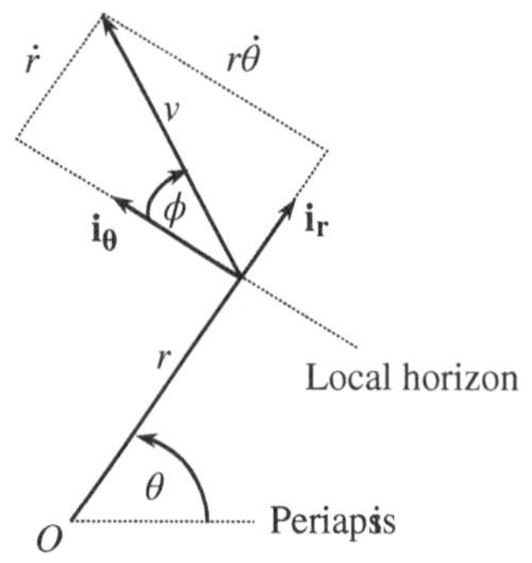

Figure 3.7 The orbital velocity vector in the local-horizon frame.

The radial and normal velocity components give rise to the orbital speed as follows:

$$v = \sqrt{\dot{r}^2 + r^2\dot{\theta}^2} = \sqrt{\frac{\mu}{p}(1 + 2e\cos\theta + e^2)} \tag{3.80}$$

or, substituting Eq. (3.40), we have the following expression for the orbital velocity, which agrees with Eq. (3.32):

$$v = \sqrt{\frac{\mu}{p}\left(\frac{2p}{r} + e^2 - 1\right)} = \sqrt{\frac{2\mu}{r} - \frac{\mu}{a}} \tag{3.81}$$

When the velocity vector is resolved in the local-horizon frame, it is seen to make an angle with the local horizon, as shown in Fig. 3.7. This angle, ϕ, called the *flight-path angle*, is defined to be the angle made by the velocity vector measured *above* the local horizon, and is thus derived from the two orbital velocity components as follows:

$$\tan\phi = \frac{\dot{r}}{r\dot{\theta}} = \frac{e\sin\theta}{1 + e\cos\theta}\,. \tag{3.82}$$

The flight-path angle provides the crucial information about the instantaneous change of the radius with the true anomaly, $\mathrm{d}r/\mathrm{d}\theta = \dot{r}/\dot{\theta} = r\tan\phi$, as the spacecraft progresses in its orbit. By definition, since $\dot{r} > 0$ when $0 < \theta < \pi$, we have $\phi > 0$ in this range of true anomalies, and $\phi < 0$ for $\pi < \theta < 2\pi$. The flight-path angle vanishes at $\theta = 0$, which denotes the point of the minimum radius (periapsis). In an elliptic orbit, we also have $\phi = 0$ at $\theta = \pi$, i.e., at the apoapsis, the point of the maximum radius. For a parabolic orbit, both the velocity components, $\dot{r}, r\dot{\theta}$, vanish at $\theta = \pi$, making $v = 0$; therefore ϕ is undefined at that point. In a hyperbolic orbit, the flight path angle approaches the limiting value $\phi \to \pi/2$ as $\theta \to \theta_\infty$ (i.e., for $r \to \infty$) along the asymptotes.

To relate the flight-path angle to the angular-momentum vector, the velocity vector is expressed in the local-horizon frame as follows:

$$\mathbf{v} = v\sin\phi\mathbf{i}_\mathrm{r} + v\cos\phi\mathbf{i}_\theta\,. \tag{3.83}$$

By taking the vector product of the angular-momentum vector with $\mu\mathbf{e}$, we have

$$\mu\mathbf{h}\times\mathbf{e} = \mathbf{h}\times(\mathbf{v}\times\mathbf{h}) - \frac{\mu}{r}\mathbf{h}\times\mathbf{r}\,, \tag{3.84}$$

or

$$h^2\mathbf{v} = \mu\mathbf{h}\times(\mathbf{e} + \mathbf{i}_\mathrm{r})\,, \tag{3.85}$$

which leads to

$$\mathbf{v} = \frac{\mu}{h}\mathbf{i}_\mathrm{h}\times(\mathbf{e} + \mathbf{i}_\mathrm{r})\,. \tag{3.86}$$

The angular-momentum vector is thus expressed as follows by substituting Eqs. (3.83) and (3.86):

$$\begin{aligned}\mathbf{h} = \mathbf{r} \times \mathbf{v} &= rv\cos\phi\mathbf{i_h}\\ &= \frac{\mu}{h}[(\mathbf{r}\cdot\mathbf{e})\mathbf{i_h} - (\mathbf{r}\cdot\mathbf{i_h})\mathbf{e} + (\mathbf{r}\cdot\mathbf{i_r})\mathbf{i_h} - (\mathbf{r}\cdot\mathbf{i_h})\mathbf{i_r}]\\ &= \frac{\mu}{h}[(\mathbf{r}\cdot\mathbf{e})\mathbf{i_h} + (\mathbf{r}\cdot\mathbf{i_r})\mathbf{i_h}]\\ &= \frac{\mu r}{h}(1+e\cos\theta)\mathbf{i_h}\,, \end{aligned} \tag{3.87}$$

which yields

$$\cos\phi = \frac{\mu}{hv}(1+e\cos\theta)\,. \tag{3.88}$$

Furthermore, from the radial velocity component in Eq. (3.83), $\dot{r} = v\sin\phi$, which is expressed in terms of the true anomaly by Eq. (3.78), we have

$$\sin\phi = \frac{\mu e\sin\theta}{hv}\,. \tag{3.89}$$

Hence, the orbital velocity, (v, ϕ), can be derived from the planar position, (r, θ), if the orbital shape is prescribed by any two constants, e.g., (a, e), (a, p), or (h, e), etc.

Example 3.6.1 *Calculate the orbital speed and the flight-path angle at the two points in the orbit given in Example 3.5.1,* $r_1 = 6700$ *km,* $\theta_1 = 25°$*, and* $r_2 = 6900$ *km,* $\theta_2 = 120°$*, respectively, assuming that the orbit is around Earth (*$\mu = 398600.4\ \mathrm{km^3/s^2}$*).*

The values of the orbital elements calculated in Example 3.5.1 are $a = 6830.5515$ km *and* $e = 0.021$*. The flight-path angles at the two points,* ϕ_1, ϕ_2*, respectively, do not depend upon* μ*, and are given by*

$$\begin{aligned}\phi_1 &= \tan^{-1}\left(\frac{e\sin\theta_1}{1+e\cos\theta_1}\right) = 0.4991°\\ \phi_2 &= \tan^{-1}\left(\frac{e\sin\theta_2}{1+e\cos\theta_2}\right) = 1.0531°\,.\end{aligned}$$

The orbital speeds, v_1, v_2*, at the two given locations are the following:*

$$\begin{aligned}v_1 &= \sqrt{2\mu/r_1 - \mu/a} = 7.7865\ \mathrm{km/s}\\ v_2 &= \sqrt{2\mu/r_2 - \mu/a} = 7.5618\ \mathrm{km/s}\,.\end{aligned}$$

The accuracy of the results is verified against the angular momentum as follows:

$$h = \sqrt{\mu a(1-e^2)} = r_1 v_1\cos\phi_1 = r_2 v_2\cos\phi_2 = 52167.6094\ \mathrm{km^2/s}\,.$$

3.7 Perifocal Frame and Lagrange's Coefficients

The position and velocity vectors in a two-body orbit can be resolved in a fixed coordinate frame with its origin at the focus, O, with an axis along the eccentricity vector ($\theta = 0$) denoted by the unit vector, $\mathbf{i_e} = \mathbf{e}/e$, the second axis along the $\theta = \pi/2$ direction denoted by the unit vector, $\mathbf{i_p}$, and the third axis along the angular momentum vector, $\mathbf{i_h} = \mathbf{h}/h = \mathbf{i_e} \times \mathbf{i_p}$. The right-handed coordinate frame $(\mathbf{i_e}, \mathbf{i_p}, \mathbf{i_h})$ is called the *perifocal frame*, and the two-body orbit expressed in this frame is the following:

$$\begin{aligned}\mathbf{r} &= r\cos\theta\mathbf{i_e} + r\sin\theta\mathbf{i_p} \\ \mathbf{v} &= v(\sin\phi\cos\theta - \cos\phi\sin\theta)\mathbf{i_e} + v(\sin\phi\sin\theta + \cos\phi\cos\theta)\mathbf{i_p}\,. \end{aligned} \tag{3.90}$$

The first of Eqs. (3.90) yields the following coordinate transformation between the perifocal frame and the local-horizon frame, $(\mathbf{i_r}, \mathbf{i_\theta}, \mathbf{i_h})$:

$$\begin{aligned}\mathbf{i_r} &= \cos\theta\mathbf{i_e} + \sin\theta\mathbf{i_p} \\ \mathbf{i_\theta} &= \mathbf{i_h} \times \mathbf{i_r} = \cos\theta\mathbf{i_p} - \sin\theta\mathbf{i_p}\,. \end{aligned} \tag{3.91}$$

The second of Eqs. (3.90) is directly obtained from Eq. (3.79). On eliminating the flight-path angle with the use of Eqs. (3.88) and (3.89), the following expression for the orbital velocity is obtained:

$$\mathbf{v} = -\frac{\mu}{h}\sin\theta\mathbf{i_e} + \frac{\mu}{h}(e + \cos\theta)\mathbf{i_p}\,. \tag{3.92}$$

Given the position and velocity at some time, τ, one would like to determine the position and velocity at some other time, t. To do so, the known position and velocity are expressed in terms of the true anomaly, θ_0, as follows:

$$\begin{aligned}\mathbf{r_0} &= \mathbf{r}(\tau) = r_0\cos\theta_0\mathbf{i_e} + r_0\sin\theta_0\mathbf{i_p} \\ \mathbf{v_0} &= \mathbf{v}(\tau) = -\frac{\mu}{h}\sin\theta_0\mathbf{i_e} + \frac{\mu}{h}(e + \cos\theta_0)\mathbf{i_p}\,, \end{aligned} \tag{3.93}$$

which in the matrix form becomes the following:

$$\begin{Bmatrix} \mathbf{r_0} \\ \mathbf{v_0} \end{Bmatrix} = \begin{pmatrix} r_0\cos\theta_0 & r_0\sin\theta_0 \\ -\frac{\mu}{h}\sin\theta_0 & \frac{\mu}{h}(e+\cos\theta_0) \end{pmatrix} \begin{Bmatrix} \mathbf{i_e} \\ \mathbf{i_p} \end{Bmatrix}. \tag{3.94}$$

Since the square matrix in Eq. (3.94) is non-singular (its determinant is $h \neq 0$), it can be inverted to obtain $\mathbf{i_e}$ and $\mathbf{i_p}$ as follows:

$$\begin{Bmatrix} \mathbf{i_e} \\ \mathbf{i_p} \end{Bmatrix} = \begin{pmatrix} \frac{1}{p}(e+\cos\theta_0) & -\frac{r_0}{h}\sin\theta_0 \\ \frac{1}{p}\sin\theta_0 & \frac{r_0}{h}(e+\cos\theta_0) \end{pmatrix} \begin{Bmatrix} \mathbf{r_0} \\ \mathbf{v_0} \end{Bmatrix}. \tag{3.95}$$

On substituting Eq. (3.95) into Eqs. (3.90) and (3.92), we have

$$\begin{Bmatrix} \mathbf{r}(t) \\ \mathbf{v}(t) \end{Bmatrix} = \begin{pmatrix} r\cos\theta & r\sin\theta \\ -\frac{\mu}{h}\sin\theta & \frac{\mu}{h}(e+\cos\theta) \end{pmatrix} \begin{pmatrix} \frac{1}{p}(e+\cos\theta_0) & -\frac{r_0}{h}\sin\theta_0 \\ \frac{1}{p}\sin\theta_0 & \frac{r_0}{h}(e+\cos\theta_0) \end{pmatrix} \begin{Bmatrix} \mathbf{r_0} \\ \mathbf{v_0} \end{Bmatrix}, \tag{3.96}$$

or

$$\begin{Bmatrix} \mathbf{r} \\ \mathbf{v} \end{Bmatrix} = \begin{pmatrix} f & g \\ \dot{f} & \dot{g} \end{pmatrix} \begin{Bmatrix} \mathbf{r}_0 \\ \mathbf{v}_0 \end{Bmatrix}, \tag{3.97}$$

where

$$\begin{aligned} f &= 1 + \frac{r}{p}[\cos(\theta - \theta_0) - 1] \\ g &= \frac{rr_0}{h}\sin(\theta - \theta_0) \\ \dot{f} &= \frac{\mathrm{d}f}{\mathrm{d}t} = -\frac{h}{p^2}[\sin(\theta - \theta_0) + e(\sin\theta - \sin\theta_0)] \\ \dot{g} &= \frac{\mathrm{d}g}{\mathrm{d}t} = 1 + \frac{r_0}{p}[\cos(\theta - \theta_0) - 1]. \end{aligned} \tag{3.98}$$

The functions f and g are called *Lagrange's coefficients*, and are useful in determining the trajectory from a known position and velocity. The square matrix

$$\Phi(t, \tau) = \begin{pmatrix} f & g \\ \dot{f} & \dot{g} \end{pmatrix} \tag{3.99}$$

is called the *state-transition matrix* and has a special significance, because it uniquely determines the current state, $(\mathbf{r},\ \mathbf{v})$, from the *initial state*, $(\mathbf{r}_0,\ \mathbf{v}_0)$, as follows:

$$\begin{Bmatrix} \mathbf{r}(t) \\ \mathbf{v}(t) \end{Bmatrix} = \Phi(t, \tau) \begin{Bmatrix} \mathbf{r}(\tau) \\ \mathbf{v}(\tau) \end{Bmatrix}. \tag{3.100}$$

Such a relationship between initial and final states of a system is rarely possible for the solution to a non-linear differential equation, such as Eq. (3.7), and is therefore a valuable property of the two-body problem. The state-transition matrix $\Phi(t, \tau)$ has the following properties:

(a) From the conservation of angular momentum,

$$\mathbf{h} = \mathbf{r} \times \mathbf{v} = (f\dot{g} - g\dot{f})\mathbf{r}_0 \times \mathbf{v}_0 = \mathbf{r}_0 \times \mathbf{v}_0, \tag{3.101}$$

it follows that

$$|\,\mathbf{\Phi}\,| = f\dot{g} - g\dot{f} = 1. \tag{3.102}$$

A consequence of the unity determinant is that the inverse of the state-transition matrix is given by

$$\mathbf{\Phi}^{-1} = \begin{pmatrix} \dot{g} & -g \\ -\dot{f} & f \end{pmatrix}. \tag{3.103}$$

A matrix which obeys Eq. (3.103) is said to be *symplectic*.

(b) Given any three points $(\tau,\ t_1,\ t_2)$ along the trajectory, it is true that

$$\mathbf{\Phi}(t_2, \tau) = \mathbf{\Phi}(t_2, t_1)\mathbf{\Phi}(t_1, \tau). \tag{3.104}$$

The state-transition matrix, $\Phi(\tau, t)$, and the corresponding Lagrange's coefficients require the separate knowledge of the true anomalies, θ_0 and θ. However, the complete solution of the two-body problem governed by Eq. (3.7) requires a determination of $\theta(t)$ from $\theta_0 = \theta(\tau)$. Therefore, Eq. (3.100) does not solve the two-body problem, and a separate procedure for evolving $\theta(t)$ from $\theta(\tau)$ is still required to complete the solution, as will be explored in Chapter 4.

Example 3.7.1 *Calculate the Lagrange's coefficients for the orbit in Example (3.5.3) relating the position and velocity at the point (r_2, θ_2) to those at (r_1, θ_1).*

From Example (3.5.3) we have $r_1 = 6700$ km, $\theta_1 = 5°$, and $r_2 = 7200$ km, $\theta_2 = 30°$, from which the orbital elements are calculated to be $a = -48229.789$ km and $e = 1.13864$, with $p = 14299.837$. Since the value of the gravitational constant, μ, is not provided, we begin by computing the following Lagrange's coefficients which do not require μ:

$$f = 1 + \frac{r_2}{p}[\cos(\theta_2 - \theta_1) - 1] = 0.952826$$

$$\dot{g} = 1 + \frac{r_1}{p}[\cos(\theta_2 - \theta_1) - 1] = 0.956102 \ .$$

The remaining two coefficients are then expressed as follows in terms of the unknown parameter, μ:

$$g = \frac{r_1 r_2}{h}\sin(\theta_2 - \theta_1) = \frac{170486.5094}{\sqrt{\mu}}$$

$$\dot{f} = -\frac{h}{p^2}[\sin(\theta_2 - \theta_1) + e(\sin\theta_2 - \sin\theta_1)] = -5.22045 \times 10^{-7}\sqrt{\mu} \ .$$

These computations are verified by ensuring that $f\dot{g} - g\dot{f} = 1$.

The constant vectors **h** and **e** (or $\mathbf{r}_0$ and $\mathbf{v}_0$) define the shape and plane of a two-body trajectory, but provide only five of the six scalar integrals required to obtain a complete solution to Eq. (3.7). The missing scalar constant pertains to the information about the location of m_2 along the trajectory at a particular time, τ, and can be expressed as the value of the true anomaly at a specific time, $\theta(\tau) = \theta_0$, which could be employed to solve for the true anomaly at any other time, $\theta(t)$. The derivation of the true anomaly at a given time, $\theta(t)$, from a known value at some other time, $\theta(\tau)$, which completes the solution to the two-body problem, is addressed in the Chapter 4.

Exercises

1. Derive the equation of motion of mass m_2 with respect to the position from the centre of mass of the two-body system. What is the result for $m_1 \gg m_2$?

2. What is the minimum velocity required for escaping from Earth's gravity ($\mu = 398600.4$ km^3/s^2) at: (a) Earth's surface. (Earth's mean equatorial radius is 6378.14 km.) (b) Geosynchronous orbit. (*Ans.* 4.3482 km/s.)

3. Find the maximum and minimum orbital speeds of Earth around the sun ($\mu = 1.327 \times 10^{11}$ km^3/s^2), if the eccentricity of Earth's orbit is 1/60. What is the mean orbital speed if the mean radius is 1 astronomical unit (AU)? (1 AU = 1.495978×10^8 km.)

4. *Voyager 1*'s current radius and speed relative to the sun are 148.61 AU and 16.9 km/s, respectively. What is the semi-major axis of the spacecraft's orbit?

5. The position and velocity vectors of a spacecraft in an Earth centred, equatorial inertial frame, $(\mathbf{I}, \mathbf{J}, \mathbf{K})$, are the following:

$$r = 6045\mathbf{I} + 3490\mathbf{J} \text{ (km)}$$

$$v = -2.457\mathbf{I} + 6.618\mathbf{J} + 2.533\mathbf{K} \text{ (km/s)}.$$

   (a) Determine the angular momentum vector and its magnitude. What is the angle made by the angular momentum vector with the **K** axis?
   (b) Calculate the total relative specific energy of the spacecraft.

6. The orbital period of an Earth satellite is 106 min. Find the apogee altitude if the perigee altitude is 200 km. (*Ans.* 1882.36 km).

7. Find the orbital period of an Earth satellite if the perigee and apogee altitudes are 250 km and 300 km, respectively.

8. Given the orbital period of Mars around the sun as 687 mean solar (24 hour) days, find the semi-major axis of Mars' orbit in AU. (*Ans.* 1.524 AU.)

9. Estimate the solar gravitational constant, μ, using Earth's orbital parameters (semi-major axis 1 AU, period 1 sidereal year =365.25636 mean solar days).

10. Two positions of a spacecraft in an Earth orbit are measured to be $r_1 = 7923$ km, $\theta_1 = 126°$ and $r_2 = 7230$ km, $\theta_2 = 58°$, respectively. Find the eccentricity, the semi-major axis, and the period of the orbit. (*Ans.* $e = 0.08164$; $a = 7593$ km; $T = 6585$ s.)

11. An Earth satellite has orbital eccentricity of 0.6 and a perigee altitude of 400 km. Calculate the maximum radial velocity and the corresponding true anomaly. (*Ans.* $\dot{r} = 3.6375$ km/s; $\theta = 90°$.)

12. What is the true anomaly in an elliptic orbit at which the speed is equal to the speed in a circular orbit of the same radius? (*Ans.* $\theta = 326.437°$.)

13. Show that the flight-path angle, ϕ, for a parabolic orbit is half the value of the true anomaly, θ.

14. Calculate the orbital speed and the flight-path angle at the two points in the orbit given in Example 3.5.3 – $r_1 = 6700$ km, $\theta_1 = 5°$, and $r_2 = 7200$ km, $\theta_2 = 30°$, respectively – assuming that the orbit is around Earth ($\mu = 398600.4$ km^3/s^2).

15. A hyperbolic Earth departure trajectory has a perigee speed of 15 km/s at an altitude of 300 km. Calculate: (a) the speed at infinite radius (the hyperbolic excess speed), and (b) the radius and speed when the true anomaly is 100°.

16. *Voyager 1*'s closest approach to Saturn was at a periapsis radius of 124000 km and the hyperbolic excess speed was 7.51 km/s. What was the angle through which the spacecraft's velocity vector was turned by Saturn ($\mu = 37.931 \times 10^6$ km^3/s^2)?

17. What is the maximum flight path angle of a spacecraft in a hyperbolic orbit?

18. A meteor is seen to approach Earth at a radius 400000 km and true anomaly 120° with speed 2.2 km/s. Compute: (a) Eccentricity of the orbit. (b) Altitude at the closest approach. (c) Speed at the closest approach.

19. A spacecraft is observed to have a radius of 10000 km, speed of 10 km/s, and flight path angle of −20° relative to Earth. What is the true anomaly at that point?

20. A satellite in a circular earth orbit of 200 km altitude is *de-orbited* by firing rocket thrusters tangentially, such that the speed instantly decreases by 500 m/s. Neglecting the effects of Earth's atmosphere, calculate the inertial speed and flight-path angle when the altitude of 100 km is reached. (*Ans.* $v = 7.41156$ km/s, $\phi = -3.6281°$.)

21. A spacecraft is tracked at an altitude of 5000 km and flight path angle of 10°, to be moving at speed 10 km/s relative to an Earth-centred inertial frame. Determine the orbital elements, a and e, of the spacecraft. (*Ans.* $a = -13315.1949$ km, $e = 1.834585$.)

22. Calculate the perifocal radius and velocity vectors of the de-orbited satellite in Exercise 20 when reaching the 100 km altitude. (*Ans.*

$$\mathbf{r} = -5773.884\mathbf{i_e} - 2937.441\mathbf{i_p} \text{ (km)}$$

$$\mathbf{v} = 3.77197\mathbf{i_e} - 6.37993\mathbf{i_p} \text{ (km/s))}.$$

4

Time in Orbit

You saw in Chapter 3 that the position and velocity in a two-body orbit at a point corresponding to the true anomaly, $\theta(t)$, can be uniquely determined from the position and velocity at any other location, $\theta(\tau)$, provided that both $\theta(t)$ and $\theta(\tau)$ are independently known. However, a complete solution to the two-body problem requires a procedure of evolving the true anomaly in time, $\theta(t)$, which would allow one to derive θ from a known $\theta(\tau)$. The missing scalar integral constant relating $\theta(t)$ to $\theta(\tau)$ would complete the set of six scalar constants required for the complete solution to the two-body problem. The sixth constant of orbital motion can be regarded as the time τ at which the orbit reaches a specific value of the true anomaly, $\theta(\tau)$. A convenient choice of this constant is the *time of periapsis*, t_0, defined to be the time when the true anomaly vanishes, $\theta(t_0) = 0$. The time of periapsis, t_0, completes the set of constants required to determine the Keplerian orbit. Along with the constants, a and e, the time of periapsis, t_0, can be used to calculate the position and velocity in the orbital plane. Hence, (a, e, t_0) constitute the planar part of the set of six constants required to completely determine the orbit, called the *classical orbital elements*. The three other classical orbital elements, which determine the orientation of the orbital plane, are described in Chapter 5.

Except in the special cases of the rectilinear motion ($h = 0$) and the circular orbit ($e = 0$) (see Chapter 3), the solution for the position and velocity as functions of time cannot be derived in a closed form. For the general case of orbital motion ($h \neq 0$), a substitution of the orbit equation, Eq. (3.40), into the expression for the constant angular momentum, $h = r^2\dot{\theta}$, yields the following:

$$\frac{\mathrm{d}\theta}{(1 + e\cos\theta)^2} = \sqrt{\frac{\mu}{p^3}}\mathrm{d}t\,, \tag{4.1}$$

which is a scalar differential equation in time to be solved for $\theta(t)$, beginning from the initial condition, $\theta(t_0) = 0$. The integration of Eq. (4.1) provides the integration constant, t_0, which would complete the solution to the two-body problem. Attempts to carry out such an integration in a closed form have occupied some of the greatest mathematicians (Euler, Lagrange, Laplace, Bessel, Fourier, Gauss, Cauchy, Leibnitz, and Newton), and have led to significant developments in all the branches of mathematics. The difficulty in obtaining a solution to Eq. (4.1) is due to the fact that it is a non-linear differential equation in terms of the unknown variable, θ. Closed-form,

Foundations of Space Dynamics, First Edition. Ashish Tewari.

exact solutions to non-linear differential equations are rare. In the following sections, analytical solutions to Eq. (4.1) are presented for the various conic-section orbits.

4.1 Position and Velocity in an Elliptic Orbit

The integrated form of Eq. (4.1) for an elliptic orbit ($0 \leq e < 1$) is called *Kepler's equation*. For an elliptic orbit, Eq. (4.1) is expressed as follows:

$$\frac{(1-e^2)^{\frac{3}{2}}\mathrm{d}\theta}{(1+e\cos\theta)^2} = n\mathrm{d}t\,, \tag{4.2}$$

where

$$n = \sqrt{\frac{\mu}{a^3}} \tag{4.3}$$

is referred to as the *mean motion*, which is the average angular frequency, $\dot{\theta}$, over a complete orbit, because the orbital period is given by Eq. (2.118) as follows:

$$T = \frac{2\pi}{\sqrt{\mu}}a^{3/2} = \frac{2\pi}{n}\,. \tag{4.4}$$

It is evident from Eq. (4.2) that the true anomaly, θ, does *not* vary uniformly with time, i.e., the angular rate of the motion, $\dot{\theta}$, is not constant, but changes with the position, θ. At the periapsis ($\theta = 0$), the rate $\dot{\theta}$ is the maximum, while at the apoapsis ($\theta = \pi$), $\dot{\theta}$ is the minimum. For the special case of a circular orbit ($e = 0$), the angular rate is constant at $\dot{\theta} = n$.

The constant frequency of a circular orbit of radius a leads to the definition of the *eccentric anomaly*, E, as an alternative measurement of the angular position of an object in an elliptic orbit. The projection of the position vector, $\mathbf{r}$, normal to the major axis intersects the auxilliary circle of radius, a, centred at the centre, C, of the ellipse (Fig. 4.1). The radial line from C to the intersection point, P, makes the angle, E, with the major axis, as shown in Fig. 4.1.

The relationship between the eccentric anomaly, E, and the true anomaly, θ, is derived from the geometry of Fig. 4.1 to be the following:

$$r\cos(\pi - \theta) + a\cos E = ae\,, \tag{4.5}$$

or

$$-r\cos(\theta) + a\cos E = ae\,, \tag{4.6}$$

which implies that

$$\cos E = \frac{ae + r\cos\theta}{a}\,. \tag{4.7}$$

Substituting the orbit equation,

$$r = \frac{a(1-e^2)}{1+e\cos\theta}\,, \tag{4.8}$$

into Eq. (4.7), we have

$$\cos E = \frac{e+\cos\theta}{1+e\cos\theta}\,. \tag{4.9}$$

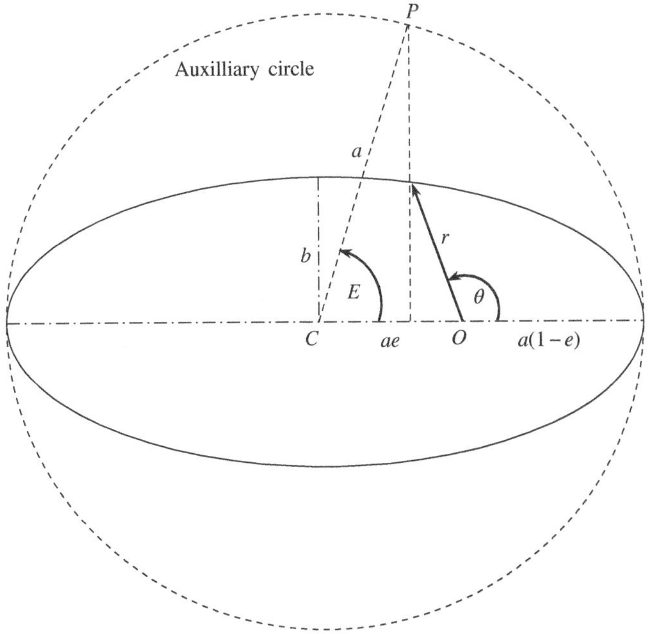

Figure 4.1 The auxilliary circle and the eccentric anomaly.

or alternatively,

$$\cos\theta = \frac{\cos E - e}{1 - e\cos E}, \tag{4.10}$$

which yields the following form of the orbit equation:

$$r = a(1 - e\cos E). \tag{4.11}$$

Equating Eqs. (4.11) and (4.8), we have

$$1 + e\cos\theta = \frac{1 - e^2}{1 - e\cos E}. \tag{4.12}$$

The trigonometric identity, $\sin E = \sqrt{1 - \cos^2 E}$, yields the following expression for $\sin E$:

$$\sin E = \frac{\sin\theta\sqrt{1 - e^2}}{1 + e\cos\theta}. \tag{4.13}$$

The substitution of Eq. (4.12) into Eq. (4.13) yields

$$\sin\theta = \frac{\sin E\sqrt{1 - e^2}}{1 - e\cos E}. \tag{4.14}$$

Combining Eqs. (4.8), (4.9), and (4.13) yields the following half-angle trigonometric identities:

$$\sin^2\frac{E}{2} = \frac{r}{a(1+e)}\sin^2\frac{\theta}{2}$$
$$\cos^2\frac{E}{2} = \frac{r}{a(1-e)}\cos^2\frac{\theta}{2}, \quad (4.15)$$

which produces the following useful relationship between θ and E:

$$\tan\frac{E}{2} = \sqrt{\frac{1-e}{1+e}}\tan\frac{\theta}{2}. \quad (4.16)$$

Equation (4.16) implies that as the true anomaly, θ, vanishes at the periapsis, so does the eccentric anomaly, E. Furthermore, when $\theta = \pi$, we have $E = \pi$, and when $\theta = \pm\pi/2$, we have $E = \cos^{-1}e$. Equation (4.16) can be used to determine E from θ (and vice versa) without any quadrant ambiguity, because the angles $E/2$ and $\theta/2$ are always in the same quadrant.

The differentiation of Eq. (4.9) results in the following differential relationship between E and θ:

$$\sin E\mathrm{d}E = \frac{(1-e^2)\sin\theta}{(1+e\cos\theta)^2}\mathrm{d}\theta. \quad (4.17)$$

The substitution of Eq. (4.13) into Eq. (4.17) produces the following:

$$\mathrm{d}E = \frac{(1-e^2)^{3/2}}{(1+e\cos\theta)^2(1-e\cos E)}\mathrm{d}\theta, \quad (4.18)$$

from which $\mathrm{d}\theta$ is eliminated by Eq. (4.1), leading to

$$(1-e\cos E)\mathrm{d}E = (1-e^2)^{3/2}\frac{\mu}{p^3}\mathrm{d}t, \quad (4.19)$$

or

$$(1-e\cos E)\mathrm{d}E = \frac{\mu}{a^3}\mathrm{d}t. \quad (4.20)$$

Integrating Eq. (4.20) from $t = t_0$, $E = 0$ results in the following *Kepler's equation*:

$$E - e\sin E = n(t-t_0). \quad (4.21)$$

Kepler's equation, Eq. (4.21), is a transcendental equation to be solved for the eccentric anomaly, E, given the present time, t, and the *time of periapsis*, t_0. The latter is the sixth integral constant required to completely solve the two-body problem for the elliptic orbit. After E is determined from Kepler's equation, it can be used to calculate the true anomaly, θ, by Eq. (4.16), which completely determines the position and velocity in the elliptic orbit. Therefore, Kepler's equation, Eq. (4.21), provides the necessary information for the evolution of the true anomaly, from the time of periapsis to the current time. The right-hand side of Eq. (4.21) is another angle, called the *mean anomaly*, defined by $M = n(t-t_0)$, which yields the following form of Kepler's equation:

$$E - e\sin E = M. \quad (4.22)$$

Once the solution E to Kepler's equation is known, it leads to the current position and velocity. In order to derive the current position and velocity, the perifocal position and velocity vectors in an elliptic orbit are expressed directly in terms of the eccentric anomaly, E, thereby eliminating the step of calculating the true anomaly. This is achieved by the elimination of the true anomaly from Eqs. (3.90) and (3.92) by the substitution of Eqs. (4.10) and (4.14), resulting in the following expressions for $\mathbf{r}$ and $\mathbf{v}$:

$$\begin{aligned} \mathbf{r} &= a(\cos E - e)\mathbf{i_e} + \sqrt{ap}\sin E\mathbf{i_p} \\ \mathbf{v} &= -\frac{\sqrt{\mu a}}{r}\sin E\mathbf{i_e} + \frac{\sqrt{\mu p}}{r}\cos E\mathbf{i_p}\,. \end{aligned} \tag{4.23}$$

An alternative form of Kepler's equation can be derived to determine $E(t)$ from a given eccentric anomaly, E_0, at time $t = 0$, which corresponds to the position and velocity, $\mathbf{r_0}$ and $\mathbf{v_0}$, respectively, by expressing the current t, as follows:

$$\begin{aligned} t &= \sqrt{\frac{a^3}{\mu}}[(E - E_0) - e(\sin E - \sin E_0)] \\ &= \sqrt{\frac{a^3}{\mu}}\left\{(E - E_0) + \frac{\mathbf{r_0}\cdot\mathbf{v_0}}{\sqrt{\mu a}}[1 - \cos(E - E_0)] - \left(1 - \frac{r_0}{a}\right)\sin(E - E_0)\right\}, \end{aligned} \tag{4.24}$$

where the following identities have been employed:

$$\begin{aligned} \sin E - \sin E_0 &= \sin E_0[\cos(E - E_0) - 1] + \cos E_0 \sin(E - E_0) \\ e\cos E_0 &= 1 - \frac{r_0}{a} \\ e\sin E_0 &= \frac{\mathbf{r_0}\cdot\mathbf{v_0}}{\sqrt{\mu a}}\,. \end{aligned} \tag{4.25}$$

When Eq. (4.24) is solved for $E - E_0$, it allows the expression of the current position and velocity in terms of those at $t = 0$ by using the Lagrange's coefficients [Eq. (3.98)] as follows:

$$\begin{aligned} \mathbf{r}(t) &= f\mathbf{r_0} + g\mathbf{v_0} \\ &= \left\{1 - \frac{a}{r_0}[1 - \cos(E - E_0)]\right\}\mathbf{r_0} \\ &\quad + \left\{t - \frac{(E - E_0) - \sin(E - E_0)}{\sqrt{\mu/a^3}}\right\}\mathbf{v_0}, \end{aligned} \tag{4.26}$$

and

$$\begin{aligned} \mathbf{v}(t) &= \dot{f}\mathbf{r_0} + \dot{g}\mathbf{v_0} \\ &= -\frac{\sqrt{\mu a}}{rr_0}\sin(E - E_0)\mathbf{r_0} + \left\{1 - \frac{a}{r}[1 - \cos(E - E_0)]\right\}\mathbf{v_0}\,. \end{aligned} \tag{4.27}$$

Thus we have

$$
\begin{aligned}
f &= 1 - \frac{a}{r_0}[1 - \cos(E - E_0)] \\
g &= t - \frac{(E - E_0) - \sin(E - E_0)}{\sqrt{\mu/a^3}} \\
&= \sqrt{\frac{a^3}{\mu}} \left\{ \frac{\mathbf{r_0} \cdot \mathbf{v_0}}{\sqrt{\mu a}}[1 - \cos(E - E_0)] + \frac{r_0}{a} \sin(E - E_0) \right\} \\
\dot{f} &= -\frac{\sqrt{\mu a}}{r r_0} \sin(E - E_0) \\
\dot{g} &= 1 - \frac{a}{r}[1 - \cos(E - E_0)] .
\end{aligned}
\tag{4.28}
$$

The time, $t = 0$, at which the initial condition, $\mathbf{r_0}(E_0)$, $\mathbf{v_0}(E_0)$, is specified is called the *epoch*.

Example 4.1.1 *Given the following perifocal position and velocity vectors around Earth at the epoch, $t = 0$:*

$$\mathbf{r} = -28655.995\mathbf{i_e} - 33178.52\mathbf{i_p} \text{ km}$$

$$\mathbf{v} = 3.556\mathbf{i_e} + 1.158\mathbf{i_p} \text{ km/s}$$

calculate the orbital elements, (a, e, t_0), and the true anomaly, θ.

This problem can be solved as follows. First, the orbital radius, r, and speed, v, are calculated from the components of **r** *and* **v** *to yield*

$$r =| \mathbf{r} |= 43840.3952 \text{ km} ; \qquad v =| \mathbf{v} |= 3.7401 \text{ km/s}.$$

Then the semi-major axis, a, is computed from the energy integral, ε, using $\mu =$ 398600.4 km^3/s^2 for Earth as follows:

$$\varepsilon = \frac{v^2}{2} - \frac{\mu}{r} = -2.0979 \text{ km}^2/\text{s}^2 = -\frac{\mu}{2a} ; \qquad a = -\frac{\mu}{2\varepsilon} = 95000 \text{ km} .$$

Next, the orbital angular momentum vector, **h**, *is computed:*

$$\mathbf{h} = \mathbf{r} \times \mathbf{v} = 84821.797\mathbf{i_h} \text{ km}^2/\text{s}.$$

Hence, $h =| \mathbf{h} |=$ 84821.797 km^2/s, and the orbital parameter is calculated to be $p = h^2/\mu =$ 18050 km, from which the orbital eccentricity is computed as follows:

$$e = \sqrt{1 - \frac{p}{a}} = 0.9 .$$

To find the time of perigee, t_0, Kepler's equation is utilized. To do so, the mean motion is calculated as follows:

$$n = \sqrt{\frac{\mu}{a^3}} = 2.156 \times 10^{-5} \text{ rad./s}$$

and the eccentric anomaly, E, at the present time, t $= 0$*, can be computed from Eq. (4.22) using the components of the position vector,* $\mathbf{r}$*, as follows:*

$$a(\cos E - e) = -28655.995 \text{ km} ; \qquad \sqrt{ap}\sin E = -33178.52 \text{ km},$$

which yields

$$\cos E = 0.59836 ; \qquad \sqrt{ap}\sin E = -0.80123 \text{ km} ,$$

or $E = 5.353839$ *rad. Substituting these values into Eq. (4.21) and solving for* t_0*, we have*

$$t_0 = \frac{e\sin E - E}{n} = -281746.741 \text{ s} ,$$

which implies that the perigee was crossed 281746.741 seconds (or 78.26 hours) ago. In comparison, the orbital period is the following:

$$T = \frac{2\pi}{n} = 291404.605 \text{ s} ,$$

or 80.95 hours. Finally, the present value of the true anomaly is calculated by Eq. (4.16) as follows:

$$\frac{\theta}{2} = \tan^{-1}\left(\sqrt{\frac{1+e}{1-e}}\tan\frac{E}{2}\right) = 2.0 \text{ rad.} ,$$

or $\theta = 4.0$ *rad. In carrying out this calculation,* $\theta/2$ *has been converted to the correct quadrant, which is the same quadrant as of* $E/2$*.*

4.2 Solution to Kepler's Equation

Kepler's equation is a transcendental equation in E, whose solution in a closed form has evaded mathematicians in the last three centuries. Given the time, t (or the mean anomaly, M), obtaining the solution, E, is a non-trivial task, requiring a numerical approximation. The solution to Kepler's equation can be cast as the problem of searching for the roots, E, of the following equation, given the constants $0 \le e < 1$ and M:

$$f(E) = E - e\sin E - M = 0 . \tag{4.29}$$

At the periapsis, both E and M vanish, and at the apoapsis we have $E = M = \pi$. For $E > 0$, we have $M > 0$, and whenever $E < 0$ then $M < 0$. Therefore, it can be verified that the function $f(E)$ in Eq. (4.29) is anti-symmetric; that is, it has the following property:

$$f(-E) = -E + e\sin E + M = -f(E) , \tag{4.30}$$

which can be extended by multiples of 2π as follows:

$$f(2k\pi - E) = (2k\pi - E) - e\sin(2k\pi - E) - (2k\pi - M) = -f(E) ; \quad (0 \le E \le \pi) , \tag{4.31}$$

where $k = 0, 1, 2, \ldots$. Furthermore, $f(E)$ satisfies the following conditions for $k\pi \leq M < (k+1)\pi$, where $k = 0, 1, 2, \ldots$:

$$f(k\pi) = k\pi - M \leq 0$$
$$f[(k+1)\pi] = (k+1)\pi - M > 0\,, \tag{4.32}$$

which implies that $f(E)$ vanishes at least once for every interval, $k\pi \leq E \leq (k+1)\pi$, $k = 0, 1, 2, \ldots$. However, since the derivative, $\mathrm{d}f(E)/\mathrm{d}E = 1 - e\cos E$ is always positive, it follows that $f(E) = 0$ has a unique solution E in each interval, $k\pi \leq E \leq (k+1)\pi$, $k = 0, 1, 2, \ldots$. From these properties of $f(E)$, it is evident that one need only seek the solution E in the interval $0 \leq E \leq \pi$, and adjust the result by a sign and a multiple of 2π as required by the given value of M. Hence, for a given M, Kepler's solution has a unique solution, E, which can be sought by a numerical procedure.

Two numerical solution methods for Kepler's equation are described next.

4.2.1 Newton's Method

Newton gave a numerical solution procedure for finding the roots of algebraic and transcendental equations, $f(x) = 0$, employing a Taylor series expansion of the function, $f(x)$, as follows:

$$f(x + \Delta x) = \sum_{k=0}^{\infty} f^{(k)}(x)\frac{(\Delta x)^k}{k!}\,, \tag{4.33}$$

where $f^{(k)} = \frac{\mathrm{d}^k f(x)}{\mathrm{d}x^k}$ is the kth-order derivative of $f(x)$. When this infinite series is approximated by a finite number of terms, the accuracy of the approximation depends upon the *step size*, Δx, as well as on the number of retained terms. When only the first two terms of the infinite series are retained, the approximation is given by

$$f(x + \Delta x) \simeq f(x) + f^{(1)}(x)(\Delta x)\,. \tag{4.34}$$

With the help of Eq. (4.34), *Newton's method* applied to the solution of Kepler's equation is summarized as follows:

1. Given a mean anomaly, M, guess an initial value for the eccentric anomaly, E. A good starting guess is $E = M + e\sin M$, although other, more refined estimates are possible, such as the following:
$$E = M + \frac{e\sin M}{\cos e - \left(\frac{\pi}{2} - e\right)\sin e + M\sin e}\,.$$
2. Calculate the modification required in the value of E such that $f(E + \Delta E) = 0$, using
$$\Delta E = -\frac{f(E)}{f^{(1)}(E)} = \frac{-E + e\sin E + M}{1 - e\cos E}\,. \tag{4.35}$$
3. Update E, using $E = E + \Delta E$.
4. Calculate $f(E) = E - e\sin E - M$.

Table 4.1 Iteration steps for the solution of Kepler's equation by Newton's method for Example 4.2.1 (all angles in rad.).

E	$f(E)$	$f^{(1)}(E)$	ΔE
$M = 1.20845848553264$	−0.187014141123649	0.929107750636731	0.201283587393912
1.40974207292656	0.00387183280517411	0.967928218786664	−0.00400012390384445
1.40574194902271	1.57904764×10^{-6}	0.967138806003956	-1.63270012×10^{-6}
1.40574031632259	2.62900812×10^{-13}	0.967138483901844	-2.7183368×10^{-13}
1.40574031632232	—	—	—

5. If the magnitude of $f(E)$ is less than or equal to a pre-selected small number, δ, called the *tolerance*, then the value of E calculated in step 4 is acceptable. Otherwise, go back to step 2, and determine a new change in E.

Newton's method normally converges in a few iterations to a very small tolerance, $|\,f(E)\,| \leq \delta$. A larger number of iterations may be necessary when the orbital eccentricity, e, is near unity. The number of iterations also depends upon the initial guess for E in the first step. It is rare for the scheme to require more than six iterations, even in the extreme cases. One can easily write a computer programme for solving Kepler's equation by Newton's method.

Example 4.2.1 *For an orbit of $a = 8500$ km and $e = 0.2$ around Earth, calculate the radius, the true anomaly, the orbital speed, and the flight-path angle 25 min. after passing the perigee.*

We begin by calculating the mean anomaly 25 min. after perigee as follows:

$$n = \sqrt{\frac{\mu}{a^3}} = 0.00080564 \text{ rad./s} \; ; \quad M = n(t - t_0) = n \times 25 \times 60 = 1.20845848553264 \text{ rad.}$$

The solution to Kepler's equation is next obtained using Newton's method for a tolerance, $\delta = 10^{-10}$ rad., starting with the rough initial guess, $E = M$, and the iteration steps are tabulated in Table 4.1.

Hence, the specified tolerance, $\delta = 10^{-10}$ rad., is met in three iterations, with the eccentric anomaly calculated to be $E = 1.405740316322$ rad. $(80.54299°)$. The true anomaly is then calculated by Eq. (4.16) to be $\theta = 1.60770988903$ rad. $(92.11499°)$. Then, using the orbit equation, the radius is computed by

$$r = \frac{p}{1 + e\cos\theta} = a(1 - e\cos E) = 8220.677 \text{ km} \; ,$$

the orbital speed by

$$v = \sqrt{\frac{2\mu}{r} - \frac{\mu}{a}} = 7.0768 \text{ km/s} \; ,$$

and the flight-path angle as follows:

$$\phi = \tan^{-1}\frac{e\sin\theta}{1 + e\cos\theta} = 0.198693 \text{ rad. } (11.3843°) \; .$$

To verify the accuracy of the calculations, the angular momentum, h, is computed by

$$h = rv\cos\phi = 57031.3884\ \text{km}^2/\text{s}$$

and compared with $h = \sqrt{\mu a(1 - e^2)} = 57031.3884\ \text{km}^2/\text{s}$.

4.2.2 Solution by Bessel Functions

Kepler's equation, Eq. (4.22), contains an odd periodic function of E given by

$$E - M = e\sin E\,, \tag{4.36}$$

which can be expressed as the following Fourier sine series:

$$E - M = e\sin E = 2\sum_{k=1}^{\infty} b_k \sin kM\,, \tag{4.37}$$

where the Fourier coefficients, b_k, are evaluated by integration by parts as follows:

$$\begin{aligned}
b_k &= \frac{1}{\pi}\int_0^{\pi} e\sin E \sin kM \mathrm{d}M \\
&= \frac{1}{\pi}\left[-e\sin E\int \frac{\cos kM}{k}\mathrm{d}M + \frac{1}{k}\int e\cos E\frac{\mathrm{d}E}{\mathrm{d}M}\cos kM\mathrm{d}M\right]_0^{\pi} \\
&= \frac{1}{k\pi}\int_0^{\pi}\frac{e\cos E}{1 - e\cos E}\cos kM\mathrm{d}M \\
&= \frac{1}{k\pi}\int_0^{\pi} e\cos E\cos kM\mathrm{d}E = \frac{1}{k\pi}\int_0^{\pi}\cos kM\mathrm{d}E - \frac{1}{k\pi}\int_0^{\pi}\cos kM\mathrm{d}M\,.
\end{aligned} \tag{4.38}$$

The last integral on the extreme right-hand side of Eq. (4.38) vanishes, leading to

$$b_k = \frac{1}{k\pi}\int_0^{\pi}\cos kM\mathrm{d}E\,. \tag{4.39}$$

The integral in Eq. (4.39) is evaluated by substituting $M = E - e\sin E$ as follows:

$$b_k = \frac{1}{k\pi}\int_0^{\pi}\cos k(E - e\sin E)\mathrm{d}E\,, \tag{4.40}$$

which is related to the *Bessel functions*, $J_k(.)$, of the first kind and order k, given by

$$J_k(ke) = \frac{1}{\pi}\int_0^{\pi}\cos k(E - e\sin E)\mathrm{d}E\,. \tag{4.41}$$

Thus we have

$$b_k = \frac{1}{k}J_k(ke)\,. \tag{4.42}$$

The substitution of Eq. (4.42) into Eq. (4.37) yields

$$e\sin E = 2\sum_{k=1}^{\infty}\frac{1}{k}J_k(ke)\sin kM \tag{4.43}$$

and

$$E = M + 2 \sum_{k=1}^{\infty} \frac{1}{k} J_k(ke) \sin kM . \tag{4.44}$$

Equation (4.44) thus gives the solution to Kepler's equation as an infinite series containing Bessel functions of the first kind, $J_k(se)$, $k = 1, 2, \dots, \infty$.

An alternative derivation of Eq. (4.44) is given as follows. A differentiation of Kepler's equation leads to

$$\mathrm{d}E = \frac{\mathrm{d}M}{1 - e \cos E} . \tag{4.45}$$

Here it is recognized that the function $1/(1 - e \cos E)$ is an even periodic function of M with the period 2π, and hence can be expanded in the following Fourier cosine series:

$$\frac{1}{1 - e \cos E} = a_0 + \sum_{k=1}^{\infty} a_k \cos kM , \tag{4.46}$$

where $a_0, a_1, a_2, \dots$ are the Fourier coefficients evaluated as follows:

$$a_0 = \frac{1}{2\pi} \int_{-\pi}^{\pi} \frac{\mathrm{d}M}{1 - e \cos E} = \frac{1}{2\pi} \int_{-\pi}^{\pi} \mathrm{d}E = 1 , \tag{4.47}$$

$$a_k = \frac{1}{\pi} \int_{-\pi}^{\pi} \frac{\cos kM \mathrm{d}M}{1 - e \cos E} = \frac{1}{\pi} \int_{-\pi}^{\pi} \cos[k(E - e \sin E)]\mathrm{d}E = 2J_k(ke) . \tag{4.48}$$

The substitution of Eqs. (4.46) and (4.48) into Eq. (4.45), and integration, results in Eq. (4.44).

The evaluation of $J_k(v)$ is possible by using the following series expansion:

$$J_k(v) = \sum_{n=0}^{\infty} \frac{(-1)^n \left(\frac{v}{2}\right)^{k+2n}}{n!(n+k)!} . \tag{4.49}$$

The Bessel functions, $J_k(v)$, are listed in Table 4.2 for the first few terms of the series in Eq. (4.49). The series in Eq. (4.44) converges within a small tolerance with a small number of terms, for small to moderate values of the orbital eccentricity, e.

Example 4.2.2 *Solve Kepler's equation for* $e = 0.1$, $M = 1.2$ *rad., by the method of Bessel functions.*

Let a tolerance of $\delta = 10^{-10}$ *rad. be selected for the solution. The solution is then approximated as follows by a finite series in Eq. (4.44):*

$$E \simeq M + 2 \sum_{k=1}^{\ell} \frac{1}{k} J_k(ke) \sin kM ,$$

where the largest neglected term is given by

$$\frac{2}{\ell + 1} J_{\ell+1}[(\ell + 1)e] \sin(\ell + 1)M .$$

With the given values of e *and* M*, the magnitude of this term is smaller than the specified tolerance,* δ*, for* $\ell = 9$*. Hence, the computed value of* E *for* $\ell = 9$ *meets the required accuracy, and is given by* $E = 1.29625496381$ *rad.*

Table 4.2 Bessel functions of the first kind and order k.

$$J_1(\nu) = \frac{\nu}{2}\left(1 - \frac{\nu^2}{8} + \frac{\nu^4}{192} - \frac{\nu^6}{9216} + \frac{\nu^8}{737280} - \cdots\right)$$

$$J_2(\nu) = \frac{\nu^2}{8}\left(1 - \frac{\nu^2}{12} + \frac{\nu^4}{384} - \frac{\nu^6}{23040} + \frac{\nu^8}{2211840} - \cdots\right)$$

$$J_3(\nu) = \frac{\nu^3}{48}\left(1 - \frac{\nu^2}{16} + \frac{\nu^4}{640} - \frac{\nu^6}{46080} + \cdots\right)$$

$$J_4(\nu) = \frac{\nu^4}{384}\left(1 - \frac{\nu^2}{20} + \frac{\nu^4}{960} - \cdots\right)$$

$$J_5(\nu) = \frac{\nu^5}{3840}\left(1 - \frac{\nu^2}{24} + \frac{\nu^4}{1344} - \cdots\right)$$

$$J_6(\nu) = \frac{\nu^6}{46080}\left(1 - \frac{\nu^2}{28} + \frac{\nu^4}{1792} - \cdots\right)$$

$$J_7(\nu) = \frac{\nu^7}{645120}\left(1 - \frac{\nu^2}{32} + \frac{\nu^4}{2304} - \cdots\right)$$

4.3 Position and Velocity in a Hyperbolic Orbit

An equivalent form of Kepler's equation can be derived for a hyperbolic orbit ($e > 1$) by introducing the *hyperbolic anomaly*, H, via hyperbolic functions. The parametric equation of a hyperbola centred at C (see Fig. 3.6), with semi-major axis $a < 0$, and semi-minor axis, $b = -a\sqrt{e^2 - 1} = \sqrt{-ap} > 0$, can be expressed as follows:

$$\frac{x^2}{a^2} - \frac{y^2}{b^2} = 1\,, \tag{4.50}$$

which, in terms of the hyperbolic functions, is given by

$$\begin{aligned} x &= a\cosh H \\ y &= b\sinh H\,. \end{aligned} \tag{4.51}$$

Thus it follows that

$$\begin{aligned} r\cos\theta &= x - ae = a(\cosh H - e) \\ r\sin\theta &= y = b\sinh H\,, \end{aligned} \tag{4.52}$$

and the orbit equation, Eq. (3.40), is the following in terms of the hyperbolic anomaly:

$$r = a(1 - e\cosh H)\,. \tag{4.53}$$

Equations (4.52) and (4.53) yield the following relationships between the hyperbolic anomaly, H, and the true anomaly, θ:

$$\begin{aligned} \cos\theta &= \frac{\cosh H - e}{1 - e\cosh H} \\ \sin\theta &= \frac{\sqrt{e^2 - 1}\sinh H}{1 - e\cosh H}\,. \end{aligned} \tag{4.54}$$

Relating the single-valued hyperbolic functions to multi-valued trigonometric functions can cause a quadrant ambiguity for the angle, θ. For the quadrant determination of the angle θ, it is necessary to have a transformation which relates H to the correct quadrant of θ. This requires a trigonometric bridge between H and θ.

The conversion of hyperbolic functions into trigonometric functions in Eqs. (4.53) and (4.54) is made possible by the following transformation of the parametric equations, Eq. (4.51):

$$\begin{aligned} x &= a \sec \zeta \\ y &= b \tan \zeta \,, \end{aligned} \tag{4.55}$$

where the angle ζ is called the *Gudermannian*. This transformation results in

$$\begin{aligned} r \cos \theta &= a(\sec \zeta - e) \\ r \sin \theta &= b \tan \zeta \,, \end{aligned} \tag{4.56}$$

and the orbit equation becomes the following:

$$r = a(1 - e \sec \zeta) \,. \tag{4.57}$$

A substitution of Eq. (4.57) into Eq. (4.56) produces the following relationships between the Gudermannian, ζ, and the true anomaly, θ:

$$\begin{aligned} \cos \theta &= \frac{e - \sec \zeta}{e \sec \zeta - 1} \\ \sin \theta &= \frac{\sqrt{e^2 - 1} \tan \zeta}{e \sec \zeta - 1} \end{aligned} \tag{4.58}$$

and

$$\begin{aligned} \sin^2 \frac{\theta}{2} &= -\frac{a(1+e)}{r \cos \zeta} \sin^2 \frac{\zeta}{2} \\ \cos^2 \frac{\theta}{2} &= \frac{a(1-e)}{r \cos \zeta} \cos^2 \frac{\zeta}{2} \,, \end{aligned} \tag{4.59}$$

which yields

$$\tan \frac{\theta}{2} = \sqrt{\frac{1+e}{e-1}} \tan \frac{\zeta}{2} \,. \tag{4.60}$$

Thus it is necessary that $\theta/2$ and $\zeta/2$ should be in the same quadrant.

The transformation between H and ζ is given by

$$\begin{aligned} \tan \zeta &= \sinh H \\ \sec \zeta &= \cosh H \,, \end{aligned} \tag{4.61}$$

or in half-angle identities, the following:

$$\tan \frac{\zeta}{2} = \tanh \frac{H}{2} \,. \tag{4.62}$$

A substitution of Eq. (4.62) into Eq. (4.60) yields the following identity:

$$\tan\frac{\theta}{2} = \sqrt{\frac{1+e}{e-1}}\tanh\frac{H}{2}\,, \tag{4.63}$$

where the correct quadrant for θ is derived from Eq. (4.60).

The differentiation of Eqs. (4.52) and (4.53) results in the following:

$$\begin{aligned} d\theta &= \frac{\sin\theta}{\sinh H}dH = \frac{b}{r}dH \\ dr &= -ae\sinh H dH\,, \end{aligned} \tag{4.64}$$

which yields the following velocity components:

$$\begin{aligned} r\dot{\theta} &= b\dot{H} \\ \dot{r} &= -ae\sinh H\dot{H}\,, \end{aligned} \tag{4.65}$$

where the rate of variation of H with time, $\dot{H}$, needs to be evaluated. The substitution of the first of Eq. (4.64) into the expression for the constant angular momentum, $h = r^2\dot{\theta}$, yields the following differential equation for the change of the hyperbolic anomaly, H with time, t:

$$(e\cosh H - 1)dH = \sqrt{\frac{-\mu}{a^3}}dt\,, \tag{4.66}$$

which, upon integration from $H = 0,\ t = t_0$ yields

$$e\sinh H - H = n(t - t_0)\,, \tag{4.67}$$

where the *hyperbolic mean motion*, n, is given by

$$n = \sqrt{\frac{-\mu}{a^3}}\,. \tag{4.68}$$

Equation (4.67) is the equivalent of Kepler's equation for a hyperbolic orbit, and can be numerically solved using Newton's method in a manner similar to Kepler's equation.

The substitution of Eq. (4.66) into Eq. (4.65) produces the following velocity components:

$$\begin{aligned} r\dot{\theta} &= \frac{\sqrt{\mu p}}{a(1 - e\cosh H)} \\ \dot{r} &= \frac{e\sqrt{-\mu a}\sinh H}{a(1 - e\cosh H)}\,. \end{aligned} \tag{4.69}$$

Employing Eqs. (4.52) and (4.69), the perifocal position and velocity in a hyperbolic orbit is expressed as follows:

$$\begin{aligned} \mathbf{r} &= a(\cosh H - e)\mathbf{i_e} + \sqrt{-ap}\sinh H\mathbf{i_p} \\ \mathbf{v} &= -\frac{\sqrt{-\mu a}}{r}\sinh H\mathbf{i_e} + \frac{\sqrt{\mu p}}{r}\cosh H\mathbf{i_p}\,. \end{aligned} \tag{4.70}$$

An alternative form of the hyperbolic Kepler's equation, Eq. (4.67), can be derived to deter mine $H(t)$ from a given hyperbolic anomaly, H_0, at epoch, $t = 0$, which corresponds to the epochal position and velocity, $\mathbf{r_0}$ and $\mathbf{v_0}$, respectively, by expressing the current time, t, as follows:

$$\begin{aligned} t &= \sqrt{\frac{-a^3}{\mu}}[-(H - H_0) + e(\sinh H - \sinh H_0)] \\ &= \sqrt{\frac{-a^3}{\mu}}\left\{-(H - H_0) - \frac{\mathbf{r_0} \cdot \mathbf{v_0}}{\sqrt{-\mu a}}[1 - \cosh(H - H_0)] + \left(1 - \frac{r_0}{a}\right)\sinh(H - H_0)\right\}. \end{aligned} \tag{4.71}$$

When Eq. (4.71) is solved for $H - H_0$, it allows the expression of the current position and velocity in terms of those at $t = 0$ by using the Lagrange's coefficients (Eq. (3.98)) as follows:

$$\begin{aligned} \mathbf{r}(t) &= f\mathbf{r_0} + g\mathbf{v_0} \\ &= \left\{1 - \frac{a}{r_0}[1 - \cosh(H - H_0)]\right\}\mathbf{r_0} \\ &+ \left\{t - \frac{\sinh(H - H_0) - (H - H_0)}{\sqrt{-\mu/a^3}}\right\}\mathbf{v_0} \end{aligned} \tag{4.72}$$

and

$$\begin{aligned} \mathbf{v}(t) &= \dot{f}\mathbf{r_0} + \dot{g}\mathbf{v_0} \\ &= -\frac{\sqrt{-\mu a}}{r r_0}\sinh(H - H_0)\mathbf{r_0} + \left\{1 - \frac{a}{r}[1 - \cosh(H - H_0)]\right\}\mathbf{v_0}. \end{aligned} \tag{4.73}$$

Lagrange's coefficients for the perifocal position and velocity in a hyperbolic orbit are thus the following:

$$\begin{aligned} f &= 1 - \frac{a}{r_0}[1 - \cosh(H - H_0)] \\ g &= t - \frac{\sinh(H - H_0) - (H - H_0)}{\sqrt{-\mu/a^3}} \\ \dot{f} &= -\frac{\sqrt{-\mu a}}{r r_0}\sinh(H - H_0) \\ \dot{g} &= 1 - \frac{a}{r}[1 - \cosh(H - H_0)]. \end{aligned} \tag{4.74}$$

Example 4.3.1 *A spacecraft is observed with a radius $r_1 = 8000$ km, orbital speed, $v_1 = 10$ km/s, and flight-path angle, $\phi_1 = -5°$ around Earth. Calculate the radius, orbital speed, and flight-path angle 1 hour after the observation is taken.*

We begin by calculating the semi-major axis, eccentricity, and time of perigee from the given data as follows:

$$\varepsilon = 100/2 - 398600.4/8000 = 0.17495 = -\frac{\mu}{2a}$$

$$a = -\mu/\varepsilon = -1139183.767 \text{ km}$$

$$h = (8000)(10)\cos(-5°) = 79695.576 \text{ km}^2/\text{s}$$

$$p = h^2/\mu = 15934.216 \text{ km}$$

$$e = \sqrt{1 - \frac{p}{a}} = 1.0069694$$

$$\theta_1 = \cos^{-1}\left[\frac{1}{e}\left(\frac{p}{r_1} - 1\right)\right] = 6.109258 \text{ rad.}$$

$$\zeta_1 = 2\tan^{-1}\left(\tan\frac{\theta}{2}\sqrt{\frac{e-1}{e+1}}\right) = 6.27291 \text{ rad.}$$

$$H_1 = 2\tanh^{-1}\left(\tan\frac{\zeta}{2}\right) = -0.01027534 \text{ rad.}$$

$$n = \sqrt{-\mu/a^3} = 5.19252 \times 10^{-7} \text{ rad./s}$$

$$t_0 = (H_1 - e\sinh H_1)/n = 138.2665 \text{ s} .$$

Next, the hyperbolic Kepler's equation is solved with

$$e\sinh H - H = n(t - t_0) = n(3600 - 138.2665) = 0.001797512 \text{ rad.}$$

using Newton's method with a tolerance of $\delta = 10^{-8}$ *rad., whose computation steps are tabulated in Table 4.3. The final value of the hyperbolic anomaly is thus obtained after five iterations to be* $H = 0.15970301$ *rad. Hence, the radius, orbital speed, and flight-path angle at the new position are the following:*

$$r = a(1 - e\cosh H) = 22599.279 \text{ km}$$

$$v = \sqrt{2\mu/r - \mu/a} = 5.9687 \text{ km/s}$$

$$\phi = \tan^{-1}(e\sqrt{-a/p}\sinh H) = 0.93871377 \text{ rad. } (53.784°) .$$

It can be verified that the computed values correspond to the same a, e, and t_0*, as for the initial position and velocity,* (r_1, H_1, v_1, ϕ_1)*.*

4.4 Position and Velocity in a Parabolic Orbit

The boundary between an elliptic orbit ($\varepsilon < 0$) and a hyperbolic escape trajectory ($\varepsilon > 0$) is that of the parabolic trajectory, which has zero relative energy, $\varepsilon = 0$. A parabolic orbit is thus the minimum energy trajectory for escaping the gravitational influence of a spherical central body, m_1. While being impractical for interplanetary travel due to its zero velocity at infinite

Table 4.3 Iteration steps for the solution of the hyperbolic Kepler's equation by Newton's method for Example 4.3.1 (all angles in rad.).

H	$f(H)$	$f^{(1)}(H)$	ΔH
$M = 0.001797512047$	−0.0017849834699	0.0069710390377	0.2560570182187
0.2578545302658	0.00288649085495	0.0406314806940	−0.0710407498236
0.1868137804422	0.00060056863064	0.0245918824596	−0.0244214175806
0.1623923628616	5.393903147×10^{-5}	0.0202761526328	−0.0026602202325
0.1597321426291	5.7795708×10^{-7}	0.0198428375334	-2.91267355×10^{-5}
0.1597030158936	6.85142097×10^{-11}	0.0198381331176	-3.453662×10^{-9}
0.15970301244	—	—	—

radius, the parabolic trajectory is often a useful device for quickly determining the minimum fuel requirements for a given mission. Equation (4.1) for a parabolic orbit is the following:

$$\frac{1}{4}\sec^4\frac{\theta}{2}\mathrm{d}\theta = \sqrt{\frac{\mu}{p^3}}\mathrm{d}t\,, \tag{4.75}$$

which on integration from $t = t_0, \theta = 0$ yields

$$\tan^3\frac{\theta}{2} + 3\tan\frac{\theta}{2} = 6\sqrt{\frac{\mu}{p^3}}(t - t_0)\,. \tag{4.76}$$

Equation (4.76) is the equivalent of Kepler's equation for a parabolic orbit, and is called *Barker's equation*. Fortunately, there exists a unique, closed-form, and real solution to Barker's equation, which is derived by substituting

$$\tan\frac{\theta}{2} = \alpha - \frac{1}{\alpha}\,, \tag{4.77}$$

and solving the resulting quadratic equation for α^3, given by

$$3\alpha - \frac{3}{\alpha} + \left(\alpha - \frac{1}{\alpha}\right)^3 = \alpha^3 - \frac{1}{\alpha^3} = 2C\,, \tag{4.78}$$

where

$$C = 3\sqrt{\frac{\mu}{p^3}}(t - \tau) \tag{4.79}$$

is an equivalent mean anomaly. The only real solution to Eq. (4.78) is the following:

$$\alpha = (C + \sqrt{C^2 + 1})^{\frac{1}{3}}\,, \tag{4.80}$$

which, substituted into Eq. (4.77), yields the true anomaly, θ, at the current time, t:

$$\tan\frac{\theta}{2} = (C + \sqrt{1 + C^2})^{\frac{1}{3}} - (C + \sqrt{1 + C^2})^{-\frac{1}{3}}\,. \tag{4.81}$$

After solving Barker's equation for the current true anomaly, θ, the perifocal position and velocity in a parabolic orbit are calculated as follows:

$$\begin{aligned}\mathbf{r} &= r\cos\theta\mathbf{i_e} + r\sin\theta\mathbf{i_p}\\ \mathbf{v} &= -\frac{\mu}{h}\sin\theta\mathbf{i_e} + \frac{\mu}{h}(1 + \cos\theta)\mathbf{i_p}\,.\end{aligned} \tag{4.82}$$

Lagrange's coefficients for the perifocal position and velocity in a parabolic orbit are derived from Eq. (3.98) as follows:

$$
\begin{aligned}
f &= 1 + \frac{r}{p}[\cos(\theta - \theta_0) - 1] \\
g &= \frac{r r_0}{h} \sin(\theta - \theta_0) \\
\dot{f} &= -\frac{2h}{p^2} \sin(\theta - \theta_0) \\
\dot{g} &= 1 + \frac{r_0}{p}[\cos(\theta - \theta_0) - 1] ,
\end{aligned}
\tag{4.83}
$$

where (r_0, θ_0) is the known initial position (not necessarily the periapsis) and the corresponding time, t_0, from which the current position, (r, θ), is determined at time, t.

4.5 Universal Variable for Keplerian Motion

To generalize the variation of the true anomaly in time for Keplerian motion, a universal variable analogous to the eccentric anomaly can be defined, such that orbits of all shapes (elliptic, parabolic, and hyperbolic) are described by the same formulae. Let a constant, α, be defined to be the reciprocal of the semi-major axis, $\alpha = 1/a$. Then the epochal form of Kepler's equation for the elliptic orbit, Eq. (4.24), and that for the hyperbolic orbit, Eq. (4.71), reduce to the following *universal Kepler's equation* (Battin, 1964):

$$
t\sqrt{\mu} = x r_0 + \frac{\mathbf{r}_0 \cdot \mathbf{v}_0}{\sqrt{\mu}} x^2 C(\alpha x^2) + (1 - r_0 \alpha) x^3 S(\alpha x^2) . \tag{4.84}
$$

Here x is the *universal variable* which relates as follows to the difference between the current and the epochal eccentric anomalies for the elliptic orbit:

$$
x = \frac{E - E_0}{\sqrt{\alpha}} , \tag{4.85}
$$

and with the hyperbolic anomaly for the hyperbolic orbit as follows:

$$
x = \frac{H - H_0}{\sqrt{-\alpha}} , \tag{4.86}
$$

and $C(.)$ and $S(.)$ are the following *Stumpff functions*:

$$
C(y) = \frac{1}{2!} - \frac{y}{4!} + \frac{y^2}{6!} - \frac{y^3}{8!} + \cdots = \begin{cases} \dfrac{1 - \cos\sqrt{y}}{y} & (y > 0) \\ \dfrac{\cosh\sqrt{-y} - 1}{-y} & (y < 0) \end{cases} \tag{4.87}
$$

$$
S(y) = \frac{1}{3!} - \frac{y}{5!} + \frac{y^2}{7!} - \frac{y^3}{9!} + \cdots = \begin{cases} \dfrac{\sqrt{y} - \sin\sqrt{y}}{(y)^{3/2}} & (y > 0) \\ \dfrac{\sinh\sqrt{-y} - \sqrt{-y}}{(-y)^{3/2}} & (y < 0) \end{cases} . \tag{4.88}
$$

The Stumpff functions satisfy the following property:

$$[1 - yS(y)]^2 = C(y)[2 - yC(y)] = 2C(4y)\,. \tag{4.89}$$

The derivatives of the Stumpff functions are the following:

$$C'(y) = \frac{1}{2y}[1 - yS(y) - 2C(y)]$$
$$S'(y) = \frac{1}{2y}[C(y) - 3S(y)]\,. \tag{4.90}$$

The constant, α, defining the shape of the orbit is determined from the position and velocity, $\mathbf{r}_0, \mathbf{v}_0$, at epoch, $t = 0$, as follows:

$$\alpha = \frac{2}{r_0} - \frac{v_0^2}{\mu}\,. \tag{4.91}$$

The universal Kepler's equation, Eq. (4.84), is therefore valid for elliptic ($\alpha > 0$), parabolic ($\alpha = 0$), and hyperbolic ($\alpha < 0$) orbits, and can be iteratively solved by Newton's method for the elliptic and hyperbolic cases. For the parabola ($\alpha = 0$), we have the following from Eq. (4.84):

$$t\sqrt{\mu} = xr_0 + \frac{\mathbf{r}_0 \cdot \mathbf{v}_0}{2\sqrt{\mu}}x^2 + \frac{x^3}{6}\,, \tag{4.92}$$

which can be solved in a closed form for the only real root, x – as in the Barker's equation, Eq. (4.76) – and hence does not require a numerical solution.

When Newton's method is employed for the solution of Eq. (4.84), the following correction is applied at $t = t_k, x = x_k$:

$$x_{k+1} = x_k + \Delta x_k = x_k - \frac{t_k\sqrt{\mu} - t\sqrt{\mu}}{\left(\sqrt{\mu}\frac{\mathrm{d}t}{\mathrm{d}x}\right)_{x=x_k}}\,, \tag{4.93}$$

where

$$\sqrt{\mu}\frac{\mathrm{d}t}{\mathrm{d}x} = r_0 + \frac{\mathbf{r}_0 \cdot \mathbf{v}_0}{\sqrt{\mu}}[x - \alpha x^3 S(\alpha x^2)] + (1 - r_0\alpha)x^2 C(\alpha x^2)\,. \tag{4.94}$$

The position and velocity vectors are evolved from the epoch, $t = 0$, $\mathbf{r}_0, \mathbf{v}_0$, to the current time, t, in terms of the universal variable, x, as follows:

$$\mathbf{r}(t) = \left[1 - \frac{x^2}{r_0}C(\alpha x^2)\right]\mathbf{r}_0 + \left[t - \frac{x^3}{\sqrt{\mu}}S(\alpha x^2)\right]\mathbf{v}_0\,, \tag{4.95}$$

$$\mathbf{v}(t) = \frac{\sqrt{\mu}}{rr_0}[\alpha x^3 S(\alpha x^2) - x]\mathbf{r}_0 + \left[1 - \frac{x^2}{r}C(\alpha x^2)\right]\mathbf{v}_0\,. \tag{4.96}$$

These relations can be used to derive the Lagrange's coefficients in terms of the universal variable.

Example 4.5.1 *Repeat the computation for Earth orbit of $a = 8500$ km and $e = 0.2$ by the universal variable method, which was carried out earlier, in Example 4.2.1.*

Table 4.4 Iteration steps for the solution of the universal Kepler's equation by Newton's method for Example 4.5.1.

x_k (km$^{1/2}$)	$t_k\sqrt{\mu}$ (km$^{3/2}$)	$(\sqrt{\mu}\mathrm{d}t/\mathrm{d}x)_{x=x_k}$ (km)	Δx_k (km$^{1/2}$)
$t\sqrt{\mu}/r_0 = 139.267959$	1027329.53223703	8397.68240058123	−9.56304444161643
129.704914710397	947861.229905016	8222.53362434529	−0.102049770747353
129.602864939649	947022.216954166	8220.67732273702	-1.1522223×10^{-5}
129.602853417426	947022.12223369	8220.67711316522	-1.557741×10^{-13}
129.602853417426	—	—	—

The value of the constant, α, is calculated as follows:

$$\alpha = 1/a = 0.00011764706 \text{ km}^{-1}$$

The following quantities related to the initial position (perigee) are specified in Example 4.2.1: $E_0 = 0$, $r_0 = a(1-e) = 6800$ km, which, substituted into Eq. (4.22), yields:

$$\mathbf{r_0} = 6800 \ \mathbf{i_e} \ (\text{km})$$
$$\mathbf{v_0} = 8.386969 \ \mathbf{i_p} \ (\text{km/s}) \ .$$

The universal form of Kepler's equation, Eq. (4.84), to be solved for the universal variable, x, is thus the following:

$$t\sqrt{\mu} = xr_0 + (1 - r_0\alpha)x^3 S(\alpha x^2) \ .$$

Given $t = 25 \times 60 = 1500$ s, this equation is solved by Newton's method with a tolerance, $\delta = 10^{-10}$ km$^{1/2}$, to yield $x = 129.602853417426$ km$^{1/2}$ in three iterations, which are listed in Table 4.4. Equation (4.85) gives the following value of the eccentric anomaly:

$$E = x\sqrt{\alpha} = 1.40574031632232 \text{ rad.} \ ,$$

which agrees with that computed in Table 4.1.

Exercises

1. For an Earth-orbiting spacecraft with an orbital period of 205 min., eccentricity $e = 0.4$, and true anomaly 60°, calculate the time since periapsis. (*Ans.* 896.895 s.)

2. Solve the problem given in Example 4.2.1 by the method of Bessel functions. What is the number of terms in the series expansion required for the same accuracy as achieved by Newton's method in Example 4.2.1? (*Ans.* 18.)

3. A spacecraft in a 600 km high circular Earth orbit fires a retro-rocket, instantly reducing the spacecraft's speed by 600 m/s. How long does the spacecraft take to impact Earth's surface ($r = 6378.14$ km)? (Neglect atmospheric effects.) (*Ans.* 1011.9 s.)

4. For an Earth orbit with a semi-major axis of 12500 km and an eccentricity of 0.472, calculate the time since perigee to a position with a true anomaly of 198°. (*Ans.* 8615.25 s.)

5. For an Earth orbit with a semi-major axis of 12500 km and an eccentricity of 0.472, what are the position (radius, true anomaly) and velocity (speed, flight path angle) 2 hours after crossing perigee? (*Ans.* $r = 18383.208$ km; $\theta = 182.59°$; $v = 3.3879$ km/s; $\phi = -2.3115°$.)

6. An Earth spacecraft has apogee radius of 21000 km and perigee radius of 9600 km. What are the position and velocity 3 hours after crossing perigee?

7. An Earth spacecraft has perigee speed of 10 km/s and perigee radius of 7972.008 km. Find the position and velocity 6 hours after crossing perigee.

8. An Earth spacecraft has perigee speed of 15 km/s and perigee altitude of 300 km. What are the position and velocity 3 hours after reaching a true anomaly of 100°? (*Ans.* $r = 163184.814$ km; $\theta = 107.77968°$; $v = 10.5124$ km/s; $\phi = 86.65239°$.)

9. Using the universal variable, determine the current position and velocity with reference to the sun of comet *West*, which last reached a perihelion distance of 0.197 AU on Feb. 25, 1976, with an orbital eccentricity of 0.99997.

10. Using the universal variable, determine the current position and velocity with reference to the sun of comet Kahoutek, which last reached its perihelion distance of 0.1424 AU on Dec. 28, 1973, with an orbital eccentricity of 1.000008.

11. Determine the current position and velocity with reference to the sun of *Voyager 1*, given the following data from January 18, 1993:
    Heliocentric radius: 51 AU.
    Semi-major axis: −480926000 km
    Orbital eccentricity: 3.724716

12. Halley's comet last passed perihelion on February 9, 1986. It has a semi-major axis $a = 17.9564$ AU and eccentricity $e = 0.967298$. What is the current position of the comet relative to the sun?

13. A spacecraft was observed with altitude, speed, and flight path angle relative to Earth of 1000 km, 10 km/s, and −25°, respectively. Determine the position and velocity 10 hours after the observation was made. (*Ans.* $r = 85241.3111$ km; $\theta = 170.8973°$; $v = 1.1416$ km/s; $\phi = 46.5944°$.)

References

Battin RH 1964. *Astronautical Guidance*. McGraw-Hill, New York.

5

Orbital Plane

Chapter 4 showed that a two-body trajectory is completely described by six scalar constants, which are called *orbital elements*. Several different sets of orbital elements can be chosen, depending upon the application. An obvious choice is the set formed by the elements of the initial position and velocity vectors, $\mathbf{r}_0, \mathbf{v}_0$, which, once specified, determine the orbit without ambiguity. Since the orbital motion is planar, one can fix the position and velocity by defining the plane of the motion, and then specifying the two-dimensional (perifocal) position and velocity in the orbital plane. For example, the set of constant vectors, $\mathbf{h}, \mathbf{e}$, fixes the orbital plane by the orientation of the right-handed, perifocal unit vectors, $\mathbf{i_e} = \mathbf{e}/e$, $\mathbf{i_h} = \mathbf{h}/h$, and $\mathbf{i_p} = \mathbf{i_h} \times \mathbf{i_e}$, while the perifocal position and velocity in the orbital plane are specified by their magnitudes, h and e, as well as the time of periapsis, t_0. Various alternative sets of scalar constants that prescribe the position and velocity in the orbital plane are (r, θ, t_0), (v, ϕ, t_0), (a, e, t_0) or $(r_0, \theta - \theta_0, t - t_0)$, etc. The fixed orientation of the orbital plane, $(\mathbf{i_e}, \mathbf{i_p}, \mathbf{i_h})$ still needs to be described by at least three more scalar constants. The *orientation*, or *attitude*, of a coordinate frame is defined relative to a reference frame, and requires a set of *kinematical parameters* for its complete description. This chapter is devoted to the representation of a coordinate frame's orientation by such a set of parameters, which is useful not only in fixing the orbital plane, but also in describing the rotational motion of a rigid body in later chapters.

5.1 Rotation Matrix

A non-rotating (inertial) reference frame, represented by a right-handed triad $(\mathbf{I}, \mathbf{J}, \mathbf{K})$, is necessary for describing the orientation of the perifocal orbital frame, $(\mathbf{i_e}, \mathbf{i_p}, \mathbf{i_h})$, by the following coordinate transformation:

$$\begin{Bmatrix} \mathbf{i_e} \\ \mathbf{i_p} \\ \mathbf{i_h} \end{Bmatrix} = \mathbf{C} \begin{Bmatrix} \mathbf{I} \\ \mathbf{J} \\ \mathbf{K} \end{Bmatrix}, \tag{5.1}$$

Foundations of Space Dynamics, First Edition. Ashish Tewari.

where $\mathbf{C}$ is the *rotation matrix* defining the transformation. A rotation matrix denotes a special transformation wherein the magnitudes of the transformed vectors are preserved, but their orientation relative to a fixed reference frame is changed.

The rotation matrix, $\mathbf{C}$, has nine elements, which are dependent on one another. Many different sets of parameters, called the *kinematical parameters*, can be employed to uniquely represent the rotation matrix. The smallest number of such parameters required to represent a rotation matrix is said to constitute a *minimal set*. In general, the rotation required to transform a right-handed reference frame, $(\mathbf{a}, \mathbf{b}, \mathbf{c})$, to another frame, $(\mathbf{a}', \mathbf{b}', \mathbf{c}')$, where $\mathbf{a}, \mathbf{b}, \mathbf{c}$, etc., are unit vectors, is given by

$$\begin{Bmatrix} \mathbf{a}' \\ \mathbf{b}' \\ \mathbf{c}' \end{Bmatrix} = \mathbf{C} \begin{Bmatrix} \mathbf{a} \\ \mathbf{b} \\ \mathbf{c} \end{Bmatrix}, \tag{5.2}$$

and $\mathbf{C}$ is the rotation matrix of the concerned transformation.

It is expected that a rotation matrix should be *orthogonal* (i.e., it must preserve the orthogonality of the unit vectors transformed by it). Furthermore, a rotation matrix should have a *unity determinant*, which implies that the unit vectors remain of unit magnitude after being multiplied by the rotation matrix, and maintain their right-handedness. Therefore, $\mathbf{C}$ must have the property of *orthogonality* of a square matrix, given by

$$\mathbf{C}^{-1} = \mathbf{C}^T, \tag{5.3}$$

with $\det(\mathbf{C}) = 1$. To derive these two properties of a rotation matrix, consider a vector, $\mathbf{V}$, alternately resolved in the original and rotated frames as follows:

$$\mathbf{V} = V_a\mathbf{a} + V_b\mathbf{b} + V_c\mathbf{c} = V_a'\mathbf{a}' + V_b'\mathbf{b}' + V_c'\mathbf{c}', \tag{5.4}$$

which can be expressed in a matrix form as follows:

$$\mathbf{V} = (\mathbf{a},\ \mathbf{b},\ \mathbf{c}) \begin{Bmatrix} V_a \\ V_b \\ V_c \end{Bmatrix} = (\mathbf{a}',\ \mathbf{b}',\ \mathbf{c}') \begin{Bmatrix} V_a' \\ V_b' \\ V_c' \end{Bmatrix}. \tag{5.5}$$

Taking the scalar products, $\mathbf{V} \cdot \mathbf{a}'$, $\mathbf{V} \cdot \mathbf{b}'$, and $\mathbf{V} \cdot \mathbf{c}'$, we have

$$\begin{Bmatrix} V_a' \\ V_b' \\ V_c' \end{Bmatrix} = \begin{pmatrix} \mathbf{a}' \cdot \mathbf{a} & \mathbf{a}' \cdot \mathbf{b} & \mathbf{a}' \cdot \mathbf{c} \\ \mathbf{b}' \cdot \mathbf{a} & \mathbf{b}' \cdot \mathbf{b} & \mathbf{b}' \cdot \mathbf{c} \\ \mathbf{c}' \cdot \mathbf{a} & \mathbf{c}' \cdot \mathbf{b} & \mathbf{c}' \cdot \mathbf{c} \end{pmatrix} \begin{Bmatrix} V_a \\ V_b \\ V_c \end{Bmatrix}. \tag{5.6}$$

The scalar products of $\mathbf{V}$ with $\mathbf{a}, \mathbf{b}$, and $\mathbf{c}$ are the following:

$$\begin{Bmatrix} V_a \\ V_b \\ V_c \end{Bmatrix} = \begin{pmatrix} \mathbf{a} \cdot \mathbf{a}' & \mathbf{a} \cdot \mathbf{b}' & \mathbf{a} \cdot \mathbf{c}' \\ \mathbf{b} \cdot \mathbf{a}' & \mathbf{b} \cdot \mathbf{b}' & \mathbf{b} \cdot \mathbf{c}' \\ \mathbf{c} \cdot \mathbf{a}' & \mathbf{c} \cdot \mathbf{b}' & \mathbf{c} \cdot \mathbf{c}' \end{pmatrix} \begin{Bmatrix} V_a' \\ V_b' \\ V_c' \end{Bmatrix}. \tag{5.7}$$

Equation (5.6) can also be expressed as follows in terms of the rotation matrix, $\mathbf{C}$:

$$\begin{Bmatrix} V'_a \\ V'_b \\ V'_c \end{Bmatrix} = \mathbf{C} \begin{Bmatrix} V_a \\ V_b \\ V_c \end{Bmatrix} , \tag{5.8}$$

where the rotation matrix, $\mathbf{C}$, is seen to be the following matrix consisting of the cosines of angles between the axes of the two coordinate frames, and is therefore also called the *direction-cosine matrix* of the transformation:

$$\mathbf{C} = \begin{pmatrix} \mathbf{a}' \cdot \mathbf{a} & \mathbf{a}' \cdot \mathbf{b} & \mathbf{a}' \cdot \mathbf{c} \\ \mathbf{b}' \cdot \mathbf{a} & \mathbf{b}' \cdot \mathbf{b} & \mathbf{b}' \cdot \mathbf{c} \\ \mathbf{c}' \cdot \mathbf{a} & \mathbf{c}' \cdot \mathbf{b} & \mathbf{c}' \cdot \mathbf{c} \end{pmatrix} . \tag{5.9}$$

Furthermore, Eq. (5.7), expressed in terms of the rotation matrix, becomes the following:

$$\begin{Bmatrix} V_a \\ V_b \\ V_c \end{Bmatrix} = \mathbf{C}^{-1} \begin{Bmatrix} V'_a \\ V'_b \\ V'_c \end{Bmatrix} , \tag{5.10}$$

and a comparison of Eqs. (5.7), (5.9), and (5.10) implies that $\mathbf{C}^{-1} = \mathbf{C}^T$ – that is, the rotation matrix, $\mathbf{C}$, is orthogonal. This property can also be deduced from Eq. (5.9) as the vectors formed out of the columns of the rotation matrix are orthogonal. The orthogonality of $\mathbf{C}$ can be expressed by the following equation:

$$\mathbf{C}^T\mathbf{C} = \mathbf{C}\mathbf{C}^T = \mathbf{I} , \tag{5.11}$$

where $\mathbf{I}$ is the identity matrix. Taking the determinant of both the sides of Eq. (5.11) yields

$$\det(\mathbf{C}^T\mathbf{C}) = \det(\mathbf{C}\mathbf{C}^T) = \det(\mathbf{C})^2 = \det(\mathbf{I}) = 1 , \tag{5.12}$$

from which it follows that $\det(\mathbf{C}) = \pm 1$. By taking the positive sign, we define a *proper* rotation to be the one for which $\det(\mathbf{C}) = 1$.

Another property of the rotation matrix can be derived from Eq. (5.2) by expressing the frame, $(\mathbf{a}', \mathbf{b}', \mathbf{c}')$, to be transformed to another frame, $(\mathbf{a}'', \mathbf{b}'', \mathbf{c}'')$, via the rotation matrix, $\mathbf{C}'$, as follows:

$$\begin{Bmatrix} \mathbf{a}'' \\ \mathbf{b}'' \\ \mathbf{c}'' \end{Bmatrix} = \mathbf{C}' \begin{Bmatrix} \mathbf{a}' \\ \mathbf{b}' \\ \mathbf{c}' \end{Bmatrix} . \tag{5.13}$$

A substitution of Eq. (5.2) into Eq. (5.13) yields

$$\begin{Bmatrix} \mathbf{a}'' \\ \mathbf{b}'' \\ \mathbf{c}'' \end{Bmatrix} = \mathbf{C}'\mathbf{C} \begin{Bmatrix} \mathbf{a}' \\ \mathbf{b}' \\ \mathbf{c}' \end{Bmatrix} = \mathbf{C}'' \begin{Bmatrix} \mathbf{a} \\ \mathbf{b} \\ \mathbf{c} \end{Bmatrix} , \tag{5.14}$$

from which it follows that two successive rotations of a coordinate frame can be represented simply by multiplying the rotation matrices of individual rotations in the correct sequence as follows:

$$\mathbf{C}'' = \mathbf{C}'\mathbf{C}. \tag{5.15}$$

Here the orientation represented by $\mathbf{C}''$ is obtained by first undergoing a rotation $\mathbf{C}$, followed by a rotation $\mathbf{C}'$.

5.2 Euler Axis and Principal Angle

The *eigenvalues* and *eigenvectors* of the rotation matrix, $\mathbf{C}$, reveal an important property of rotational transformation. Let $\mathbf{s}$ be an eigenvector associated with the eigenvalue, λ, of $\mathbf{C}$:

$$\mathbf{C}\mathbf{s} = \lambda\mathbf{s}. \tag{5.16}$$

On pre-multiplication of Eq. (5.4) by the *Hermitian conjugate* of each side, we have

$$(\mathbf{C}\mathbf{s})^H(\mathbf{C}\mathbf{s}) = \bar{\lambda}\lambda\mathbf{s}^H\mathbf{s}, \tag{5.17}$$

where $\overline{(.)}$ deotes the complex conjugate of (.), and $(.)^H = \overline{(.)}^T$. Since $\mathbf{C}$ is real and satisfies Eq. (5.11), it follows that

$$(\bar{\lambda}\lambda - 1)\mathbf{s}^H\mathbf{s} = 0, \tag{5.18}$$

which implies that

$$\bar{\lambda}\lambda = 1 \tag{5.19}$$

because $\mathbf{s}$ is nonzero. Equation (5.19) states the fact that all three eigenvalues of $\mathbf{C}$ have the magnitude unity. Since complex eigenvalues occur in conjugate pairs, it follows that one of the eigenvalues of $\mathbf{C}$ must be real ($\lambda_1 = 1$), for which Eq. (5.16) becomes

$$\mathbf{C}\mathbf{s}_1 = \mathbf{s}_1. \tag{5.20}$$

Equation (5.20) implies that the eigenvector, $\mathbf{s}_1$, associated with the eigenvalue, $\lambda_1 = 1$, is unchanged by the rotation. It also follows from Eqs. (5.11) and (5.20) that $\mathbf{s}_1{}^T\mathbf{s}_1 = 1$, i.e., $\mathbf{s}_1$ is a unit vector. Therefore, the unit vector, $\mathbf{s}_1$, being invariant under the coordinate transformation given by Eq. (5.2), represents the fixed axis about which the rotation takes place. This axis of rotation, $\mathbf{s}_1$, is termed the *Euler axis*, and passes through the common origin of the coordinate frames, $(\mathbf{a}, \mathbf{b}, \mathbf{c})$ and $(\mathbf{a}', \mathbf{b}', \mathbf{c}')$, as shown in Fig. 5.1.

The other two eigenvalues of $\mathbf{C}$ are a pair of complex conjugates with the unit magnitude, and can be expressed as $\lambda_{2,3} = e^{\pm i\Phi} = \cos\Phi \pm i\sin\Phi$. Therefore, Eq. (5.16) yields

$$\mathbf{C}\mathbf{s}_{2,3} = e^{\pm i\Phi}\mathbf{s}_{2,3}. \tag{5.21}$$

In the complex plane representation of a vector, the factor, $e^{i\Phi}$, multiplying a vector implies a positive rotation of the vector by the angle, Φ. Hence, the eigenvectors, $\mathbf{s}_{2,3}$, which are complex conjugates, undergo rotations by angle $\pm\Phi$, respectively, when the coordinate frame is rotated about the axis $\mathbf{s}_1$. A consequence of $\mathbf{C}$ being orthogonal [Eq. (5.11)] is that its eigenvectors are mutually perpendicular. Since the vectors $\mathbf{s}_{2,3}$ are perpendicular to $\mathbf{s}_1$, their rotation must be

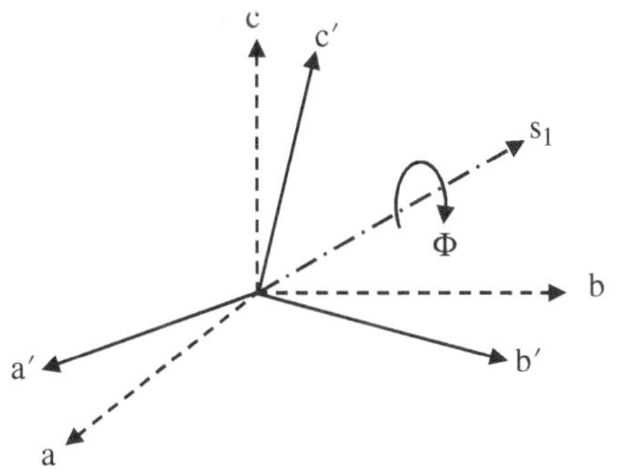

Figure 5.1 The Euler axis, $\mathbf{s}_1$, and the principal angle, Φ, defining the orientation of the coordinate frame, $(\mathbf{a}', \mathbf{b}', \mathbf{c}')$, with respect to $(\mathbf{a}, \mathbf{b}, \mathbf{c})$.

equal to the angle of coordinate frame rotation. Therefore, Φ is the angle by which the frame $(\mathbf{a}, \mathbf{b}, \mathbf{c})$ is rotated about the Euler axis, $\mathbf{s}_1$, to produce the frame, $(\mathbf{a}', \mathbf{b}', \mathbf{c}')$. The angle of rotation, Φ, is called the *principal angle*, and is shown in Fig. 5.1.

A simple method of obtaining the principal angle is through the *trace* of $\mathbf{C}$:

$$\text{trace}(\mathbf{C}) = \lambda_1 + \lambda_2 + \lambda_3 = 1 + e^{i\Phi} + e^{-i\Phi} = 1 + 2\cos\Phi\,, \tag{5.22}$$

or

$$\cos\Phi = \frac{1}{2}[\text{trace}(\mathbf{C}) - 1]\,. \tag{5.23}$$

There are two values of Φ, differing only in sign, which satisfy Eq. (5.23). However, Eq. (5.20) is satisfied for both $\mathbf{s}_1$ and $-\mathbf{s}_1$; hence a rotation by angle Φ about $\mathbf{s}_1$ is the same as a rotation by $-\Phi$ about $-\mathbf{s}_1$. Thus there is no loss of generality by taking the positive value of the principal angle, Φ, from Eq. (5.23).

The rotation matrix can be expressed in terms of the Euler axis and the principal angle as follows by the use of Eqs. (5.20) and (5.23):

$$\mathbf{C} = \mathbf{I}\cos\Phi + (1 - \cos\Phi)\mathbf{s}_1{\mathbf{s}_1}^T - \sin\Phi\mathbf{S}(\mathbf{s}_1)\,, \tag{5.24}$$

where $\mathbf{S}(\mathbf{s}_1)$ is the following skew-symmetric matrix formed out of the components of $\mathbf{s}_1 = s_a\mathbf{a} + s_b\mathbf{b} + s_c\mathbf{c}$:

$$\mathbf{S}(\mathbf{s}_1) = \begin{pmatrix} 0 & -s_c & s_b \\ s_c & 0 & -s_a \\ -s_b & s_a & 0 \end{pmatrix}. \tag{5.25}$$

This relationship between the rotation matrix and the Euler-axis/principal-angle combination is called *Euler's formula.*

The Euler-axis/principal-angle representation of the rotation described by Eq. (5.2) is depicted in Fig. 5.1, where it can be seen that it consists of a right-handed rotation by the principal angle, Φ, about the Euler axis, $\mathbf{s}_1$, passing through the common origin of the two frames. Such a combination of the Euler axis and principal angle, $(\mathbf{s}_1, \Phi)$, required to represent the rotation matrix, $\mathbf{C}$, in Eq. (5.2) thus consists of four scalar parameters, i.e., three components (or the direction cosines) constituting $\mathbf{s}_1 = s_a\mathbf{a} + s_b\mathbf{b} + s_c\mathbf{c}$, and the angle, Φ, as the fourth parameter. However, out of these four kinematical parameters, (s_a, s_b, s_c, Φ), only three are independent because $\mathbf{s}_1$, being a unit vector, must satisfy

$$s_a^2 + s_b^2 + s_c^2 = 1\,. \tag{5.26}$$

Hence the minimum number of scalar kinematical parameters required to represent the rotation matrix, $\mathbf{C}$, is three. A set of three kinematical parameters used for representing the rotation matrix is therefore a *minimal set*. When a set of more than three scalar parameters is used to represent $\mathbf{C}$, such as the set $(\mathbf{s}_1, \Phi)$, it is said to be a *redundant set*, and all the kinematical parameters of such a set are not independent of one another.

A four-parameter set related to the Euler-axis/principal-angle representation is the set of *Euler symmetric parameters*, which is also termed the *quaternion*. A quaternion is a special set composed of four mutually dependent scalar parameters, q_1, q_2, q_3, q_4, such that the first three form a vector, called the *vector part*,

$$\mathbf{q} = \begin{Bmatrix} q_1 \\ q_2 \\ q_3 \end{Bmatrix}, \tag{5.27}$$

and the fourth, q_4, represents the *scalar part*. The quaternion for attitude representation can be derived from the Euler axis, $\mathbf{s}_1 = s_a\mathbf{a} + s_b\mathbf{b} + s_c\mathbf{c}$, and the principal rotation angle, Φ, as follows:

$$\begin{aligned} q_1 &= s_a \sin\frac{\Phi}{2} \\ q_2 &= s_b \sin\frac{\Phi}{2} \\ q_3 &= s_c \sin\frac{\Phi}{2} \\ q_4 &= \cos\frac{\Phi}{2}\,. \end{aligned} \tag{5.28}$$

It follows from Eq. (5.28) that q_1, q_2, q_3, q_4 must satisfy the constraint equation,

$$q_1^2 + q_2^2 + q_3^2 + q_4^2 = 1\,. \tag{5.29}$$

Hence, as expected, the parts of the quaternion, $(\mathbf{q}, q_4)$, are not independent of one another. The quaternion satisfies a special set of mathematical operations called the *quaternion algebra*, which was invented by Hamilton.

The rotation matrix, $\mathbf{C}$, can be expressed in terms of the quaternion by substituting Eq. (5.28) into Euler's formula, Eq. (5.24), as follows:

$$\mathbf{C} = (q_4^2 - \mathbf{q}^T\mathbf{q})\mathbf{I} + 2\mathbf{q}\mathbf{q}^T - 2q_4\mathbf{S}(\mathbf{q})\,, \tag{5.30}$$

where $\mathbf{S}(\mathbf{q})$ is the following skew-symmetric matrix function formed out of the elements of vector $\mathbf{q}$:

$$\mathbf{S}(\mathbf{q}) = \begin{pmatrix} 0 & -q_3 & q_2 \\ q_3 & 0 & -q_1 \\ -q_2 & q_1 & 0 \end{pmatrix}. \tag{5.31}$$

A minimal (three parameter) set of kinematical parameters related to the Euler-axis/principal-angle representation is the set of *modified Rodrigues parameters*, which is defined by

$$\mathbf{p} = \mathbf{s}_1 \tan \frac{\Phi}{4}\,. \tag{5.32}$$

Clearly, the set $\mathbf{p} = (p_1, p_2, p_3)^T$ is non-singular for principal rotations of $\Phi < 360°$. However, there is a singularity at $\Phi = 360°$, where $\mathbf{p}$ is undefined. Such singularities are inevitable in minimal representations.

The rotation matrix can be expressed in terms of the modified Rodrigues parameters as follows:

$$\mathbf{C} = \mathbf{I} + \frac{4(\mathbf{p}^T\mathbf{p} - 1)}{(1 + \mathbf{p}^T\mathbf{p})^2}\mathbf{S}(\mathbf{p}) + \frac{8}{(1 + \mathbf{p}^T\mathbf{p})^2}\mathbf{S}^2(\mathbf{p})\,, \tag{5.33}$$

where $\mathbf{S}(\mathbf{p})$ is the following skew-symmetric matrix formed out of the elements of $\mathbf{p} = (p_1, p_2, p_3)^T$:

$$\mathbf{S}(\mathbf{p}) = \begin{pmatrix} 0 & -p_3 & p_2 \\ p_3 & 0 & -p_1 \\ -p_2 & p_1 & 0 \end{pmatrix}. \tag{5.34}$$

5.3 Elementary Rotations and Euler Angles

The simplest coordinate transformation is a rotation about one of the axes of a given coordinate frame, and is called an *elementary rotation*. In an elementary rotation of a coordinate frame, $(\mathbf{a}, \mathbf{b}, \mathbf{c})$, the Euler axis, $\mathbf{s}_1$, is aligned with one of the axes of the original frame, ($\mathbf{a}$, $\mathbf{b}$, or $\mathbf{c}$), thereby producing no change in the corresponding axis of the transformed frame. A positive rotation of the frame, $(\mathbf{a}, \mathbf{b}, \mathbf{c})$, about $\mathbf{a}$ (that is, the first axis of the frame) by an angle Φ is depicted in Fig. 5.2, and is described by the following rotation matrix:

$$\begin{aligned} \mathbf{C}_1(\Phi) &= \begin{pmatrix} \mathbf{a}'\cdot\mathbf{a} & \mathbf{a}'\cdot\mathbf{b} & \mathbf{a}'\cdot\mathbf{c} \\ \mathbf{b}'\cdot\mathbf{a} & \mathbf{b}'\cdot\mathbf{b} & \mathbf{b}'\cdot\mathbf{c} \\ \mathbf{c}'\cdot\mathbf{a} & \mathbf{c}'\cdot\mathbf{b} & \mathbf{c}'\cdot\mathbf{c} \end{pmatrix} \\ &= \begin{pmatrix} \cos(0) & \cos\left(\frac{\pi}{2}\right) & \cos\left(\frac{\pi}{2}\right) \\ \cos\left(\frac{\pi}{2}\right) & \cos\Phi & \cos\left(\frac{\pi}{2} - \Phi\right) \\ \cos\left(\frac{\pi}{2}\right) & \cos\left(\frac{\pi}{2} + \Phi\right) & \cos\Phi \end{pmatrix} \\ &= \begin{pmatrix} 1 & 0 & 0 \\ 0 & \cos\Phi & \sin\Phi \\ 0 & -\sin\Phi & \cos\Phi \end{pmatrix}, \end{aligned} \tag{5.35}$$

whereas the rotation matrix for a positive rotation about $\mathbf{b}$ (the second axis of the frame; see Fig 5.3) by the same angle is

$$\mathbf{C}_2(\Phi) = \begin{pmatrix} \mathbf{a}'\cdot\mathbf{a} & \mathbf{a}'\cdot\mathbf{b} & \mathbf{a}'\cdot\mathbf{c} \\ \mathbf{b}'\cdot\mathbf{a} & \mathbf{b}'\cdot\mathbf{b} & \mathbf{b}'\cdot\mathbf{c} \\ \mathbf{c}'\cdot\mathbf{a} & \mathbf{c}'\cdot\mathbf{b} & \mathbf{c}'\cdot\mathbf{c} \end{pmatrix}$$

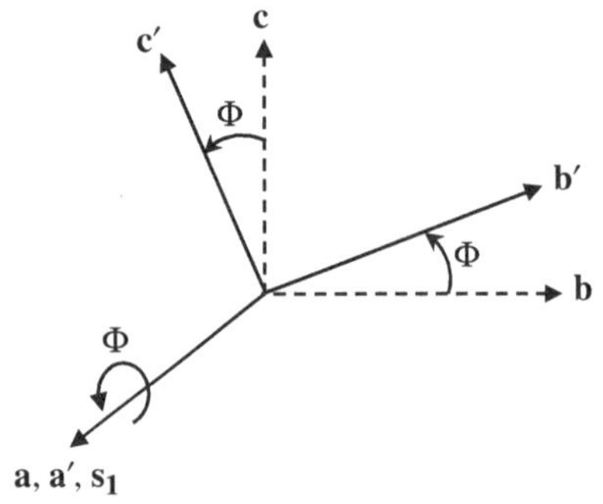

Figure 5.2 Elementary rotation about the first axis, **a**, of the coordinate frame, (**a**, **b**, **c**).

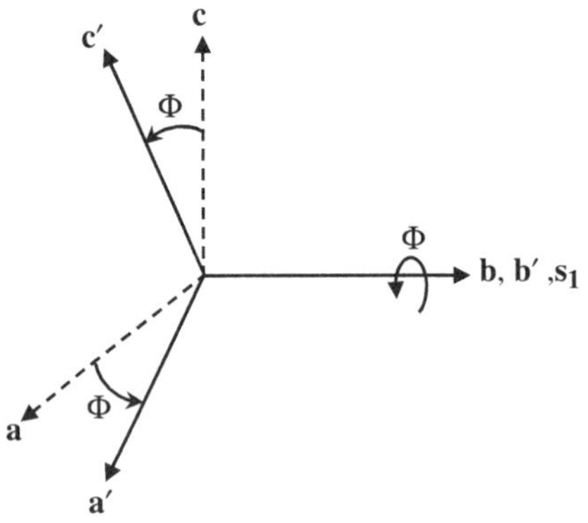

Figure 5.3 Elementary rotation about the second axis, **b**, of the coordinate frame, (**a**, **b**, **c**).

$$= \begin{pmatrix} \cos\Phi & \cos\left(\frac{\pi}{2}\right) & \cos\left(\frac{\pi}{2}+\Phi\right) \\ \cos\left(\frac{\pi}{2}\right) & \cos(0) & \cos\left(\frac{\pi}{2}\right) \\ \cos\left(\frac{\pi}{2}-\Phi\right) & \cos\left(\frac{\pi}{2}\right) & \cos\Phi \end{pmatrix} \tag{5.36}$$

$$= \begin{pmatrix} \cos\Phi & 0 & -\sin\Phi \\ 0 & 1 & 0 \\ \sin\Phi & 0 & \cos\Phi \end{pmatrix}.$$

Similarly, a positive rotation about the third axis, **c**, by Φ (Fig. 5.4) is denoted by

$$\mathbf{C}_3(\Phi) = \begin{pmatrix} \mathbf{a}'\cdot\mathbf{a} & \mathbf{a}'\cdot\mathbf{b} & \mathbf{a}'\cdot\mathbf{c} \\ \mathbf{b}'\cdot\mathbf{a} & \mathbf{b}'\cdot\mathbf{b} & \mathbf{b}'\cdot\mathbf{c} \\ \mathbf{c}'\cdot\mathbf{a} & \mathbf{c}'\cdot\mathbf{b} & \mathbf{c}'\cdot\mathbf{c} \end{pmatrix}$$

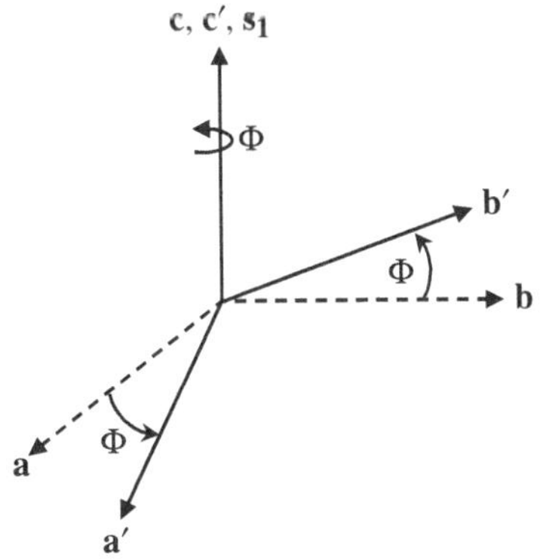

Figure 5.4 Elementary rotation about the third axis, **c**, of the coordinate frame, (**a**, **b**, **c**).

$$
= \begin{pmatrix} \cos\Phi & \cos\left(\frac{\pi}{2}-\Phi\right) & \cos\left(\frac{\pi}{2}\right) \\ \cos\left(\frac{\pi}{2}+\Phi\right) & \cos\Phi & \cos\left(\frac{\pi}{2}\right) \\ \cos\left(\frac{\pi}{2}\right) & \cos\left(\frac{\pi}{2}\right) & \cos(0) \end{pmatrix} \tag{5.37}
$$

$$
= \begin{pmatrix} \cos\Phi & \sin\Phi & 0 \\ -\sin\Phi & \cos\Phi & 0 \\ 0 & 0 & 1 \end{pmatrix}.
$$

A general orientation can be described by a sequence of successive elementary rotations, as discussed later.

While the Euler-axis/principal angle combination (and its associated four parameter sets, such as the quaternion) can be utilized for describing the orientation of a coordinate frame, a three-parameter representation is desired for greater simplicity. Being a minimal representation, a three-parameter kinematical set yields the smallest number of variables required to describe a frame's rotation. An example of a minimal attitude representation is the set of modified Rodrigues parameters defined by Eq. (5.32). However, a disadvantage of a minimal representation is that it suffers from singularity at a particular value of the principal angle, Φ. Another minimal representation is derived from the elementary rotations. A general orientation can be obtained by using successive elementary rotations. The largest number of such rotations needed to uniquely specify a given orientation, called the rotational *degrees of freedom*, is three. Therefore, a set of three elementary rotations, each about a specific coordinate axis, forms a minimal set of kinematical parameters for describing a given orientation. Such a representation of the attitude by three angles is called an *Euler-angle* representation, and the concerned angles are known as the *Euler angles*. The sequence of successive axial rotations is of utmost importance in the Euler angle representation, which can be easily verified. For example, rotating a book by three elementary rotations of 90°, each about a different axis, and then carrying out the rotations about the same axes in a different sequence, produces a different final attitude of the book in each case. The general orientation resulting from a set of three elementary rotations is described by the following rotation matrix:

$$\mathbf{C} = \sum_{i=1}^{3} \mathbf{C}_i(\Phi_i), \tag{5.38}$$

where the Euler angles, Φ_i, $i = 1, 2, 3$, represent a particular sequence of elementary rotations about the axes, $i = 1, 2, 3$, of the coordinate frame. Each rotation modifies the coordinate axes, whose current orientation is used to produce the next elementary rotation, as clarified by the following example.

Example 5.3.1 *Consider the set of sequential elementary rotations, $(\psi_3, \theta_2, \phi_1)$, as an example of employing Euler angles to represent a general orientation. As shown in Fig. 5.5, this sequence consists of a rotation by angle ψ about the third axis, $\mathbf{c}$, of the original coordinate frame, $(\mathbf{a}, \mathbf{b}, \mathbf{c})$, resulting in the intermediate orientation, $(\mathbf{a}', \mathbf{b}', \mathbf{c}')$. Since the third axis of the frame is invariant in this elementary rotation, we have $\mathbf{c}' = \mathbf{c}$. The next elementary rotation in this sequence is carried out by angle θ about the second axis, $\mathbf{b}'$, of the intermediate frame, which produces the intermediate orientation, $(\mathbf{a}'', \mathbf{b}'', \mathbf{c}'')$. Clearly, the second axis of the frame is invariant in this elementary rotation; hence $\mathbf{b}'' = \mathbf{b}'$. The final elementary rotation in the given sequence is by angle ϕ about the first axis of the frame (which is invariant under this rotation), $\mathbf{a}''' = \mathbf{a}''$, thereby resulting in the final orientation of the coordinate frame, $(\mathbf{a}''', \mathbf{b}''', \mathbf{c}''')$. This specific Euler angle representation is commonly used to describe the general attitude of a rigid body, and the concerned Euler angles are termed the* yaw angle, *ψ, the* pitch angle, *θ, and the* roll angle, *ϕ.*

The rotation matrix corresponding to the Euler-angle representation, $(\psi_3, \theta_2, \phi_1)$, is the following:

$$\mathbf{C} = \mathbf{C}_1(\phi)\mathbf{C}_2(\theta)\mathbf{C}_3(\psi) = \begin{pmatrix} 1 & 0 & 0 \\ 0 & \cos\phi & \sin\phi \\ 0 & -\sin\phi & \cos\phi \end{pmatrix}$$

$$\times \begin{pmatrix} \cos\theta & 0 & -\sin\theta \\ 0 & 1 & 0 \\ \sin\theta & 0 & \cos\theta \end{pmatrix} \begin{pmatrix} \cos\psi & \sin\psi & 0 \\ -\sin\psi & \cos\psi & 0 \\ 0 & 0 & 1 \end{pmatrix} \tag{5.39}$$

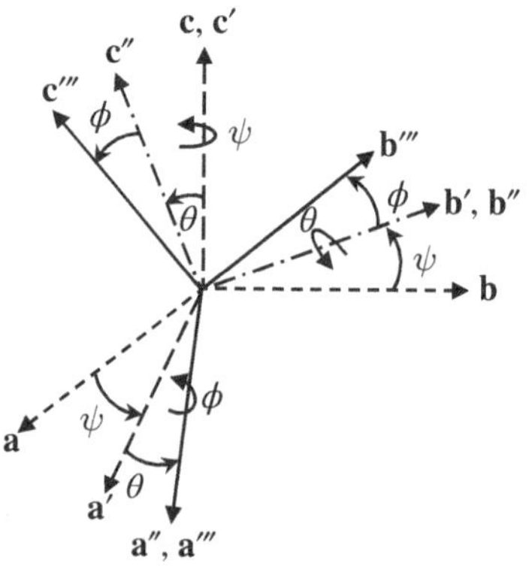

Figure 5.5 The Euler-angle representation, ψ_3, θ_2, ϕ_1.

$$= \begin{pmatrix} \cos\theta\cos\psi & \cos\theta\sin\psi & -\sin\theta \\ (\sin\phi\sin\theta\cos\psi - \cos\phi\sin\psi) & (\sin\phi\sin\theta\sin\psi + \cos\phi\cos\psi) & \sin\phi\cos\theta \\ (\cos\phi\sin\theta\cos\psi + \sin\phi\sin\psi) & (\cos\phi\sin\theta\sin\psi - \sin\phi\cos\psi) & \cos\phi\cos\theta \end{pmatrix}.$$

It can be verified that this rotation matrix has the properties of orthogonality and a unity determinant. To represent the attitude by the given Euler angles, (ϕ, θ, ψ), the latter should be uniquely determined from the rotation matrix, **C**. *Equation (5.39) indicates that the Euler angles can be determined by the following inverse transformation:*

$$\begin{aligned} \phi &= \tan^{-1}\frac{c_{23}}{c_{33}} \\ \theta &= -\sin^{-1} c_{13} \\ \psi &= \tan^{-1}\frac{c_{12}}{c_{11}}, \end{aligned} \tag{5.40}$$

where c_{ij} represents the element (i,j) of **C**. *Of course, neither $c_{11} = \cos\theta\cos\psi$ nor $c_{33} = \cos\phi\cos\theta$ must vanish. Otherwise the angles ϕ and ψ become undefined. However, this cannot be avoided at the specific orientations given by $\theta = \pm\pi/2$, for which $c_{11} = c_{33} = 0$; hence the angles ϕ and ψ cannot be determined. Such an orientation where an Euler-angle representation fails to be useful is called a* singularity *of the particular representation. Hence, $\theta = \pm\pi/2$ are the singular points of the representation, $(\psi_3, \theta_2, \phi_1)$, and the latter can be utilized only in the range $-\pi/2 < \theta < \pi/2$, where the representation is said to be non-singular.*

It is important to mention that Euler-angle representations are not unique, i.e., a particular orientation can be described by any number of different sequences of elementary rotations, provided they are non-singular at the given orientation. Furthermore, it is not necessary that all the three coordinate axes of the successive frames should be involved in describing a specific orientation. For example, the classical Euler-angle representation, $(\Omega_3, i_1, \omega_3)$, is traditionally used to describe the orientation of an orbital plane (as discussed in Section 5.4), which employs elementary rotations about only the first and the third axes. Such a sequence of Euler angles which begins and ends with the same current axis of the successive frames is said to be *symmetric*. The symmetric and asymmetric sets of Euler-angle representations are qualitatively different, as seen from the structure of their corresponding rotation matrices.

5.4 Euler-Angle Representation of the Orbital Plane

The orientation of the perifocal orbital frame, $(\mathbf{i_e}, \mathbf{i_p}, \mathbf{i_h})$, relative to a non-rotating (inertial) reference frame, $(\mathbf{I}, \mathbf{J}, \mathbf{K})$, is given by the general coordinate transformation of Eq. (5.1), which can be represented by a suitable choice of Euler angles. The traditional method of describing the orientation of the orbital plane is by the Euler-angle representation, $(\Omega_3, i_1, \omega_3)$. Combined with the Keplerian parameters for representing a conic section, a, e, and the time of periapsis, t_0, the three Euler angles, Ω, ω, i, form the set of *classical orbital elements*. The Euler-angle representation, $(\Omega_3, i_1, \omega_3)$, used to describe an orbital plane's orientation relative to an inertial frame, is shown in Fig. 5.6. As seen in the figure, this particular representation begins with

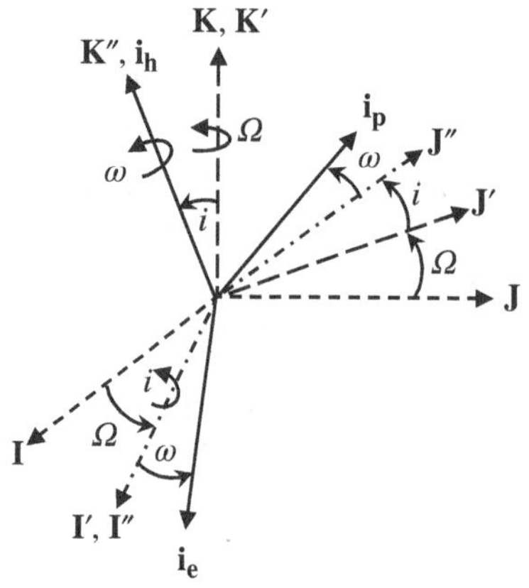

Figure 5.6 The Euler-angle representation, $(\Omega_3, i_1, \omega_3)$, of an orbital plane.

a positive rotation of the reference frame, $(\mathbf{I}, \mathbf{J}, \mathbf{K})$, about the third axis, $\mathbf{K}$, by the angle, Ω, followed next by a positive rotation by the angle, i, about the first axis, $\mathbf{I}'$, of the intermediate frame, $(\mathbf{I}', \mathbf{J}', \mathbf{K}')$, which results in the subsequent intermediate frame, $(\mathbf{I}'', \mathbf{J}'', \mathbf{K}'')$. The final rotation in this sequence is by the angle, ω, about the third axis, $\mathbf{K}''$, which results in the perifocal frame, $(\mathbf{i}_e, \mathbf{i}_p, \mathbf{i}_h)$. Certain axes of the intermediate frames, $(\mathbf{I}', \mathbf{J}', \mathbf{K}')$ and $(\mathbf{I}'', \mathbf{J}'', \mathbf{K}'')$, are themselves useful because they point in important directions, as discussed later in this chapter.

The successive elementary rotations involved in the Euler-angle representation, $(\Omega_3, i_1, \omega_3)$, are given by the following coordinate transformations:

$$\begin{Bmatrix} \mathbf{I}' \\ \mathbf{J}' \\ \mathbf{K}' \end{Bmatrix} = \mathbf{C}_3(\Omega) \begin{Bmatrix} \mathbf{I} \\ \mathbf{J} \\ \mathbf{K} \end{Bmatrix}$$
$$\begin{Bmatrix} \mathbf{I}'' \\ \mathbf{J}'' \\ \mathbf{K}'' \end{Bmatrix} = \mathbf{C}_1(i) \begin{Bmatrix} \mathbf{I}' \\ \mathbf{J}' \\ \mathbf{K}' \end{Bmatrix} \tag{5.41}$$
$$\begin{Bmatrix} \mathbf{i}_e \\ \mathbf{i}_p \\ \mathbf{i}_h \end{Bmatrix} = \mathbf{C}_3(\omega) \begin{Bmatrix} \mathbf{I}'' \\ \mathbf{J}'' \\ \mathbf{K}'' \end{Bmatrix},$$

or

$$\begin{Bmatrix} \mathbf{i}_e \\ \mathbf{i}_p \\ \mathbf{i}_h \end{Bmatrix} = \mathbf{C}_3(\omega)\mathbf{C}_1(i)\mathbf{C}_3(\Omega) \begin{Bmatrix} \mathbf{I} \\ \mathbf{J} \\ \mathbf{K} \end{Bmatrix}. \tag{5.42}$$

The rotation matrix representing the orientation of $(\mathbf{i}_e, \mathbf{i}_p, \mathbf{i}_h)$ relative to $(\mathbf{I}, \mathbf{J}, \mathbf{K})$ is therefore the following:

$$\mathbf{C} = \mathbf{C}_3(\omega)\mathbf{C}_1(i)\mathbf{C}_3(\Omega) = \begin{pmatrix} c_{11} & c_{12} & c_{13} \\ c_{21} & c_{22} & c_{23} \\ c_{31} & c_{32} & c_{33} \end{pmatrix}, \tag{5.43}$$

where the elements, c_{ij}, are the following:

$$c_{11} = \cos\Omega\cos\omega - \sin\Omega\sin\omega\cos i, \tag{5.44}$$

$$c_{12} = \sin\Omega\cos\omega + \cos\Omega\sin\omega\cos i, \tag{5.45}$$

$$c_{13} = \sin\omega\sin i, \tag{5.46}$$

$$c_{21} = -\cos\Omega\sin\omega - \sin\Omega\cos\omega\cos i, \tag{5.47}$$

$$c_{22} = -\sin\Omega\sin\omega + \cos\Omega\cos\omega\cos i, \tag{5.48}$$

$$c_{23} = \cos\omega\sin i, \tag{5.49}$$

$$c_{31} = \sin\Omega\sin i, \tag{5.50}$$

$$c_{32} = -\cos\Omega\sin i, \tag{5.51}$$

$$c_{33} = \cos i. \tag{5.52}$$

It is evident from these expressions that the Euler-angle representation, $(\Omega_3, i_1, \omega_3)$, has singularities at $i = 0, \pm\pi$. The inverse transformation

$$\begin{Bmatrix} \mathbf{I} \\ \mathbf{J} \\ \mathbf{K} \end{Bmatrix} = \mathbf{C}^* \begin{Bmatrix} \mathbf{i}_e \\ \mathbf{i}_p \\ \mathbf{i}_h \end{Bmatrix} \tag{5.53}$$

is represented by the rotation matrix, $\mathbf{C}^* = \mathbf{C}^T = \mathbf{C}_3{}^T(\Omega)\mathbf{C}_1{}^T(i)\mathbf{C}_3{}^T(\omega)$.

5.4.1 Celestial Reference Frame

The inertial, Cartesian reference frame, $(\mathbf{I}, \mathbf{J}, \mathbf{K})$, must be fixed relative to objects far beyond the solar system so that it can be used as a reference for the orientation of orbital planes within the solar system. Such a reference frame is called a *celestial frame*. For an orbit around Earth, the axes of a celestial frame can be taken as pointing towards either specific distant stars or special points in Earth's orbit around the sun, as seen from Earth. A common choice of an Earth-based celestial frame is the one whose **I**-axis points towards the *vernal equinox*, which indicates the location of the sun against a background of distant stars, as it crosses the equatorial plane from the southern to the northern hemisphere. The vernal equinox thus indicates a direction along the intersection of Earth's equatorial plane and the plane of the orbit of Earth around the sun (called the *ecliptic* plane). The vernal equinox occurs at noon on the first day of spring, around March 21. In the early days of astronomy about 2200 years ago, the vernal equinox coincided with the *first point of Aries*, which is a special star in the constellation Aries. However, due to a precession of Earth's spin axis (to be discussed later) with a period of about 26000 years, the vernal equinox shifts slowly with time, and no longer coincides with the first point of Aries. When $\mathbf{I}$ is taken as pointing towards the vernal equinox (which indicates the line of intersection of the equatorial and ecliptic planes), it leaves us with two possible choices for the plane, $(\mathbf{I}, \mathbf{J})$ – that is, either the equatorial plane or the ecliptic plane. In the former choice the $\mathbf{K}$ axis is the rotational (polar) axis of Earth, and in the latter an axis normal to the ecliptic plane. The $\mathbf{J}$-axis completes the right-handed, orthogonal triad, $\mathbf{J} = \mathbf{K} \times \mathbf{I}$.

5.4.2 Local-Horizon Frame

The Cartesian coordinates of the position vector are resolved in the celestial frame as follows:

$$\mathbf{r} = r_X\mathbf{I} + r_Y\mathbf{J} + r_Z\mathbf{K}. \tag{5.54}$$

Alternatively, the position vector, $\mathbf{r}$, can be referred to the celestial frame using the spherical coordinates, (r, δ, λ), as follows:

$$\mathbf{r} = r(\cos\delta\cos\lambda\mathbf{I} + \cos\delta\sin\lambda\mathbf{J} + \sin\delta\mathbf{K}), \tag{5.55}$$

where the angle, δ, called the *declination* (or the *celestial latitude*), is the angle between the position vector, $\mathbf{r}$, and the plane $(\mathbf{I}, \mathbf{J})$, and is defined to be positive above the plane. The angle λ, called the *right ascension*, or the *celestial longitude*, is the angle made by the projection of $\mathbf{r}$ on the plane $(\mathbf{I}, \mathbf{J})$ with the vernal equinox direction, $\mathbf{I}$, and is defined to be positive towards the east. The spherical coordinates are determined from the Cartesian coordinates as follows:

$$r = |\,\mathbf{r}\,| = \sqrt{r_X{}^2 + r_Y{}^2 + r_Z{}^2}, \tag{5.56}$$

$$\delta = \sin^{-1}\frac{r_Z}{r} = \sin^{-1}\frac{\mathbf{r}\cdot\mathbf{K}}{r}, \tag{5.57}$$

and

$$\begin{aligned}\sin\lambda &= \frac{r_Y}{r\cos\delta} = \frac{\mathbf{r}\cdot\mathbf{J}}{r\cos\delta}\\ \cos\lambda &= \frac{r_X}{r\cos\delta} = \frac{\mathbf{r}\cdot\mathbf{I}}{r\cos\delta}.\end{aligned} \tag{5.58}$$

To specify the velocity vector, $\mathbf{v}$, a moving *local-horizon* frame, $(\mathbf{i_r}, \mathbf{i}_\lambda, \mathbf{i}_\delta)$, is employed, as depicted in Fig. 5.7. The axis $\mathbf{i_r} = \mathbf{r}/r$ is along $\mathbf{r}$, while $\mathbf{i}_\lambda$ and $\mathbf{i}_\delta$ point towards the east and north respectively, as shown in Fig. 5.7. The unit vector $\mathbf{i}_\lambda$ points towards the direction of increasing λ, and $\mathbf{i}_\delta$ towards that of increasing δ, and the local-horizon frame is right-handed such that $\mathbf{i_r} \times \mathbf{i}_\lambda = \mathbf{i}_\delta$ (Fig. 5.7). The velocity vector, $\mathbf{v}$, can be resolved in spherical coordinates, (v, ϕ, ψ), referred to the local-horizon frame, where v is the orbital speed, and ϕ is the flight-path angle made by $\mathbf{v}$ with the local horizon, $(\mathbf{i}_\lambda, \mathbf{i}_\delta)$, defined to be positive towards $\mathbf{i_r}$. The angle ψ is the *velocity azimuth* angle between the projection of $\mathbf{v}$ on the local-horizon plane, $(\mathbf{i}_\lambda, \mathbf{i}_\delta)$, and the axis $\mathbf{i}_\delta$, measured positive towards $\mathbf{i}_\lambda$, as shown in Fig. 5.7. The parameters $(r, \lambda, \delta, v, \phi, \psi)$ form a set of orbital variables for determining the position and velocity at any instant of time.

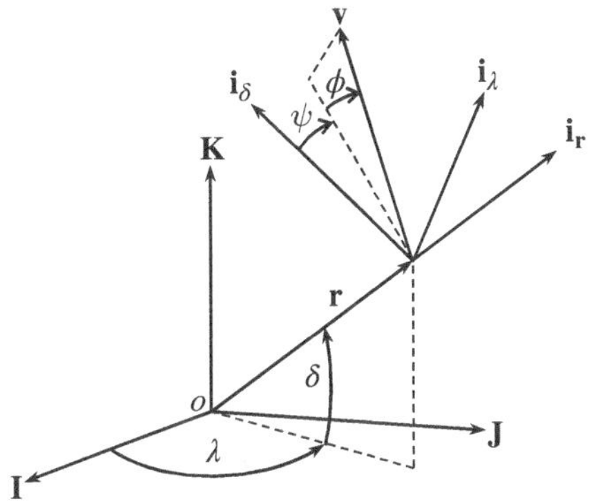

Figure 5.7 Spherical celestial coordinates, (r, δ, λ), and the local-horizon frame, $(\mathbf{i}_r, \mathbf{i}_\lambda, \mathbf{i}_\delta)$.

The orientation of the local-horizon frame, $(\mathbf{i}_r, \mathbf{i}_\lambda, \mathbf{i}_\delta)$, relative to the Cartesian, celestial frame, $(\mathbf{I}, \mathbf{J}, \mathbf{K})$, can be represented by the Euler-angle representation, $(\lambda)_3, (\frac{\pi}{2} - \delta)_2, (\frac{-\pi}{2})_2$. Hence, the rotation matrix relating the two frames is obtained as

$$\begin{Bmatrix} \mathbf{i}_r \\ \mathbf{i}_\lambda \\ \mathbf{i}_\delta \end{Bmatrix} = \mathbf{C}_{\mathbf{LH}} \begin{Bmatrix} \mathbf{I} \\ \mathbf{J} \\ \mathbf{K} \end{Bmatrix}, \tag{5.59}$$

where

$$\begin{aligned} \mathbf{C}_{\mathbf{LH}} &= \mathbf{C}_2\left(\frac{-\pi}{2}\right)\mathbf{C}_2\left(\frac{\pi}{2} - \delta\right)\mathbf{C}_3(\lambda) \\ &= \begin{pmatrix} \cos\delta\cos\lambda & \cos\delta\sin\lambda & \sin\delta \\ -\sin\lambda & \cos\lambda & 0 \\ -\sin\delta\cos\lambda & -\sin\delta\sin\lambda & \cos\delta \end{pmatrix}. \end{aligned} \tag{5.60}$$

The spherical coordinates, (v, ϕ, ψ), are derived from the celestial, Cartesian coordinates defining the velocity vector, $\mathbf{v} = v_X\mathbf{I} + v_Y\mathbf{J} + v_Z\mathbf{K}$, by expressing the velocity vector in the local-horizon frame as follows:

$$\begin{aligned} \mathbf{v} &= \mathbf{C}_{\mathbf{LH}} \begin{Bmatrix} v_X \\ v_Y \\ v_Z \end{Bmatrix} \\ &= v(\sin\phi\mathbf{i} + \cos\phi\sin\psi\mathbf{j} + \cos\phi\cos\psi\mathbf{k}), \end{aligned} \tag{5.61}$$

which results in $v = |\mathbf{v}| = \sqrt{v_X{}^2 + v_Y{}^2 + v_Z{}^2}$,

$$\phi = \sin^{-1}\frac{\mathbf{v}\cdot\mathbf{i}}{v}, \tag{5.62}$$

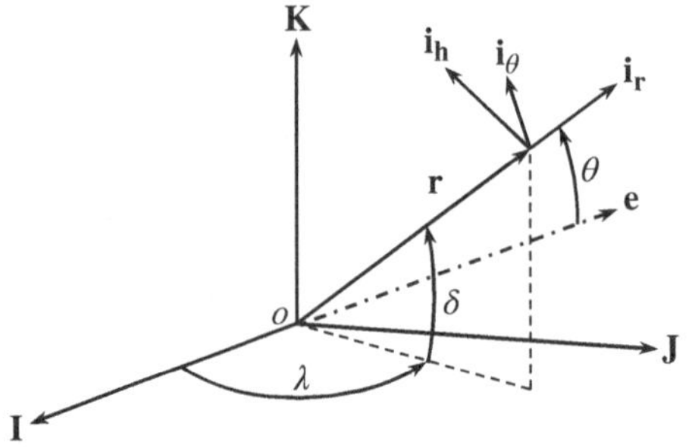

Figure 5.8 The local-horizon frame, $(\mathbf{i}_r, \mathbf{i}_\theta, \mathbf{i}_h)$.

and

$$\sin\psi = \frac{\mathbf{v}\cdot\mathbf{j}}{v\cos\phi} \qquad (5.63)$$
$$\cos\psi = \frac{\mathbf{v}\cdot\mathbf{k}}{v\cos\phi}\,.$$

Another local-horizon frame which is useful in representing the position and velocity in a two-body orbit is the frame $(\mathbf{i}_r, \mathbf{i}_\theta, \mathbf{i}_h)$, where the orientation of $\mathbf{i}_r$ in the celestial frame, $(\mathbf{I}, \mathbf{J}, \mathbf{K})$, is given (as previously) via the angles δ and λ (see Fig. 5.8) as follows:

$$\mathbf{i}_r = \cos\delta\cos\lambda\mathbf{I} + \cos\delta\sin\lambda\mathbf{J} + \sin\delta\mathbf{K}\,, \qquad (5.64)$$

while $\mathbf{i}_h = \mathbf{h}/h$ and $\mathbf{i}_\theta = \mathbf{i}_h \times \mathbf{i}_r$ are normal to $\mathbf{i}_r$; hence the plane $(\mathbf{i}_\theta, \mathbf{i}_h)$ is the local horizon. Such a local-horizon frame is often employed in resolving the orbital perturbations, as seen in Chapter 9.

5.4.3 *Classical Euler Angles*

The orientation of the orbital frame, $(\mathbf{i}_e, \mathbf{i}_p, \mathbf{i}_h)$, relative to a celestial reference frame with $\mathbf{I}$ towards the vernal equinox is shown in Fig. 5.9. Since the axis $\mathbf{i}_e = \mathbf{e}/e$ indicates the direction of the periapsis, its direction is important in describing the orbital plane's orientation. Another fix on the orbital plane is supplied by the direction of the angular-momentum vector, $\mathbf{i}_h = \mathbf{h}/h$. The Euler-angle representation, $(\Omega_3, i_1, \omega_3)$, is traditionally used to represent the orientation of the perifocal frame according to Eq. (5.42); hence the angles Ω, i, ω are called the *classical Euler angles*. It is now described how these angles offer an insight into the orientation of the orbital plane.

The intersection of the perifocal orbital plane, $(\mathbf{i}_e, \mathbf{i}_p)$, with the reference plane, $(\mathbf{I}, \mathbf{J})$, is called the *line of nodes*. The *ascending node* is the point on the line of nodes where the orbit crosses the plane $(\mathbf{I}, \mathbf{J})$ in the $\mathbf{K}$ direction. A unit vector $\mathbf{n}$ pointing towards the ascending node makes an angle Ω with the $\mathbf{I}$-axis, as shown in Fig. 5.9. Thus the angle Ω is measured on the plane $(\mathbf{I}, \mathbf{J})$ in an anti-clockwise direction (Fig. 5.9), and represents a positive rotation of $\mathbf{I}$ about $\mathbf{K}$ according to the right-hand rule to produce $\mathbf{i}_n$; hence Ω is termed the *right-ascension of the ascending*

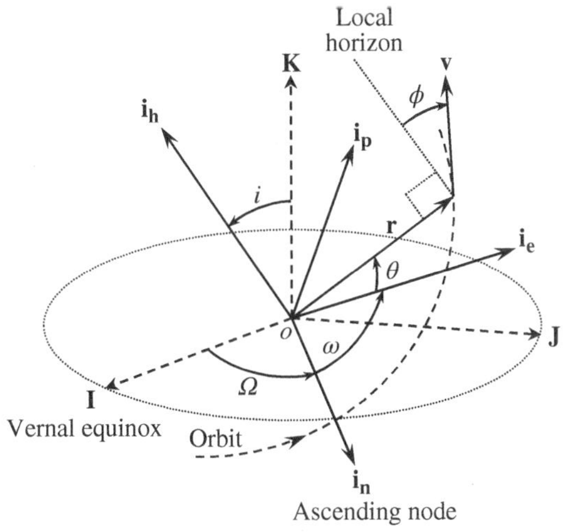

Figure 5.9 The classical Euler angles, (Ω, i, ω).

node. Since $\mathbf{i}_n$ is the same as the direction $\mathbf{I}'$ produced by the first elementary rotation of the sequence $(\Omega_3, i_1, \omega_3)$, one can use Eq. (5.41) to derive $\mathbf{i}_n$ (and Ω) as follows:

$$\mathbf{i}_n = \mathbf{I}' = \mathbf{I}\cos\Omega + \mathbf{J}\sin\Omega \,. \tag{5.65}$$

Furthermore, since the ascending node vector, $\mathbf{i}_n$, being normal to both $\mathbf{K}$ and $\mathbf{i}_h$, is the axis of the second elementary rotation of the sequence $(\Omega_3, i_1, \omega_3)$ by the Euler angle, i, it can be computed as follows:

$$\mathbf{i}_n = \frac{\mathbf{K} \times \mathbf{i}_h}{|\,\mathbf{K} \times \mathbf{i}_h\,|} \,. \tag{5.66}$$

The right ascension of the ascending node, Ω, is then computed from:

$$\cos\Omega = \mathbf{I} \cdot \mathbf{i}_n \,, \tag{5.67}$$

$$\sin\Omega = \mathbf{J} \cdot \mathbf{i}_n \,. \tag{5.68}$$

The *orbital inclination*, i, is the second classical Euler angle describing the orbital plane and, being the positive rotation about $\mathbf{i}_n$ required to produce $\mathbf{i}_h$ from $\mathbf{K}$ (Fig. 5.9), is computed as follows:

$$\cos i = \mathbf{K} \cdot \mathbf{i}_h \,, \tag{5.69}$$

$$\sin i = |\,\mathbf{K} \times \mathbf{i}_h\,| \,. \tag{5.70}$$

An orbit with $0 \le i < \pi/2$ is called a *direct* (or *prograde*) orbit, while that with $\pi/2 < i \le \pi$ is said to be a *retrograde* orbit. An orbit with $i = \pi/2$ directly passes over the geographical poles; hence it is termed a *polar* orbit.

Finally, the third classical Euler angle, ω, which represents a positive rotation of $\mathbf{i_n}$ about $\mathbf{i_h}$ to produce $\mathbf{i_e}$ in the orbital plane, is called the *argument of periapsis*. From the last elementary rotation of the sequence $(\Omega_3, i_1, \omega_3)$ about $\mathbf{i_h}$, Eq. (5.41) produces the following:

$$\begin{aligned} \cos\omega &= \mathbf{I}'' \cdot \mathbf{i_e} = \mathbf{i_n} \cdot \mathbf{i_e} \\ \sin\omega &= \mathbf{i_h} \cdot (\mathbf{I}'' \times \mathbf{i_e}) = \mathbf{i_h} \cdot (\mathbf{i_n} \times \mathbf{i_e}), \end{aligned} \tag{5.71}$$

which yield a unique value of ω.

The classical Euler angles, (Ω, i, ω), uniquely determine the orbital plane, except for $i = 0, \pm\pi$, for which the representation $(\Omega_3, i_1, \omega_3)$ becomes singular, and the angles Ω and ω are undefined. The classical Euler angles, along with the orbital elements, (a, e, t_0), form the set of *classical orbital elements*, $(a, e, t_0, \Omega, \omega, i)$, and uniquely determine the plane and the shape of a Keplerian orbit. Apart from the singularities of the Euler-angle representation $(\Omega_3, i_1, \omega_3)$ at $i = 0, \pm\pi$, the argument of periapsis, ω, is undefined for a circular orbit, $e = 0$. Thus, the utility of the classical orbital elements, $(a, e, t_0, \Omega, \omega, i)$, is limited to non-circular orbits, which do not lie on the plane $(\mathbf{I}, \mathbf{J})$. A way of avoiding the singularities of the Euler angles is to choose the plane $(\mathbf{I}, \mathbf{J})$ such that the orbital inclination falls in the range $0 < i < \pi$, which is possible by selecting $(\mathbf{I}, \mathbf{J})$ to be either the equatorial or the ecliptic plane.

The relationships between the classical Euler angles and the celestial spherical coordinates for the position and velocity vectors can be derived from the rotation matrices for the perifocal and local-horizon frames. For example, a useful relation among the declination, δ, the orbital inclination, i, and the velocity azimuth, ψ, is derived by substituting Eqs. (5.59) and (5.61) into Eq. (5.69), with $\mathbf{h} = \mathbf{r} \times \mathbf{v}$:

$$\cos i = \cos\delta \sin\psi\,. \tag{5.72}$$

Another similar relationship between the classical Euler angles and the celestial spherical coordinates is the following:

$$\tan i \sin(\lambda - \Omega) = \tan\delta\,. \tag{5.73}$$

Equations (5.72) and (5.73) relate to a special branch of mathematics called *spherical trigonometry*, where three spherical arcs having a right angle between them satisfy a specific relationship between the other two angles.

It is evident from Eq. (5.72) that the minimum possible orbital inclination, i, for a given declination, δ, occurs at $\psi = 90^\circ$, which denotes an eastward flight path, and equals the local declination, $i = \delta$. Employing an equatorial plane for $(\mathbf{I}, \mathbf{J})$, this implies that it is impossible to launch a spacecraft into an orbit with an inclination smaller than the latitude of the launch site, without carrying out a change in the orbital plane. As will be discussed in Chapter 6, the orbital inclination can be changed using propulsive manoeuvres, but these are very expensive and significantly increase the cost of a launch. However, a zero-inclination (equatorial) orbit can be obtained from a launch site situated at the equator, without requiring a propulsive plane change. Such a launch site also gives the maximum velocity advantage due to Earth's rotation for an eastward launch. The market for geosynchronous satellite launches, which is currently the most lucrative and requires a zero orbital inclination, dictates that the launch sites be located as near the equator as physically possible.

The classical Euler angles, (Ω, i, ω), are determined from the measured position and velocity in the celestial frame once the perifocal frame is derived using the transformations presented in this section.

Example 5.4.1 *An Earth-orbiting spacecraft is observed to have the following geocentric position and velocity vectors in a celestial reference frame:*

$$\mathbf{r} = -10\mathbf{I} - 3300\mathbf{J} + 11000\mathbf{K} \text{ (km)}$$

$$\mathbf{v} = -2\mathbf{I} - 3.3\mathbf{J} + 2.5\mathbf{K} \text{ (km/s)} .$$

Determine the declination, right ascension, flight-path angle, and velocity azimuth when the observation is taken. Also calculate the classical orbital elements of the spacecraft's orbit.

The orbital radius is $r = |\mathbf{r}| = 11484.3415$ *km, and the orbital speed is* $v = |\mathbf{v}| = 4.5978$ km/s. *From Eqs. (5.57) and (5.58), we have*

$$\delta = \sin^{-1}\frac{11000}{11484.3415} = 73.301^\circ ,$$

$$\sin\lambda = \frac{-3300}{11484.3415\cos(73.301^\circ)} = -0.9999954$$

$$\cos\lambda = \frac{-10}{11484.3415\cos(73.301^\circ)} = -0.0030303 ,$$

which yield $\lambda = 269.8264^\circ$. *The rotation matrix of the local horizon frame is the following, according to Eq. (5.60):*

$$\mathbf{C_{LH}} = \begin{pmatrix} -0.0008708 & -0.2873478 & 0.9578259 \\ 0.9999954 & -0.0030303 & 0 \\ 0.0029025 & 0.9578215 & 0.2873491 \end{pmatrix} .$$

The velocity components in the local horizon frame are thus the following:

$$\mathbf{v} = \mathbf{C_{LH}} \begin{Bmatrix} -2 \\ -3.3 \\ 2.5 \end{Bmatrix}$$

$$= 3.344554\mathbf{i} - 1.989991\mathbf{j} - 2.448243\mathbf{k} \text{ (km/s)} .$$

Finally, by employing Eqs. (5.62) and (5.63) the flight-path angle and the velocity azimuth are obtained as follows:

$$\phi = \sin^{-1}\frac{3.344554}{4.5978} = 46.671^\circ ,$$

$$\sin\psi = \frac{-1.989991}{4.5978\cos(46.671^\circ)} = -0.63074405$$

$$\cos\psi = \frac{-2.448243}{4.5978\cos(46.671^\circ)} = -0.77599094 ,$$

which yield $\psi = 219.105^\circ$.

The determination of the orbital elements begins with the semi-major axis and orbital eccentricity as follows:

$$\varepsilon = v^2/2 - \mu/r = -24.138163\ \text{km}^2/\text{s}^2$$

$$a = -mu/(2\varepsilon) = 8256.643\ \text{km}$$

$$\mathbf{h} = \mathbf{r} \times \mathbf{v} = 28050\mathbf{I} - 21975\mathbf{J} - 6567\mathbf{K}\ \text{km}^2/\text{s}$$

$$h = |\,\mathbf{h}\,| = 36232.9769 \text{km}^2/\text{s}$$

$$\mathbf{e} = \frac{\mathbf{v} \times \mathbf{h}}{\mu} - \mathbf{i_r}$$

$$= 0.193065\mathbf{I} + 0.430326\mathbf{J} - 0.61534\mathbf{K}$$

$$e = |\,\mathbf{e}\,| = 0.7753048\ .$$

Next, the nodal vector, $\mathbf{i_n}$*, the right ascension of the ascending node,* Ω*, and the orbital inclination,* i*, are computed according to Eqs. (5.66)–(5.70):*

$$\mathbf{i_n} = \frac{\mathbf{K} \times \mathbf{i_h}}{|\,\mathbf{K} \times \mathbf{i_h}\,|}$$

$$= 0.6167055\mathbf{I} + 0.78719399\mathbf{J}$$

$$\Omega = \sin^{-1}(0.78719399) = 0.9062457\ \text{rad.}\ (51.924°)$$

$$\cos i = \mathbf{K} \cdot \mathbf{i_h} = -0.181244\ ; \quad \sin i = |\,\mathbf{K} \times \mathbf{i_h}\,| = 0.983438$$

$$i = \cos^{-1}(-0.181244) = 1.753047\ \text{rad.}\ (100.442°)\ .$$

Finally, the argument of perigee, ω*, is computed according to Eq. (5.71):*

$$\cos\omega = \mathbf{i_n} \cdot \mathbf{i_e} = 0.590495\ ; \quad \sin\omega = \mathbf{i_h} \cdot (\mathbf{i_n} \times \mathbf{i_e}) = -0.807041$$

$$\omega = \sin^{-1}(-0.807041) = 5.344061\ \text{rad.}\ (306.192°)\ .$$

It is now verified that the computed values of the spherical coordinates of the position and velocity – (r, δ, λ) *and* (v, ϕ, ψ)*, respectively – satisfy the spherical trigonometric relations, Eqs. (5.72) and (5.73):*

$$\cos i = -0.181244 = (0.287349)(-0.630744) = \cos\delta \sin\psi$$

$$\tan i \sin(\lambda - \Omega) = (-5.426054)(-0.614317) = 3.333318 = \tan\delta\ .$$

Example 5.4.2 *For an orbit around Earth with* $a = 9000$ km, $e = 0.2$, $t_0 = -1200\ (s)$, $\Omega = 30°$, $\omega = 235°$, $i = 100°$, *determine the position and velocity at* $t = 30$ min *in the geocentric celestial reference frame.*

The perifocal velocity and position are calculated by first computing the mean motion, $n = \sqrt{\frac{\mu}{a^3}} = 0.0007394$ rad./s, *and the mean anomaly,* $M = n(t - t_0) = 2.21833$ rad. *and then solving Kepler's equation (Chapter 4) to yield* $E = 2.35931$ rad., *resulting in*

$$\mathbf{r} = a(\cos E - e)\mathbf{i_e} + a\sqrt{1 - e^2}\sin E\mathbf{i_p}$$
$$= -8183.763\mathbf{i_e} + 6215.92\mathbf{i_p} \text{ km}$$
$$\mathbf{v} = \frac{an(-\sin E\mathbf{i_e} + \sqrt{1 - e^2}\cos E\mathbf{i_p})}{1 - e\cos E}$$
$$= -4.10829\mathbf{i_e} - 4.0505\mathbf{i_p} \text{ km/s} .$$

The rotation matrix giving the transformation from the perifocal frame to the celestial frame is computed using Eq. (5.43) as follows:

$$\mathbf{C}^* = \mathbf{C}^T = \begin{pmatrix} -0.56785 & 0.65961 & 0.4924 \\ -0.1636 & 0.49583 & -0.85287 \\ -0.80671 & -0.56486 & -0.17365 \end{pmatrix} ,$$

from which the celestial position and velocity are obtained to be

$$\mathbf{r} = \mathbf{C}^* \begin{Bmatrix} -8183.763 \\ 6215.92 \\ 0 \end{Bmatrix} = \begin{Bmatrix} 8747.242 \\ 4420.928 \\ 3090.761 \end{Bmatrix} \text{ km}$$
$$= 8747.242\mathbf{I} - 3221.591\mathbf{J} + 11420.412\mathbf{K} \text{ (km)} ,$$

$$\mathbf{v} = \mathbf{C}^* \begin{Bmatrix} -4.10829 \\ -4.0505 \\ 0 \end{Bmatrix} = \begin{Bmatrix} -0.338796 \\ -1.336228 \\ 5.602143 \end{Bmatrix} \text{ km/s}$$
$$= -0.338796\mathbf{I} - 1.336228\mathbf{J} + 5.602143\mathbf{K} \text{ (km/s)} .$$

5.5 Planet-Fixed Coordinate System

The methods of coordinate transformation due to rotation can be applied to resolving the orbital motion in a reference frame firmly attached to a rotating central body, such as a planet. A planet-fixed frame is useful in determining the position and velocity with reference to a ground station, which is located at a fixed point on the surface and rotates with the planet. Such a planet-centered, rotating frame, $(SX'Y'Z)$, with origin S at the centre, and which shares the axis SZ with the celestial frame, $(SXYZ)$, is shown in Fig. 5.10. The axes of the planet-fixed coordinate frame, SX', SY', and SZ', are represented by the unit vectors, $\mathbf{I}', \mathbf{J}'$, and $\mathbf{K}$, respectively. The axes of the celestial frame, SX, SY, and SZ, are denoted by the unit vectors, $\mathbf{I}, \mathbf{J}$, and $\mathbf{K}$, respectively.

A set of spherical coordinates, (r, δ, l), denoting the radius, the latitude (or declination), and the longitude, respectively, are chosen to represent the position vector, $\mathbf{r}$, in the planet-fixed frame with the origin, S, at the planet's centre, as shown in Fig. 5.10. The same vector in the celestial frame is resolved by (r, δ, λ), where λ is the right ascension. The orientation of the

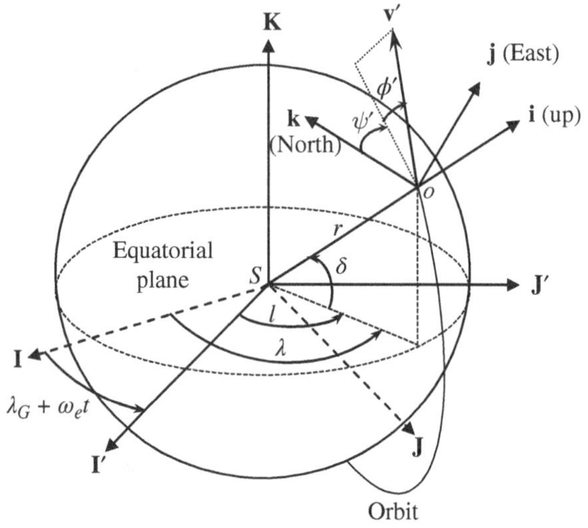

Figure 5.10 Planet-fixed, rotating coordinates and the local horizon.

planet-fixed coordinate frame, $(\mathbf{I}', \mathbf{J}', \mathbf{K})$, relative to the celestial frame, $(\mathbf{I}, \mathbf{J}, \mathbf{K})$, is given by

$$\begin{Bmatrix} \mathbf{I}' \\ \mathbf{J}' \\ \mathbf{K} \end{Bmatrix} = \mathbf{C_{pf}} \begin{Bmatrix} \mathbf{I} \\ \mathbf{J} \\ \mathbf{K} \end{Bmatrix}, \tag{5.74}$$

where

$$\mathbf{C_{pf}} = \mathbf{C}_3(\lambda - l) = \begin{pmatrix} \cos(\lambda - l) & \sin(\lambda - l) & 0 \\ -\sin(\lambda - l) & \cos(\lambda - l) & 0 \\ 0 & 0 & 1 \end{pmatrix}. \tag{5.75}$$

For the velocity relative to the planet-fixed frame, $\mathbf{v}'$, the spherical coordinates v', ϕ', ψ', are employed, representing the relative speed, the relative flight-path angle, and the relative velocity azimuth resolved in the local-horizon frame, $(\mathbf{i}, \mathbf{j}, \mathbf{k})$, with the origin at the centre of mass, o, of the orbiting body. The axes of the local-horizon frame are denoted by the unit vectors, $\mathbf{i} = \mathbf{r}/r$ (up), $\mathbf{j}$ (east), and $\mathbf{k} = \mathbf{i} \times \mathbf{j}$ (north), respectively, as shown in Fig. 5.10. The coordinate transformation between the planet-fixed and the local-horizon frames is given by

$$\begin{Bmatrix} \mathbf{i} \\ \mathbf{j} \\ \mathbf{k} \end{Bmatrix} = \mathbf{C_{LH}} \mathbf{C_{pf}}^T \begin{Bmatrix} \mathbf{I}' \\ \mathbf{J}' \\ \mathbf{K} \end{Bmatrix}, \tag{5.76}$$

where $\mathbf{C_{LH}}$ is given by Eq. (5.60), repeated here as follows:

$$\mathbf{C_{LH}} = \begin{pmatrix} \cos\delta\cos\lambda & \cos\delta\sin\lambda & \sin\delta \\ -\sin\lambda & \cos\lambda & 0 \\ -\sin\delta\cos\lambda & -\sin\delta\sin\lambda & \cos\delta \end{pmatrix}. \tag{5.77}$$

The longitude, l, is an important planet-fixed coordinate. The lines of constant longitude on the planet's surface are called *meridians*. The distant stars cross each meridian at the same local time; hence a meridian and its corresponding local time have been synonymous since the earliest days of navigation.

To utilize the planet-fixed coordinates, the longitude, l, must be calculated from the current right ascension, λ, at the given time, t, as well as at the right ascension of the zero longitude line, $\mathbf{I}'$, known at some previous time, $t = 0$. The zero longitude on Earth is the Greenwich meridian, whose right ascension is denoted by λ_G. The calculation of the current longitude is then performed as follows, by taking into account the planet's rotation in the given time, t, follows (Fig. 5.10):

$$l = \lambda - \lambda_G - \omega_e t. \tag{5.78}$$

The right ascension of the Greenwich meridian, $\mathbf{I}'$, is available from periodically published astronomical (or *ephemeris*) charts, such as *the American Ephemeris and Nautical Almanac.*

The relationship between the inertial velocity, $\mathbf{v}$, and the relative velocity, $\mathbf{v}'$, is given by the following chain rule of vector differentiation (Chapter 2):

$$\mathbf{v} = \mathbf{v}' + \boldsymbol{\omega} \times \mathbf{r}\,, \tag{5.79}$$

where $\boldsymbol{\omega} = \omega_e\mathbf{K}$ is the angular velocity of rotation of the planet, and $\mathbf{r} = r\mathbf{i}$. The relative velocity is expressed in terms of the spherical coordinates as follows:

$$\mathbf{v}' = v'(\sin\phi'\mathbf{i} + \cos\phi'\sin\psi'\mathbf{j} + \cos\phi'\cos\psi'\mathbf{k})\,. \tag{5.80}$$

This yields the following from Eq. (5.79):

$$v(\sin\phi\mathbf{i} + \cos\phi\sin\psi\mathbf{j} + \cos\phi\cos\psi\mathbf{k}) = v'(\sin\phi'\mathbf{i} + \cos\phi'\sin\psi'\mathbf{j} + \cos\phi'\cos\psi'\mathbf{k}) + \omega_e r\cos\delta\mathbf{j}\,. \tag{5.81}$$

On comparing the respective vector components on both the sides of Eq. (5.81), we have

$$\begin{aligned} v\sin\phi &= v'\sin\phi' \\ v\cos\phi\sin\psi &= v'\cos\phi'\sin\psi' + \omega_e r\cos\delta \\ v\cos\phi\cos\psi &= v'\cos\phi'\cos\psi'\,. \end{aligned} \tag{5.82}$$

Therefore, the relationships between the inertial and relative velocity coordinates are given by the following:

$$\begin{aligned} \tan\psi' &= \tan\psi - \frac{\omega_e r\cos\delta}{v\cos\phi\cos\psi} \\ \tan\phi' &= \tan\phi\frac{\cos\psi'}{\cos\psi} \\ v' &= v\frac{\sin\phi}{\sin\phi'}\,. \end{aligned} \tag{5.83}$$

Example 5.5.1 *A spacecraft is to be launched into an Earth orbit of inclination* 30° *such that the inertial speed of the ascending orbit at an altitude of 300 km and latitude* 20° *is 7 km/s, with* *a flight-path angle of 2°. What is the velocity of the spacecraft relative to Earth at the given point?*

The inertial velocity azimuth at the given point is calculated by Eq. (5.72) as follows:

$$\psi = \sin^{-1}\frac{\cos(30°)}{\cos(20°)} = 61.568° \text{ or } 118.432°.$$

Since the given point is on the ascending node of the latitude crossing, the correct velocity azimuth is $\psi = 61.568°$. The rotational speed of Earth is calculated from the sidereal day (23 hours, 56 minutes, and 4.09 seconds). We recall from Chapter 1 that the sidereal day is the time taken by Earth to complete one rotation, and is obtained by adding $\frac{360°}{365.242}$ to the 360° angle rotated in 24 hours relative to the sun (the mean solar day). Thus we have

$$\omega_e = \frac{2\pi}{23 \times 3600 + 56 \times 60 + 4.09} = 7.29211 \times 10^{-5} \text{ rad./s}.$$

Next, the relative velocity coordinates are calculated from Eq. (5.83) by substituting $r = 6678.14$ km, $v = 7$ km/s, $\phi = 2°$:

$$\psi' = \tan^{-1}\left(\tan\psi - \frac{\omega_e r \cos\delta}{v\cos\phi\cos\psi}\right) = 59.676°$$

$$\phi' = \tan^{-1}\left(\tan\phi\frac{\cos\psi'}{\cos\psi}\right) = 2.1208°$$

$$v' = v\frac{\sin\phi}{\sin\phi'} = 6.6014 \text{ km/s}.$$

Hence the change in the speed and the velocity azimuth caused by Earth's rotation is significant, while the change in the flight-path angle is relatively small.

Example 5.5.2 *A spacecraft currently has the right ascension $\lambda = 270°$, relative to a geocentric celestial frame. Calculate the longitude of the spacecraft if the right ascension of the Greenwich meridian 50 minutes earlier was 23°.*

On substituting $\lambda_G = 23°$, $\omega_e = 0.004178°/\text{s}$, and $t = (50)(60)$ s into Eq. (5.78), we have

$$l = 270° - 23° - (0.004178)(3000) = 234.466°.$$

The decrease in the longitude due to Earth's rotation is $\omega_e t = 12.5342°$.

Exercises

1. Derive the rotation matrix representing the transformation of a right-handed frame, $(\mathbf{i}, \mathbf{j}, \mathbf{k})$, to the frame $(\mathbf{j}, \mathbf{i}, -\mathbf{k})$.
2. For the rotation of Exercise 1, find the Euler axis and the principal angle.
3. For the rotation of Exercise 1, find an Euler-angle representation.

4. The rotation matrix representing a rotation of a right-handed frame, $(\mathbf{i}, \mathbf{j}, \mathbf{k})$, is the following:

$$\mathbf{C} = \begin{pmatrix} -0.1710 & 0.4698 & 0.8660 \\ 0.9397 & 0.3420 & 0 \\ -0.2962 & 0.8138 & -0.5 \end{pmatrix}.$$

Derive:

(a) The Euler-axis/ principal angle representation. (*Ans.* $\mathbf{s}_1 = 0.5445\mathbf{i} + 0.7776\mathbf{j} + 0.3144\mathbf{k}$, $\Phi = 131.6436°$.)
(b) An Euler-angle representation.
(c) The modified Rodrigues parameters.
(d) The quaternion. [*Ans.* $\mathbf{q} = (0.4967,\ 0.7094,\ 0.2868)^T$, $q_4 = 0.4096$.]

5. Derive the rotation matrix of the Euler sequence $(\Omega_3, i_1, -\Omega_3)$. The orientation of the reference frame resulting from this Euler sequence is called the *equinoctial axes*.

6. Given the following position and velocity vectors in an Earth-centered, equatorial celestial frame:

$$r = 6045\mathbf{I} + 3490\mathbf{J} \text{ (km)}$$

$$v = -2.457\mathbf{I} + 6.618\mathbf{J} + 2.533\mathbf{K} \text{ (km/s)},$$

determine the classical orbital elements. (*Ans.* $a = 6877.25076$ km, $e = 0.158171$, $t_0 = -1362.287$ s, $i = 20°$, $\Omega = 30°$, $\omega = 255.581°$.)

7. Given the following position and velocity vectors in an Earth-centered, equatorial celestial frame:

$$r = -5888.9727\mathbf{J} - 3400\mathbf{K} \text{ (km)}$$

$$v = 10.691338\mathbf{I} \text{ (km/s)},$$

determine the classical orbital elements.

8. The orbital elements of an Earth spacecraft are $a = 7016$ km, $e = 0.05$, $i = 45°$, $\Omega = 0$, $\omega = 20°$. Determine the declination, right ascension, speed, flight-path angle, and velocity azimuth when the true anomaly is $10°$. (*Ans.* $\delta = 20.705°$, $\lambda = 22.208°$, $v = 7.91877$ km/s, $\phi = 0.4741°$, $\psi = 310.893°$.)

9. Show that the maximum assistance from planetary rotation at any latitude is obtained for an eastward launch.

10. At what point in an orbit is the inclination equal to the latitude?

11. An Earth satellite is in an orbit of inclination $89.1°$ relative to the equatorial plane, eccentricity 0.00132, argument of perigee $261°$, and perigee altitude of 917 km. (a) What are the maximum declination of the spacecraft and the corresponding radius? (*Ans.* $89.1°$, 7314.303 km.) (b) What is the maximum height of the spacecraft above the equatorial plane? (*Ans.* 7313.401 km.)

12. A spacecraft in a 200 km-high circular Earth orbit fires its rocket, instantly increasing the speed by 600 m/s. Assuming zero atmospheric drag and using an equatorial inertial frame, estimate the spherical coordinates of position and velocity 30 minutes after the firing of the rocket, if the declination, right ascension, and azimuth at the time of firing are 45°, 20°, and 45°, respectively. (*Ans.* $r = 8111.59275$ km, $\delta = 11.7139°$, $\lambda = 157.8603°$, $v = 6.88372$ km/s, $\phi = 8.9846°$, $\psi = 149.2939°$.)

13. Accounting for Earth's rotation is crucial for accuracy in naval gunnery. For example, in the Second World War the German cruiser *Scharnhorst* repeatedly hit the aircraft carrier, *HMS Glorious*, with 11-inch shells at the horizontal range of 24 km, thereby sinking her. Assuming that the engagement took place at the latitude of 68.633°, that the firing was in the north-south direction, and that the flight time of the shells was 36 s, calculate the error caused by neglecting Earth's rotation during firing. (*Ans.* ± 22.96 m.)

14. Referring to Exercise 13, calculate the targeting error caused by neglecting Earth's rotation if the battle had taken place at the equator. (*Ans.* ± 63 m.)

6

Orbital Manoeuvres

Spacecraft can change their orbits by applying thrust generated by rocket engines. The deliberate variation of the orbit by thrust application is called an *orbital manoeuvre*. A rocket engine works on the principle of generating thrust by expelling a stream of particles at a high velocity relative to the spacecraft. The net rate of change of momentum of the exhausted mass appears as a thrust force acting in a direction opposite to the rocket exhaust by Newton's second law of motion, because the net momentum of the system (spacecraft plus the exhaust) is conserved. The principle of conservation of linear momentum is expressed as follows:

$$(m + \Delta m)\mathbf{v} = m(\mathbf{v} + \Delta\mathbf{v}) + \Delta m(\mathbf{v} + \mathbf{v}_e)\,, \tag{6.1}$$

or

$$m\Delta\mathbf{v} = -\Delta m\mathbf{v}_e\,, \tag{6.2}$$

where $\mathbf{v}_e$ is the *exhaust velocity*, the net mass of the spacecraft is m, and the spacecraft's velocity is $\mathbf{v}$, just before an incremental mass Δm is exhausted by the rocket engine with a total velocity $\mathbf{v} + \mathbf{v}_e$, thereby causing a change in the velocity of the spacecraft by $\Delta\mathbf{v}$. Equation (6.2) is the ideal rocket equation, because it assumes that the entire change of momentum of the exhaust is converted into a change in the spacecraft's velocity. It also implies that the maximum acceleration of the spacecraft is achieved when the exhaust velocity, $\mathbf{v}_e$, is opposite in direction to the spacecraft's velocity, $\mathbf{v}$. Assuming this optimum exhaust direction is maintained, i.e., $\mathbf{v}_e = -v_e\mathbf{v}/v$, and dividing Eq. (6.2) by the time interval, Δt, during which the incremental change takes place, we have the following scalar equation:

$$m\frac{\Delta v}{\Delta t} = v_e\frac{\Delta m}{\Delta t}\,, \tag{6.3}$$

which, in the limit $\Delta t \to 0$, results in the following differential equation:

$$m\frac{\mathrm{d}v}{\mathrm{d}t} = -v_e\frac{\mathrm{d}m}{\mathrm{d}t} = f\,, \tag{6.4}$$

where $f(t)$ is the thrust generated at the time instant t when the spacecraft's mass is $m(t)$ and its velocity is $v(t)$. The rate of change of the spacecraft's mass is given by

$$\dot{m} = \mathrm{d}m/\mathrm{d}t = \lim_{\Delta t \to 0}\left(\frac{-\Delta m}{\Delta t}\right)\,, \tag{6.5}$$

Foundations of Space Dynamics, First Edition. Ashish Tewari.

which is a negative quantity. If the exhaust velocity, v_e, is constant, then Eq. (6.4), can be integrated from the initial condition, $t = 0$, $m(0) = m_0$, $v(0) = v_0$, to yield

$$v(t) - v_0 = v_e \ln\left[\frac{m_0}{m(t)}\right] . \tag{6.6}$$

Equation (6.6) is referred to as the *rocket equation*, and gives the ideal increase in the velocity by the consumption of mass, $m_0 - m$, of a rocket propellant of a constant exhaust velocity, v_e. The rocket equation is commonly expressed as follows in terms of the *specific impulse* defined by $I_{sp} = v_e/g_0$, where $g_0 = 9.81\ \mathrm{m/s^2}$:

$$v(t) - v_0 = g_0 I_{sp} \ln\left[\frac{m_0}{m(t)}\right] . \tag{6.7}$$

For a rocket of ideal construction, the specific impulse is solely the property of the propellant. The larger the specific impulse, the higher is the exhaust speed, and hence the higher is the velocity increment produced per unit mass of the propellant. Thus I_{sp} is a measure of a rocket's efficiency. Rockets can be broadly divided into two categories: (a) *chemical* rockets and (b) *ion-propulsion* (or plasma) rockets. A chemical rocket's propellant consists of the combination of a fuel and an oxidizer. These are burned together in a combustion chamber, and result in a mixture of gases which are exhausted in a nozzle to produce thrust. Table 6.1 lists some chemical propellants and their specific impulse in seconds. An ion-propulsion rocket accelerates a mass of charged particles through a magnetic field. Since the exhaust speed achieved in an ion-propulsion rocket is 15–20 times that of a chemical rocket, its propulsive efficiency is much higher. However, the mass exhaust rate, $-\dot{m}$, of such a rocket is less than a thousandth of a chemical rocket. Hence, an ion-propulsion rocket can generate only a small thrust, f (typically $0.10 - 0.25$ Newton). An orbital manoeuvre carried out by an ion-propulsion rocket requires that the small thrust should be applied continuously for a very long time. The ion-propulsion technology is currently confined to small robotic spacecraft on lunar and interplanetary missions, where a long time is available for performing the necessary low-thrust manoeuvres. In contrast, due to its high mass exhaust rate (therefore thrust), a chemical rocket needs to be operated only briefly for a given orbital manoeuvre. Due to these reasons, chemical rockets are the only practical option for most spacecraft (as well as launch vehicles).

Table 6.1 The specific impulse, I_{sp}, of some chemical rocket propellants

Propellant	I_{sp} (s)
Cold gas	50
Hydrazine	230
Solid ammonium perchlorate	268
Liquid oxygen/kerosene	304
Liquid NO_4/MMH	313
Liquid oxygen/liquid hydrogen	460
Liquid lithium/fluorine	542

As opposed to chemical rockets which have a constant exhaust velocity, v_e, ion-propulsion engines have a constant power output, $P = f(t)v_e(t) = -\dot{m}v_e^2(t) = \text{const.}$, which results in a time-variable specific impulse, $I_{sp}(t)$. Such an engine also has a variable mass exhaust rate, $-\dot{m}$, which is inversely proportional to the square of the exhaust speed. The acceleration produced by an ion-propulsion engine can therefore be expressed as follows:

$$a(t) = \frac{f(t)}{m(t)} = \frac{1}{m(t)}\sqrt{-P\dot{m}}\,. \tag{6.8}$$

Most orbital manoeuvres are performed by chemical rocket engines, which apply a large thrust in a relatively small duration, which is negligible in comparison with the orbital time scale, $\frac{2\pi\sqrt{p^3}}{\sqrt{\mu}}$. Therefore, it is reasonable to assume that the velocity change occurs instantaneously at the point of thrust application. Such an approximation involves the assumption of an impulsive thrust acceleration, f/m, which results in a step (instantaneous) change in the velocity, Δv; hence the concerned manoeuvre is said to be *impulsive*. Using the rocket equation, Eq. (6.7), this sudden change in the velocity at a particular instant is approximated by

$$\Delta v = v_f - v_i = g_0 I_{sp} \ln\left(1 + \frac{\Delta m}{m_f}\right), \tag{6.9}$$

where v_i, v_f are the initial and final velocities immediately before and after the application of the thrust, respectively; m_f is the final mass of the spacecraft, and $\Delta m = m_i - m_f$ is the mass of the propellant spent in the manoeuvre. Thus the velocity change for an impulsive manoeuvre translates into the required propellant mass as follows:

$$\Delta m = m_f[e^{\Delta v/(g_0 I_{sp})} - 1] = m_i[1 - e^{-\Delta v/(g_0 I_{sp})}]\,. \tag{6.10}$$

Since the cost of a space mission increases nearly exponentially with the required propellant mass, it is necessary to design a mission in which the sum of the magnitudes of all the velocity changes is the minimum. This chapter discusses how such optimum impulsive manoeuvres can be designed. Analysis of some simple continuous thrust manoeuvres is also presented.

6.1 Single-Impulse Orbital Manoeuvres

The simplest impulsive manoeuvre is the case of intersecting initial and final trajectories, as shown in Fig. 6.1, where a single velocity impulse, $\Delta\mathbf{v}$, is sufficient to produce a velocity change from $\mathbf{v_i}$ to $\mathbf{v_f}$ at a given orbital position, $\mathbf{r}$. The final velocity, $\mathbf{v_f}$, makes an angle α with the initial velocity, $\mathbf{v_i}$. From the vector triangle in Fig. 6.1, it is evident that the relationship among the magnitudes, $v_i, v_f, \Delta v$, is given by the following cosine law:

$$\Delta v = \sqrt{v_i^2 + v_f^2 - 2v_i v_f \cos\alpha}\,. \tag{6.11}$$

A general single-impulse manoeuvre involves a change in both the shape and the plane of the orbit. When only the orbital shape needs to be changed without affecting the orbital plane, the initial and final orbits are coplanar. A coplanar manoeuvre involves a change in the orbital speed, $(v_f - v_i)$, and the flight-path angle, α, which results in an orbit of modified shape, e, and size, a. A manoeuvre involving only a plane change is called a *plane-change manoeuvre*, and

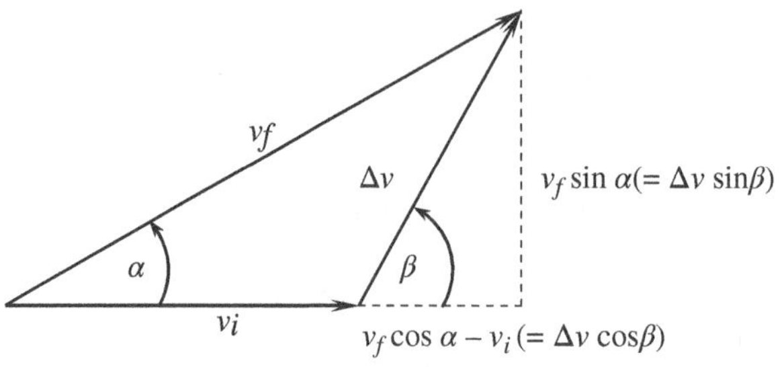

Figure 6.1 A single-impulse manoeuvre for intersecting initial and final orbits.

preserves the shape and size of the orbit. Equation (6.11) leads to the following expression for the velocity impulse required for a plane change by angle, α, at a constant speed, v_i:

$$\Delta v = 2v_i \sin\frac{\alpha}{2}\,. \tag{6.12}$$

The plane-change impulse must be applied at the angle $\beta = (\alpha + \pi)/2$ from the initial velocity direction denoted by $\mathbf{v_i}/v_i$, as shown in Fig. 6.1. Since the velocity impulse magnitude for a plane change is directly proportional to the speed at which the change is performed, even a small-angle change can require a large velocity impulse, which translates into a high propellant expenditure. Hence, a plane-change manoeuvre should be carried out at the smallest possible speed (for example, at the apogee of an elliptic orbit). For a plane change by 60°, the impulse magnitude equals the orbital speed.

Example 6.1.1 *A spacecraft in a circular orbit of altitude* 600 km *and inclination* 30° *around Earth ($\mu = 398600.4$ km^3/s^2, $r_0 = 6378.14$ km) must be transferred to an elliptic orbit with perigee altitude* 300 km, *apogee altitude* 1000 km, *and inclination* 0°. *Calculate the magnitudes and directions of the velocity impulses required for the transfer.*

An impulse can be applied only at the intersection of the initial and the final orbits. Since the final orbital plane is not uniquely specified, it is sufficient to send the spacecraft to any plane of 0° *inclination. Hence a choice is available of where to apply the required inclination change. The most desirable point for making the plane change is at the apogee of the elliptic orbit, because it requires the least impulse magnitude. The eccentricity and the semi-major axis of the final orbit, and the first velocity impulse at the intersection point, $r = r_i = r_0 + 600 = 6978.14$ km, are calculated as follows:*

$$e = \frac{r_a - r_p}{r_a + r_p} = 0.0498$$

$$a = \frac{r_p}{1 - e} = 7028.14 \text{ km}$$

$$v_p = \sqrt{\frac{2\mu}{r_p} - \frac{\mu}{a}} = 7.9158 \text{ km/s}$$

$$v_i = \sqrt{\frac{\mu}{r_i}} = 7.5579 \text{ km/s}$$

$$v_f = \sqrt{\frac{2\mu}{r_i} - \frac{\mu}{a}} = 7.5847 \text{ km/s}$$

$$\cos\alpha = \cos\phi = \frac{r_p v_p}{r_i v_f} = 0.998785$$

$$\Delta v_1 = \sqrt{v_i^2 + v_f^2 - 2 v_i v_f \cos\alpha} = 0.37427 \text{ km/s} .$$

The transfer angle, α, is determined from $\alpha = \cos^{-1}(0.998785) = \pm 2.825°$, each sign denoting one of the two possible intersection points. The direction of the first impulse in the orbital plane is given by $\beta = \sin^{-1}(v_f \sin\alpha / \Delta v_1) = \pm 87.302°$ (Fig. 6.1).

The second velocity impulse is applied at the apogee, $r = r_a = a(1 + e)$, of the elliptic orbit resulting from the first impulse, in order to change the orbital inclination by 30°:

$$v_a = \sqrt{\frac{2\mu}{r_a} - \frac{\mu}{a}} = 7.1648 \text{ km/s}$$

$$\Delta v_2 = 2 v_a \sin\frac{30°}{2} = 3.70876 \text{ km/s} .$$

The second impulse makes an angle $\beta = (30° + 180°)/2 = 105°$ with the initial orbital plane. The total velocity impulse magnitude for the orbital change is $\Delta v = \Delta v_1 + \Delta v_2 = 4.083$ km/s.

When the initial and final orbital planes are uniquely defined by the given sets of classical Euler angles, Ω, ω, i, the point of plane change cannot be chosen arbitrarily. In the general plane change manoeuvre, the impulse must wait until the spacecraft reaches the line of nodes formed by the intersection of the two orbital planes. If the manoeuvre must be performed by a single velocity impulse, the two orbits should also intersect at the line of nodes, which is true only for circular orbits of equal radii. Figure 6.2 describes the geometry of the general plane-change manoeuvre in terms of the initial and final orbital angular momenta, $\mathbf{h_i}$ and $\mathbf{h_f}$, respectively. The nodal vector between the two planes denoting the point of application of the velocity impulse is given by

$$\mathbf{n_{if}} = \frac{\mathbf{h_i} \times \mathbf{h_f}}{|\, \mathbf{h_i} \times \mathbf{h_f} \,|} . \tag{6.13}$$

The impulsive manoeuvre is carried out at $\mathbf{r_i} = r_i \mathbf{n_{if}}$, from which the speed of the plane change, v_i, is calculated. The cosine of the angle, α, between the two planes is obtained as follows:

$$\cos\alpha = \frac{\mathbf{h_i} \cdot \mathbf{h_f}}{h_i h_f} . \tag{6.14}$$

The direction of the plane-change impulse relative to the initial velocity vector is given by $\beta = \frac{\alpha}{2} + 90°$, which results in the following expression for the velocity impulse:

$$\Delta\mathbf{v} = \Delta v \left(\cos\beta \frac{\mathbf{v_i}}{v_i} + \sin\beta \frac{\mathbf{h_i}}{h_i} \right) , \tag{6.15}$$

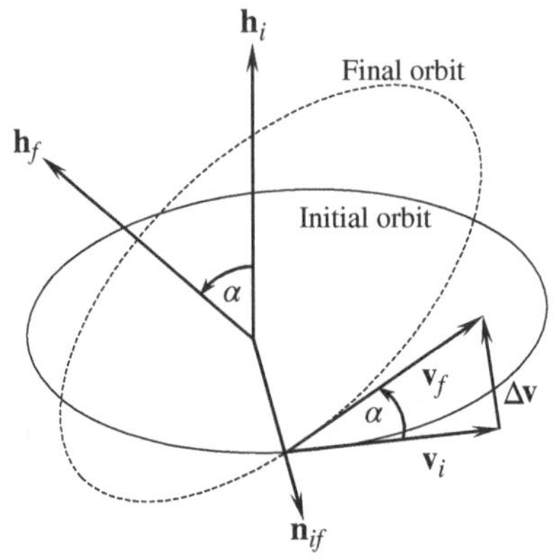

Figure 6.2 A general plane-change manoeuvre.

leading to

$$\mathbf{v_f} = \mathbf{v_i} + \Delta\mathbf{v} = \mathbf{v_i} + \Delta v\left(\cos\beta\frac{\mathbf{v_i}}{v_i} + \sin\beta\frac{\mathbf{h_i}}{h_i}\right). \tag{6.16}$$

Since the plane-change impulse is applied out of the orbital plane at an angle greater than 90°, it does not affect the orbital speed and the flight-path angle. Hence a and e are unchanged. Furthermore, since the initial and final orbits share the same radius, r_i, there cannot be any change in the true anomaly, θ, due to a plane-change impulse. Hence, the time of periapsis, t_0, is also unchanged.

Example 6.1.2 *A spacecraft is in an Earth orbit of* $a = 7000$ km, $e = 0.5$, $\Omega = 80°$, $\omega = 250°$, $i = -20°$. *When the spacecraft is at its apogee, a velocity impulse is applied at an angle,* $\beta = 110°$*, to the orbital plane, measured from the initial velocity vector,* $\mathbf{v_i}$*, as shown in Fig. 6.1, such that there is no change in the orbital speed, v. Determine the new orbit of the spacecraft.*

The magnitude of the impulse is obtained from Eq. (6.12) by substituting $\alpha = 2\beta - 180° = 40°$. *A major component of the velocity impulse is in the direction of the initial angular momentum vector,* $\mathbf{h_i}$*; thus it would tend to increase the orbital inclination, i, and decrease the right ascension of the ascending node, Ω. We begin the solution by deriving the angular momentum vector of the initial orbital planes from the perifocal apogee position and velocity, transformed into the celestial frame, by the rotation matrix,* $\mathbf{C}^*$. *The nodal vector,* $\mathbf{n_{if}}$*, essentially lies towards the apogee. The final velocity at the orbit intersection is obtained by vectorially adding the velocity impulse to the initial velocity. Thus, the final angular momentum can be calculated, and the orbital elements defining the new orbital plane obtained. These calculation steps are listed as follows:*

$$n = \sqrt{\mu/a^3} = 0.001078 \text{ rad/s}$$

$$\mathbf{r}_a = -a(1+e)\mathbf{i_e} = -10500\mathbf{i_e} \text{ (km)}$$

$$\mathbf{v}_a = -an\frac{\sqrt{1-e^2}}{1+e}\mathbf{i_p} = -4.3567\mathbf{i_p} \text{ (km/s)}$$

$$\mathbf{C}^* = \mathbf{C}_3(250^\circ)\mathbf{C}_1(-20^\circ)\mathbf{C}_3(80^\circ) = \begin{pmatrix} 0.81022 & 0.47969 & -0.33682 \\ -0.49016 & 0.86961 & 0.05939 \\ 0.32139 & 0.11698 & 0.93969 \end{pmatrix}$$

$$\mathbf{r}_\mathrm{i} = \mathbf{C}^*\mathbf{r}_a = -8507.268\mathbf{I} + 5146.673\mathbf{J} - 3374.635\mathbf{K} \text{ (km)}$$

$$\mathbf{v}_\mathrm{i} = \mathbf{C}^*\mathbf{v}_a = -2.0899\mathbf{I} - 3.7886\mathbf{J} - 0.5096\mathbf{K} \text{ (km/s)}$$

$$\mathbf{h}_\mathrm{i} = \mathbf{r}_\mathrm{i} \times \mathbf{v}_\mathrm{i} = -15408.1913\mathbf{I} + 2716.8798\mathbf{J} + 42986.7224\mathbf{K} \text{ (km}^2\text{/s)}$$

$$\Delta v = 2v_a \sin\frac{40^\circ}{2} = 2.9802 \text{ km/s}$$

$$\Delta\mathbf{v} = \Delta v\left(\cos\beta\frac{\mathbf{v}_\mathrm{i}}{v_i} + \sin\beta\frac{\mathbf{h}_\mathrm{i}}{h_i}\right) = -0.4543\mathbf{I} + 1.0527\mathbf{J} + 2.7508\mathbf{K} \text{ (km/s)}$$

$$\mathbf{v}_\mathrm{f} = \mathbf{v}_\mathrm{i} + \Delta\mathbf{v} = -2.5442\mathbf{I} - 2.7359\mathbf{J} + 2.2412\mathbf{K} \text{ (km/s)}$$

$$\mathbf{h}_\mathrm{f} = \mathbf{r}_\mathrm{i} \times \mathbf{v}_\mathrm{f} = 2301.6697\mathbf{I} + 27651.7440\mathbf{J} + 36369.4305\mathbf{K} \text{ (km}^2\text{/s)}$$

$$\mathbf{n} = \mathbf{h}_\mathrm{f} \times \mathbf{K}/h_f = -0.996554\mathbf{I} + 0.082951\mathbf{J}$$

$$\Omega = \cos^{-1}(-0.996554) = 175.2418^\circ$$

$$i = \cos^{-1}\left(\mathbf{K}\cdot\frac{\mathbf{h}_\mathrm{f}}{h_f}\right) = 37.3411^\circ$$

$$\mathbf{e} = \frac{\mathbf{v}_\mathrm{f} \times \mathbf{h}_\mathrm{f}}{\mu} - \frac{\mathbf{r}_\mathrm{i}}{r_i} = 0.810216\mathbf{I} - 0.49016\mathbf{J} + 0.32139\mathbf{K}$$

$$\cos\omega = \mathbf{n}\cdot\mathbf{e}/e = -0.84808 \;; \quad \sin\omega = (\mathbf{h}_\mathrm{f}/h_f)\cdot(\mathbf{n}\times\mathbf{e}/e) = 0.5298637$$

$$\omega = \cos^{-1}(-0.84808) = 148.0038^\circ \,.$$

Hence, the new orbit after the application of the impulse is given by the classical orbital elements, $a = 7000$ km, $e = 0.5$, $\Omega = 175.2418^\circ$, $\omega = 148.0038^\circ$, $i = 37.3411^\circ$. *The reader can verify this calculation by computing the celestial rotation matrix,* $\mathbf{C}^*$, *with the new Euler angles, whose multiplication with the perifocal position,* $\mathbf{r}_a$, *should yield the same celestial position,* $\mathbf{r}_\mathrm{i}$.

When both the shape and the plane of an orbit are to be changed, the sequence of the coplanar and plane-change manoeuvres can be selected depending upon the speeds at which they are to be performed (plane change is more efficient at a smaller speed). However, when the coplanar and the plane-change manoeuvres are to be performed at the same radius, the two separate impulses required for each are combined into a single-impulse manoeuvre, rather than carrying out each separately. In such a case, the velocity impulses required for the separately performed individual manoeuvres form two sides of a vector triangle, whose resultant is the impulse required for the combined manoeuvre (Fig. 6.1). Since the resultant is always smaller than the sum of the individual magnitudes, such a combined impulse can save a significant amount of propellant.

6.2 Multi-impulse Orbital Transfer

When the initial and final orbits do not intersect, the orbital transfer cannot be made by a single impulse; hence a multi-impulse manoeuvre becomes necessary. Equation (6.11) implies that the minimum velocity impulse magnitude required in a coplanar manoeuvre for a given pair of initial and final speeds, v_i, v_f, corresponds to the case of $\alpha = 0$, which is the case of a *tangential* impulse. Hence a coplanar manoeuvre has the minimum possible total velocity change when all the velocity impulses required for the manoeuvre are applied tangentially. Such a manoeuvre requires the least possible propellant mass, and is therefore said to be an *optimal* manoeuvre between the given initial and final orbits. The trajectory connecting the initial and final orbits in such a manoeuvre must essentially be elliptic, because only an elliptic orbit has two points where the velocity is tangential to the orbit (periapsis and apoapsis). The tangential velocity impulses must be applied at the periapsis and the apoapsis of the *transfer ellipse*.

Example 6.2.1 *Calculate the smallest possible total velocity change required in a two-impulse orbital transfer from the circular Earth orbit of 600 km altitude to the intersecting elliptic orbit of Example 6.1.2.*

The optimal manoeuvres between coplanar circular and elliptic orbits require that the impulses be applied tangentially at both the initial and final orbits. Since the initial orbit is circular, any point can be chosen to apply a tangential impulse. However, in the final elliptic orbit, there are only two points where tangential velocity changes can take place, namely the perigee and the apogee. The transfer ellipse with the smallest possible semi-major axis, a, has the least orbital energy, ϵ, and is therefore the optimal one. Such an ellipse ends at either the perigee or the apogee of the final elliptic orbit, and begins at the point in the circular orbit that is diametrically opposite the terminating point.

6.2.1 Hohmann Transfer

Hohmann showed that the optimal transfer between two circular orbits consists of two velocity impulses applied tangentially to the respective orbits. Hohmann transfer between two circular orbits of radii r_1 and r_2, respectively, is depicted in Fig. 6.3. Since both the velocity impulses, Δv_1 and Δv_2, are tangential to the orbits, the semi-major axis of the transfer ellipse is the average of the two radii, $a = (r_1 + r_2)/2$. Using the multi-variable optimization, it can be shown that Hohmann transfer satisfies the necessary conditions for optimality.

Let the two velocity impulses be applied at angles ϕ_1 and ϕ_2, respectively, from the local horizon, $(\mathbf{i}_r, \mathbf{i}_\theta)$:

$$\begin{aligned}
\Delta \mathbf{v}_1 &= \Delta v_{1\theta} \mathbf{i}_{\theta 1} + \Delta v_{1r} \mathbf{i}_{r1} \\
&= (v_1' \cos \phi_1 - v_1)\mathbf{i}_{\theta 1} + v_1' \sin \phi_1 \mathbf{i}_{r1} \\
\Delta \mathbf{v}_2 &= \Delta v_{2\theta} \mathbf{i}_{\theta 2} + \Delta v_{2r} \mathbf{i}_{r2} \\
&= (v_2 - v_2' \cos \phi_2)\mathbf{i}_{\theta 2} - v_2' \sin \phi_2 \mathbf{i}_{r2},
\end{aligned} \tag{6.17}$$

where $v_1 = \sqrt{\mu/r_1}$ and $v_2 = \sqrt{\mu/r_2}$ are the two circular orbital speeds, v_1' denotes the speed immediately after the application of the first impulse, and v_2' denotes the speed immediately before the application of the second impulse. The impulse magnitudes are thus the following:

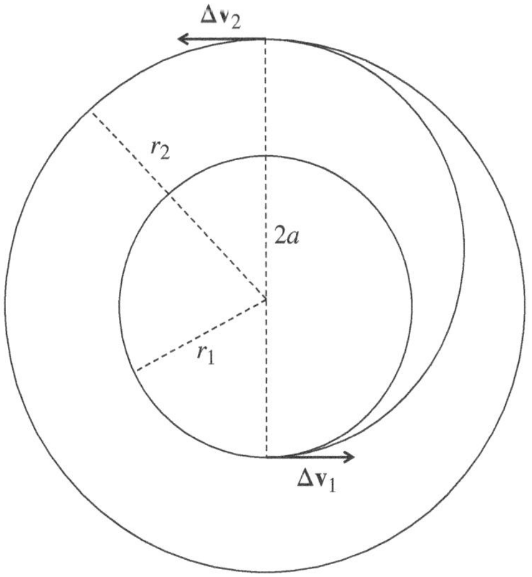

Figure 6.3 Geometry of Hohmann transfer between two circular orbits.

$$\Delta v_1 = |\Delta \mathbf{v}_1| = \sqrt{(v_1' - v_1)^2 + 2v_1 v_1'(1 - \cos\phi_1)}$$

$$\Delta v_2 = |\Delta \mathbf{v}_2| = \sqrt{(v_2' - v_2)^2 + 2v_2 v_2'(1 - \cos\phi_2)} \tag{6.18}$$

The total velocity impulse magnitude, $J = \Delta v_1 + \Delta v_2$, is the performance index to be minimized with respect to the control variables, $v_1', v_2', \phi_1, \phi_2$, subject to the following constraints on the transfer orbit:

(i) The orbital energy, ϵ, is constant along the transfer ellipse. This implies that

$$(v_1')^2 - (v_2')^2 + 2(v_2^2 - v_1^2) = 0. \tag{6.19}$$

(ii) The orbital angular momentum magnitude, h, is conserved along the transfer ellipse, thereby implying that

$$v_2^2 v_1' \cos\phi_1 - v_1^2 v_2' \cos\phi_2 = 0. \tag{6.20}$$

The Hamiltonian of the optimal control problem is thus given by

$$H = \sqrt{(v_1' - v_1)^2 + 2v_1 v_1'(1 - \cos\phi_1)} + \sqrt{(v_2' - v_2)^2 + 2v_2 v_2'(1 - \cos\phi_2)}$$
$$+\lambda_1[(v_1')^2 - (v_2')^2 + 2(v_2^2 - v_1^2)] + \lambda_2(v_2^2 v_1' \cos\phi_1 - v_1^2 v_2' \cos\phi_2), \tag{6.21}$$

where λ_1, λ_2 are the Lagrange multipliers associated with the two equality constraints.

The stationary conditions for the Hamiltonian yield the following:

$$\frac{\partial H}{\partial v_1'} = \frac{v_1' - v_1 \cos\phi_1}{\Delta v_1} + 2\lambda_1 v_1' + \lambda_2 v_2^2 \cos\phi_1 = 0$$

$$\frac{\partial H}{\partial v_2'} = \frac{v_2' - v_2 \cos\phi_2}{\Delta v_2} - 2\lambda_1 v_2' - \lambda_2 v_1^2 \cos\phi_2 = 0$$

$$\frac{\partial H}{\partial \phi_1} = \frac{v_1 v_1'}{\Delta v_1} \sin\phi_1 - \lambda_2 v_2^2 v_1' \sin\phi_1 = 0 \tag{6.22}$$

$$\frac{\partial H}{\partial \phi_2} = \frac{v_2 v_2'}{\Delta v_2} \sin\phi_2 + \lambda_2 v_1^2 v_2' \sin\phi_2 = 0,$$

the last two of which require $\phi_1 = \phi_2 = 0$, which, when substituted into the constraints [Eqs. (6.19) and (6.20)],

$$v_1' = \frac{v_1^2}{\sqrt{\frac{v_1^2+v_2^2}{2}}}$$

$$v_2' = \frac{v_2^2}{\sqrt{\frac{v_1^2+v_2^2}{2}}}. \tag{6.23}$$

This proves that a Hohmann transfer ellipse is indeed an extremal trajectory, satisfying the necessary conditions of optimality.

The magnitudes of the two velocity impulses in a Hohmann transfer are obtained by merely subtracting the initial speed from the final speed at respective instants as follows:

$$\Delta v_1 = v_{f1} - v_{i1} = \sqrt{\frac{2\mu}{r_1} - \frac{\mu}{a}} - \sqrt{\frac{\mu}{r_1}} \tag{6.24}$$

and

$$\Delta v_2 = v_{f2} - v_{i2} = \sqrt{\frac{\mu}{r_2}} - \sqrt{\frac{2\mu}{r_2} - \frac{\mu}{a}}, \tag{6.25}$$

where r_1, r_2 are the initial and final circular orbit radii, respectively, and a is the semi-major axis of the transfer ellipse. The net velocity impulse of a Hohmann transfer is the sum of the two impulses, given by

$$\begin{aligned} \Delta v_H &= \Delta v_1 + \Delta v_2 \\ &= \sqrt{\frac{2\mu}{r_1} - \frac{2\mu}{r_1 + r_2}} - \sqrt{\frac{\mu}{r_1}} + \sqrt{\frac{\mu}{r_2}} - \sqrt{\frac{2\mu}{r_2} - \frac{2\mu}{r_1 + r_2}}. \end{aligned} \tag{6.26}$$

The time required for a Hohmann transfer, t_H, is half the period of the transfer ellipse, and is given by

$$t_H = \pi\sqrt{\frac{a^3}{\mu}} = \frac{\pi(r_1 + r_2)^{\frac{3}{2}}}{\sqrt{8\mu}}. \tag{6.27}$$

Example 6.2.2 *Calculate the velocity impulses and the time required for a Hohmann transfer from a circular, earth orbit of altitude 250 km (parking orbit) to a* geosynchronous *orbit.*

The radius of the geosynchronous orbit was calculated in Chapter 3 from Earth's sidereal period, $T_s = 23$ hr., 56 min., 4.09 s, *to be the following:*

$$r_2 = \left(\frac{T_s\sqrt{\mu}}{2\pi}\right)^{\frac{2}{3}} = 42164.17 \text{ km} .$$

The semi-major axis of the transfer ellipse is $a = \frac{r_1+r_2}{2} = 24396.155$ km. *The two impulse magnitudes are then calculated by*

$$\Delta v_1 = \sqrt{\frac{2\mu}{r_1} - \frac{\mu}{a}} - \sqrt{\frac{\mu}{r_1}} = 2.44 \text{ km/s} .$$

$$\Delta v_2 = \sqrt{\frac{\mu}{r_2}} - \sqrt{\frac{\mu}{r_2} - \frac{\mu}{a}} = 1.472 \text{ km/s} .$$

Both the impulses are applied in the direction of motion. The Hohmann transfer time is computed as follows:

$$t_H = \pi\sqrt{\frac{a^3}{\mu}} = 18961.08 \text{ s (5 hr., 16 min., 1.08 s)} .$$

6.2.2 Rendezvous in Circular Orbit

When two spacecraft are in circular orbits of different radii around the same central body, a rendezvous between them can take place by Hohmann transfer. Let one of the spacecraft called the *target* be maintained in the orbit of radius r_f, while the other spacecraft, initially in an orbit of radius r_i, is manoeuvred by applying the two tangential velocity impulses, such that the rendezvous between the spacecraft takes place at the radius r_f. A rendezvous between two (or more) spacecraft requires that their positions and velocities should match at the time of rendezvous, $t = t_f$. Without a loss of generality in the discussion, let $r_f > r_i$. Then since the manoeuvring spacecraft travels more rapidly than the target, the target must be slightly advanced in its angular position when compared to the manoeuvring spacecraft at $t = 0$ when the first impulse is applied. This situation is depicted in Fig. 6.4, when the target leads the manoeuvring spacecraft by a *phase angle*, θ_H, at the instant of the first impulse, $t = 0$. The value of the correct phase angle for rendezvous at $t = t_f$ by Hohmann transfer is derived as follows:

$$\frac{\pi - \theta_H}{\sqrt{\mu/r_f^3}} = \frac{\pi(r_i + r_f)^{\frac{3}{2}}}{\sqrt{8\mu}} , \tag{6.28}$$

or

$$\theta_H = \pi\left[1 - \left(\frac{1 + r_i/r_f}{2}\right)^{3/2}\right] . \tag{6.29}$$

Considering the two extreme possibilities for the ratio, $0 \le r_i/r_f \le 1$, one arrives at the range of the correct phase angle to be $0 \le \theta_H \le 0.64645\pi$. If the two orbits are in different planes, then

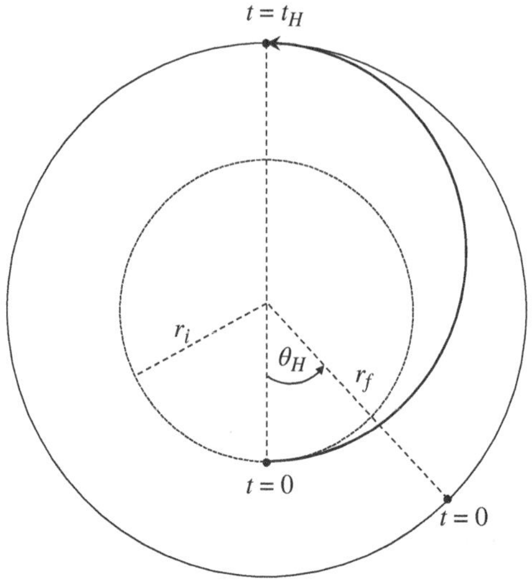

Figure 6.4 Rendezvous by Hohmann transfer.

the rendezvous point must be located at the intersection of the two planes, and a plane-change manoeuvre must be carried out at the final time, $t = t_f$.

The rendezvous problem is thus a matter of timing the velocity impulses to be applied to the manoeuvring spacecraft. If the initial phase angle between the two spacecraft is different from the correct phase angle, θ_H, then a *waiting time*, t_w, is necessary before the first velocity impulse can be applied. This time is calculated from the error in the phase angle, $\Delta\theta$, as follows:

$$t_w = \frac{\Delta\theta}{\sqrt{\mu/r_i^3} - \sqrt{\mu/r_f^3}}. \tag{6.30}$$

For the extreme case of $\Delta\theta = 2\pi$, the waiting time becomes the *synodic period* of the two orbits, T_s, which is given by

$$T_s = \frac{2\pi}{\sqrt{\mu/r_i^3} - \sqrt{\mu/r_f^3}}. \tag{6.31}$$

Therefore, an alternative expression for the waiting time is $t_w = T_s\Delta\theta/(2\pi)$. The synodic period, and hence the waiting time, can be very large when the two radii are close to each other, i.e., $r_f/r_i \simeq 1$. In such a case, Hohmann transfer is no longer a viable option for making the rendezvous, and other (non-optimal) transfer trajectories, which have a reasonable waiting time, must be chosen.

Example 6.2.3 *A spacecraft* A *is in a circular orbit of inclination* 30° *and 300 km altitude around Earth. It has to rendezvous with a space station in a zero-inclination orbit of 400 km altitude by using Hohmann transfer. The initial positions of the two spacecraft relative to the line of intersection of the two orbital planes are shown in Fig. 6.5. Calculate the total time and*

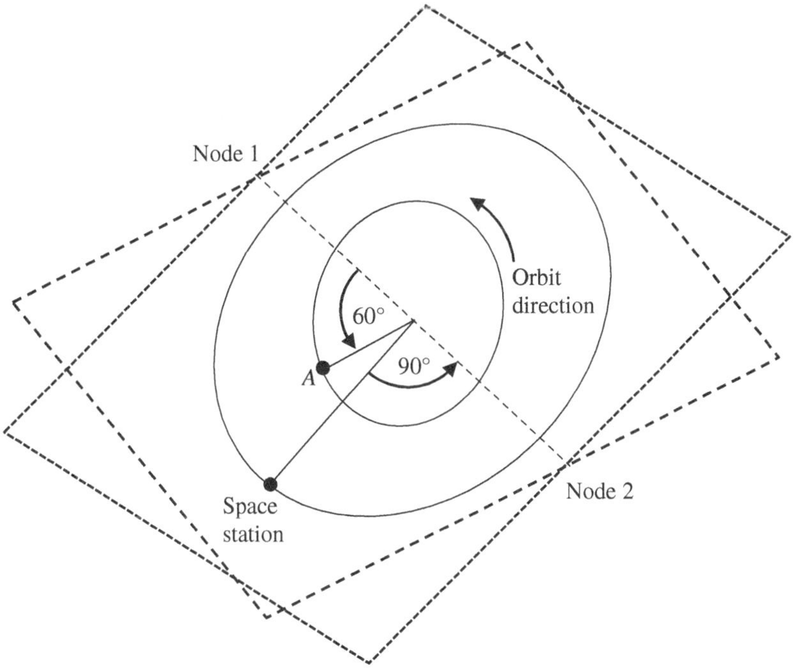

Figure 6.5 Rendezvous by Hohmann transfer between spacecraft *A* and a space station originally in non-coplanar circular orbits.

the total propellant mass required by A *for the rendezvous, assuming a specific impulse of* 300 *s and the final spacecraft mass of* 500 *kg.*

Given the initial and final radii of $r_i = 6378.14 + 300 = 6678.14$ *km and* $r_i = 6378.14 + 400 = 6778.14$ *km, we have the following correct phase angle required for the rendezvous by Hohmann transfer:*

$$\theta_H = \pi \left[1 - \left(\frac{1 + r_i/r_f}{2} \right)^{3/2} \right] = 0.034698 \text{ rad. } (1.988^\circ).$$

Figure 6.5 shows that the space station (target) leads the manoeuvring spacecraft, A, *by a phase angle of* 30°. *The error in the phase angle is thus given by* $\Delta\theta = 30(\pi/180) - 0.034698 = 0.4889$ *rad., (or* 28.012°*). The waiting time is therefore calculated to be the following:*

$$t_w = \frac{\Delta\theta}{\sqrt{\mu/r_i^3} - \sqrt{\mu/r_f^3}} = \frac{0.4889}{0.0011569 - 0.0011314} = 19167.4188 \text{ s.}$$

Hence, the first impulse is applied at $t = t_w$, *and the second at* $t = t_w + t_H$, *where* t_H, *the time during Hohmann transfer, is calculated as follows:*

$$t_H = \frac{\pi - \theta_H}{\sqrt{\mu/r_f^3}} = 2746.1455 \text{ s.}$$

The positions of the two spacecraft just prior to the initiation of Hohmann transfer at $t = t_w =$ 19167.4188 *s measured from the line of intersection at node 1 (see Fig. 6.5) are given by* $\theta_i = 23.22146 = 6\pi + 4.371906$ *rad., and* $\theta_f = 23.25616 = 6\pi + 4.40660$ *rad., which indicates that both the spacecraft have completed more than three orbits before the Hohmann transfer is initiated. The phase angle at* $t = t_w$ *is* $\theta_f - \theta_i = 0.034698$ *rad., which precisely equals the correct phase angle,* θ_H*. After completing the Hohmann transfer, the spacecraft* A *is located at* $\theta_i = 4.371906 + \pi = 7.5134985$ *rad., from node 1, while the space station is located at* $\theta_f = 4.40660 + t_H\sqrt{\mu/r_f^3} = 7.5134985$ *rad. thereby confirming that the two spacecraft have achieved the same radial and angular position measured from node 1, and the manoeuvring spacecraft, A, has reached a circular orbit of radius,* r_f*. However, the rendezvous has yet to be made, because the two spacecraft are in different orbital planes. Since node 1 has been crossed at this point, the earliest opportunity of rendezvous is at node 2, which requires both the spacecraft to traverse an additional angle of* $\theta' = 3\pi - \theta_f = 1.91128$ *rad., as shown in Fig. 6.6. This requires an additional time of*

$$t' = \frac{\theta'}{\sqrt{\mu/r_f^3}} = 1689.3558 \text{ s}$$

before rendezvous is made at node 2. Therefore, the total time until rendezvous is $t_w + t_H + t' =$ 23602.9201 s, *which amounts to 6 hr., 33 min., 22.9201 s.*

The two velocity impulses required for the Hohmann transfer are calculated by

$$\Delta v_1 = \sqrt{\frac{2\mu}{r_i} - \frac{2\mu}{r_i + r_f}} - \sqrt{\frac{\mu}{r_i}} = 0.028654 \text{ km/s.}$$

$$\Delta v_2 = \sqrt{\frac{\mu}{r_f}} - \sqrt{\frac{2\mu}{r_f} - \frac{2\mu}{r_i + r_f}} = 0.028548 \text{ km/s}$$

The third velocity impulse required for plane change at the rendezvous point is given by

$$\Delta v_3 = 2\sqrt{\frac{\mu}{r_f}} \sin(15\pi/180) = 3.969537 \text{ km/s.}$$

Hence, the total velocity impulse for rendezvous is $\Delta v = \Delta v_1 + \Delta v_2 + \Delta v_3 = 4.026738$ *km/s.*

Finally, the propellant spent during the manoeuvre is calculated by

$$\Delta m = m_f[e^{\Delta v/(g_0 I_{sp})} - 1] = 500[e^{(4.026738\times 1000)/(9.81\times 300)} - 1] = 1464.2204 \; kg \; .$$

6.2.3 Outer Bi-elliptic Transfer

A generalization of Hohmann transfer for the three-impulse case is the *outer bi-elliptic transfer* depicted in Fig. 6.7. Here the transfer consists of two ellipses, and takes the spacecraft to an intermediate radius, $r_3 > r_2 > r_1$, with all the three impulses applied tangentially. The third impulse, Δv_3, is applied opposite to the direction of the orbit, as shown in Fig. 6.7. The time of flight is determined by r_3, which itself is a non-linear function of r_1 and r_2. Thus the sole design parameter for this transfer is the intermediate radius, r_3.

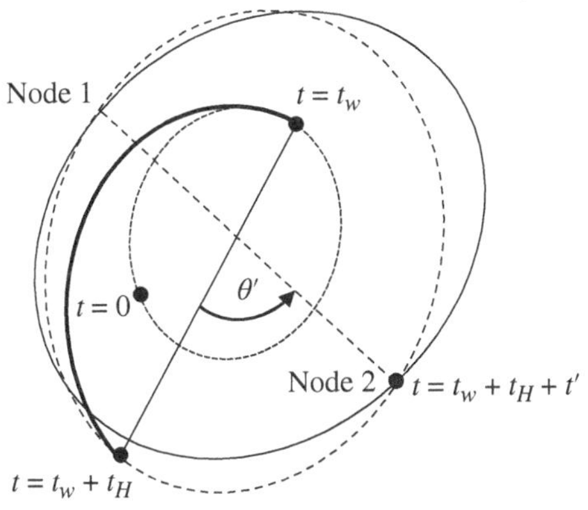

Figure 6.6 Positions of the manoeuvring spacecraft, A, at different times during rendezvous with a space station by Hohmann transfer.

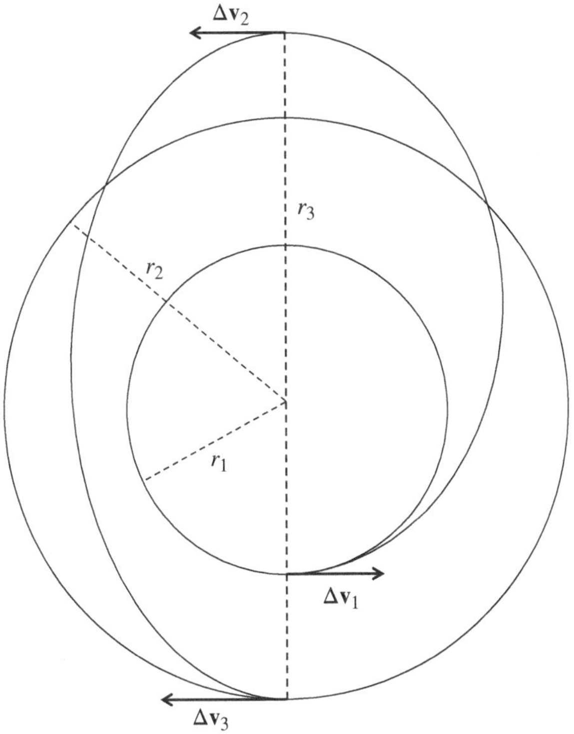

Figure 6.7 Geometry of the outer bi-elliptic transfer between two circular orbits.

The net velocity impulse of the outer bi-elliptic (OB) transfer is the sum of the following three impulses (Fig. 6.7):

$$\begin{aligned}\Delta v_B &= \Delta v_1 + \Delta v_2 + \Delta v_3 \\ &= \sqrt{\frac{2\mu}{r_1} - \frac{2\mu}{r_1 + r_3}} - \sqrt{\frac{\mu}{r_1}} + \sqrt{\frac{2\mu}{r_3} - \frac{2\mu}{r_2 + r_3}} - \sqrt{\frac{2\mu}{r_3} - \frac{2\mu}{r_1 + r_3}} \\ &+ \sqrt{\frac{2\mu}{r_2} - \frac{2\mu}{r_2 + r_3}} - \sqrt{\frac{\mu}{r_2}}.\end{aligned} \tag{6.32}$$

When the ratio r_2/r_1 is large, the outer bi-elliptic transfer can be more efficient than Hohmann transfer. To find the critical value, r_2/r_1, consider $x = r_2/r_1 \geq 1$ and $y = r_3/r_1 \geq 1$ as the two non-dimensional parameters of the OB transfer. The total velocity change required for the Hohmann and OB transfers for the same value of r_1 and r_2 are then respectively expressed as follows:

$$\Delta v_H = \sqrt{\frac{\mu}{r_1}} \left[-1 + \frac{1}{\sqrt{x}} + \sqrt{\frac{2x}{1+x}} - \sqrt{\frac{2}{x(1+x)}} \right] \tag{6.33}$$

$$\Delta v_B = \sqrt{\frac{\mu}{r_1}} \left[-1 - \frac{1}{\sqrt{x}} + \sqrt{\frac{2y}{1+y}} + \sqrt{\frac{2x}{y(x+y)}} - \sqrt{\frac{2}{y(1+y)}} + \sqrt{\frac{2y}{x(x+y)}} \right].$$

The difference between the net non-dimensional velocity change of the two transfers is then expressed by the following function:

$$f(x, y) = \frac{\Delta v_B - \Delta v_H}{\sqrt{\frac{\mu}{r_1}}}, \tag{6.34}$$

which vanishes for $x = y$ (that is, when the two transfers become exactly the same). The value of $x = y$ for which the OB transfer becomes more efficient than Hohmann transfer is then obtained from the stationary condition of $f(x, y)$ given by

$$\begin{aligned}\frac{\partial f}{\partial x} &= \frac{\partial f}{\partial y} = 0 \\ &= \frac{3y+1}{(y+1)^{3/2}} - \sqrt{\frac{x}{x+y}} \\ &= \frac{3x+1}{(x+1)^{3/2}} - \frac{1}{\sqrt{2}}\end{aligned} \tag{6.35}$$

or

$$x^3 - 15x^2 - 9x - 1 = 0, \tag{6.36}$$

whose only positive real root is $x = 15.5817$. Since the second derivative of $f(x, y)$ is always negative, given by

$$\frac{\partial^2 f}{\partial x \partial y} = -\frac{1}{2\sqrt{2xy}(x+y)^{3/2}} < 0, \tag{6.37}$$

it follows that $f(x, y) = 0$ is a maximum point for $x = y$. Therefore, $f(x, y) < 0$ for $x > 15.5817$, thereby implying that OB transfer is more efficient than Hohmann transfer for $r_2/r_1 > 15.5817$.

The OB transfer is advantageous in a rendezvous problem where the initial and final radii are very close, thereby resulting in a large waiting time for Hohmann transfer.

6.3 Continuous Thrust Manoeuvres

The motion of a spacecraft around a spherical body of gravitational constant, μ, powered by a continuous thrust engine, which produces an acceleration input, $\mathbf{a}(t)$, is represented by the following vector differential equations:

$$\dot{\mathbf{v}} + \mu\frac{\mathbf{r}}{r^3} = \mathbf{a} \tag{6.38}$$

and

$$\dot{\mathbf{r}} = \mathbf{v}, \tag{6.39}$$

where $\mathbf{r}(t)$ (with $r(t) = |\mathbf{r}(t)|$) is the position vector measured from the centre of mass of the central body, and $\mathbf{v}(t)$ is the velocity vector. In this section, the symbol $a =| \mathbf{a} |$ denotes the magnitude of the thrust acceleration, and should not be confused with the semi-major axis Sections 6.1 and 6.2. Consider a local-horizon frame, $(\mathbf{i_r}, \mathbf{i}_\theta, \mathbf{i_h})$, for resolving Eqs. (4.1) and (4.2), where $\mathbf{i_r} = \mathbf{r}/r$, $\mathbf{i_h} = \mathbf{r} \times \mathbf{v}/|\mathbf{r} \times \mathbf{v}|$, and $\mathbf{i}_\theta = \mathbf{i_h} \times \mathbf{i_r}$. The plane $(\mathbf{i}_\theta, \mathbf{i_h})$ is the local horizon. The thrust-acceleration vector is resolved into the radial, circumferential, and normal components, a_r, a_θ, and a_n, respectively, as follows:

$$\mathbf{a} = a_r\mathbf{i_r} + a_\theta\mathbf{i}_\theta + a_n\mathbf{i_h} \,. \tag{6.40}$$

The rotation of the local-horizon frame caused by the applied thrust can be represented by the Euler angles, (θ_3, α_2), as follows (Chapter 5):

$$\begin{Bmatrix} \mathbf{i}'_\mathbf{r} \\ \mathbf{i}'_\theta \\ \mathbf{i}'_\mathbf{h} \end{Bmatrix} = \mathbf{C}_3(\theta) \begin{Bmatrix} \mathbf{i_r} \\ \mathbf{i}_\theta \\ \mathbf{i_h} \end{Bmatrix}, \tag{6.41}$$

$$\begin{Bmatrix} \mathbf{i}''_\mathbf{r} \\ \mathbf{i}''_\theta \\ \mathbf{i}''_\mathbf{h} \end{Bmatrix} = \mathbf{C}_2(\alpha) \begin{Bmatrix} \mathbf{i}'_\mathbf{r} \\ \mathbf{i}'_\theta \\ \mathbf{i}'_\mathbf{h} \end{Bmatrix}, \tag{6.42}$$

or

$$\begin{Bmatrix} \mathbf{i}''_\mathbf{r} \\ \mathbf{i}''_\theta \\ \mathbf{i}''_\mathbf{h} \end{Bmatrix} = \mathbf{C}_2(\alpha)\mathbf{C}_3(\theta) \begin{Bmatrix} \mathbf{i_r} \\ \mathbf{i}_\theta \\ \mathbf{i_h} \end{Bmatrix}, \tag{6.43}$$

where

$$\mathbf{C}_2(\alpha)\mathbf{C}_3(\theta) = \begin{pmatrix} \cos\alpha\cos\theta & \cos\alpha\sin\theta & -\sin\alpha \\ -\sin\theta & \cos\theta & 0 \\ \sin\alpha\cos\theta & \sin\alpha\sin\theta & \cos\alpha \end{pmatrix}.$$

The angular velocity of the rotation from $(\mathbf{i_r}, \mathbf{i}_\theta, \mathbf{i_h})$ to $(\mathbf{i}''_\mathbf{r}, \mathbf{i}''_\theta, \mathbf{i}''_\mathbf{h})$ is given by a vector sum of the two elementary rotation velocities:

$$\begin{aligned}\boldsymbol{\omega} &= \dot{\theta}\mathbf{i_h} + \dot{\alpha}\mathbf{i}'_\theta \\ &= -\dot{\alpha}\sin\theta\mathbf{i_r} + \dot{\alpha}\cos\theta\mathbf{i}_\theta + \dot{\theta}\mathbf{i_h}\,.\end{aligned} \tag{6.44}$$

The time derivative of the position vector is then given by

$$\begin{aligned}\dot{\mathbf{r}} = \mathbf{v} &= \dot{r}\mathbf{i_r} + \boldsymbol{\omega}\times\mathbf{r} \\ &= \dot{r}\mathbf{i_r} + (-\dot{\alpha}\sin\theta\mathbf{i_r} + \dot{\alpha}\cos\theta\mathbf{i}_\theta + \dot{\theta}\mathbf{i_h})\times\mathbf{r} \\ &= \dot{r}\mathbf{i_r} + r\dot{\theta}\mathbf{i}_\theta - r\dot{\alpha}\cos\theta\mathbf{i_h}\,.\end{aligned} \tag{6.45}$$

Similarly, the time derivative of the velocity vector is derived as follows:

$$\begin{aligned}\dot{\mathbf{v}} &= \ddot{r}\mathbf{i_r} + \dot{r}\dot{\theta}\mathbf{i}_\theta + r\ddot{\theta}\mathbf{i}_\theta - r\ddot{\alpha}\cos\theta\mathbf{i_h} - \dot{r}\dot{\alpha}\cos\theta\mathbf{i_h} + r\dot{\alpha}\dot{\theta}\sin\theta\mathbf{i_h} + \boldsymbol{\omega}\times\mathbf{v} \\ &= (\ddot{r} - r\dot{\theta}^2 - r\dot{\alpha}^2\cos^2\theta)\mathbf{i_r} + (r\ddot{\theta} + 2\dot{r}\dot{\theta} - r\dot{\alpha}^2\sin\theta\cos\theta)\mathbf{i}_\theta - (r\ddot{\alpha} + 2\dot{r}\dot{\alpha})\cos\theta\mathbf{i_h}\,.\end{aligned} \tag{6.46}$$

The substitution of Eqs. (6.45) and (6.46) into Eq. (6.38) results in the following scalar constituents of the equations of motion:

$$\ddot{r} - r\dot{\theta}^2 - r\dot{\alpha}^2\cos^2\theta + \frac{\mu}{r^2} = a_r\,, \tag{6.47}$$

$$r\ddot{\theta} + 2\dot{r}\dot{\theta} - r\dot{\alpha}^2\sin\theta\cos\theta = a_\theta\,, \tag{6.48}$$

and

$$-(r\ddot{\alpha} + 2\dot{r}\dot{\alpha})\cos\theta = a_n\,. \tag{6.49}$$

The set of non-linear, coupled ordinary differential equations, Eqs. (6.47)–(6.49), requires a numerical solution, such as by the Runge-Kutta method (Appendix A). However, in certain special cases, a closed-form solution can be derived, as described next.

6.3.1 Planar Manoeuvres

When the motion is confined to a plane we have $\alpha = \text{const.}$, and Eqs. (6.47)–(6.49) become the following:

$$\ddot{r} - r\dot{\theta}^2 + \frac{\mu}{r^2} = a_r\,, \tag{6.50}$$

$$r\ddot{\theta} + 2\dot{r}\dot{\theta} = a_\theta\,, \tag{6.51}$$

and

$$a_n = 0\,. \tag{6.52}$$

By defining $h = r^2\dot{\theta}$ to be the angular momentum, Eqs. (6.50) and (6.51) are alternatively expressed as follows:

$$\ddot{r} - \frac{h^2}{r^3} + \frac{\mu}{r^2} = a_r\,, \tag{6.53}$$

$$\dot{h} = ra_\theta\,. \tag{6.54}$$

6.3.2 Constant Radial Acceleration from Circular Orbit

Equations (6.53) and (6.54) are simplified for a constant radial acceleration, $a_r = a =$ const., and $a_\theta = 0$. Then it follows that $h =$ const., and the differential equations are to be solved subject to an initial condition, $r(0) = r_0$, $\dot{r}(0) = \dot{r}_0$, $\theta(0) = \theta_0$, and for a known constant acceleration, a. The solution requires an integration of the following differential equation in time:

$$\ddot{r} - \frac{h^2}{r^3} + \frac{\mu}{r^2} = a\,, \tag{6.55}$$

to obtain $r(t)$, followed by the integration of

$$\dot{\theta} = h/r^2\,, \tag{6.56}$$

to obtain $\theta(t)$. This is generally possible only by a numerical method.

Equation (6.55) is amenable to an exact closed-form solution for the velocity, $v = \sqrt{\dot{r} + r\dot{\theta}}$, in terms of the radius, r, in the case of the initial condition being a circular orbit of radius r_0, which implies $\dot{r}(0) = 0$ and $h = \sqrt{\mu r_0}$. The second-order time derivative on the left-hand side of Eq. (6.55) is expressed as follows:

$$\ddot{r} = \frac{1}{2\dot{r}}\frac{\mathrm{d}(\dot{r}^2)}{\mathrm{d}t} = \frac{1}{2}\frac{\mathrm{d}(\dot{r}^2)}{\mathrm{d}r}\,. \tag{6.57}$$

Substituting Eq. (6.57) into Eq. (6.55) and integrating with respect to r from $r(0) = r_0$, $\dot{r}(0) = 0$ to r, $\dot{r}$, we have

$$\dot{r}^2 = 2a(r - r_0) + \frac{\mu}{r}\left(2 - \frac{r}{r_0} - \frac{r_0}{r}\right) = (r - r_0)\left[2a - \frac{\mu}{r_0 r^2}(r - r_0)\right]\,. \tag{6.58}$$

The circumferential velocity component is given by

$$r\dot{\theta} = h/r = \frac{\sqrt{\mu r_0}}{r}\,. \tag{6.59}$$

Hence the net velocity is the following:

$$\begin{aligned} v = \sqrt{\dot{r}^2 + r^2\dot{\theta}^2} &= \sqrt{(r - r_0)\left[2a - \frac{\mu}{r_0 r^2}(r - r_0)\right] + \frac{\mu r_0}{r^2}} \\ &= \sqrt{2a(r - r_0) + \mu\left(\frac{2}{r} - \frac{1}{r_0}\right)}\,. \end{aligned} \tag{6.60}$$

The radius, r_e, where the escape velocity, $v = \sqrt{2\mu/r_e}$, is reached is calculated from Eq. (6.60) as follows:

$$r_e = r_0 + \frac{\mu}{2ar_0}\,. \tag{6.61}$$

By examining the two velocity components we find that the circumferential velocity component, $r\dot{\theta}$, decreases as the radius, r, increases. Therefore, the escape velocity is reached due to the increase of the radial component, $\dot{r}$, with the radius, which requires that the acceleration, a, should be sufficiently large. To find the smallest possible value of a for which the escape velocity is reached, consider the following expression for the square of the radial velocity component derived from Eqs. (6.60) and (6.61):

$$\dot{r}^2 = \frac{2a(r - r_0)}{r^2}[r^2 - (r_e - r_0)(r - r_0)] . \tag{6.62}$$

Equation (6.62) indicates that as the escape radius, r_e, increases the radial velocity decreases. Therefore, $\dot{r}$ can become zero (or negative) for large values of the escape radius, r_e. To derive the condition when $\dot{r}$ vanishes, the term within brackets on the right-hand side of Eq. (6.62) is equated to zero, which results in the following quadratic solution for r:

$$\begin{aligned} r &= \frac{1}{2}\left[(r_e - r_0) \pm \sqrt{(r_e - r_0)^2 - 4r_0(r_e - r_0)}\right] \\ &= \frac{1}{2}\left[(r_e - r_0) - \sqrt{(r_e - r_0)(r_e - 5r_0)}\right] , \end{aligned} \tag{6.63}$$

where the negative sign is taken to correspond to the case $r_0 < r < r_e$. For a real root, r, Eq. (6.63) requires that $r_e > 5r_0$. If this is the case, then Eq. (6.61) yields the requirement that $a > \mu/(8r_0^2)$.

To complete the solution, it is necessary to integrate Eqs. (6.58) and (6.59) in time to obtain $r(t)$ and $\theta(t)$. This requires numerical integration. However, the time required to achieve the escape velocity when $a > \mu/(8r_0^2)$ can be expressed in a closed form as follows:

$$t_e = \int_{r_0}^{r_e} \frac{\mathrm{d}r}{\dot{r}} = \frac{1}{\sqrt{2a}} \int_{r_0}^{r_e} \frac{r\mathrm{d}r}{\sqrt{(r - r_0)[r^2 - \frac{\mu}{2ar_0}(r - r_0)]}} . \tag{6.64}$$

The integral in Eq. (6.64) can be expressed in terms of elliptic integrals (Battin, 1999), which, however, must be numerically evaluated.

6.3.3 Constant Circumferential Acceleration from Circular Orbit

The case of a constant thrust acceleration in the circumferential direction, $a_\theta = a =$ const. and $a_r = 0$, yields the following equations of planar motion:

$$\ddot{r} - \frac{h^2}{r^3} + \frac{\mu}{r^2} = 0 , \tag{6.65}$$

$$\dot{h} = ra . \tag{6.66}$$

The substitution of Eq. (6.65) into Eq. (6.66) yields

$$\dot{h} = \frac{\mathrm{d}}{\mathrm{d}t}\sqrt{r^3\ddot{r} + \mu r} = ra . \tag{6.67}$$

Equation (6.67), being a third-order non-linear differential equation, needs numerical integration in time to find the solution $r(t)$ for a given constant a, and subject to the initial condition, $r(0)$, $\dot{r}(0)$, and $\ddot{r}(0)$.

Let the applied acceleration be small, $| a | \ll \mu/r^2$, and the orbit be initially circular prior to the thrust application, which corresponds to the initial condition $r(0) = r_0$, $\dot{r}(0) = 0$, $\ddot{r}(0) = 0$, and implies $h(0) = \sqrt{\mu r_0}$. The small thrust acceleration amounts to the approximation, $|\ddot{r}| \ll \mu/r^2$; hence the following is obtained from Eq. (6.67):

$$\frac{\mathrm{d}}{\mathrm{d}t}\sqrt{r^3\ddot{r} + \mu r} \simeq \frac{\mathrm{d}}{\mathrm{d}t}\sqrt{\mu r} = ra\,. \tag{6.68}$$

The assumption $|\ddot{r}| \ll \mu/r^2$ must be carefully employed; otherwise it may lead to the erroneous result that the radial velocity, $\dot{r}$, is constant. Taking the time derivative on the left-hand side of the Eq. (6.68) results in the following:

$$\frac{1}{2}\sqrt{\frac{\mu}{r}}\dot{r} \simeq ra\,,$$

or

$$\dot{r} \simeq \frac{2a}{\sqrt{\mu}}r^{3/2}\,. \tag{6.69}$$

Integrating Eq. (6.69) from $r(0) = r_0$ to $r(t) = r$, we have

$$\frac{1}{\sqrt{r}} \simeq \frac{1}{\sqrt{r_0}} - \frac{a}{\sqrt{\mu}}t\,,$$

or

$$r \simeq \frac{r_0}{\left(1 - at\frac{\sqrt{r_0}}{\mu}\right)^2}\,. \tag{6.70}$$

The approximation of Eq. (6.68) applied to Eq. (6.65) results in $h \simeq \sqrt{\mu r}$, and yields the following approximate solution for the square of the orbital speed:

$$v^2 = \dot{r}^2 + \frac{h^2}{r^2} \simeq \frac{4a^2}{\mu}r^3 + \frac{\mu}{r}\,. \tag{6.71}$$

To calculate the escape radius, r_e, we put $v = \sqrt{2\mu/r}$ in Eq. (6.71), resulting in

$$r_e \simeq \left(\frac{\mu^2}{4a^2}\right)^{1/4} = \sqrt{\frac{\mu}{2a}}\,. \tag{6.72}$$

Hence the escape radius is independent of the initial radius, r_0, and depends only on the thrust acceleration, a.

The time until escape, t_e, is approximated by integrating Eq. (6.69) from $r(0) = r_0$ to $r(t_e) = r_e$ to yield

$$t_e \simeq \frac{\sqrt{\mu}}{a}\left(\frac{1}{\sqrt{r_0}} - \frac{1}{\sqrt{r_e}}\right) = \frac{\sqrt{\mu}}{a}\left[\frac{1}{\sqrt{r_0}} - \left(\frac{2a}{\mu}\right)^{1/4}\right]\,. \tag{6.73}$$

Example 6.3.1 *To check the accuracy of the approximation of Eq. (6.68), simulate the escape radius for Earth ($\mu = 398600.4$ km^3/s^2) spacecraft powered by an engine of constant circumferential thrust acceleration, $a = 10^{-6}$ km/s^2 in an initial circular orbit of radius $r_0 =$ 100000 km. The Runge-Kutta simulation (Appendix A) of the exact equations, Eqs. (6.65) and*

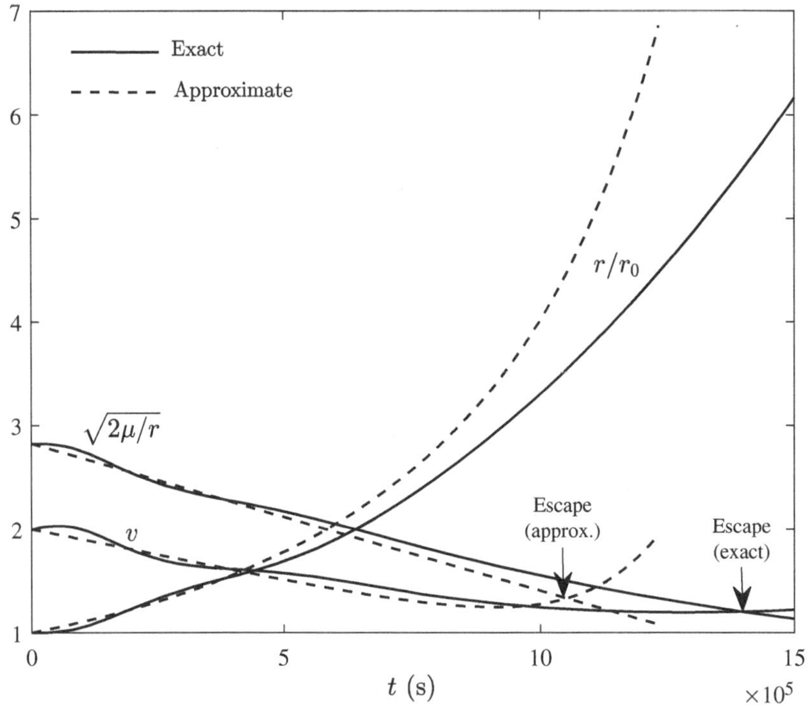

Figure 6.8 Plots of simulated (exact) and approximate escape trajectory for a constant circumferential acceleration, $a = 10^{-6}$ km/s^2, applied to an initially circular orbit of radius $r_0 = 100{,}000$ km around Earth.

(6.66), is plotted in Fig. 6.8 as solid lines, and compared with the approximate solution (dashed lines) obtained by applying the assumption of Eq. (6.68). The figure reveals an escape time of $t_e = 1397351.5$ *s and the corresponding escape radius as* $r_e = 5.465 \times 10^5$ *km, for which the escape velocity is* $v_e = \sqrt{2\mu/r_e} = 1.208$ *km/s. In comparison, the approximate values calculated by Eqs. (6.72) and (6.73) are as follows:*

$$r_e = \sqrt{\frac{\mu}{2a}} = 446430.51 \text{ km} \; ; \qquad t_e = \frac{\sqrt{\mu}}{a}\left[\frac{1}{\sqrt{r_0}} - \left(\frac{2a}{\mu}\right)^{1/4}\right] = 1051584.7 \text{ s}.$$

The approximate escape velocity is $v_e = \sqrt{2\mu/r_e} = 1.336$ *km/s. Hence, there is a significant error (about 18–25%) in values calculated by the approximation of Eq. (6.68). The reason for this error is that whereas the approximation* $a \ll \mu/r^2$ *is valid for* $r = r_0$*, with* $\mu/r_0^2 = 3.986 \times 10^{-5}$ km/s^2*, it becomes increasingly invalid as the escape radius is approached, for which we have* $\mu/r_e^2 = 2 \times 10^{-6}$ km/s^2*, which is only twice the value of a.*

6.3.4 Constant Tangential Acceleration from Circular Orbit

If a constant thrust acceleration a is applied along the instantaneous velocity direction, $\mathbf{i}_v = \mathbf{v}/v$, then Eq. (6.38) becomes the following:

$$\frac{d\mathbf{v}}{dt} + \frac{\mu}{r^3}\mathbf{r} = a\mathbf{i}_v . \tag{6.74}$$

Select a direction $\mathbf{i}_n$ normal to $\mathbf{i}_v$, such that $\mathbf{i}_n \times \mathbf{i}_v = \mathbf{i}_h = \mathbf{h}/h$, as shown in Fig. 6.9. Thus the right-handed coordinate frame, $(\mathbf{i}_n, \mathbf{i}_v, \mathbf{i}_h)$, is used to resolve the acceleration vector as follows:

$$\begin{aligned}\frac{d\mathbf{v}}{dt} &= \dot{v}\mathbf{i}_v + \boldsymbol{\omega} \times \mathbf{v} \\ &= \dot{v}\mathbf{i}_v + (\dot{\theta} - \dot{\phi})\mathbf{i}_h \times \mathbf{i}_v = \dot{v}\mathbf{i}_v - v\dot{\theta}\mathbf{i}_n .\end{aligned} \tag{6.75}$$

The radius vector is resolved in this velocity-normal frame as follows in terms of the flight-path angle, ϕ (Fig. 6.9):

$$\mathbf{r} = r\sin\phi\mathbf{i}_v + r\cos\phi\mathbf{i}_n . \tag{6.76}$$

The substitution of Eqs. (6.75) and (6.76) into Eq. (6.74) yields the following scalar components:

$$\dot{v} + \frac{\mu}{r^2}\sin\phi = a \tag{6.77}$$

and

$$-v(\dot{\theta} - \dot{\phi}) + \frac{\mu}{r^2}\cos\phi = 0 . \tag{6.78}$$

Here $v(\dot{\theta} - \dot{\phi})$ is termed the *centripetal acceleration* as it is directed towards the central body along $-\mathbf{i}_n$, and $\dot{v}$ is called the *tangential acceleration*, being directed along $\mathbf{i}_v$.

Now consider an arc length, s, measured along the flight direction (Fig. 6.9), such that $v = \dot{s}$, and

$$ds = dr + rd\theta . \tag{6.79}$$

Thus it is noted that

$$\frac{dr}{ds} = \sin\phi ; \qquad r\frac{d\theta}{ds} = \cos\phi , \tag{6.80}$$

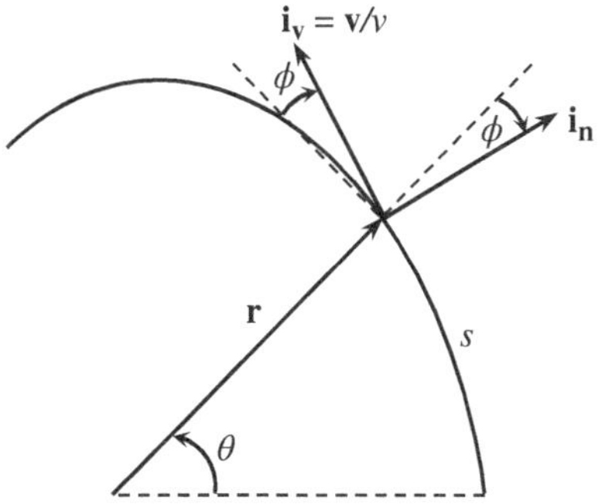

Figure 6.9 The tangential ($\mathbf{i}_v$) and normal ($\mathbf{i}_n$) directions for resolving the motion variables, with arc length s measured along the orbit.

and we have from Eq. (6.77):

$$v\frac{\mathrm{d}v}{\mathrm{d}s} + \frac{\mu}{r^2}\frac{\mathrm{d}r}{\mathrm{d}s} = a\,. \tag{6.81}$$

The variation of the flight-path angle, ϕ, with s is derived as follows by differentiating the first of Eq. (6.80):

$$\frac{\mathrm{d}^2 r}{\mathrm{d}s^2} = \cos\phi\frac{\mathrm{d}\phi}{\mathrm{d}s}\,, \tag{6.82}$$

which, substituted into Eq. (6.78), along with Eq. (6.80), yields the following:

$$rv^2\frac{\mathrm{d}^2 r}{\mathrm{d}s^2} + \left(\frac{\mu}{r} - v^2\right)\cos^2\phi = 0\,. \tag{6.83}$$

or

$$rv^2\frac{\mathrm{d}^2 r}{\mathrm{d}s^2} + \left(\frac{\mu}{r} - v^2\right)\left[1 - \left(\frac{\mathrm{d}r}{\mathrm{d}s}\right)^2\right] = 0\,. \tag{6.84}$$

An examination of Eq. (6.84) reveals that the radial acceleration vanishes when either $\mathrm{d}r/\mathrm{d}s$ becomes unity (i.e., $\phi = \pm\pi/2$) or the orbit becomes circular, i.e., $v^2 = \mu/r$. If neither of these conditions is satisfied, and $v^2 < \mu/r$, then the flight path continuously curves towards the central body, i.e., $\mathrm{d}^2r/\mathrm{d}s^2 < 0$. Since the speed continuously increases due to the tangential acceleration according to Eq. (6.81), the inward curvature is maintained until the condition $v^2 = \mu/r$ is reached. This natural inward curvature is a feature of all trajectories when a tangential thrust is applied, and is termed a *gravity turn*. A launch to a circular orbit is an example of a gravity-turn trajectory.

Equations (6.81) and (6.84) are differential equations in the independent variable, s, written in an abbreviated form as follows:

$$vv' + \frac{\mu}{r^2}r' = a \tag{6.85}$$

and

$$rv^2r'' + \left(v^2 - \frac{\mu}{r}\right)[(r')^2 - 1] = 0\,, \tag{6.86}$$

where the prime denotes the derivative with respect to s; that is $(.)' = \mathrm{d}(.)/\mathrm{d}s$. As in the case of a constant radial thrust, these equations are amenable to a closed-form solution for the velocity, v, as a function of r, beginning from a circular orbit which corresponds to the initial condition $t = 0$, $r(0) = r_0$, $v(0) = v_0 = \sqrt{\mu/r_0}$, $r'(0) = 0$. In that case, the solution to Eq. (6.85) is obtained to be the following:

$$v^2 = 2as + \mu\left(\frac{2}{r} - \frac{1}{r_0}\right)\,. \tag{6.87}$$

For the escape velocity, $v = \sqrt{2\mu/r}$, to be reached, Eq. (6.86) requires that

$$2rr'' + (r')^2 - 1 = 0\,. \tag{6.88}$$

Equation (6.85) is a non-linear differential equation to be integrated for $r(s)$, and generally requires a numerical solution even for the case of starting from a circular orbit. However, for a small thrust acceleration (such as that provided by an ion-propulsion rocket), $|\,a\,| \ll \mu/r^2$, we have $r'' \simeq 0$; hence the following approximation can be applied in Eq. (6.86):

$$v^2 \simeq \frac{\mu}{r}\,, \tag{6.89}$$

which implies that an approximately circular shape of the orbit is maintained. The substitution of Eq. (6.89) into Eq. (6.85) produces the following approximate solution for the radius:

$$r \simeq \frac{r_0}{1 - 2as/v_0^2} = \frac{r_0}{1 - 2ar_0 s/\mu}, \tag{6.90}$$

whose substitution into Eq. (6.88) and solving for s yields the following arc length until escape, s_e:

$$s_e \simeq \frac{v_0^2}{2a}\left[1 - \frac{(20a^2 r_0^2)^{1/4}}{v_0}\right], \tag{6.91}$$

which produces the following escape radius, r_e, by Eq. (6.90):

$$r_e \simeq \frac{r_0 v_0}{(20a^2 r_0^2)^{1/4}}. \tag{6.92}$$

The approximate velocity for the small thrust acceleration is given by substituting Eq. (6.90) into Eq. (6.87):

$$v^2 \simeq v_0^2 - 2as. \tag{6.93}$$

The approximate time until escape is calculated as follows:

$$t_e = \int_0^{s_e} \frac{ds}{v} = \frac{v_0}{a}\left[1 - \left(\frac{20a^2 r_0^2}{v_0^4}\right)^{1/8}\right]. \tag{6.94}$$

Exercises

1. A spacecraft is in an Earth orbit of eccentricity, $e = 0.7$. When the altitude is 500 km and the speed is 7.4 km/s, a velocity impulse of magnitude 500 m/s is applied at an angle 60° from the initial velocity direction in order to increase the speed and flight-path angle simultaneously, but without changing the orbital plane. Determine the new orbit of the spacecraft. (*Ans.* $a = 6969.325$ km, $e = 0.738246$.)

2. A launch site is located at the latitude of 30°. What should be the direction of launch (ψ) for achieving an orbital inclination of 45°, without requiring a plane-change manoeuvre in orbit?

3. A satellite is to be launched from Cape Canaveral (USA) ($\delta = 28.5°$) to an equatorial orbit of altitude 200 km.
   (a) Assuming an impulsive launch with $e = 0.1$ for the launch trajectory, and neglecting the atmospheric effects, estimate the minimum velocity change to be imparted by the rocket engine, as well as the direction of launch in Earth-fixed frame. (Non-impulsive thrust and atmospheric drag will cause an increase of about 1.5 km/s in the actual value). (*Ans.* 10.98416 km/s.)
   (b) If the satellite weighs 1000 kg, calculate the mass of the liquid oxygen/liquid hydrogen propellant required for the launch. (*Ans.* 10405.66 kg.)

4. Repeat Exercise 3 for a launch from Plesetsk (Russia) ($\delta = 62.8°$).

5. What is the minimum velocity impulse magnitude required for capturing a spacecraft from a hyperbolic Earth orbit of $a = -18110$ km, $e = 1.369$, to an elliptic orbit of $a = 10441.547$ km, $e = 0.36$? (*Ans.* 2.88048 km/s.)

6. Repeat Exercise 5 for the same orbital manoeuvre around Mars ($\mu = 42828.3$ km^3/s^2).

7. A spacecraft is in a circular Earth orbit of radius 15000 km. Calculate the velocity impulse required to send it to an elliptic orbit of perigee altitude 500 km and apogee radius 22000 km. (*Ans.* 2.77333 km/s.)

8. A spacecraft is in a circular orbit of radius 8000 km around Earth. Calculate the velocity impulse required to send it to a hyperbolic orbit of perigee radius 8000 km and eccentricity $e = 1.5$. (*Ans.* 1.5864 km/s.)

9. A spacecraft is required to undergo a plane change by 20°, as well as a speed change from 7.3 km/s to 8.2 km/s, with a 5° increase in the flight-path angle. Determine the total velocity impulse required when:
   (a) The speed and the flight-path angle are changed together, while the plane change is carried out separately. (*Ans.* 3.66024 km/s.)
   (b) All the three changes are carried out simultaneously. (*Ans.* 2.9083 km/s.)

10. In Exercise 9, determine the mass of the propellant saved by option (b) when compared to option (a), if the final mass of the spacecraft is 2000 kg and its rocket propellant is liquid oxygen/kerosene.

11. A spacecraft's velocity vector in a Moon-centered celestial, equatorial frame is observed to be $\mathbf{v} = -2\mathbf{I} + \mathbf{J} - \mathbf{K}$ km/s when its altitude, declination, and right ascension are 500 km, 30°, and 300°, respectively. A velocity impulse of $\Delta\mathbf{v} = \mathbf{I} - \mathbf{J} + \mathbf{K}$ km/s is now applied to the spacecraft. Calculate the position and velocity vectors 10 minutes after the impulse is applied. (For the Moon: $\mu = 4902.8$ km^3/s^2, $r_0 = 1737.1$ km.)

12. Spacecraft A has landed on the Moon, while spacecraft B is orbiting the moon in a circular orbit of altitude 185 km. Spacecraft A must now make a rendezvous with B. Assuming that the two spacecraft are in the same orbital plane, and that the initial launch speed of A from the Moon's surface is 1.5 km/s, determine spacecraft A's flight-path angle at launch from the Moon's surface such that it reaches B's orbit tangentially. (*Ans.* 15.487°.)

13. Calculate the total velocity change, waiting time, and flight time for a Hohmann transfer from a parking orbit of 200 km altitude inclined at 13.5° to a geostationary target position which is currently 50° ahead of the spacecraft. (*Ans.* $\Delta v = 4.6546$ km/s, $t_w = 4858.288$ s, $t_H = 18931.96$ s.)

14. An Earth communications satellite has attained an equatorial orbit. However, the apogee altitude is 41756 km and eccentricity is 0.0661. What is the minimum velocity change required to place the satellite in the geostationary orbit? (*Ans.* 0.10002 km/s.)

15. Estimate the total time and the total velocity change required for a Mars mission if the initial angular positions of Earth and Mars from the rendezvous point located at an ecliptic node of the Mars orbit are 90° and 140°, respectively, as measured along the respective

orbits. Assume that the Mars orbital plane is inclined at 2° to the ecliptic, Mars's sidereal orbital period is 1.881 times that of Earth, the planetary orbits are circular, and the planetary gravitation on the spacecraft is negligible. (*Ans.* $t_H + t_w = 267.198$ mean solar days, $\Delta v =$ 6.43635 km/s.)

16. Estimate the total time and total velocity change required for a Venus mission if the angular positions of Earth and Venus from the rendezvous point at launch are 60° and 120°, respectively, measured along the respective orbits. Assume that Venus's orbital plane is inclined at 3.4° to the ecliptic, Venus's sidereal orbital period is 0.6152 times that of Earth, the planetary orbits are circular, and planetary gravitation on the spacecraft is negligible.

17. For an outer bi-elliptic (OB) transfer, derive an expression for the total time of transfer, t_{OB}. Furthermore, assume that a rendezvous is to be made with a target in the circular orbit of the larger radius, r_2 (Fig. 6.7), and the target initially ($t = 0$) leads the manoeuvring spacecraft by an angle, $(\theta_H + \Delta\theta)$, with θ_H being the correct phase angle for rendezvous by Hohmann transfer. Suppose instead of waiting until $\Delta\theta$ becomes zero before initiating the Hohmann transfer, the spacecraft is immediately launched into an OB transfer such that the rendezvous is made at time $t = t_{OB}$. Derive an expression for the intermediate radius, r_3, and show that if the synodic period is greater than half the period of the final orbit, $T_s \geq \pi\sqrt{r_2^3/\mu}$, the total time for rendezvous by OB transfer, t_{OB}, is smaller than that required by Hohmann transfer, $(t_w + t_H)$, for phase error larger than a critical value, $\Delta\theta > \Delta\theta_i$. (*Hint*: $\Delta\theta_i = 3\pi[1 - (r_1/r_2)^{3/2}]$.)

18. A spacecraft originally in a circular orbit of radius 8000 km around Earth begins to apply a constant radial acceleration of $a = 10^{-4}$ km/s^2. Determine the maximum possible radius reached by the spacecraft. (*Ans.* 8274.855 km.)

19. In Exercise 18, if the engine applying the radial acceleration is switched off at the instant of reaching the maximum radius, determine the orbital elements a and e of the resulting orbit.

References

Battin RH 1999. *An Introduction to the Mathematics and Methods of Astrodynamics*. American Institute of Aeronautics and Astronautics (AIAA) Education Series, Reston, VA.

7

Relative Motion in Orbit

Relative motion between two spacecraft around a central body is an important problem of space guidance and navigation. Whenever a rendezvous, docking, and undocking of two (or more) spacecraft, targeting and interception of satellites, or the ejection of small items from spacecraft are considered, the relative motion between objects in different orbits about the same central body becomes relevant. While the relative motion can be modelled by separately solving the Keplerian equations of motion of any two spacecraft, and then subtracting their respective inertial position and velocity vectors to obtain the relative position and velocity vectors, the problem is considerably simplified when the separation between the spacecraft is relatively small. In such a case, the equations of relative motion in orbit can be approximated by a set of linear differential equations called the *Hill-Clohessy-Wiltshire equations* (covered in Section 7.1). In this chapter, these approximate, linearized equations are derived and solved for special problems, such as interception and rendezvous of spacecraft around a spherical, central body.

Consider two spacecraft idealized as points A and B orbiting a spherical, central body of gravitational constant μ, as shown in Fig. 7.1. Spacecraft A is in a known Keplerian orbit while spacecraft B can be manoeuvred by the application of a thrust acceleration, $\mathbf{u}$. Since spacecraft A is used as a reference for measuring the relative position of spacecraft B, spacecraft A is called the *target*, and the orbit of A is called the reference (or target) orbit. Spacecraft B is termed the *manoeuvring spacecraft*, whose relative position and velocity with respect to A can be controlled. The positions of the two spacecraft (which are approximated as being of negligible masses and dimensions compared to those of the central body) in an inertial reference frame are described by the radius vectors, $\mathbf{r}_A(t)$ and $\mathbf{r}_B(t)$, respectively, which are governed by the following equations of motion:

$$\ddot{\mathbf{r}}_A + \frac{\mu}{r_A^3}\mathbf{r}_A = \mathbf{0}\,, \tag{7.1}$$

$$\ddot{\mathbf{r}}_B + \frac{\mu}{r_B^3}\mathbf{r}_B = \mathbf{u}\,. \tag{7.2}$$

The position of spacecraft B relative to the centre of mass of spacecraft A is described by the *relative position* vector, $\boldsymbol{\rho} = \mathbf{r}_B - \mathbf{r}_A$ (Fig. 7.1). The target's orbit is defined by the classical orbital elements, $a, e, i, \Omega, \omega, t_0$, calculated from the initial position, $\mathbf{r}_A(0)$, and velocity, $\dot{\mathbf{r}}_A(0)$, and resolved in an inertial reference frame, $(\mathbf{I}, \mathbf{J}, \mathbf{K})$, with its origin, o, at the centre of the attracting body. The relative position of the manoeuvring spacecraft B is resolved in a local-horizon

Foundations of Space Dynamics, First Edition. Ashish Tewari.

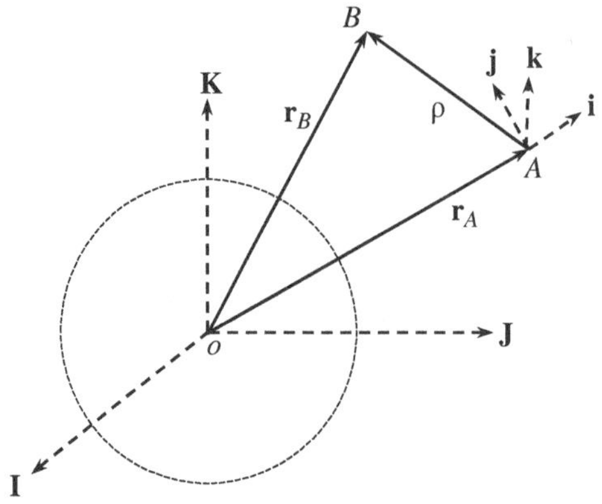

Figure 7.1 Geometry of relative motion in orbit.

frame, $(\mathbf{i}, \mathbf{j}, \mathbf{k})$, which has its origin at the target, A (Fig. 7.1), as follows:

$$\boldsymbol{\rho} = \mathbf{r}_B - \mathbf{r}_A = x\mathbf{i} + y\mathbf{j} + z\mathbf{k} , \tag{7.3}$$

where $\mathbf{i} = \mathbf{r}_A/r_A$, $\mathbf{j}$ is along the circumferential orbital direction of the target, and $\mathbf{k} = \mathbf{i} \times \mathbf{j}$ is normal to the target's orbital plane and parallel to its angular momentum vector.

A *relative velocity* vector, $\boldsymbol{v}$, is defined to be the velocity of B as seen by an observer on A, and is resolved in the moving frame, $(\mathbf{i}, \mathbf{j}, \mathbf{k})$, as follows:

$$\boldsymbol{v} = \frac{\delta \boldsymbol{\rho}}{\delta t} = \dot{x}\mathbf{i} + \dot{y}\mathbf{j} + \dot{z}\mathbf{k} , \tag{7.4}$$

where $\delta(.)/\delta t$ represents taking the time derivative of (.) by considering $(\mathbf{i}, \mathbf{j}, \mathbf{k})$ to be a stationary frame.

The inertial velocity difference between B and A is obtained by taking into account the velocity of the frame, $(\mathbf{i}, \mathbf{j}, \mathbf{k})$, as follows:

$$\begin{aligned} \dot{\mathbf{r}}_B &= \frac{\delta \mathbf{r}_B}{\delta t} + \boldsymbol{\omega} \times \mathbf{r}_B \\ \dot{\mathbf{r}}_A &= \frac{\delta \mathbf{r}_A}{\delta t} + \boldsymbol{\omega} \times \mathbf{r}_A \\ \dot{\mathbf{r}}_B - \dot{\mathbf{r}}_A &= \frac{\delta(\mathbf{r}_B - \mathbf{r}_A)}{\delta t} + \boldsymbol{\omega} \times (\mathbf{r}_B - \mathbf{r}_A) , \end{aligned} \tag{7.5}$$

or

$$\dot{\mathbf{r}}_B - \dot{\mathbf{r}}_A = \frac{\delta \boldsymbol{\rho}}{\delta t} + \boldsymbol{\omega} \times \boldsymbol{\rho} = \boldsymbol{v} + \boldsymbol{\omega} \times \boldsymbol{\rho} , \tag{7.6}$$

where $\boldsymbol{\omega}$ is the angular velocity of a moving frame, $(\mathbf{i}, \mathbf{j}, \mathbf{k})$, given by $\boldsymbol{\omega} = \dot{\theta}_A \mathbf{k}$, with θ_A being the true anomaly of the target in its orbit.

The differential equation governing the relative motion is derived by subtracting Eq. (7.1) from Eq. (7.2) to yield the following:

$$\ddot{\boldsymbol{\rho}} + \frac{\mu}{r_B^3}\mathbf{r}_B - \frac{\mu}{r_A^3}\mathbf{r}_A = \mathbf{u} . \tag{7.7}$$

The solution to Eq. (7.7), $\rho(t)$, subject to an initial condition, $[\rho(0), \dot{\rho}(0)]$, and a known acceleration input, $\mathbf{u}$, requires knowledge of the individual position vectors of the two spacecraft, $\mathbf{r}_A(t)$ and $\mathbf{r}_B(t)$. Therefore, the exact solution to the problem of relative motion requires separately solving the equations of motion for the individual spacecraft, Eqs. (7.1) and (7.2), subject to an initial condition, $(\mathbf{r}_A(0), \dot{\mathbf{r}}_A(0), \mathbf{r}_B(0), \dot{\mathbf{r}}_B(0))$, and a known input, $\mathbf{u}$. Being non-linear, coupled differential equations, the simultaneous solution to Eqs. (7.1) and (7.2) is difficult, requiring a numerical solution. However, if the separation of the two spacecraft is relatively small at all times – that is, $|\rho(t)| \ll r_A(t)$ and $|\rho(t)| \ll r_B(t)$, $t \geq 0$ – then Eq. (7.7) is greatly simplified, resulting in the following approximation:

$$\ddot{\rho} \simeq \frac{\partial \mathbf{f}}{\partial \rho}\rho + \mathbf{u} , \tag{7.8}$$

where

$$\mathbf{f} = \frac{\mu}{r_A^3}\mathbf{r}_A - \frac{\mu}{r_B^3}\mathbf{r}_B . \tag{7.9}$$

The square matrix, $\partial \mathbf{f}/\partial \rho$, evaluated at the target (or the reference) orbit, $\rho = \mathbf{0}$, is called the *gravity-gradient matrix*, and using the definition, $\mathbf{r}_B = \mathbf{r}_A + \rho$, is evaluated as follows:

$$\begin{aligned}
\frac{\partial \mathbf{f}}{\partial \rho} &= \lim_{\rho \to 0} \mu \frac{\partial}{\partial \rho}\left(\frac{\mathbf{r}_A}{r_A^3} - \frac{\mathbf{r}_A + \rho}{|\mathbf{r}_A + \rho|^3}\right) \\
&= -\lim_{\rho \to 0}\left[\frac{\mu}{|\mathbf{r}_A + \rho|^3}\mathbf{I} + \mu(\mathbf{r}_A + \rho)\frac{\partial |\mathbf{r}_A + \rho|^{-3}}{\partial \rho}\right] \\
&= -\frac{\mu}{r_A^3}\mathbf{I} + \lim_{\rho \to 0}\frac{3\mu(\mathbf{r}_A + \rho)(\mathbf{r}_A + \rho)^T}{|\mathbf{r}_A + \rho|^5} \\
&= -\frac{\mu}{r_A^3}\mathbf{I} + \frac{3\mu \mathbf{r}_A \mathbf{r}_A^T}{\mathbf{r}_A^5} .
\end{aligned} \tag{7.10}$$

The substitution of Eq. (7.10) into Eq. (7.8) results in the following approximate equation of relative motion:

$$\ddot{\rho} \simeq -\left(\frac{\mu}{r_A^3}\mathbf{I} - \frac{3\mu \mathbf{r}_A \mathbf{r}_A^T}{\mathbf{r}_A^5}\right)\rho + \mathbf{u} , \tag{7.11}$$

with $\mathbf{r}_A(t)$ available as the reference Keplerian orbit of the target spacecraft A satisfying Eq. (7.1).

When the reference frame of the approximate motion is chosen such that one of the axes is always aligned with the radius vector of the target, $\mathbf{r}_A(t)$, a further simplification occurs in Eq. (7.11). Let such a reference frame, $(\mathbf{i}, \mathbf{j}, \mathbf{k})$, be selected such that $\mathbf{i} = \mathbf{r}_A/r_A$ is pointed radially outward, $\mathbf{k}$ is normal to the target's orbital plane and aligned with its angular momentum vector, and $\mathbf{j} = \mathbf{k} \times \mathbf{i}$ completes the right-handed triad. Therefore, we have

$$\begin{aligned}
\mathbf{r}_A &= r_A \mathbf{i} \\
\rho &= x\mathbf{i} + y\mathbf{j} + z\mathbf{k} ,
\end{aligned} \tag{7.12}$$

where (x, y, z) are the components of $\boldsymbol{\rho}$ resolved in $(\mathbf{i}, \mathbf{j}, \mathbf{k})$. A substitution of Eq. (7.12) into Eqs. (7.10) and (7.11) yields the following:

$$\frac{\partial \mathbf{f}}{\partial \boldsymbol{\rho}} = \frac{\mu}{r_A^3}\begin{pmatrix} 2 & 0 & 0 \\ 0 & -1 & 0 \\ 0 & 0 & -1 \end{pmatrix}, \tag{7.13}$$

$$\ddot{\boldsymbol{\rho}} \simeq \frac{\mu}{r_A^3}\begin{pmatrix} 2 & 0 & 0 \\ 0 & -1 & 0 \\ 0 & 0 & -1 \end{pmatrix}\begin{Bmatrix} x \\ y \\ z \end{Bmatrix} = \frac{\mu}{r_A^3}\begin{Bmatrix} 2x \\ -y \\ -z \end{Bmatrix} + \begin{Bmatrix} u_x \\ u_y \\ u_z \end{Bmatrix}, \tag{7.14}$$

where $\mathbf{u} = u_x\mathbf{i} + u_y\mathbf{j} + u_z\mathbf{k}$ is the applied thrust acceleration. The left-hand side of Eq. (7.14) must be evaluated in an inertial frame by taking into account the angular velocity of the moving reference frame, $(\mathbf{i}, \mathbf{j}, \mathbf{k})$. The Section 7.1 considers a special case, where the target A moves in a circular orbit.

7.1 Hill-Clohessy-Wiltshire Equations

Consider the special case of a target spacecraft A in a circular orbit of radius $r_A =$ const. and frequency $n = \sqrt{\mu/r_A^3}$. The relative position of the manoeuvring spacecraft B from the target is given by Eq. (7.12) as $\boldsymbol{\rho} = (x, y, z)^T$ in the moving, Cartesian reference frame, $(\mathbf{i}, \mathbf{j}, \mathbf{k})$, as described above, and is governed by Eq. (7.14), which in the present case has a constant gravity-gradient matrix, $\partial \mathbf{f}/\partial \boldsymbol{\rho}$, and becomes the following:

$$\ddot{\boldsymbol{\rho}} \simeq n^2\begin{Bmatrix} 2x \\ -y \\ -z \end{Bmatrix} + \begin{Bmatrix} u_x \\ u_y \\ u_z \end{Bmatrix}. \tag{7.15}$$

To evaluate the left-hand side of Eq. (7.14), the time derivative of $\boldsymbol{\rho}$ is taken as follows:

$$\begin{aligned} \dot{\boldsymbol{\rho}} &= \dot{x}\mathbf{i} + \dot{y}\mathbf{j} + \dot{z}\mathbf{k} + n\mathbf{k} \times (x\mathbf{i} + y\mathbf{j} + z\mathbf{k}) \\ &= (\dot{x} - ny)\mathbf{i} + (\dot{y} + nx)\mathbf{j} + \dot{z}\mathbf{k}\,. \end{aligned} \tag{7.16}$$

Then the second time derivative of $\boldsymbol{\rho}$ is derived to be the following:

$$\begin{aligned} \ddot{\boldsymbol{\rho}} &= (\ddot{x} - n\dot{y})\mathbf{i} + (\ddot{y} + n\dot{x})\mathbf{j} + \ddot{z}\mathbf{k} + n\mathbf{k} \times [(\dot{x} - ny)\mathbf{i} + (\dot{y} + nx)\mathbf{j} + \dot{z}\mathbf{k}] \\ &= (\ddot{x} - 2n\dot{y} - n^2x)\mathbf{i} + (\ddot{y} + 2n\dot{x} - n^2y)\mathbf{j} + \ddot{z}\mathbf{k}\,. \end{aligned} \tag{7.17}$$

Finally, the substitution of Eq. (7.17) into Eq. (7.15) produces the following equations of approximate relative motion, called the *Hill-Clohessy-Wiltshire* (HCW) equations:

$$\begin{Bmatrix} \ddot{x} - 2n\dot{y} - n^2x \\ \ddot{y} + 2n\dot{x} - n^2y \\ \ddot{z} \end{Bmatrix} = n^2\begin{Bmatrix} 2x \\ -y \\ -z \end{Bmatrix} + \begin{Bmatrix} u_x \\ u_y \\ u_z \end{Bmatrix}, \tag{7.18}$$

which are expressed in the following scalar form:

$$\begin{aligned}\ddot{x} - 2n\dot{y} - 3n^2x &= u_x \\ \ddot{y} + 2n\dot{x} &= u_y \\ \ddot{z} + n^2z &= u_z \,.\end{aligned} \tag{7.19}$$

The HCW equations, Eq. (7.19), are linear, and hence can be solved in a closed form for a known acceleration input vector, $(u_x(t), u_y(t), u_z(t))$, $t \geq 0$, and with a given initial condition, $(x(0), y(0), z(0), \dot{x}(0), \dot{y}(0), \dot{z}(0))$. The relative positional coordinates of the HCW model are termed the *radial displacement*, x, the *in-track displacement*, y, and the *out-of-plane displacement*, z.

The HCW equations can be alternatively derived using the binomial theorem for the approximation, $x \ll r_A$, $y \ll r_A$, and $z \ll r_A$, in Eq. (7.2) as follows. The velocity of the manoeuvring spacecraft B is the following:

$$\begin{aligned}\dot{\mathbf{r}}_B &= \dot{\mathbf{r}}_A + \dot{\rho} \\ &= n\mathbf{k} \times r_A\mathbf{i} + \dot{x}\mathbf{i} + \dot{y}\mathbf{j} + \dot{z}\mathbf{k} + n\mathbf{k} \times (x\mathbf{i} + y\mathbf{j} + z\mathbf{k}) \\ &= (\dot{x} - ny)\mathbf{i} + (\dot{y} + nr_A + nx)\mathbf{j} + \dot{z}\mathbf{k} \,,\end{aligned} \tag{7.20}$$

and the acceleration of the manoeuvring spacecraft is given by

$$\begin{aligned}\ddot{\mathbf{r}}_B &= (\ddot{x} - n\dot{y})\mathbf{i} + (\ddot{y} + n\dot{x})\mathbf{j} + \ddot{z}\mathbf{k} + n\mathbf{k} \times [(\dot{x} - ny)\mathbf{i} + (\dot{y} + nr_A + nx)\mathbf{j} + \dot{z}\mathbf{k}] \\ &= [(\ddot{x} - 2n\dot{y} - n^2(r_A + x)]\mathbf{i} + (\ddot{y} + 2n\dot{x} - n^2y)\mathbf{j} + \ddot{z}\mathbf{k} \,.\end{aligned} \tag{7.21}$$

The acceleration due to gravity of the manoeuvring spacecraft is approximated using the binomial theorem as follows:

$$\begin{aligned}-\frac{\mu}{r_B^3}\mathbf{r}_B &= -\mu\frac{(r_A + x)\mathbf{i} + y\mathbf{j} + z\mathbf{k}}{(r_A^2 + 2xr_A + x^2 + y^2 + z^2)^{3/2}} \\ &\simeq -r_A\frac{\mu}{r_A^3}\mathbf{i} - \frac{\mu}{r_A^3}(-2x\mathbf{i} + y\mathbf{j} + z\mathbf{k}) \\ &= n^2(2x - r_A)\mathbf{i} - n^2y\mathbf{j} - n^2z\mathbf{k} \,.\end{aligned} \tag{7.22}$$

Equations (7.21) and (7.22) substituted into Eq. (7.2) produce the HCW equations, Eq. (7.19).

It is evident from Eq. (7.19) that the out-of-plane motion, $z(t)$, is independent of the motion projected in the target's orbital plane, $[(x(t), y(t))]$. For the unforced motion, $u_z = 0$, the last HCW equation in Eq. (7.19) can be directly integrated in time to yield

$$z(t) = z(0)\cos nt + \frac{\dot{z}(0)}{n}\sin nt \,, \tag{7.23}$$

where $z(0), \dot{z}(0)$ is the out-of-plane initial condition. Similarly, the in-plane relative motion can be solved for the unforced system, $u_x = u_y = 0$, independently of $z(t)$, resulting in

$$x(t) = x(0)(4 - 3\cos nt) + \frac{\dot{x}(0)}{n}\sin nt + \frac{2\dot{y}(0)}{n}(1 - \cos nt), \tag{7.24}$$

$$y(t) = y(0) + 6x(0)(\sin nt - nt) + \frac{2\dot{x}(0)}{n}(\cos nt - 1) + \frac{\dot{y}(0)}{n}(4\sin nt - 3nt) \,. \tag{7.25}$$

Example 7.1.1 *A spacecraft in a circular orbit of period 90 min. around Earth releases a small instrument package with an out-of-plane velocity 1 m/s. Determine the position and the velocity of the package relative to the spacecraft after 30 min.*

Assuming the package is not large enough to affect the spacecraft's orbital motion by its release, one can employ the HCW model for this calculation. The orbital frequency is $n = 2\pi/(90 \times 60) = 0.00116355$ *rad./s, which makes the final angular dispacement of the target* $nt = 2\pi/3$. *The initial condition is given by* $z(0) = 0, \dot{z}(0) = 1$ *m/s, with all the remaining state variables being zero at* $t = 0$. *Because in the HCW equations the out-of-plane motion is decoupled from the coplanar motion, Eq. (7.23) gives us the following:*

$$z(t) = z(0)\cos nt + \frac{\dot{z}(0)}{n}\sin nt = 744.294 \text{ m} \,,$$

$$\dot{z}(t) = -nz(0)\sin nt + \dot{z}(0)\cos nt = \cos(2\pi/3) = -0.5 \text{ m/s} \,.$$

The radius of the target's orbit is $r_A = (\mu/n^2)^{\frac{1}{3}} = 6652.556$ *km. Since the displacement is small relative to the radius of the target's orbit,* $z(t)/r_A \simeq 0.0001$, *the calculation by the HCW model is accurate.*

Example 7.1.2 *Spacecraft* B *is observed to have the following initial position and velocity relative to a space station* A, *which is in a circular orbit of radius 7000 km around Earth:*

$$x(0) = -5 \text{ km} \,; \quad \dot{x}(0) = 0.01 \text{ km/s} \,; y(0) = 1 \text{ km} \,; \quad \dot{y}(0) = -0.02 \text{ km/s} \,,$$

and $z(0) = \dot{z}(0) = 0$. *Estimate the position and velocity of* B *relative to the space station 20 minutes after the observation is taken.*

With $r_A = 7000$ *km, the frequency of the space station's orbit is* $n = \sqrt{\mu/r_A^3} = 0.001078$ *rad./s, and the final angular displacement of the station at* $t = 1200$ *s is given by* $nt = 1.29361$ *rad. Since the out-of-plane displacement and velocity component are zero at* $t = 0$, *they will remain zero for* $t \geq 0$ *by the HCW model. Using the HCW model, spacecraft* B *has the following coplanar orbital position and velocity relative to* A *at* $t = 20$ *minutes:*

$$x(t) = x(0)(4 - 3\cos nt) + \frac{\dot{x}(0)}{n}\sin nt + \frac{2\dot{y}(0)}{n}(1 - \cos nt) = -33.925 \text{ km}$$

$$\dot{x}(t) = 3nx(0)\sin nt\dot{x}(0)\cos nt + 2\dot{y}(0)\sin nt = -0.0513 \text{ km/s}$$

$$y(t) = y(0) + 6x(0)(\sin nt - nt) + \frac{2\dot{x}(0)}{n}(\cos nt - 1) + \frac{\dot{y}(0)}{n}(4\sin nt - 3nt)$$
$$= -1.9006 \text{ km}$$

$$\dot{y}(t) = 6nx(0)(\cos nt - 1) - 2\dot{x}(0)\sin nt + \dot{y}(0)(4\cos nt - 3) = 0.0424 \text{ km/s} \,.$$

The relative displacement, $\rho(t) = \sqrt{x^2(t) + y^2(t)} = 33.978$ *km, is quite small (less than 0.5%) compared to the target's orbital radius,* r_A*; hence the calculation by the HCW model is valid.*

7.2 Linear State-Space Model

The HCW model given by Eq. (7.19) is linear, and can be represented by the following linear time-invariant state-space model :

$$\frac{\delta \boldsymbol{\rho}}{\delta t} = \mathbf{v}$$
$$\frac{\delta \mathbf{v}}{\delta t} = \mathbf{A}_1 \boldsymbol{\rho} + \mathbf{A}_2 \mathbf{v} + \mathbf{a} \,, \tag{7.26}$$

where $\boldsymbol{\rho} = (x, y, z)^T$, $\mathbf{v} = (\dot{x}, \dot{y}, \dot{z})^T$, $\mathbf{a} = (a_x, a_y, a_z)^T$, and

$$\mathbf{A}_1 = \begin{pmatrix} 3n^2 & 0 & 0 \\ 0 & 0 & 0 \\ 0 & 0 & -n^2 \end{pmatrix}$$
$$\mathbf{A}_2 = \begin{pmatrix} 0 & 2n & 0 \\ -2n & 0 & 0 \\ 0 & 0 & 0 \end{pmatrix} . \tag{7.27}$$

By defining the state vector to be $\mathbf{X} = (\boldsymbol{\rho}^T, \mathbf{v}^T)^T = (x, y, z, \dot{x}, \dot{y}, \dot{z})^T$, Eq. (7.26) can be expressed as follows:

$$\frac{\delta \mathbf{X}}{\delta t} = \mathbf{A}\mathbf{X} + \mathbf{B}\mathbf{a} \,, \tag{7.28}$$

where $\delta \mathbf{X}/\delta t = (\delta \boldsymbol{\rho}^T/\delta t, \delta \mathbf{v}^T/\delta t)^T = (\dot{x}, \dot{y}, \dot{z}, \ddot{x}, \ddot{y}, \ddot{z})^T$, and

$$\mathbf{A} = \begin{pmatrix} \mathbf{0} & \mathbf{I} \\ \mathbf{A}_1 & \mathbf{A}_2 \end{pmatrix} ; \qquad \mathbf{B} = \begin{pmatrix} \mathbf{0} \\ \mathbf{I} \end{pmatrix} . \tag{7.29}$$

The state solution to Eq. (7.29), $\mathbf{X}(t)$, $t \geq 0$, for a given initial state, $\mathbf{X}(t_0) = (\boldsymbol{\rho}^T(t_0), \mathbf{v}^T(t_0))^T$, and with a known input profile, $\mathbf{a}(t)$, $t \geq 0$, can be derived in a closed form to be the following:

$$\mathbf{X}(t) = \boldsymbol{\Phi}(t, t_0)\mathbf{X}(t_0) + \int_{t_0}^{t} \boldsymbol{\Phi}(t - \tau, 0)\mathbf{B}\mathbf{a}(\tau)\mathrm{d}\tau \,, \tag{7.30}$$

where $\boldsymbol{\Phi}(t, t_0)$ is the *state-transition matrix* representing the evolution of the *unforced* linear system of Eq. (7.29), i.e.,

$$\frac{\delta \mathbf{X}}{\delta t} = \mathbf{A}\mathbf{X} \,, \tag{7.31}$$

from the time t_0 to time t as follows:

$$\mathbf{X}(t) = \boldsymbol{\Phi}(t, t_0)\mathbf{X}(t_0) \,. \tag{7.32}$$

Since $\boldsymbol{\Phi}(t, t_0)$ is a property of the unforced linear system, it depends solely on the state dynamics matrix, $\mathbf{A}$, and is expressed as the following *matrix exponential*:

$$\boldsymbol{\Phi}(t, t_0) = e^{\mathbf{A}(t-t_0)} = \mathbf{I} + \mathbf{A}(t - t_0) + \frac{1}{2}\mathbf{A}^2(t - t_0)^2 + \cdots + \frac{1}{k!}\mathbf{A}^k(t - t_0)^k + \cdots \,. \tag{7.33}$$

It is evident from its definition given by Eq. (7.32) that a state-transition matrix must have the following properties:

$$\mathbf{\Phi}(t_0, t_0) = \mathbf{I}, \tag{7.34}$$

$$\frac{\delta \mathbf{\Phi}}{\delta t}(t, t_0) = \mathbf{A}\mathbf{\Phi}(t, t_0), \tag{7.35}$$

$$\mathbf{\Phi}(t_0, t) = \mathbf{\Phi}^{-1}(t, t_0), \tag{7.36}$$

$$\mathbf{\Phi}(t_0, t) = \mathbf{\Phi}(t_0, t_1)\mathbf{\Phi}(t_1, t)\ ; \qquad (t_0 \leq t_1 \leq t). \tag{7.37}$$

The application of Laplace transform, $\mathcal{L}(.)$, to Eq. (7.31) with the initial condition $\mathbf{X}(0) = \mathbf{X}_0$ provides a method for calculating the state-transition matrix, $\mathbf{\Phi}(t, 0)$, as follows:

$$\mathcal{L}(\frac{\delta \mathbf{X}}{\delta t}) = s\mathbf{X}(s) - \mathbf{X}_0 = \mathbf{A}\mathbf{X}(s)\ , \tag{7.38}$$

where s denotes the Laplace variable and $\mathbf{X}(s) = \mathcal{L}(\mathbf{X}(t))$. Solving for $\mathbf{X}(s)$ in Eq. (7.38) and taking the inverse Laplace transform, $\mathcal{L}^{-1}(.)$, we have

$$\mathbf{X}(t) = \mathcal{L}^{-1}(s\mathbf{I} - \mathbf{A})^{-1})\mathbf{X}_0\ , \tag{7.39}$$

or

$$\mathbf{\Phi}(t, 0) = e^{\mathbf{A}t} = \mathcal{L}^{-1}(s\mathbf{I} - \mathbf{A})^{-1})\ . \tag{7.40}$$

Once $\mathbf{\Phi}(t, 0) = e^{\mathbf{A}t}$ is known, the general state-transition matrix, $\mathbf{\Phi}(t, t_0) = e^{\mathbf{A}(t-t_0)}$ is easily obtained by replacing t with $t - t_0$.

The state-transition matrix for the HCW model is calculated to be the following:

$$\mathbf{\Phi}(t, 0) = e^{\mathbf{A}t} = \begin{pmatrix} \mathbf{\Phi}_{\rho\rho} & \mathbf{\Phi}_{\rho v} \\ \mathbf{\Phi}_{v\rho} & \mathbf{\Phi}_{vv} \end{pmatrix}, \tag{7.41}$$

where

$$\mathbf{\Phi}_{\rho\rho} = \begin{bmatrix} 4 - 3\cos nt & 0 & 0 \\ 6(\sin nt - nt) & 1 & 0 \\ 0 & 0 & \cos nt \end{bmatrix}, \tag{7.42}$$

$$\mathbf{\Phi}_{\rho v} = \begin{bmatrix} \frac{1}{n}\sin nt & \frac{2}{n}(1 - \cos nt) & 0 \\ \frac{2}{n}(\cos nt - 1) & \frac{1}{n}(4\sin nt - 3nt) & 0 \\ 0 & 0 & \frac{1}{n}\sin nt \end{bmatrix}, \tag{7.43}$$

$$\mathbf{\Phi}_{v\rho} = \frac{\delta \mathbf{\Phi}_{\rho\rho}}{\delta t} = \begin{bmatrix} 3n\sin nt & 0 & 0 \\ 6n(\cos nt - 1) & 0 & 0 \\ 0 & 0 & -n\sin nt \end{bmatrix}, \tag{7.44}$$

$$\mathbf{\Phi}_{vv} = \frac{\delta \mathbf{\Phi}_{\rho v}}{\delta t} = \begin{bmatrix} \cos nt & 2\sin nt & 0 \\ -2\sin nt & (4\cos nt - 3) & 0 \\ 0 & 0 & \cos nt \end{bmatrix}. \tag{7.45}$$

This expression of the state-transition matrix agrees with the solution to the unforced HCW equations given by Eqs. (7.23)–(7.25), which can be expressed as follows:

$$\boldsymbol{\rho}(t) = \boldsymbol{\Phi}_{\rho\rho}(t)\boldsymbol{\rho}(0) + \boldsymbol{\Phi}_{\rho v}(t)\boldsymbol{v}(0) \tag{7.46}$$

and

$$\boldsymbol{v}(t) = \boldsymbol{\Phi}_{v\rho}(t)\boldsymbol{\rho}(0) + \boldsymbol{\Phi}_{vv}(t)\boldsymbol{v}(0)\,. \tag{7.47}$$

7.3 Impulsive Manoeuvres About a Circular Orbit

The solution to the linearized model of relative motion referenced to a circular orbit given by the Hill-Clohessy-Wiltshire equations in Section 7.2 is useful when considering the response to a given initial condition, $\mathbf{X}(0) = \boldsymbol{\rho}(0), \dot{\boldsymbol{\rho}}(0)$, when the applied inputs, $\mathbf{a}$, are zero. Such an unforced response is the relative trajectory between any two instances where impulsive inputs are applied on spacecraft B. Called a *zero-thrust* (or null-thrust) trajectory, such a trajectory can be used to devise an impulsive control strategy for achieving a desired objective for a given set of initial and final conditions.

7.3.1 *Orbital Rendezvous*

A common manoeuvre involving relative orbital motion is the *rendezvous* between two (or more) spacecraft which are initially in different orbits. A rendezvous requires that beginning from an initial separation, $\boldsymbol{\rho}(0)$, and relative velocity, $\boldsymbol{v}(0)$, at time $t = 0$, the spacecraft must arrive at the same orbital position and velocity at a future time, $t = t_f$, where the relative separation and velocity vectors vanish, i.e., $\boldsymbol{\rho}(t_f) = \boldsymbol{v}(t_f) = \mathbf{0}$. A two-impulse rendezvous manoeuvre is the simplest to design, consisting of the first impulse applied to the manoeuvring spacecraft at $t = 0$ to change its initial velocity such that the relative position vector vanishes at $t = t_f$, and a second impulse applied at $t = t_f$ to make the relative velocity vector zero.

When the rendezvous involves a target in a circular orbit and the initial separation is small, one can employ the HCW model to derive the two impulses required for the manoeuvre. Let the target be in a circular orbit of frequency n. For the given (small) initial separation and relative velocity, $\boldsymbol{\rho}(0)$ and $\boldsymbol{v}(0)$, the first velocity impulse at $t = 0$ is the difference between the required initial velocity, $\boldsymbol{v}_r(0)$, and the actual initial velocity given by

$$\Delta\mathbf{v}_1 = \boldsymbol{v}_r(0) - \boldsymbol{v}(0)\,, \tag{7.48}$$

where the required initial relative velocity is calculated by Eq. (7.46) for making $\boldsymbol{\rho}(t_f) = \mathbf{0}$ as follows:

$$\boldsymbol{v}_r(0) = -\boldsymbol{\Phi}_{\rho v}{}^{-1}(t_f)\boldsymbol{\Phi}_{\rho\rho}(t_f)\boldsymbol{\rho}(0)\,. \tag{7.49}$$

The second velocity impulse is applied at $t = t_f$ to make $\boldsymbol{v}(t_f)$ vanish, and is calculated by Eq. (7.47) to be the following:

$$\Delta\mathbf{v}_2 = -\boldsymbol{v}(t_f) = -\boldsymbol{\Phi}_{v\rho}\boldsymbol{\rho}(0) - \boldsymbol{\Phi}_{vv}\boldsymbol{v}_r(0)\,. \tag{7.50}$$

The matrix inversion, $\mathbf{\Phi}_{\rho v}{}^{-1}(t)$, required in Eq. (7.49) is easily carried out by recognizing that the last row and column represent the out-of-plane motion, which is decoupled from the coplanar motion given by the first two rows and columns. Therefore we have

$$\mathbf{\Phi}_{\rho v}{}^{-1}(t) = \begin{bmatrix} \frac{n}{\Delta}(4\sin nt - 3nt) & \frac{2n}{\Delta}(\cos nt - 1) & 0 \\ \frac{2n}{\Delta}(1 - \cos nt) & \frac{n}{\Delta}\sin nt & 0 \\ 0 & 0 & \frac{n}{\sin nt} \end{bmatrix}, \tag{7.51}$$

where

$$\Delta = 8(1 - \cos nt) - 3nt \sin nt .$$

The inverse exists for $nt \neq (k-1)\pi$, $k = 1, 2, 3, \ldots$. Thus the final time, t_f, cannot be taken to be an integral multiple of half the target's orbital period, π/n. However, if the relative motion is planar – i.e., the out-of-plane motion is absent, $z(0) = \dot{z}(0) = 0$ – then odd multiples of π/n are allowed for t_f.

Example 7.3.1 *Spacecraft* A *is in a circular orbit of frequency, n, and radius, c. Another spacecraft,* **B**, *has an initial separation and relative velocity with respect to* A *at* $t = 0$ *given by*

$$x(0) = 0.001c\ ; \quad \dot{x}(0) = -0.003nc\ ; \quad z(0) = -0.002c\ ; \quad \dot{z}(0) = 0.001nc\ ,$$

and $y(0) = \dot{y}(0) = 0$. *Estimate the two velocity impulses applied on* B *such that it makes a rendezvous with* A *at* $t_f = \frac{\pi}{2n}$.

For the given rendezvous time, we have $\cos nt_f = 0$ *and* $\sin nt_f = 1$. *Substituting these into Eq. (7.51) yields the following matrix inverse,* $\mathbf{\Phi}_{\rho v}{}^{-1}(t_f)$:

$$\mathbf{\Phi}_{\rho v}{}^{-1}(t) = \begin{bmatrix} \frac{n}{\Delta}(4 - 3\pi/2) & -\frac{2n}{\Delta} & 0 \\ \frac{2n}{\Delta} & \frac{n}{\Delta} & 0 \\ 0 & 0 & n \end{bmatrix},$$

where $\Delta = 8 - 3\pi/2$. *Furthermore, we have*

$$\mathbf{\Phi}_{\rho\rho} = \begin{bmatrix} 4 & 0 & 0 \\ 6(1 - \pi/2) & 1 & 0 \\ 0 & 0 & 0 \end{bmatrix},$$

and $\boldsymbol{\rho}(0) = (0.001c,\ 0, -0.002c)^T$. *Therefore, the required initial relative velocity is the following:*

$$\begin{aligned} \mathbf{v}_r(0) &= -\mathbf{\Phi}_{\rho v}{}^{-1}(t_f)\mathbf{\Phi}_{\rho\rho}(t_f)\boldsymbol{\rho}(0) \\ &= -\frac{0.001nc}{\Delta}\begin{Bmatrix} 4 \\ 14 - 3\pi \\ 0 \end{Bmatrix} . \end{aligned}$$

Since the initial relative velocity is $\mathbf{v}(0) = (-0.003nc,\ 0,\ 0.001nc)^T$*, the first velocity impulse is given by*

$$\Delta\mathbf{v}_1 = \mathbf{v}_r(0) - \mathbf{v}(0) = 0.001nc \begin{Bmatrix} -\frac{4}{\Delta} + 3 \\ -\frac{(14-3\pi)}{\Delta} \\ -1 \end{Bmatrix}.$$

To calculate the second velocity impulse, the following transition matrices are required:

$$\mathbf{\Phi}_{v\rho} = \begin{pmatrix} 3n & 0 & 0 \\ -6n & 0 & 0 \\ 0 & 0 & -n \end{pmatrix}$$

$$\mathbf{\Phi}_{vv} = \begin{pmatrix} 0 & 2 & 0 \\ -2 & -3 & 0 \\ 0 & 0 & 0 \end{pmatrix}.$$

Substituting these into Eq. (7.50) we have

$$\Delta\mathbf{v}_2 = -\mathbf{\Phi}_{v\rho}\rho(0) - \mathbf{\Phi}_{vv}\mathbf{v}_r(0) = -0.001nc \begin{Bmatrix} 3 + \frac{(6\pi-28)}{\Delta} \\ -6 + \frac{(50-9\pi)}{\Delta} \\ 2 \end{Bmatrix}.$$

7.4 Keplerian Relative Motion

The exact relative motion of spacecraft B relative to the target A around a spherical body is described by separately solving the equations of Keplerian motion, Eqs. (7.1) and (7.2), for the two spacecraft, and then finding the difference in their position and inertial velocity vectors, $\rho = \mathbf{r}_B - \mathbf{r}_A$, and $\mathbf{v} = \dot{\mathbf{r}}_B - \dot{\mathbf{r}}_A$. It is assumed that the initial conditions of the two spacecraft at time $t = 0$ are known, and given by the relative position and relative inertial velocity, $\rho(0)$ and $\mathbf{v}(0)$, respectively. The propagation in time of the positions of the two spacecraft in their respective orbits is computed by an appropriate method, such as the Lagrange's coefficients (Chapter 3):

$$\begin{Bmatrix} \mathbf{r}_A(t) \\ \dot{\mathbf{r}}_A(t) \end{Bmatrix} = \begin{pmatrix} f_A(t) & g_A(t) \\ \dot{f}_A(t) & \dot{g}_A(t) \end{pmatrix} \begin{Bmatrix} \mathbf{r}_A(0) \\ \dot{\mathbf{r}}_A(0) \end{Bmatrix} \tag{7.52}$$

and

$$\begin{Bmatrix} \mathbf{r}_B(t) \\ \dot{\mathbf{r}}_B(t) \end{Bmatrix} = \begin{pmatrix} f_B(t) & g_B(t) \\ \dot{f}_B(t) & \dot{g}_B(t) \end{pmatrix} \begin{Bmatrix} \mathbf{r}_B(0) \\ \dot{\mathbf{r}}_B(0) \end{Bmatrix}. \tag{7.53}$$

The transformation from the respective orbital perifocal frames to a common celestial reference frame requires the rotation matrices, $\mathbf{C}*_A$ and $\mathbf{C}*_B$, for the two spacecraft. Finally, the relative position and relative inertial velocity, $\rho(t)$ and $\mathbf{v}(t)$, are transformed from the celestial frame,

$(\mathbf{I}, \mathbf{J}, \mathbf{K})$, to the target's local-horizon frame, $(\mathbf{i}, \mathbf{j}, \mathbf{k})$, using the rotation matrix, $\mathbf{C_{LH}}$, of the target. Such a computational procedure can be automated in a suitable programming environment.

Example 7.4.1 *A satellite, **B**, has deviated from its original position, A, in a geosynchronous circular orbit. This deviation is resolved at $t = 0$ in a frame attached to A by $x = -100$ km, $y = 250$ km, $z = 30$ km, and $\dot{x} = \dot{y} = \dot{z} = 0$. Find the relative position and velocity of the satellite with respect to A after 10 hours.*

We will first carry out the calculation using the approximate HCW model, and then compare the result with that obtained exactly from the Keplerian solution. As calculated in Chapter 3, the radius and frequency of a geosynchronous Earth orbit are $r_A = 42164.168$ km and $n = 7.292116 \times 10^{-5}$ rad./s, respectively. Then the HCW model predicts the following relative position and velocity at $t = 36000$ s:

$$\rho(t) = \mathbf{\Phi}_{\rho\rho}(t)\rho(0) + \mathbf{\Phi}_{\rho v}(t)\mathbf{v}(0) = -660.876\mathbf{i} + 1528.829\mathbf{j} - 26.088\mathbf{k} \text{ km}$$

$$\mathbf{v}(t) = \mathbf{\Phi}_{v\rho}(t)\rho(0) + \mathbf{\Phi}_{vv}(t)\mathbf{v}(0) = -0.0108\mathbf{i} + 0.0818\mathbf{j} - 0.0011\mathbf{k} \text{ km/s} .$$

Next, the Keplerian motion of the two spacecraft is determined. Since the target, A, is in a circular orbit, its true anomaly, θ_A, can be taken to be zero when $t = 0$. Furthermore, we select the celestial frame, $(\mathbf{I}, \mathbf{J}, \mathbf{K})$, such that it coincides with the orientation of the target-fixed frame, $(\mathbf{i}, \mathbf{j}, \mathbf{k})$, at $t = 0$. The initial radius and velocity vectors of B are therefore given by

$$\begin{aligned}\mathbf{r}_B(0) &= \rho(0) + \mathbf{r}_A(0) = [x(0) + r_A]\mathbf{I} + y(0)\mathbf{J} + z(0)\mathbf{K} \\ &= 42064.168\mathbf{I} + 250\mathbf{J} + 30\mathbf{K} \text{ km}\end{aligned}$$

$$\begin{aligned}\dot{\mathbf{r}}_B(0) &= \frac{\delta \mathbf{r}_B}{\delta t}(0) + \boldsymbol{\omega} \times \mathbf{r}_B(0) = n\mathbf{k} \times \mathbf{r}_B(0) \\ &= n(42064.168\mathbf{J} - 250\mathbf{I}) \text{ km/s}.\end{aligned}$$

Then the orbital elements of B are calculated as follows using the methods from Chapter 5:

$$r_0 = |\, \mathbf{r}_B(0) \,| = 42064.9216 \text{ km} ; \quad v_0 = |\, \dot{\mathbf{r}}_B(0) \,| = 3.067422 \text{ km/s}$$

$$a = \frac{-\mu}{v_0^2 - 2\mu/r_0} = 41770.634 \text{ km}$$

$$\mathbf{h} = \mathbf{r}_B(0) \times \dot{\mathbf{r}}_B(0) = -92.021\mathbf{I} - 0.5469\mathbf{J} + 129030.835\mathbf{K} \text{ (km}^2\text{/s)}$$

$$h = |\, \mathbf{h} \,| = 129030.868 \text{ km}^2\text{/s} ; \quad p = h^2/\mu = 41768.5606 \text{ km}$$

$$e = \sqrt{1 - p/a} = 0.0070453$$

$$i = \cos^{-1}[(\mathbf{h}/h) \cdot \mathbf{K}] = 0.000713 \text{ rad.}$$

$$\mathbf{e} = \frac{1}{\mu}[\dot{\mathbf{r}}_B(0) \times \mathbf{h}] - \mathbf{r}_B(0)/r_0 = -0.0070452\mathbf{I} - 4.187 \times 10^{-5}\mathbf{J} - 5.025 \times 10^{-6}\mathbf{K}$$

$$\mathbf{i}_n = \frac{\mathbf{K} \times (\mathbf{h}/h)}{|\, \mathbf{K} \times (\mathbf{h}/h) \,|} = 0.0059432\mathbf{I} - 0.99998\mathbf{J} ; \quad \Omega = -1.56485 \text{ rad.}$$

$$\cos\omega = \mathbf{i}_n \cdot (\mathbf{e}/e) = 0 ; \quad \sin\omega = [\mathbf{i}_n \times (\mathbf{e}/e)] \cdot \mathbf{h}/h = -1 ; \quad \omega = 3\pi/2$$

$$\cos\theta(0) = \frac{\mathbf{r}_B(0)\cdot\mathbf{e}}{er_0} = -1\ ; \quad \theta(0) = \pi = E(0)$$

$$t_0 = -\pi/n_B = -\frac{\pi}{\sqrt{\mu/a^3}} = -42480.303 \text{ s}\ .$$

Given the orbital elements of B *calculated above, the eccentric anomaly at* $t = 36000$ *s is computed by solving Kepler's equation as follows:*

$$M = n_B(t - t_0) = 5.8039403 \text{ rad.} = E - e\sin E\ ; \quad E = 5.80067122 \text{ rad.}$$

Therefore, we have the following perifocal position and velocity of B *at* $t = 36000$ *s:*

$$r_B(t) = a(1 - e\cos E) = 41509.9448 \text{ km}\ ; \quad v_B(t) = \sqrt{\frac{2\mu}{r_B} - \frac{\mu}{a}} = 3.1084492 \text{ km/s}$$

$$\mathbf{r}_B(t) = a(\cos E - e)\mathbf{i_e} + \sqrt{ap}\sin E\mathbf{i_p} = 36707.442\mathbf{i_e} - 19381.415\mathbf{i_p} \text{ (km)}$$

$$\dot{\mathbf{r}}_B(t) = -\frac{\sqrt{\mu a}}{r}\sin E\mathbf{i_e} + \frac{\sqrt{\mu p}}{r}\cos E\mathbf{i_p} = 1.44237\mathbf{i_e} + 2.75355\mathbf{i_p} \text{ (km/s)}.$$

To transform the perifocal position and velocity to the celestial frame, it is necessary to calculate the rotation matrix, $\mathbf{C}*$, *as follows:*

$$\mathbf{C}* = \mathbf{C_3}^T(\Omega)\mathbf{C_1}^T(i)\mathbf{C_3}^T(\omega) = \begin{pmatrix} -0.9999821 & 0.0059432 & -0.00071317 \\ -0.0059432 & -0.9999823 & -4.2386\times10^{-6} \\ -0.0007132 & 0 & 1 \end{pmatrix}.$$

Multiplication of the perifocal vectors with this rotation matrix yields the following celestial position and velocity vectors:

$$\mathbf{r}_B(t) = -36821.972\mathbf{I} + 19162.913\mathbf{J} - 26.179\mathbf{K} \text{ (km)}$$

$$\dot{\mathbf{r}}_B(t) = -1.42598\mathbf{I} - 2.76207\mathbf{J} - 0.001029\mathbf{K} \text{ (km/s)}\ .$$

The position and velocity vectors of the target, A, *at* $t = 36000$ *s are easily calculated as follows because of its circular orbit with* $\mathbf{k} = \mathbf{K}$*:*

$$\mathbf{r}_A(t) = r_A\mathbf{i} = r_A(\cos\theta_A\mathbf{I} + \sin\theta_A\mathbf{J})$$

$$\dot{\mathbf{r}}_A(t) = nr_A\mathbf{j} = nr_A(-\sin\theta_A\mathbf{I} + \cos\theta_A\mathbf{J})\ ,$$

where $\theta_A = nt = 2.625162$ *rad. is the angle moved by* A *in its orbit. The relative distance and velocity of* B *relative to* A *at* $t = 36000$ *s are thus the following:*

$$\rho(t) = \mathbf{r}_B(t) - \mathbf{r}_A(t) = -156.557\mathbf{I} - 1656.896\mathbf{J} - 26.179\mathbf{K} \text{ (km)}$$

$$\dot{\mathbf{r}}_B(t) - \dot{\mathbf{r}}_A(t) = 0.092223\mathbf{I} - 0.088386\mathbf{J} - 0.0010287\mathbf{K} \text{ (km/s)}\ .$$

The transformation to the local-horizon frame attached to A *involves multiplication of these vectors with the following rotation matrix:*

$$\mathbf{C_{LH}} = \begin{pmatrix} \cos\theta_A & \sin\theta_A & 0 \\ -\sin\theta_A & \cos\theta_A & 0 \\ 0 & 0 & 1 \end{pmatrix},$$

Table 7.1 Comparison of relative position and velocity in a geosynchronous orbit calculated by the approximate HCW model and the exact Keplerian equations in Example 7.4.1

Method	$\rho(t)$ (km)	$v(t)$ (km/s)
HCW	$-660.876\mathbf{i} + 1528.829\mathbf{j} - 26.088\mathbf{k}$	$-0.0108\mathbf{i} + 0.0818\mathbf{j} - 0.0011\mathbf{k}$
Keplerian	$-682.001\mathbf{i} + 1518.1205\mathbf{j} - 26.179\mathbf{k}$	$-0.01314\mathbf{i} + 0.08105\mathbf{j} - 0.001029\mathbf{k}$

which yields

$$\rho(t) = -682.001\mathbf{i} + 1518.1205\mathbf{j} - 26.179\mathbf{k} \text{ (km)}$$

$$\dot{\mathbf{r}}_B(t) - \dot{\mathbf{r}}_A(t) = -0.12384\mathbf{i} + 0.03132\mathbf{j} - 0.00103\mathbf{k} \text{ (km/s)} .$$

Finally, the relative velocity of B *seen by an observer on* A, $\mathbf{v}$, *is calculated by Eq. (7.6) as follows:*

$$\begin{aligned} \mathbf{v}(t) &= \dot{\mathbf{r}}_B(t) - \dot{\mathbf{r}}_A(t) - \boldsymbol{\omega} \times \rho(t) \\ &= -0.013136\mathbf{i} + 0.081054\mathbf{j} - 0.001029\mathbf{k} . \end{aligned}$$

A comparison of $\boldsymbol{\rho}(t)$ *and* $\mathbf{v}(t)$ *with the values computed by the approximate HCW model reveals an agreement between the two calculations, which are listed in Table 7.1. The planar displacements,* (x, y), *calculated by the HCW model are off by tens of km due to the large time,* t, *involved.*

Exercises

1. A small free-flying experiment package is ejected radially outward with speed 0.01 times the orbital velocity from a satellite in a circular Earth orbit of period 90 minutes. Using the approximate equations of relative motion (Hill-Clohessy-Wiltshire), find the position and velocity of the package relative to the satellite after 90 minutes.
2. Repeat Exercise 1 for the same orbit around the moon ($\mu = 4902.8$ km^3/s^2).
3. Compare the solution obtained in Exercise 1 with that obtained after solving the exact Keplerian equation of motion for the package.
4. Two spacecraft are in the same circular, Earth orbit of period two hours, but are separated by an in-track displacement of $x = 5486$ m. In order to make a rendezvous, the leading spacecraft fires a retro-rocket, instantaneously decreasing its speed by 3.05 m/s. Using the Hill-Clohessy-Wiltshire model, estimate the relative distance and relative velocity after 30 minutes. Will the rendezvous be finally made? If so, what is the final time of the rendezvous?
5. To avoid some orbiting debris, a GEO satellite applies an in-track velocity impulse. Sometime later, the debris has passed and the satellite has the following relative position and velocity with respect to its initial position:

$$y = -120.6 \text{ km} , \qquad x = 71.12 \text{ km} , \qquad \dot{y} = -10.636 \text{ m/s} , \qquad \dot{x} = 20.221 \text{ m/s}.$$

The satellite now initiates a two-impulse manoeuvre to return to its original location in two hours. What are the magnitudes and directions of the two impulses obtained from the Hill-Clohessy-Wiltshire model?

6. A space station is in a circular Earth orbit of period 90 minutes. A *Soyuz* capsule is in the same orbit, but lagging behind the space station by in-track displacement of 100 km. Estimate the magnitudes and directions of velocity impulses on the *Soyuz* for a rendezvous between the two spacecraft in 30 minutes.

7. In Exercise 12 of Chapter 6, suppose spacecraft A initially fails to make a rendezvous with spacecraft B as required, but has reached a circular orbit of the same radius as B. If the initial in-track displacement of A from B is 10 km, estimate the magnitudes and directions of two velocity impulses on spacecraft A for a rendezvous in 30 minutes.

8

Lambert's Problem

Transferring a spacecraft between two given positions around a spherical body in a prescribed time requires the determination of the Keplerian orbit connecting the two positions, and is called *Lambert's problem*. Such a problem is typical in the guidance of spacecraft and ballistic missiles, as well as in the orbital determination of space objects from two observed positions separated by a specific time interval. Lambert's problem attracted the attention of the greatest mathematicians, such as Euler, Gauss, and Lagrange, and is responsible for many advances in analytical and computational mechanics. A solution to Lambert's problem was first presented by Gauss in his *Theoria Motus* (1801), where he determined the orbit of the asteroid Ceres from its angular position observed at three time instants.

8.1 Two-Point Orbital Transfer

Lambert's problem must be solved whenever an orbit between two positions separated by an angle, θ, and observed at different times, $\mathbf{r}_1(t_1)$ and $\mathbf{r}_2(t_2)$, is to be determined, as shown in Fig. 8.1. In this chapter, the convention $r_2 > r_1$ is taken without any loss of generality. The change to the case $r_1 > r_2$ is easily made by reversing the nomenclature. Furthermore, the case $r_1 = r_2$ is trivially handled by a circular transfer orbit. Lambert's problem can be alternatively stated as finding the orbit which involves a change in the position vector between $\mathbf{r}_1$ and $\mathbf{r}_2$ in time $t_{12} = t_2 - t_1$. Lambert's problem is a special *two-point boundary value problem* which requires a solution, $\mathbf{r}(t)$, $\mathbf{v}(t)$, subject to the boundary conditions, $\mathbf{r}(t_1) = \mathbf{r}_1$, $\mathbf{r}(t_2) = \mathbf{r}_2$, to the two-body problem,

$$\frac{\partial \mathbf{r}}{\partial t} = \mathbf{v}, \tag{8.1}$$

$$\frac{\partial \mathbf{v}}{\partial t} + \mu \frac{\mathbf{r}}{r^3} = \mathbf{0}, \tag{8.2}$$

where $\mathbf{r}(t)$ and $\mathbf{v}(t)$ are the position and velocity vectors, respectively, measured in an inertial reference frame with origin at the centre of the spherical body of gravitational constant, μ. The orbital plane is the plane formed by the two position vectors, $\mathbf{r}_1$ and $\mathbf{r}_2$, and hence can be determined from the two given positions unless the transfer angle is $\theta = \pi$. Assuming that $\theta \neq \pi$, the normal to the orbital plane is the following:

$$\mathbf{i}_h = \pm \frac{\mathbf{r}_1 \times \mathbf{r}_2}{|\mathbf{r}_1 \times \mathbf{r}_2|}, \tag{8.3}$$

Foundations of Space Dynamics, First Edition. Ashish Tewari.

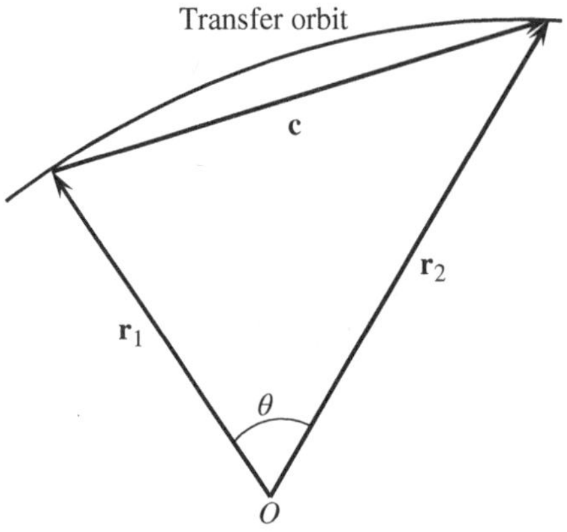

Figure 8.1 Geometry of a two-point orbital transfer.

where $\mathbf{i_h} = \mathbf{h}/h$ denotes the direction of the orbital angular momentum vector. The sign ambiguity indicates the two possible ways in which the transfer can be made, each giving a specific direction of the orbit direction, $\mathbf{h} = \mathbf{r} \times \mathbf{v}$, according to the right hand rule. When $\theta = \pi$, i.e., $\mathbf{r_1}/r_1 = -\mathbf{r_2}/r_2$, the plane of transfer is undefined, and additional information must be given to fix the orbital plane. With the orbital plane known, it remains to find the shape and size of the orbit, as well as the time of periapsis.

8.1.1 Transfer Triangle and Terminal Velocity Vectors

The *chord* connecting the two positions (Fig. 8.1) is defined by the vector, **c**, as follows:

$$\mathbf{c} = \mathbf{r_2} - \mathbf{r_1} \,. \tag{8.4}$$

The two radii, r_1 and r_2, and the chord length, c, constitute the fixed *triangle of transfer* in the orbital plane. As discussed earlier, when the orbit direction is not fixed, each transfer triangle, (r_1, r_2, c), specifies two possible ways the transfer can be made: with the transfer angle θ, and $2\pi - \theta$. When the orbit direction is fixed, the *angle of transfer* between $\mathbf{r_1}$ and $\mathbf{r_2}$, shown in Fig. 8.1, is also fixed.

Let $\mathbf{v}(t_1) = \mathbf{v_1}$ be the velocity at $t = t_1$ such that the transfer between $\mathbf{r}(t_1) = \mathbf{r_1}$ and $\mathbf{r}(t_2) = \mathbf{r_2}$ takes place in time $t_{12} = t_2 - t_1$. Then Lambert's problem requires the determination of the Lagrange's coefficients (Chapter 4), f, g, in terms of the true anomalies, θ_1 and θ_2, corresponding to the orbital radii r_1 and r_2, respectively, with $\theta = \theta_2 - \theta_1$ being the transfer angle. The given position at $t = t_2$ is given in terms of the Lagrange's coefficients by

$$\mathbf{r_2} = f\mathbf{r_1} + g\mathbf{v_1} \,, \tag{8.5}$$

where

$$f = 1 + \frac{r_2}{p}(\cos\theta - 1)$$

$$g = \frac{r_1 r_2}{h} \sin\theta$$
$$\dot{f} = \frac{\mathrm{d}f}{\mathrm{d}t} = -\frac{h}{p^2}[\sin\theta + e(\sin\theta_2 - \sin\theta_1)]$$
$$\dot{g} = \frac{\mathrm{d}g}{\mathrm{d}t} = 1 + \frac{r_1}{p}(\cos\theta - 1)\,. \tag{8.6}$$

Furthermore, we have

$$\mathbf{r_1} = f'\mathbf{r_2} + g'\mathbf{v_2}\,, \tag{8.7}$$

where

$$f' = 1 + \frac{r_1}{p}(\cos\theta - 1) = \dot{g}$$
$$g' = -\frac{r_1 r_2}{h}\sin\theta = -g\,. \tag{8.8}$$

Thus, we have

$$\mathbf{r_1} = \dot{g}\mathbf{r_2} - g\mathbf{v_2}\,. \tag{8.9}$$

By virtue of Eqs. (8.5) and (8.9), the terminal velocity vectors, $\mathbf{v_1}$ and $\mathbf{v_2}$, can be resolved into components along the chord and the local radius, v_c and v_r, respectively. It is evident from Fig. 8.1 that the chord and the local radius are not necessarily perpendicular to each other. Thus we have

$$v_r \neq \dot{r}\,,$$

and it follows that

$$\mathbf{v_1} = \frac{1}{g}[(\mathbf{r_2} - \mathbf{r_1}) + (1 - f)\mathbf{r_1}] = v_c\mathbf{i_c} + v_r\mathbf{i_{r1}}$$
$$\mathbf{v_2} = \frac{1}{g}[(\mathbf{r_2} - \mathbf{r_1}) - (1 - \dot{g})\mathbf{r_2}] = v_c\mathbf{i_c} - v_r\mathbf{i_{r2}}\,, \tag{8.10}$$

where

$$\mathbf{i_c} = \frac{\mathbf{r_2} - \mathbf{r_1}}{c}$$
$$\mathbf{i_{r1}} = \frac{\mathbf{r_1}}{r_1}$$
$$\mathbf{i_{r2}} = \frac{\mathbf{r_2}}{r_2} \tag{8.11}$$
$$v_c = \frac{c}{g} = \frac{ch}{r_1 r_2 \sin\theta}$$
$$v_r = \frac{h(1 - \cos\theta)}{p\sin\theta}\,,$$

where h denotes the orbital angular momentum and $p = h^2/\mu$ the orbital parameter. Hence the terminal velocity vectors depend not only on the transfer triangle, (r_1, r_2, c), but also on the orbital parameter, $p = a(1 - e^2)$. The last two equations of Eq. (8.11) imply that the parameter can be determined from the transfer triangle and the ratio of the chordal and radial velocity components as follows:

$$p = \frac{r_1 r_2 (1 - \cos\theta)}{c} \frac{v_c}{v_r} . \tag{8.12}$$

Furthermore, we have

$$v_c v_r = \frac{c\mu \sec^2\frac{\theta}{2}}{2 r_1 r_2} , \tag{8.13}$$

which is an important result showing that the product of the two velocity components depends only upon the transfer triangle, (r_1, r_2, c), and is independent of the shape and size of the transfer orbit. Hence, transfer orbits of various parameter values could possibly satisfy the given transfer triangle. This also implies that while (r_1, r_2, θ) are fixed, the true anomalies, θ_1 and θ_2, vary depending upon the parameter, p, of the orbit.

If the *transfer time*, t_{12}, is known, the two-point boundary-value problem given by the fixed transfer sector, (r_1, r_2, θ), becomes the Lambert's problem whose solution, (a, e, t_0), yields a transfer orbit. If the transfer time is not fixed, the two-point boundary-value problem does not have a unique solution, and orbits of different shapes and sizes can be selected to pass through the points, P_1 and P_2.

8.2 Elliptic Transfer

Let us first consider the case of an elliptic transfer orbit, and later generalize the discussion to include other shapes. An elliptic transfer is distinguished from the other types of transfer in that for a given semi-major axis, a, and a fixed transfer sector, (r_1, r_2, θ), there are two possible ways in which the transfer can be made. These are detailed in this section.

It is interesting to note an important characteristic of an ellipse of semi-major axis, a, with focus, O, on which two points, P_1 and P_2, corresponding to the radii r_1 and r_2, respectively. As shown in Fig. 8.2, the points P_1 and P_2 are located at $r_1^* = 2a - r_1$ and $r_2^* = 2a - r_2$, respectively, from the vacant focus, V. This implies that a position on the ellipse can be alternatively described as the distance from the vacant focus, given by the *virtual radius*, $r^* = 2a - r$, instead of the radius, r, measured from the focus, O. A consequence of this fact is that circles of virtual radii, $r_1^* = 2a - r_1$ and $r_2^* = 2a - r_2$, centred at the points P_1 and P_2, respectively, must intersect at the vacant focus, as shown in Fig. 8.2. However, for an arbitrary value of a, there are two possible intersection points of the circles given by V and V', as shown in Fig. 8.2, both of which qualify to be the vacant focii of the transfer orbit. While the transfer orbits marked Ellipse #1 and #2 in the figure, with the vacant focii V and V', respectively, have the same semi-major axis, a, their values of eccentricity, e, are different. This is because the distance between the focus and the vacant focus of an ellipse is $2ae$, and as shown in Fig. 8.2, the distances OV and OV' are different. Since the energy of the transfer orbit, $\varepsilon = -\mu/(2a)$, is only a function of a, the vacant focii V and V' denote the two possibilities of the equally energetic transfer ellipses.

From the geometry depicted in Fig. 8.2, we have

$$r_1 + r_1^* = 2a \, ; \qquad r_2 + r_2^* = 2a \, , \tag{8.14}$$

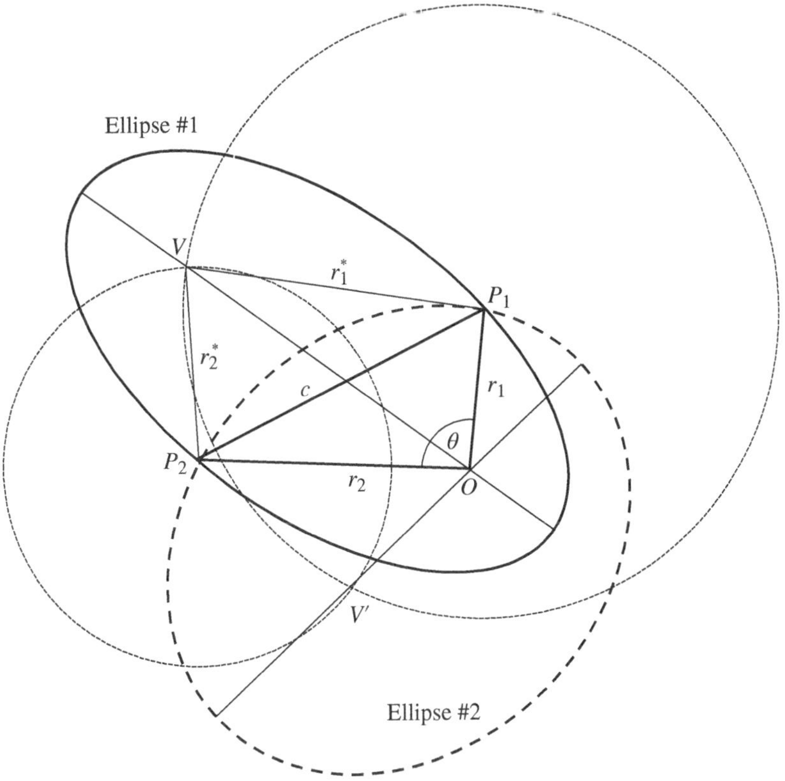

Figure 8.2 The vacant focus, V, of Ellipse #1 of the semi-major axis, a, as the point of intersection of circles of radii $r_1^* = 2a - r_1$ and $r_2^* = 2a - r_2$, centred at the points P_1 and P_2, which are located at radii, r_1 and r_2, respectively, from the focus, O. Ellipse #2 is an alternative transfer ellipse, also of semi-major axis a, but with vacant focus, V'.

or

$$r_1 + r_2 + r_1^* + r_2^* = 4a\,, \tag{8.15}$$

where r_1^* and r_2^* are the respective distances of the points, P_1 and P_2, from the vacant focus, V.

8.2.1 Locus of the Vacant Focii

When the semi-major axis of the transfer ellipse is varied while keeping the transfer triangle (r_1, r_2, θ) fixed, the vacant focii, V and V', describe a smooth curve shown in Fig. 8.3. The locus of V and the locus of V' are equidistant from the chord, **c**. At the chord, the two focii merge into a single point, and their locii at this point are perpendicular to the chord. Furthermore, the difference of the distances between the vacant focus, V, and the points P_2 and P_1, that is $r_2^* - r_1^* = r_1 - r_2$, is constant. Hence it follows that the locus of V and V' is a conic section with focii at P_1 and P_2 and the major axis along the chord. The semi-major axis of this conic section is $\mid r_1 - r_2 \mid$ (that is, the minimum difference, $r_2^* - r_1^*$). The distance between the focii, P_1 and P_2,

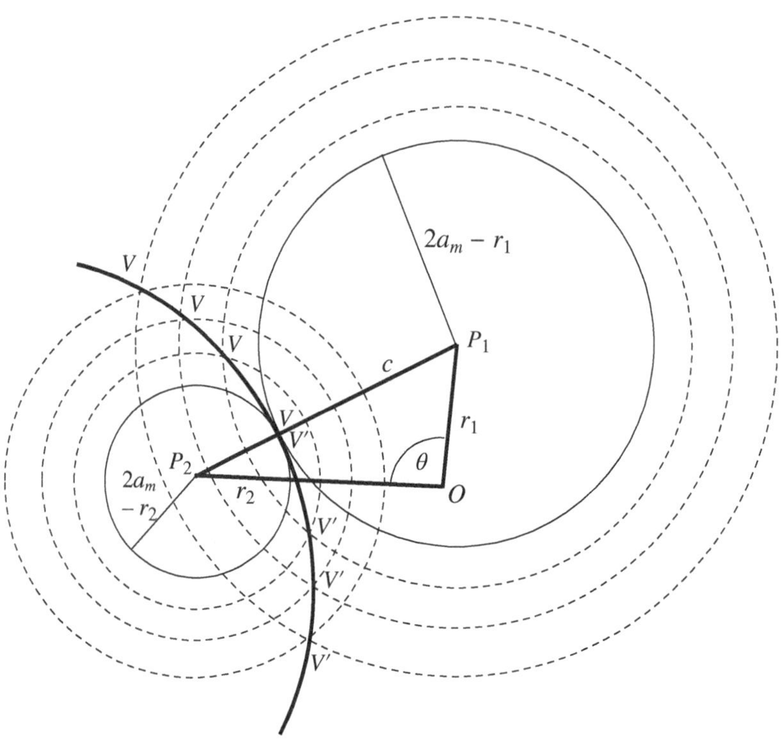

Figure 8.3 The locus of the vacant focii, V and V', of transfer ellipses of the semi-major axis, a, as the latter is varied, resulting in the minimum value of $a = a_m$ corresponding to the minimum-energy transfer between P_1 and P_2.

of the conic section being the chord length, c, implies that the eccentricity of the conic section is $c/ \mid r_1 - r_2 \mid$. From the triangle inequality, we have $c = \mid \mathbf{r_2} - \mathbf{r_1} \mid > r_2 - r_1$, which implies that the conic section is a hyperbola of eccentricity $c/(r_2 - r_1) > 1$ and semi-major axis $r_1 - r_2 < 0$.

The locus of the vacant focus, V', produces ellipses of a smaller eccentricity, e, than that of V. This is because the distance between the focii, $ae = OV'$, is smaller than the distance, $ae = OV$, for the same value of the semi-major axis, a, as shown in Fig. 8.3. We shall return to the locus of the vacant focus in Section 8.2.2.

8.2.2 *Minimum-Energy and Minimum-Eccentricity Transfers*

For the special case of $a = a_m$, the two vacant focii of the elliptic transfer, V and V', merge into a single point located on the chord, as shown in Fig. 8.3. The minimum-energy ellipse is therefore the boundary between the two possible transfers shown in Fig. 8.3 with the same semi-major axis, a. Hence $a = a_m$ is the smallest possible value of the semi-major axis for which a transfer between r_1 and r_2 can take place, and corresponds to the *minimum orbital transfer energy*, $\varepsilon_m = -\mu/(2a_m)$. For $a < a_m$, there is no intersection between the two circles, which implies that the orbital energy is insufficient to make the transfer between r_1 and r_2. As shown in Fig. 8.3, the circles centred at P_1 and P_2 of radii $r_1^* = 2a_m - r_1$ and $r_2^* = 2a_m - r_2$ (depicted as solid circles in the figure) are tangential to each other at the chord line, thereby yielding a single point of intersection, $V = V'$. It follows from Fig. 8.3 that $c = 4a_m - r_1 - r_2$; hence the smallest possible value for the distance, $r_1^* + r_2^*$, is the chord length, c. Therefore, we have

$$2a_m = \frac{1}{2}(r_1 + r_2 + c)\,. \tag{8.16}$$

The semi-perimeter, s, of the transfer triangle is half the sum of the sides of the transfer triangle:

$$s = \frac{r_1 + r_2 + c}{2}\,. \tag{8.17}$$

Hence, the major axis of the minimum-energy transfer ellipse is the same as the semi-perimeter,

$$2a_m = s\,. \tag{8.18}$$

The geometry of the minimum-energy elliptic transfer is depicted in Fig. 8.4. The figure indicates that a minimum-energy transfer is equivalent to a rectilinear transfer between r_1^* and r_2^*, for which we have, $c = r_1^* + r_2^*$. Hence the chordal velocity component, v_c, is the same as the radial velocity component, v_r, for a minimum-energy transfer (i.e., $v_c/v_r = 1$), and we have the following expressions for the minimum-energy parameter from Eq. (8.12):

$$p_m = \frac{r_1 r_2}{c}(1 - \cos\theta) = 2\frac{r_1 r_2}{c}\sin^2\frac{\theta}{2} = \frac{2}{c}(s - r_1)(s - r_2)\,. \tag{8.19}$$

Hence the parameter, p, of a general orbit can be expressed as follows:

$$p = p_m \frac{v_c}{v_r}\,. \tag{8.20}$$

As θ is varied from 0 to π, the minimum-energy orbital parameter, p_m, varies from 0 to $2r_1 r_2/c$, and the ratio v_c/v_r (thus p) varies from infinity to zero.

An important property of elliptic transfer for a given transfer triangle is derived by writing the orbit equation, $r = p/(1 + e\cos\theta)$, for the two given radii as follows:

$$\mathbf{r_1} \cdot \mathbf{e} = r_1 - p\,; \qquad \mathbf{r_2} \cdot \mathbf{e} = r_2 - p\,, \tag{8.21}$$

or

$$(\mathbf{r_2} - \mathbf{r_1}) \cdot \mathbf{e} = -(r_2 - r_1)\,, \tag{8.22}$$

and dividing by the chord length, c, we have

$$\mathbf{i_c} \cdot \mathbf{e} = -(r_2 - r_1)/c\,, \tag{8.23}$$

which implies that the eccentricity vectors of all possible transfer ellipses for the given transfer triangle, (r_1, r_2, c), have a constant projection with the chordwise direction, $\mathbf{i_c}$. This yields the following minimum possible value for the orbital eccentricity, $e_s = |\, \mathbf{e_s}\, |$, of the transfer ellipse:

$$e_s = (r_2 - r_1)/c\,, \tag{8.24}$$

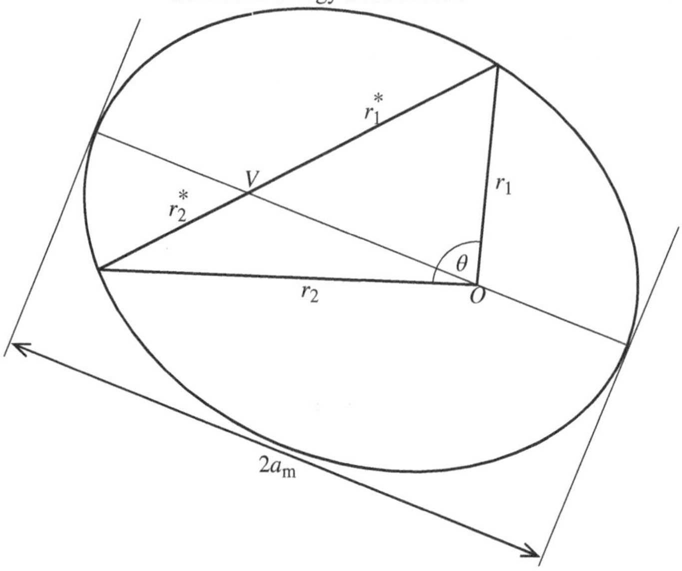

Figure 8.4 Geometry of the minimum-energy orbital transfer.

which is called the *minimum-eccentricity ellipse.* The major axis of the minimum-eccentricity ellipse is parallel to the chord, with the semi-major axis given by $a_s = (r_1 + r_2)/2$. Recall that the major axis of the hyperbola described by the vacant focus is also parallel to the chord, and has its eccentricity given by $1/e_s$. The minimum-eccentricity transfer ellipse is said to be *symmetrical* because the arc length from the periapsis to point P_1 is the same as that from the apoapsis to point P_2. From this symmetry, the geometrical relationship between point P_1 and the focus, O, is exactly the same as that between point P_2 and the vacant focus, V. Due to the symmetry of transfer, the minimum-eccentricity ellipse is also called the *fundamental ellipse.*

8.3 Lambert's Theorem

For a given transfer triangle (Fig. 8.1), *Lambert's theorem* states that the transfer time, $t_{12} = t_2 - t_1$, is a function of only the semi-major axis, a, of the transfer orbit, the sum of the two radii, $r_1 + r_2$, and the chord, $c = |\,\mathbf{r_2} - \mathbf{r_1}\,|$, joining the two given positions. Thus, we have

$$t_{12} = t_{12}(a, \mathbf{r_1} + \mathbf{r_2}, \mathbf{c})\ . \tag{8.25}$$

Since there is no dependence of the transfer time on the orbital eccentricity, one can choose any value of e for the orbit, provided $(a, r_1 + r_2, c)$ are unchanged. This is tantamount to changing the orbital shape for the same flight time by moving the focii of the transfer orbit in such a way that a, $r_1 + r_2$, and c are unaffected. A geometrical interpretation of Lambert's theorem can be seen by considering the individual shapes of the possible transfer orbits.

8.3.1 *Time in Elliptic Transfer*

The time of flight, t_{12}, in an elliptic transfer between r_1 and r_2 is derived by using Kepler's equation as follows:

$$t_{12} = \frac{2}{n}\left(\frac{E_2 - E_1}{2} - e \sin\frac{E_2 - E_1}{2}\cos\frac{E_1 + E_2}{2}\right), \tag{8.26}$$

where $n = \sqrt{\mu/a^3}$ is the mean motion of the transfer ellipse and E_1, E_2 are the eccentric anomalies corresponding to the radii, r_1 and r_2, respectively. The eccentric anomalies and the orbital eccentricity, e, can be eliminated from Eq. (8.26) by applying the following identities:

$$r_1 + r_2 = 2a\left(1 - e\cos\frac{E_2 - E_1}{2}\cos\frac{E_1 + E_2}{2}\right), \tag{8.27}$$

$$c^2 = 4a^2\left(1 - e^2\cos^2\frac{E_1 + E_2}{2}\right)\sin^2\frac{E_2 - E_1}{2}, \tag{8.28}$$

and defining

$$\psi = (E_2 - E_1)/2\,; \qquad \cos\phi = e\cos\frac{E_1 + E_2}{2}. \tag{8.29}$$

Substituting these into Eq. (8.26) results in

$$t_{12} = \frac{2}{n}(\psi - \sin\psi\cos\phi), \tag{8.30}$$

where

$$r_1 + r_2 = 2a(1 - \cos\psi\cos\phi) \tag{8.31}$$

and

$$c = 2a\sin\psi\sin\phi. \tag{8.32}$$

Since n, ψ, and ϕ are determined from a, $r_1 + r_2$, and c, this proves Lambert's theorem for an elliptic transfer.

An alternative form of Eq. (8.30) is derived by introducing *Lagrange's variables*, the angles $\alpha = \phi + \psi$ and $\beta = \phi - \psi$, such that the perimeter of the transfer triangle is given by

$$2s = r_1 + r_2 + c = 2a(1 - \cos\alpha) \tag{8.33}$$

and

$$2(s - c) = r_1 + r_2 - c = 2a(1 - \cos\beta) \tag{8.34}$$

or

$$\sin\frac{\alpha}{2} = \sqrt{\frac{s}{2a}}\,; \qquad \sin\frac{\beta}{2} = \sqrt{\frac{s - c}{2a}}. \tag{8.35}$$

Hence Eq. (8.30) becomes

$$t_{12} = \frac{1}{n}[(\alpha - \sin\alpha) - (\beta - \sin\beta)], \tag{8.36}$$

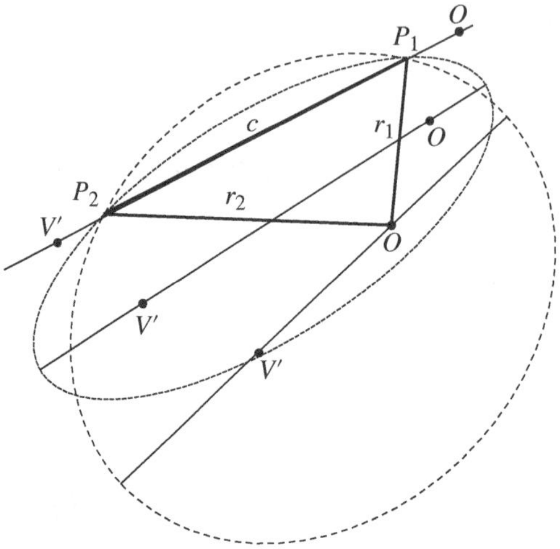

Figure 8.5 Geometrical interpretation of Lambert's theorem for calculating the time of flight in an elliptic orbit (P_1 and P_2 on the same side of the major axis).

which is called the *Lagrange's form* of the time equation for an elliptic transfer. There is a quadrant ambiguity of the angles α and β in this equation, which can be resolved after understanding the geometric significance of the equation as follows.

Consider an elliptic transfer between P_1 and P_2 in time t_{12}, as shown in Fig. 8.5. According to Lambert's theorem, if the radii, r_1 and r_2, and the chord, c, between them are fixed, and the focii, O and V (or V'), are varied such that the ellipse's semi-major axis, a, is unchanged, then the time of flight, t_{12}, is held constant during the variation. In that case, the locus described by the focus, O, is an ellipse with focii at P_1 and P_2, and major axis $r_1 + r_2$. Similarly, the locus of the vacant focus is also an ellipse with focii at P_1 and P_2, with major axis, $r_1^* + r_2^* = 4a - (r_1 + r_2)$. Now an elliptic transfer between P_1 and P_2 can be made by travelling along either the shorter or the longer arc of the transfer ellipse (i.e., with a transfer angle, θ, or $2\pi - \theta$). Since each of these transfer angles, θ and $2\pi - \theta$, corresponds to the same chord length, c, the time of flight, t_{12}, in each case must be the same if carried out along the ellipses of the same semi-major axis, a. For a given transfer angle, θ, a further distinction in the geometry of the transfer arises when considering whether the apoapsis of the ellipse is crossed during the transfer, which corresponds to the transfer ellipses of the same semi-major axis labelled Ellipse #1 and Ellipse #2, respectively, in Fig. 8.2. These two possibilities are investigated next.

Let us first consider the branch of the vacant locus, V', as shown in Fig. 8.3, which, being closer to the focus, O, depicts an ellipse of smaller eccentricity than that of the upper branch of the locus corresponding to V. Since both P_1 and P_2 lie on the same side of the major axis of the transfer ellipse (Fig. 8.3), a crossing of the apoapsis does not take place during such a transfer. By moving the focus, O, in the opposite direction to that of the vacant focus, V', while keeping the transfer triangle fixed, the eccentricity of the orbit is continuously increased. The locii of O and V are separate ellipses (not shown in Fig. 8.5) of major axes $r_1 + r_2$ and $r_1^* + r_2^*$, respectively, with focii at P_1 and P_2. The movement of the focii O and V in this manner causes the transfer ellipse to become continuously more elongated, until both O and V' fall on the chord line, as shown in Fig. 8.5. This is the limiting case of a rectilinear ellipse ($p = 0$), and represents the straight-line motion between P_1 and P_2. In a rectilinear ellipse, the distance between the two focii equals the major axis, i.e., $OV' = 2a$, and for the rectilinear transfer, we have $r_2 = r_1 + c$. The rectilinear elliptic motion is described by the following equation:

$$v = \dot{r} = \sqrt{\mu}\left(\frac{2}{r} - \frac{1}{a}\right)^{1/2}, \tag{8.37}$$

which is integrated to yield

$$t_{12} = \frac{1}{\sqrt{\mu}} \int_{s-c}^{s} \frac{r\mathrm{d}r}{\sqrt{2r - r^2/a}}. \tag{8.38}$$

The integration limits in Eq. (8.38) are the distances $r_1 = s - c$ and $r_2 = s$ measured along the straight line OP_2, representing the direct rectilinear transfer from P_1 to P_2. By introducing the Lagrange's variables, α and β, defined by Eq. (8.35) and changing the integration variable to ξ by

$$r = a(1 - \cos\xi), \tag{8.39}$$

we have

$$t_{12} = \sqrt{\frac{a^3}{\mu}} \int_{\beta}^{\alpha} (1 - \cos\xi)\mathrm{d}\xi, \tag{8.40}$$

or

$$t_{12} = \frac{1}{n}[(\alpha - \sin\alpha) - (\beta - \sin\beta)], \tag{8.41}$$

which is the same as Eq. (8.36).

The case when the vacant focus is on the branch of the locus, V, the transfer between P_1 and P_2 requires passing the apoapsis during the transfer, because P_1 and P_2 fall on the opposite sides of the major axis, as shown in Fig. 8.3. The eccentricity of the orbit is increased by moving the focii O and V in opposite directions along their respective ellipses focussed at P_1 and P_2; the situation produced is depicted in Fig. 8.6, wherein the limiting case of a rectilinear ellipse is once

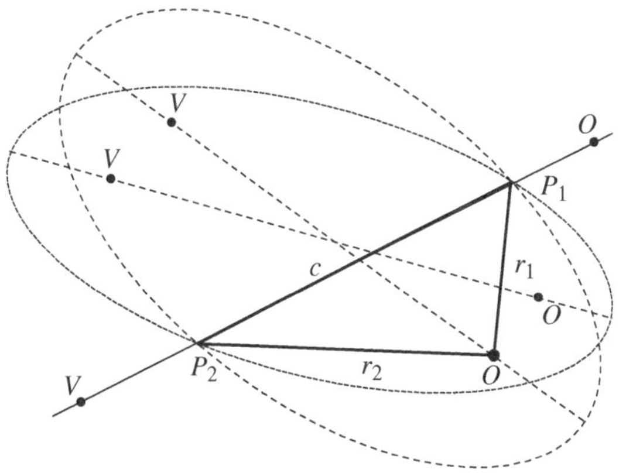

Figure 8.6 Geometrical interpretation of Lambert's theorem for calculating the time of flight in an elliptic orbit (P_1 and P_2 on the opposite sides of the major axis).

again reached for both O and V falling on the chord line. Since the rectilinear transfer between P_1 and P_2 must now cross the vacant focus, V, it is required that an additional distance of P_2 to V, and back from V to P_2, must be travelled. Hence twice the time to traverse the distance between $r = r_2 = s$ to $r = 2a$ must be added to Eq. (8.41). Therefore, we have

$$\begin{aligned} t_{12} &= \frac{1}{n}[(\alpha - \sin\alpha) - (\beta - \sin\beta)] + \frac{2}{\sqrt{\mu}}\int_s^{2a} \frac{r\mathrm{d}r}{\sqrt{2r - r^2/a}} \\ &= \frac{1}{n}[(\alpha - \sin\alpha) - (\beta - \sin\beta)] + 2\sqrt{\frac{a^3}{\mu}}\int_\alpha^{\pi} (1 - \cos\xi)\mathrm{d}\xi \\ &= \frac{2\pi}{n} - \frac{1}{n}[(\alpha - \sin\alpha) + (\beta - \sin\beta)]\,. \end{aligned} \tag{8.42}$$

Equations (8.41) and (8.42) are derived from Eq. (8.26) by regarding the angles β and α to be the eccentric anomalies along the rectilinear ellipse ($e = 1$) corresponding to the two positions in their respective quadrants, i.e., $\tilde{E}_1 = \beta$, $\tilde{E}_2 = \alpha$ for Eq. (8.41), and $\tilde{E}_1 = \beta$ and $\tilde{E}_2 = 2\pi - \alpha$ for Eq. (8.42). Therefore, assuming that the correct quadrant is taken for $\tilde{E}_2$, Eqs. (8.41) and (8.42) collapse into the following single equation:

$$t_{12} = \frac{1}{n}[(\tilde{E}_2 - \sin\tilde{E}_2) - (\tilde{E}_1 - \sin\tilde{E}_1)]\,. \tag{8.43}$$

Since an arbitrary value of a results in two different values of t_{12} for a given sector, (r_1, r_2, θ), the flight time is a double-valued function of a in an elliptic transfer.

The two different possibilities of the transfer angle, $\theta \le \pi$ and $\theta \ge \pi$, are covered by $0 \le \tilde{E}_1 \le \pi$ and $-\pi \le \tilde{E}_1 \le 0$, respectively. Hence Eq. (8.43) is extended as follows:

$$t_{12} = \frac{1}{n}[(\tilde{E}_2 - \sin\tilde{E}_2) - (\tilde{E}_1 - \sin\tilde{E}_1)]\,; \quad \begin{matrix} (0 \le \tilde{E}_1 \le \pi) & (\theta \le \pi) \\ (-\pi \le \tilde{E}_1 \le 0) & (\theta \ge \pi) \end{matrix}\,. \tag{8.44}$$

Taking $\tilde{E}_1$ and $\tilde{E}_2$ in their correct quadrants is therefore crucial in deriving the time of flight for the elliptic Lambert's transfer.

For the minimum-energy transfer, we have $a = a_m = s/2$,

$$\sin\frac{\tilde{E}_{2_m}}{2} = \sqrt{\frac{s}{2a_m}} = 1\,; \qquad \sin\frac{\tilde{E}_{1_m}}{2} = \sqrt{\frac{s-c}{s}}\,, \tag{8.45}$$

which implies $\tilde{E}_{2_m} = \pi$. Hence the time of flight in the minimum-energy transfer is expressed as follows:

$$t_m = \frac{1}{n_m}[\pi - (\tilde{E}_{1_m} - \sin\tilde{E}_{1_m})]\,, \tag{8.46}$$

where $n_m = 2\sqrt{2\mu/s^3}$. This is the limiting case where both Eqs. (8.41) and (8.42) yield the same value of $t_{12} = t_m$ for $a = a_m$. The time of flight in the minimum-energy transfer, t_m, falls between the time of transfer required by Ellipse #1 – Eq. (8.42) and that required by Ellipse #2 – Eq. (8.41). Hence, the two elliptic transfer cases can be distinguished as follows:

$$t_{12} = \frac{1}{n}[(\tilde{E}_2 - \sin\tilde{E}_2) - (\tilde{E}_1 - \sin\tilde{E}_1)]\,; \quad \begin{matrix} (\tilde{E}_2 = \alpha) & (t \le t_m) \\ (\tilde{E}_2 = 2\pi - \alpha) & (t \ge t_m) \end{matrix}\,, \tag{8.47}$$

and the range of $\tilde{E}_1$ is given by Eq. (8.44) depending upon the transfer angle, θ.

For the minimum-eccentricity transfer, we have $a = a_s - (r_1 + r_2)/2$, and

$$\sin\frac{\tilde{E}_{2_s}}{2} = \sqrt{\frac{s}{r_1 + r_2}}\,; \qquad \sin\frac{\tilde{E}_{1_s}}{2} = \sqrt{\frac{s-c}{s}} = \sqrt{\frac{r_1 + r_2}{s}}\,, \tag{8.48}$$

thereby implying that $\cos\tilde{E}_{1_s} = -\cos\tilde{E}_{2_s} = c/(r_1 + r_2)$; hence the time of flight is given by

$$t_s = \frac{1}{n_s}(\pi - 2\tilde{E}_{1_s}) = \frac{1}{n_s}(2\tilde{E}_{2_s} - \pi)\,, \tag{8.49}$$

where $n_s = 2\sqrt{2\mu/(r_1 + r_2)^3}$.

8.3.2 *Time in Hyperbolic Transfer*

For a hyperbola, the time of flight is derived to be the following from the hyperbolic Kepler's equation:

$$t_{12} = \frac{1}{n}[(e\sinh H_2 - H_2) - (e\sinh H_1 - H_1)]\,, \tag{8.50}$$

where $n = \sqrt{-\mu/a^3}$ is the mean motion of the hyperbolic transfer orbit. Introducing the variables ψ and *phi* defined by

$$\psi = (H_2 - H_1)/2\,; \qquad \cosh\phi = e\cosh\frac{H_1 + H_2}{2}\,, \tag{8.51}$$

Eq. (8.50) is expressed as follows:

$$t_{12} = \frac{2}{n}(\sinh\psi\cosh\phi - \psi)\,. \tag{8.52}$$

Using the parametric equations of the hyperbola as well as the orbit equation, $r = a(1 - e\cosh H)$, we have the following for the transfer triangle, (r_1, r_2, c):

$$r_1 + r_2 = 2a(1 - \cosh\psi\cosh\phi)\,, \tag{8.53}$$

$$c = -2a\sinh\psi\sinh\phi\,, \tag{8.54}$$

and

$$s = a(1 - \cosh\psi\cosh\phi - \sinh\psi\sinh\phi)\,. \tag{8.55}$$

Then *Lagrange's variables*, α and β, can be defined as follows in a manner similar to the elliptic transfer:

$$\sinh\frac{\alpha}{2} = \sqrt{\frac{-s}{2a}}\,; \qquad \sinh\frac{\beta}{2} = \sqrt{\frac{-(s-c)}{2a}}\,. \tag{8.56}$$

A substitution of these results in the following equation for the time of flight:

$$t_{12} = \frac{1}{n}[(\sinh\alpha - \alpha) - (\sinh\beta - \beta)]\,. \tag{8.57}$$

Equation (8.57) corresponds to the transfer where both α and β are positive quantities. This is the case of the transfer angle, $\theta < \pi$. The transfer with $\theta > \pi$ corresponds to $\alpha > 0$ and $\beta < 0$.

However, if the positive values of α and β computed by Eq. (8.56) are to be used in all cases, then the expression for the time of flight is modified by replacing β with $-\beta$, resulting in the following comprehensive expression:

$$t_{12} = \frac{1}{n}[(\sinh\alpha - \alpha) \mp (\sinh\beta - \beta)], \tag{8.58}$$

where the upper sign is to be used for $\theta < \pi$, and the lower sign for $\theta > \pi$.

The geometric significance of Eq. (8.58) is investigated in the manner similar to that of the elliptic transfer. As in the case of elliptic transfer, circles centred at P_1 and P_2 of radii $r_1 + r_2$ and $r_1^* + r_2^*$, respectively, intersect at the vacant focii, V and V', corresponding to hyperbolae of the smaller and the larger eccentricity orbits, respectively. However, since $e > 1$ for a hyperbola, the eccentricity of the transfer orbit must be decreased to reach the limiting case of the rectilinear hyperbola ($e = 1$). Furthermore, being an open orbit, a hyperbola does not curve around the vacant focus; hence the only distinction between the cases $\theta < \pi$ and $\theta > \pi$ is provided by a transfer which crosses the periapsis. Therefore, the roles played by the focus, O, and the vacant focus, V, are reversed from those of the ellipse. By moving the focii O and V' of the larger eccentricity hyperbola in opposite directions along their respective hyperbolic locii, the eccentricity of the transfer hyperbola is continuously decreased, until the case of a rectilinear hyperbola is reached, for which O and V' fall on the chord line. Then an integration of the rectilinear motion equation, Eq. (8.37) is carried out between r_1 and r_2 by changing the integration variable to η as follows:

$$r = a(1 - \cosh\eta), \tag{8.59}$$

with $r_1 = a(1 - \cosh\beta)$ and $r_2 = a(1 - \cosh\alpha)$. This integration yields

$$t_{12} = \sqrt{\frac{-a^3}{\mu}} \int_{\beta}^{\alpha} (1 - \cosh\eta)\mathrm{d}\eta, \tag{8.60}$$

or

$$t_{12} = \frac{1}{n}[(\sinh\alpha - \alpha) - (\sinh\beta - \beta)]. \tag{8.61}$$

Equation (8.57) represents a transfer with $\theta < \pi$, which implies a hyperbola with P_1 and P_2 on the same side of the major axis. When the vacant focus, V, corresponding to the smaller eccentricity transfer is chosen, its movement in the opposite direction to that of O ultimately results in a rectilinear hyperbola for which the focus, O, must be crossed in the transfer between P_1 and P_2. This represents a hyperbolic transfer with $\theta > \pi$ which requires going around the focus O via the periapsis. On the rectilinear hyperbola, the distance between points P_1 and P_2 is $c = r_2 - r_1$, while that between O and P_2 is $r_2 = s > c$. Hence, an additional time of twice that required to move between O (i.e, $r = 0$) and P_1 (i.e, $r = s - c$) must be added to Eq. (8.61) as follows:

$$\begin{aligned} t_{12} &= \frac{1}{n}[(\sinh\alpha - \alpha) - (\sinh\beta - \beta)] + \frac{2}{\sqrt{\mu}} \int_0^{s-c} \frac{r\mathrm{d}r}{\sqrt{2r - r^2/a}} \\ &= \frac{1}{n}[(\sinh\alpha - \alpha) - (\sinh\beta - \beta)] + 2\sqrt{\frac{-a^3}{\mu}} \int_0^{\beta} (1 - \cosh\eta)\mathrm{d}\eta \\ &= \frac{1}{n}[(\sinh\alpha - \alpha) + (\sinh\beta - \beta)]. \end{aligned} \tag{8.62}$$

The two results given by Eqs. (8.61) and (8.62) are combined into a single equation which covers all possible cases of transfer:

$$t_{12} = \frac{1}{n}[(\sinh \alpha - \alpha) \mp (\sinh \beta - \beta)], \tag{8.63}$$

which is the same as Eq. (8.58), denoting both $\theta < \pi$ (negative sign) and $\theta > \pi$ (positive sign).

A comparison of Eqs. (8.58) and (8.50) reveals that α and β are the hyperbolic anomalies, $\tilde{H}_1$ and $\tilde{H}_2$, respectively, for the rectilinear hyperbola ($e = 1$), corresponding to the positions r_1 and r_2. However, the correct quadrant for the Gudermannian, ζ_1, such that $\sinh \tilde{H}_1 = \tan \zeta_1$ (Chapter 4) during the transfer, requires that $\tilde{H}_1$ should be replaced by $-\tilde{H}_1$ for $\theta > \pi$.

8.3.3 Time in Parabolic Transfer

For the special case of a parabolic transfer ($a = \infty$), Barker's equation (Chapter 4) is applied to yield the time of flight as follows:

$$t_{12} = \frac{\sqrt{p^3}}{6\sqrt{\mu}}\left[\left(\tan^3\frac{\theta_2}{2} - \tan^3\frac{\theta_1}{2}\right) + 3\left(\tan\frac{\theta_2}{2} - \tan\frac{\theta_1}{2}\right)\right], \tag{8.64}$$

where θ_1 and θ_2 are the true anomalies corresponding to the radii, r_1 and r_2, respectively. Equation (8.64) is simplified in notation by introducing the variable $\sigma = \sqrt{p}\tan(\theta/2)$,

$$\sigma_1 = \sqrt{p}\tan\frac{\theta_1}{2}\,; \qquad \sigma_2 = \sqrt{p}\tan\frac{\theta_2}{2}, \tag{8.65}$$

which results in the following:

$$t_{12} = \frac{1}{6\sqrt{\mu}}(\sigma_2 - \sigma_1)[(\sigma_2 - \sigma_1)^2 + 3(p + \sigma_1\sigma_2)]. \tag{8.66}$$

The orbit equation for a parabola is expressed as follows:

$$r = \frac{1}{2}(p + \sigma^2), \tag{8.67}$$

which yields

$$p + \sigma_1\sigma_2 = p\sec\frac{\theta_1}{2}\sec\frac{\theta_2}{2}\cos\frac{(\theta_2 - \theta_1)}{2} = 2\sqrt{r_1 r_2}\cos\frac{\theta}{2}. \tag{8.68}$$

Covering the two cases, $\theta < \pi$ and $\theta > \pi$, we write

$$p + \sigma_1\sigma_2 = \pm 2\sqrt{s(s-c)}, \tag{8.69}$$

where the upper sign is used for $\theta < \pi$ and the lower for $\theta > \pi$. It is also noted from the orbit equation that

$$2(r_1 + r_2) = 2p + \sigma_1^2 + \sigma_2^2 = (\sigma_2 - \sigma_1)^2 + 2(p + \sigma_1\sigma_2). \tag{8.70}$$

Hence we have

$$(\sigma_2 - \sigma_1)^2 = 2[(r_1 + r_2) - (p + \sigma_1\sigma_2)] = 4(\sqrt{s} \mp \sqrt{s-c})^2. \tag{8.71}$$

Since $\sigma_2 \geq \sigma_1$, it follows that

$$\sigma_2 - \sigma_1 = 2(\sqrt{s} \mp \sqrt{s-c}). \tag{8.72}$$

The substitution of Eqs. (8.68) and (8.72) into Eq. (8.66) yields the following *Euler's equation* for the time of flight in a parabolic orbit:

$$t_{12} = \frac{1}{6\sqrt{\mu}}[(r_1 + r_2 + c)^{3/2} \mp (r_1 + r_2 - c)^{3/2}], \tag{8.73}$$

where the positive sign is taken for the transfer angle, $\theta > \pi$, and the negative sign is taken for $\theta < \pi$.

Another method of deriving the Euler's equation for a parabolic transfer is by transferring an elliptic orbit (or a hyperbolic orbit) to an equivalent rectilinear parabola ($p = 0$) by increasing the semi-major axis, a, to infinity while moving the focus and the vacant focus to change the eccentricity to unity. The time of flight in a rectilinear parabola for the direct transfer between P_1 and P_2, i.e., for $\theta < \pi$, is derived as follows:

$$v = \dot{r} = \sqrt{\frac{2\mu}{r}}, \tag{8.74}$$

whose integration yields

$$t_{12} = \frac{1}{\sqrt{2\mu}} \int_{s-c}^{s} \sqrt{r}\mathrm{d}r = \frac{1}{3}\sqrt{\frac{2}{\mu}}[s^{3/2} - (s-c)^{3/2}]. \tag{8.75}$$

For $\theta > \pi$, the additional time of twice that required to move between O and P_1 must be added to Eq. (8.75) as follows:

$$\begin{aligned} t_{12} &= \frac{1}{3}\sqrt{\frac{2}{\mu}}[s^{3/2} - (s-c)^{3/2}] + \frac{2}{\sqrt{2\mu}} \int_{0}^{s-c} \sqrt{r}\mathrm{d}r \\ &= \frac{1}{3}\sqrt{\frac{2}{\mu}}[s^{3/2} - (s-c)^{3/2}] + \frac{2}{3}\sqrt{\frac{2}{\mu}}(s-c)^{3/2} \\ &= \frac{1}{3}\sqrt{\frac{2}{\mu}}[s^{3/2} + (s-c)^{3/2}]. \end{aligned} \tag{8.76}$$

Combining both the cases, Eqs. (8.75) and (8.76), into a single equation yields

$$t_{12} = \frac{1}{3}\sqrt{\frac{2}{\mu}}[s^{3/2} \mp (s-c)^{3/2}], \tag{8.77}$$

which is the same as the Euler's equation, Eq. (8.73).

8.4 Solution to Lambert's Problem

In Section 8.3, the expressions for the flight time, $t_{12} = t_2 - t_1$, were derived for a transfer between two given coplanar radial positions, $r_1(t_1)$ and $r_2(t_2)$, separated by a chord length, c, by elliptic, hyperbolic, and parabolic orbits. These expressions of the flight time for the various cases of transfer are listed in Table 8.1.

The solution to Lambert's problem requires *orbit determination* between $\mathbf{r}_1(t_1)$ and $\mathbf{r}_2(t_2)$. As discussed previously, the transfer angle, θ, as well as the plane and the orbit direction are determined from the given position vectors and times. The orbital velocities required at the two points are resolved in the radial and chordal components, v_r and v_c, thereby expressing the parameter, p, in terms of v_r and v_c. However, the size, a, and the shape, e, of the orbit require an iterative solution by matching the given transfer time to that listed in Table 8.1. Since the parabolic flight time is quickly determined from the transfer triangle, (r_1, r_2, c), the nature of the transfer is identified by the required flight time. If the given value of t_{12} is greater than the parabolic flight time, we have an elliptic transfer, and if t_{12} is smaller than the parabolic flight time, the case of a hyperbolic transfer is indicated. In the case of an elliptic transfer, two different flight times are possible for the given transfer angle, θ, with the same value of the semi-major axis, a, as given in Table 8.1. Hence t_{12} is a double-valued function of the ellipse's semi-major axis, a, which implies that an iterative solution for a given t_{12} does not yield a unique value of a. Furthermore, the derivative of t_{12} with respect to the independent variable, a, given by

$$\frac{dt_{12}}{da} = \frac{3}{2}\sqrt{\frac{a}{\mu}}[(\alpha - \sin\alpha) - (\beta - \sin\beta)] - \sqrt{\frac{1}{a\mu}}\left[s\tan\frac{\alpha}{2} - (s - c)\tan\frac{\beta}{2}\right] \tag{8.78}$$

Table 8.1 Time of flight, $t_{12} = t_2 - t_1$, for a transfer between $r_1(t_1)$ and $r_2(t_2)$ with a chord length, c, between them, for various transfer orbits.

Type	t_{12}
Ellipse	$\sin\frac{\alpha}{2} = \sqrt{\frac{r_1+r_2+c}{4a}}$; $\sin\frac{\beta}{2} = \sqrt{\frac{r_1+r_2-c}{4a}}$
$(\theta \le \pi, t_{12} \le t_m)$	$\sqrt{\frac{a^3}{\mu}}[(\alpha - \sin\alpha) - (\beta - \sin\beta)]$
$(\theta \le \pi, t_{12} \ge t_m)$	$\sqrt{\frac{a^3}{\mu}}[(2\pi - \alpha + \sin\alpha) - (\beta - \sin\beta)]$
$(\theta \ge \pi, t_{12} \le t_m)$	$\sqrt{\frac{a^3}{\mu}}[(\alpha - \sin\alpha) + (\beta - \sin\beta)]$
$(\theta \ge \pi, t_{12} \ge t_m)$	$\sqrt{\frac{a^3}{\mu}}[(2\pi - \alpha + \sin\alpha) + (\beta - \sin\beta)]$
Hyperbola	$\sinh\frac{\alpha}{2} = \sqrt{\frac{r_1+r_2+c}{-4a}}$; $\sinh\frac{\beta}{2} = \sqrt{\frac{r_1+r_2-c}{-4a}}$
$(\theta \le \pi)$	$\sqrt{\frac{-a^3}{\mu}}[(\sinh\alpha - \alpha) - (\sinh\beta - \beta)]$
$(\theta \ge \pi)$	$\sqrt{\frac{-a^3}{\mu}}[(\sinh\alpha - \alpha) + (\sinh\beta - \beta)]$
Parabola	
$(\theta \le \pi)$	$\frac{1}{6\sqrt{\mu}}[(r_1 + r_2 + c)^{3/2} - (r_1 + r_2 - c)^{3/2}]$
$(\theta \ge \pi)$	$\frac{1}{6\sqrt{\mu}}[(r_1 + r_2 + c)^{3/2} + (r_1 + r_2 - c)^{3/2}]$

is infinite for a minimum-energy orbit, $a = a_m = s/2$, for which $\alpha = \pi$. This indicates a singularity at the point used to distinguish between the two types of elliptic transfers, $t_{12} < t_m$ and $t_{12} > t_m$. These features of the elliptic time equation show that a is not a convenient independent variable. A transformation to a new variable is therefore necessary for a numerical solution.

8.4.1 Parameter of Transfer Orbit

Before taking up the methods for the solution of the time equation for the semi-major axis, a, it is necessary to express the parameter, p, in terms of a as well as α and β, representing the transfer triangle. From the orbit equation, we have $p = r_1(1 + e\cos\theta_1) = r_2[1 + e\cos(\theta_1 + \theta)]$, which, substituted into the following identity,

$$\cos^2(\theta_1 + \theta) - 2\cos(\theta_1 + \theta)\cos\theta_1\cos\theta + \cos^2\theta_1 - \sin^2\theta = 0\,, \tag{8.79}$$

results into the following quadratic equation to be solved for the parameter, p:

$$ar_1^2(p - r_2)^2 - 2ar_1r_2(p - r_2)(p - r_1)\cos\theta + ar_2^2(p - r_1)^2 - r_1^2r_2^2(a - p)\sin^2\theta = 0\,, \tag{8.80}$$

which, is applicable to both the elliptic orbit ($a > 0$), and the hyperbolic orbit ($a < 0$).

A simplification occurs in Eq. (8.80) by recognizing the cosine law for the transfer triangle, $r_1^2 + r_2^2 - 2r_1r_2\cos\theta = c^2$, and on collecting terms, we have

$$ac^2p^2 - r_1r_2(1 - \cos\theta)[2a(r_1 + r_2) - r_1r_2(1 + \cos\theta)]p + ar_1^2r_2^2(1 - \cos\theta)^2 = 0\,, \tag{8.81}$$

which, on introducing the semi-perimeter, $s = (r_1 + r_2 + c)/2$, becomes

$$ac^2p^2 - r_1r_2(1 - \cos\theta)[-2s(s - c - 2a) - 2ac]p + ar_1^2r_2^2(1 - \cos\theta)^2 = 0\,. \tag{8.82}$$

The roots of the quadratic equation are derived separately for the ellipse and the hyperbola in terms of the Lagrange's variables, α and β, which are given for each case in Table 8.1, to be the following:

$$\begin{aligned} p &= \frac{4a(s - r_1)(s - r_2)}{c^2}\sin^2\frac{\alpha \pm \beta}{2} \quad \text{(ellipse)} \\ p &= \frac{-4a(s - r_1)(s - r_2)}{c^2}\sinh^2\frac{\alpha \pm \beta}{2} \quad \text{(hyperbola)}\,, \end{aligned} \tag{8.83}$$

where the upper sign is taken for $\theta < \pi$, and the lower sign for $\theta > \pi$. These can be alternatively expressed in terms of the parameter, p_m, of the minimum-energy orbit as follows:

$$\begin{aligned} \frac{p}{p_m} &= \frac{2a}{c}\sin^2\frac{\alpha \pm \beta}{2} \quad \text{(ellipse)} \\ \frac{p}{p_m} &= \frac{-2a}{c}\sinh^2\frac{\alpha \pm \beta}{2} \quad \text{(hyperbola)}\,. \end{aligned} \tag{8.84}$$

For each value of a, the hyperbola has two values of p corresponding to whether $\theta < \pi$ or $\theta > \pi$. However, for a given pair (a, θ), the ellipse has two values of p: one corresponding to the case $t < t_m$, and the other for $t > t_m$ for which α is replaced by $2\pi - \alpha$ in Eq. (8.83). Hence, a given value of a corresponds to a total of *four* ways in which an elliptic transfer can take place, each having a different value for the parameter, p. Equation (8.83) can be generalized for all cases as follows:

$$p = \frac{2s(s-r_1)(s-r_2)(s-c)}{c^2}\left[\sqrt{\frac{1}{s-c}-\frac{1}{2a}} \pm \mathrm{sgn}(t_m - t_{12})\sqrt{\frac{1}{s}-\frac{1}{2a}}\right]^2 . \tag{8.85}$$

This expression is used in Eq. (8.10) to determine the radial and chordal velocity components, v_r and v_c, and hence the velocity vectors, $\mathbf{v}_1$ and $\mathbf{v}_2$.

8.4.2 Stumpff Function Method

To numerically solve the time equations of Table 8.1 for the semi-major axis, a, a set of convenient universal formulae is required. Instead of the semi-major axis, a, it is necessary to introduce an alternative universal variable which avoids the numerical problems of the elliptic case discussed earlier. In this subsection, two alternative, yet equivalent algorithms are described, based upon the *Stumpff functions* introduced in Chapter 4.

Algorithm No. 1

Consider the following universal variables, (x, y), for a numerical solution of the time equation:

$$\begin{aligned} x &= \tilde{E}_2^2 \text{ (ellipse)}\,; \qquad x = -\tilde{H}_2^2 \text{ (hyperbola)} \\ y &= \tilde{E}_1^2 \text{ (ellipse)}\,; \qquad y = -\tilde{H}_1^2 \text{ (hyperbola)}\,, \end{aligned} \tag{8.86}$$

with the following range of the anomalies corresponding to a rectilinear orbit ($e = 1$):

$$\begin{aligned} \tilde{E}_2 &= \begin{cases} \alpha\,; & (t \le t_m) \\ 2\pi - \alpha\,; & (t \ge t_m) \end{cases} \\ \tilde{E}_1 &= \begin{cases} \beta\,; & (\theta \le \pi) \\ -\beta\,; & (\theta \ge \pi) \end{cases} \\ \tilde{H}_1 &= \begin{cases} \beta\,; & (\theta \le \pi) \\ -\beta\,; & (\theta \ge \pi) \end{cases}, \end{aligned} \tag{8.87}$$

and the Lagrange's variables, α and β, determined from the transfer triangle as follows:

$$\begin{aligned} \sin\frac{\alpha}{2} &= \sqrt{\frac{s}{2a}} \text{ (ellipse)}\,; \qquad \sinh\frac{\alpha}{2} = \sqrt{\frac{s}{-2a}} \text{ (hyperbola)} \\ \sin\frac{\beta}{2} &= \sqrt{\frac{s-c}{2a}} \text{ (ellipse)}\,; \qquad \sinh\frac{\beta}{2} = \sqrt{\frac{s-c}{-2a}} \text{ (hyperbola)}\,. \end{aligned} \tag{8.88}$$

Considering the ranges of the eccentric and hyperbolic anomalies, the universal variables are limited by

$$-\infty \le x \le 4\pi^2\,; \qquad y \le x\,; -\infty \le y \le \pi^2\,.$$

To eliminate a from the time equation, the following is employed:

$$a = \frac{s}{1-\cos\tilde{E}_2} = \frac{s-c}{1-\cos\tilde{E}_1} \quad \text{(ellipse)}$$
$$a = \frac{s}{1-\cosh\tilde{H}_2} = \frac{s-c}{1-\cosh\tilde{H}_1} \quad \text{(hyperbola)}, \tag{8.89}$$

where $s \neq c$. Hence, the time equation becomes the following:

$$t_{12}\sqrt{\mu} = \left(\frac{s}{1-\cos\tilde{E}_2}\right)^{3/2}(\tilde{E}_2 - \sin\tilde{E}_2) - \left(\frac{s-c}{1-\cos\tilde{E}_1}\right)^{3/2}(\tilde{E}_1 - \sin\tilde{E}_1) \quad \text{(ellipse)}$$
$$t_{12}\sqrt{\mu} = \left(\frac{s}{\cosh\tilde{H}_2 - 1}\right)^{3/2}(\sinh\tilde{H}_2 - \tilde{H}_2) - \left(\frac{s-c}{\cosh\tilde{H}_1 - 1}\right)^{3/2}(\sinh\tilde{H}_1 - \tilde{H}_1)$$
$$\text{(hyperbola)}. \tag{8.90}$$

The substitution of Eq. (8.86) into Eq. (8.90) yields the following time equation in terms of the universal variables:

$$t_{12}\sqrt{\mu} = \left[\frac{s}{C(x)}\right]^{3/2} S(x) \mp \left[\frac{s-c}{C(y)}\right]^{3/2} S(y)$$
$$syC(y) = (s-c)xC(x), \tag{8.91}$$

where $s \neq c$, and $C(z), S(z)$ are the following *Stumpff functions*, introduced in Chapter 4:

$$C(z) = \frac{1}{2!} - \frac{z}{4!} + \frac{z^2}{6!} - \cdots = \begin{cases} \frac{1-\cos\sqrt{z}}{z} & (z > 0) \\ \frac{\cosh\sqrt{-z}-1}{-z} & (z < 0) \end{cases} \tag{8.92}$$

$$S(z) = \frac{1}{3!} - \frac{z}{5!} + \frac{z^2}{7!} - \cdots = \begin{cases} \frac{\sqrt{z}-\sin\sqrt{z}}{(z)^{3/2}} & (z > 0) \\ \frac{\sinh\sqrt{-z}-\sqrt{-z}}{(-z)^{3/2}} & (z < 0) \end{cases}.$$

In Eq. (8.91), the upper sign is to be taken for $\theta < \pi$, and the lower sign for $\theta > \pi$. The roots (x, y) of Eq. (8.91) yield the semi-major axis by

$$a = \frac{s}{xC(x)} = \frac{s-c}{yC(y)}. \tag{8.93}$$

For the determination of the sign term, $\text{sgn}(t_m - t_{12})$, in the expression for the parameter, p, given by Eq. (8.85), this term can be replaced by $\text{sgn}(\pi^2 - x)$.

A suitable numerical procedure can be adopted to find the roots (x, y) of Eq. (8.91), wherein the flight time, t_{12}, is a monotonically varying function of x, with a finite slope at the minimum-eccentricity elliptic orbit, $x = \pi^2$. To apply Newton's method for the solution, two nested iterations are required: (a) to determine y, given x, and (b) to determine x, given t_{12} and the current value of the time function,

$$t_{12_k} = \frac{1}{\sqrt{\mu}}\left\{\left[\frac{s}{C(x_k)}\right]^{3/2} S(x_k) \mp \left[\frac{s-c}{C(y)}\right]^{3/2} S(y)\right\}. \tag{8.94}$$

The evolution of (x, y) after the kth step is given as follows using the identities for $C(z), S(z)$ presented in Chapter 4:

$$x_{k+1} = x_k \left[\frac{t_{12} - t_{12_k}}{(x\mathrm{d}t_{12}/\mathrm{d}x)_{x=x_k}} + 1 \right]$$

$$y_{k+1} = y_k - \frac{2}{1 - y_k S(y_k)} \left[y_k C(y_k) - \frac{s-c}{s} x C(x) \right] , \tag{8.95}$$

where

$$x\frac{\mathrm{d}t_{12}}{\mathrm{d}x} = \frac{1}{2\sqrt{\mu C(x)}} \left\{ s^{3/2} \mp \sqrt{2 - xC(x)} \left[\frac{(s-c)^{3/2}}{\sqrt{2 - yC(y)}} - \frac{3}{2} t_{12}\sqrt{\mu} \right] \right\} . \tag{8.96}$$

Example 8.4.1 *Calculate the orbital elements, (a, e, t_0), for a transfer from $r_1 = 6800$ km to $r_2 = 7000$ km around Earth in $t_{12} = 1000$ s, with the transfer angle, $\theta = 60°$.*

The parameters of the transfer triangle are calculated as follows:

$$c = \sqrt{r_1^2 + r_2^2 - 2r_1 r_2 \cos\theta} = 6902.1736 \text{ km}$$

$$s = \frac{1}{2}(r_1 + r_2 + c) = 10351.0868 \text{ km} .$$

Using Algorithm No. 1 to solve the Lambert's equation by Newton's method with a tolerance of $|\,\mathrm{d}x\,| \leq 10^{-9}$, the following solution is obtained:

$$x = 4.900746146 \; ; \qquad y = 1.176929405 ,$$

which implies

$$a = \frac{s}{xC(x)} = \frac{s-c}{yC(y)} = 6471.15914 \text{ km} .$$

The angles α and β are thus calculated by Eq. (8.35) to be the following:

$$\alpha = 2\sin^{-1}\sqrt{\frac{s}{2a}} = 2.2137629 \text{ rad.} \; ; \quad \beta = 2\sin^{-1}\sqrt{\frac{s-c}{2a}} = 1.0848638 \text{ rad.}$$

The minimum-energy ellipse has $a_m = s/2 = 5175.543$ km, for which the minimum-energy transfer time is obtained by Eq. (8.46) to be

$$t_m = \frac{1}{n_m}[\pi - (\tilde{E}_{1_m} - \sin\tilde{E}_{1_m})] = 1682.923 \text{ s} ,$$

where

$$\tilde{E}_{1_m} = 2\sin^{-1}\sqrt{\frac{s-c}{s}} = 1.23066244 \text{ rad.}$$

Since $t_{12} < t_m$, and $\theta < \pi$, the following expression is used in calculating the parameter, p, of the transfer orbit via Eq. (8.83):

$$p = \frac{4a(s-r_1)(s-r_2)}{c^2} \sin^2\frac{\alpha+\beta}{2} = 6425.9469 \text{ km} ,$$

which yields

$$e = \sqrt{1 - \frac{p}{a}} = 0.08358664 \, .$$

Finally, the time of perigee is derived from the eccentric anomaly at radius, r_1,

$$E_1 = \cos^{-1}\left\{\frac{1}{e}\left(1 - \frac{r_1}{a}\right)\right\} = 2.22427056 \text{ rad.}$$

to be the following:

$$t_0 = \frac{e \sin E_1 - E_1}{\sqrt{\mu/a^3}} = -1779.24815 \text{ s.}$$

It is verified that α and β satisfy Eq. (8.36) for the given time of flight, t_{12}:

$$t_{12} = \frac{1}{n}[(\alpha - \sin\alpha) - (\beta - \sin\beta)] = 1000 \text{ s} \, .$$

Algorithm No. 2

An alternative approach is to define an auxilliary variable, ξ, by

$$\xi = (E_2 - E_1)\sqrt{a} \text{ (ellipse)} \, ; \qquad \xi = (H_2 - H_1)\sqrt{-a} \text{ (hyperbola)} \, , \tag{8.97}$$

where $E_2 - E_1$ and $H_2 - H_1$ denote the difference between the eccentric and hyperbolic anomalies, respectively, corresponding to the radii, r_1 and r_2, along the transfer orbit of semi-major axis, a. The following epochal form of Kepler's equation, which was derived in Chapter 4, is employed for the time of flight:

$$t_{12} = \sqrt{\frac{a^3}{\mu}}\left\{(E_2 - E_1) + \frac{\mathbf{r}_1 \cdot \mathbf{v}_1}{\sqrt{\mu a}}[1 - \cos(E_2 - E_1)] - \left(1 - \frac{r_1}{a}\right)\sin(E_2 - E_1)\right\} \text{ (ellipse)}$$

$$t_{12} = \sqrt{\frac{-a^3}{\mu}}\left\{-(H_2 - H_1) - \frac{\mathbf{r}_1 \cdot \mathbf{v}_1}{\sqrt{-\mu a}}[1 - \cosh(H_2 - H_1)] + \left(1 - \frac{r_1}{a}\right)\sinh(H_2 - H_1)\right\} \text{ (hyperbola)} \, . \tag{8.98}$$

To evaluate the time equation common to all orbits, consider the Lagrange's coefficients given by Eq. (8.6) for the transfer sector, (r_1, r_2, θ):

$$\begin{aligned} f &= 1 + \frac{r_2}{p}(\cos\theta - 1) \\ g &= \frac{r_1 r_2}{h}\sin\theta \\ \dot{g} &= \frac{dg}{dt} = 1 + \frac{r_1}{p}(\cos\theta - 1) \, , \end{aligned} \tag{8.99}$$

where the remaining Lagrange's coefficient, $\dot{f}$, is derived from the identity, $f\dot{g} - g\dot{f} = 1$, and depends upon the particular shape of the transfer orbit. These expressions are given in terms of the following universal parameter, η,

$$\eta = \frac{r_1 r_2}{p}(1 - \cos\theta), \tag{8.100}$$

as follows:

$$f = 1 - \frac{\eta}{r_1}$$

$$g = A\sqrt{\frac{\eta}{\mu}}, \tag{8.101}$$

$$\dot{g} = 1 - \frac{\eta}{r_2}, \tag{8.102}$$

where

$$A = \sin\theta\sqrt{\frac{r_1 r_2}{1 - \cos\theta}} \tag{8.103}$$

is a constant of the transfer sector, (r_1, r_2, θ).

The term $\mathbf{r_1} \cdot \mathbf{v_1}$ in the time equation, Eq. (8.98), is evaluated by using the Lagrange's coefficients as follows:

$$\begin{aligned}\mathbf{v_1} &= \frac{1}{g}(\mathbf{r_2} - f\mathbf{r_1}) \\ \mathbf{r_1} \cdot \mathbf{v_1} &= \frac{1}{g}(r_1 r_2 \cos\theta - f r_1^2) \\ &= \frac{\sqrt{\mu}}{A\sqrt{\eta}}\left(\frac{r_2^2 - r_1^2 - c^2}{2} - y r_1\right).\end{aligned} \tag{8.104}$$

By resolving $\mathbf{r_1}$ and $\mathbf{r_2}$ in directions along and normal to $\mathbf{r_1}$, the first of Eqs. (8.104) yields the speed, v_1, and the semi-major axis, a, in terms of the Lagrange's coefficients as follows:

$$\begin{aligned}v_1 &= \frac{1}{g}\sqrt{(r_2\cos\theta - f r_1)^2 + r_2^2\sin^2\theta} \\ a &= \frac{-\mu}{v_1^2 - 2\mu/r_1}.\end{aligned} \tag{8.105}$$

In order to eliminate a from Eq. (8.98), the universal variable is transformed as follows:

$$z = \frac{\xi^2}{a} = (E_2 - E_1)^2 \text{ (ellipse)}; \qquad z = \frac{\xi^2}{a} = -(H_2 - H_1)^2 \text{ (hyperbola)}. \tag{8.106}$$

The elliptic transfer is given by $z > 0$, the hyperbolic transfer by $z < 0$, and the parabolic transfer by $z = 0$.

The following expression for η is obtained in terms of ξ and z:

$$\eta = \xi^2 C(z), \tag{8.107}$$

where $C(z)$, $S(z)$ are the Stumpff functions. A substitution of ξ, z, and η into Eq. (8.98) yields the following universal time equation:

$$t_{12}\sqrt{\mu} = S(z)\xi^3 + A\xi\sqrt{C(z)}, \tag{8.108}$$

and the Lagrange's coefficients relating the two positions are expressed as the following functions of ξ and z:

$$\begin{aligned} f &= 1 - \frac{\xi^2}{r_1}C(z) \\ \dot{f} &= \frac{\xi\sqrt{\mu}}{r_1 r_2}[zS(z) - 1] \\ g &= t_{12} - \frac{1}{\sqrt{\mu}}x^3 S(z) \\ \dot{g} &= 1 - \frac{\xi^2}{r_2}C(z), \end{aligned} \tag{8.109}$$

where the parabolic transfer, $C(0) = \frac{1}{2}$, $S(0) = \frac{1}{6}$, is analytically solved.

For solving the universal time equation, Eq. (8.108), by Newton's method, we write

$$F(\xi) = A\xi\sqrt{C(z)} + S(z)\xi^3 - t_{12}\sqrt{\mu}, \tag{8.110}$$

and take the derivative with respect to ξ,

$$F'(\xi) = A\sqrt{C(z)} + 3S(z)\xi^2 + \left(A\xi\frac{C'(z)}{2\sqrt{C(z)}} + S'(z)\xi^3\right)\frac{dz}{d\xi}, \tag{8.111}$$

where

$$\frac{dz}{d\xi} = \frac{2\xi}{a} \tag{8.112}$$

and

$$C'(z) = -\frac{1}{4!} + \frac{2z}{6!} - \frac{3z^2}{8!} - \dots = \begin{cases} \dfrac{\cos\sqrt{z} + \frac{1}{2}\sqrt{z}\sin\sqrt{z} - 1}{z^2} & (z > 0) \\ \dfrac{\cosh\sqrt{-z} - 1 - \frac{1}{2}\sqrt{-z}\sinh\sqrt{-z}}{z^2} & (z < 0) \end{cases} \tag{8.113}$$

$$S'(z) = -\frac{1}{5!} + \frac{2z}{7!} - \frac{3z^2}{9!} - \dots = \begin{cases} \dfrac{\frac{z}{2}(1 - \cos\sqrt{z}) - \frac{3}{2}(z - \sqrt{z}\sin\sqrt{z})}{z^3} & (z > 0) \\ \dfrac{\frac{z}{2}(\cosh\sqrt{-z} - 1) - 1 + \frac{3}{2}\sqrt{-z}(\sinh\sqrt{-z} - \sqrt{-z})}{(-z)^3} & (z < 0). \end{cases}$$

The correction in each step is then calculated by

$$\Delta\xi = -\frac{F(\xi)}{F'(\xi)}. \tag{8.114}$$

Example 8.4.2 *Use Algorithm No.2 to perform the computation carried out in Example 8.4.1.*

With the given values of the transfer around Earth in $t_{12} = 1000$ s, with $r_1 = 6800$ km, $r_2 = 7000$ km, and $\theta = 60°$, Algorithm No. 2 is implemented by Newton's method with a tolerance, $| F(\xi) | \leq 10^{-10}$ s, and results in the following after 18 iterations:

$$\xi = 90.8126135679$$

$$z = 1.2744132235$$

$$a = \frac{\xi^2}{z} = 6471.15914 \text{ km}$$

$$\eta = 3703.7343 \text{ km}$$

$$p = \frac{r_1 r_2}{\eta}(1 - \cos\theta) = 6425.9469 \text{ km}$$

$$e = \sqrt{1 - \frac{p}{a}} = 0.08358664$$

$$E_1 = \cos^{-1}\left\{\frac{1}{e}\left(1 - \frac{r_1}{a}\right)\right\} = 2.22427056 \text{ rad.}$$

$$t_0 = \frac{e \sin E_1 - E_1}{\sqrt{\mu/a^3}} = -1779.24815 \text{ s .}$$

These values agree with those computed in Example 8.4.1 with Algorithm No. 1.

8.4.3 Hypergeometric Function Method

The solution to Lambert's problem can be formulated in terms of the *hypergeometric* functions evaluated by continued fractions (Battin, 1999) instead of the Stumpff functions. Such a formulation improves convergence, and renders the method insensitive to the starting point of the iterations. Define a universal variable, x, such that the semi-major axis, a, of the transfer orbit is given by

$$a = \frac{a_m}{1 - x^2}, \tag{8.115}$$

where a_m is the semi-major axis of the minimum-energy ellipse connecting the initial and final radii. For an elliptic transfer, the value of x lies in the range $-1 < x < 1$ ($x = 0$ for the minimum-energy transfer), while for a parabola we have $x = 1$. A hyperbolic transfer is defined by $x > 1$. Furthermore, for a given transfer sector, (r_1, r_2, θ), a parameter, λ, is defined as follows:

$$\lambda = \frac{r_1 r_2}{s} \cos\frac{\theta}{2}, \tag{8.116}$$

where s is the semi-perimeter,

$$s = 2a_m = \frac{r_1 + r_2 + c}{2}, \tag{8.117}$$

which implies that $-1 < \lambda < 1$.

Additional variables, y, η, and z, are defined by

$$y = \sqrt{1 - \lambda^2(1 - x^2)}\,, \tag{8.118}$$

$$\eta = y - x\lambda\,, \tag{8.119}$$

and

$$z = \frac{1}{2}(1 - \lambda - x\eta)\,, \tag{8.120}$$

which transform the time equation to the following:

$$t_{12}\sqrt{\frac{\mu}{a_m^3}} = \frac{4}{3}\eta^3 F(z) + 4\lambda\eta\,. \tag{8.121}$$

Here $F(z)$ is a hypergeometric function (Battin, 1999) which can be approximated by a continued fraction as follows:

$$F(z) = \cfrac{1}{1 - \cfrac{\gamma_1 z}{1 - \cfrac{\gamma_2 z}{1 - \cfrac{\gamma_3 z}{1 - \ddots}}}} \tag{8.122}$$

and

$$\gamma_n = \begin{cases} \frac{(n+2)(n+5)}{(2n+1)(2n+3)} & n \text{ odd} \\ \frac{n(n-3)}{(2n+1)(2n+3)} & n \text{ even}\,. \end{cases} \tag{8.123}$$

The orbital parameter of the transfer orbit is derived by substituting Eqs. (8.116)–(8.119) into Eqs. (8.84) and (8.19), and is the following:

$$p = \frac{2r_1 r_2}{s\eta^2}\sin^2\frac{\theta}{2} = \frac{p_m c}{s\eta^2}\,, \tag{8.124}$$

where p_m is the parameter of the minimum-energy ellipse.

When adopting Newton's method to solve the time equation for x, we write

$$f(x) = \frac{4}{3}\eta^3 F(z) + 4\lambda\eta - t_{12}\sqrt{\frac{\mu}{a_m^3}} \tag{8.125}$$

and

$$f'(x) = 4[\eta^2 F(z) + \lambda]\frac{\mathrm{d}\eta}{\mathrm{d}x} + \frac{4}{3}\eta^3 F'(z)\frac{\mathrm{d}z}{\mathrm{d}x}\,, \tag{8.126}$$

where the prime denotes differentiation with respect to the argument. From the definitions given above, we have

$$\frac{\mathrm{d}\eta}{\mathrm{d}x} = -\frac{\lambda\eta}{y}\,, \tag{8.127}$$

$$\frac{\mathrm{d}z}{\mathrm{d}x} = -\frac{\eta^2}{2y}\,, \tag{8.128}$$

and

$$F'(z) = \frac{6 - 3\gamma_1 G(z)}{2(1 - z)[1 - z\gamma_1 G(z)]}\,, \tag{8.129}$$

where

$$G(z) = \cfrac{1}{1 - \cfrac{\gamma_2 z}{1 - \cfrac{\gamma_3 z}{1 - \cfrac{\gamma_4 z}{1 - \ddots}}}} . \tag{8.130}$$

Newton's method can be implemented by choosing a finite number, N, of the continued fractions in Eqs. (8.122) and (8.130) for convergence to a specified tolerance, $| f(x) | \leq \delta$. Such an implementation is carried out in the following example.

Example 8.4.3 *Consider the geocentric, inertial position at the time $t = 0$ of a spacecraft in an Earth orbit given by*

$$\mathbf{r}(0) = \begin{pmatrix} 6500 \\ -2000 \\ 1000 \end{pmatrix} \text{km} ; \qquad \mathbf{v}(0) = \begin{pmatrix} -1 \\ 5 \\ -8 \end{pmatrix} \text{km/s} .$$

It must have the following position and velocity at the time $t = 1500$ s:

$$\mathbf{r}(1500) = \begin{pmatrix} -2800 \\ 5000 \\ -7000 \end{pmatrix} \text{km} ; \qquad \mathbf{v}(1500) = \begin{pmatrix} -8 \\ 2 \\ -1 \end{pmatrix} \text{km/s} .$$

Calculate the two velocity impulses required by the spacecraft to reach the final position and velocity.

This is a problem of transfer orbit determination from the position measured at two different time instants, as well as determining the changes in the velocity at each instant on the transfer orbit.

The transfer triangle is given by

$$\begin{aligned}
r_1 &= | \mathbf{r}(0) | = 6873.8635 \; km \\
r_2 &= | \mathbf{r}(1500) | = 9046.5463 \; km \\
\cos\theta &= \frac{\mathbf{r}(0) \cdot \mathbf{r}(1500)}{r_1 r_2} = -0.5660554 \\
\theta &= 124.4756^\circ \\
c &= \sqrt{r_1^2 + r_2^2 - 2 r_1 r_2 \cos\theta} = 14124.0929 \text{ km} \\
s &= \frac{1}{2}(r_1 + r_2 + c) = 15022.2514 \text{ km} .
\end{aligned}$$

Using the hypergeometric function method, we arrive at the following solution by Newton's method with a tolerance of $| f(x) | \leq 10^{-10}$ in six iterations:

$$\begin{aligned}
x &= 0.839774596 \\
y &= 0.9911486348 \\
\lambda &= 0.2445169524 \\
\eta &= 0.7858095099
\end{aligned}$$

$$
\begin{aligned}
z &= 0.047790092\\
a &= \frac{s}{2(1-x^2)} = 25480.564 \text{ km}\\
p &= \frac{2r1r2}{s\eta^2}\sin^2\frac{\theta}{2} = 10498.355 \text{ km}\\
e &= \sqrt{1-\frac{p}{a}} = 0.7668022993\ .
\end{aligned}
$$

The Lagrange's coefficients required to calculate the initial and final velocities on the transfer orbit are the following:

$$
\begin{aligned}
f &= 1 + \frac{r_2}{p}(\cos\theta - 1) = -0.349486919\\
g &= \frac{r_1 r_2}{h}\sin\theta = 792.4556 \text{ s}\\
\dot{g} &= 1 + \frac{r_1}{p}(\cos\theta - 1) = -0.0253845635\ .
\end{aligned}
$$

The velocity vectors in the transfer orbit at $t = 0$ *and* $t = 1500$ *are thus the following:*

$$
\mathbf{v}_1 = \frac{1}{g}[\mathbf{r}(1500) - f\mathbf{r}(0)] = \begin{pmatrix} -0.6667062\\ 5.4274666\\ -8.3922851 \end{pmatrix} \text{km/s}
$$

$$
\mathbf{v}_2 = \frac{1}{g}[\dot{g}\mathbf{r}(1500) - \mathbf{r}(0)] = \begin{pmatrix} -8.1126607\\ 2.3636368\\ -1.0376709 \end{pmatrix} \text{km/s}\ ,
$$

which yield the two velocity impulses to be applied to the spacecraft as follows:

$$
\Delta\mathbf{v}(0) = \mathbf{v}_1 - \mathbf{v}(0) = \begin{pmatrix} -0.3332938\\ -0.4274666\\ 0.3922851 \end{pmatrix} \text{km/s}
$$

$$
\Delta\mathbf{v}(1500) = \mathbf{v}(1500) - \mathbf{v}_2 = \begin{pmatrix} -0.1126607\\ 0.3636368\\ -0.0376709 \end{pmatrix} \text{km/s}\ .
$$

Exercises

1. Derive an expression for the radial and chordal velocity components, v_r, and v_c, respectively, from the general expression for the parameter of the transfer orbit, p, given by Eq. (8.85).

2. A corollary to Lambert's theorem is the *mean value theorem* which states that the eccentric (or hyperbolic) anomaly of the *mean position* in an orbital transfer is the arithmetic mean

of the eccentric (or hyperbolic) anomalies of the initial and final positions. The mean position, $\mathbf{r_0}$, is the intermediate position where the mean-value theorem of differential calculus dictates that the tangent to the orbit should be parallel to the chord joining $\mathbf{r_1}$ and $\mathbf{r_2}$. Show that for a general orbit, the mean position, $\mathbf{r_0}$, satisfies the following:

$$\mathbf{r_0} \cdot (\mathbf{r_2} - \mathbf{r_1}) = r_0(r_2 - r_1)$$

and for an elliptic transfer, we have

$$r_0 = a\left[1 - e\cos\left(\frac{E_1 + E_2}{2}\right)\right],$$

where

$$r_1 = a(1 - e\cos E_1)\,; \qquad r_2 = a(1 - e\cos E_2)\,.$$

Similarly, show that for a hyperbolic transfer, the mean radius is given by

$$r_0 = a\left[1 - e\cosh\left(\frac{H_1 + H_2}{2}\right)\right],$$

where

$$r_1 = a(1 - e\cosh H_1)\,; \qquad r_2 = a(1 - e\cosh H_2)\,.$$

3. Determine the minimum-energy transfer trajectory for reaching altitude 300 km, latitude −45°, and right ascension 60°, beginning from an initial altitude, latitude, and right ascension of 120 km, 10°, and 256°, respectively. Also calculate the corresponding time of flight. (*Ans.* $a = 6412.33387$ km, $e = 0.166190908$, $t_0 = -1208.373661$ s measured from $t_1 = 0$, $t_{12} = 2550.12$ s.)

4. A spacecraft is to be transferred from

$$\mathbf{r_1} = -5000\mathbf{I} + 2000\mathbf{J} + 9000\mathbf{K}\ \text{(km)}$$

to

$$\mathbf{r_2} = 10000\mathbf{I} - 3000\mathbf{J} - 1000\mathbf{K}\ \text{(km)}$$

around Earth in 2000 s. Determine the classical orbital elements, $(a, e, t_0, i, \Omega, \omega)$, of the transfer orbit.

5. Repeat the computation of Example 8.4.1 by Algorithm No. 2 of the Stumpff function method, as well as by the hypergeometric function method.

6. Calculate the time of flight for a parabolic transfer from $r_1 = 6800$ km to $r_2 = 7000$ km around Earth with the transfer angle, $\theta = 60°$. (*Ans.* $t_{12} = 635.0945$ s.)

7. Calculate the orbital elements, (a, e, t_0), for a transfer from $r_1 = 6800$ km to $r_2 = 7500$ km around Earth with the transfer angle, $\theta = 5°$, in $t_{12} = 150$ s. (*Ans.* $a = 5514.1502$ km, $e = 0.7727925$, $t_0 = -739.716$ s.)

8. Determine the transfer trajectory around Earth and the initial and final velocity vectors for a ballistic missile from a burn-out point of 200 km, latitude 20°, and right ascension 80°, to a re-entry point given by altitude, latitude, and right ascension of 100 km, 45°, and −5°, respectively. The transfer must take place in a retrograde direction in exactly 30 min. (*Ans.* $a = 5197.627$ km, $e = 0.52355$, $t_0 = -980.3175$ s, $\mathbf{v}_1 = 4.56\mathbf{I} + 0.794\mathbf{J} + 4.804\mathbf{K}$ km/s, $\mathbf{v}_2 = -1.985\mathbf{I} - 5.723\mathbf{J} - 3.111\mathbf{K}$ km/s.)

9. Repeat the computation of Example 8.4.3 by Algorithm No. 1 and Algorithm No. 2 of the Stumpff function method.

10. Spacecraft A is in an orbit around Earth of $a = 6700$ km, $e = 0.01$, $\Omega = 120°$, $\omega = -50°$, $i = 91°$, $t_0 = -2000$ s. Spacecraft B is in an Earth orbit with $a = 6600$ km, $e = 0.1$, $\Omega = 110°$, $\omega = -25°$, $i = 10°$, $t_0 = -1000$ s. It is decided to make a rendezvous between two spacecraft by manoeuvring spacecraft B, such that it attains the same position and velocity as that of A after 2000 s. Find the initial and final impulse magnitudes and directions in the celestial frame. (*Ans.* $\Delta\mathbf{v}_1 = 0.1866\mathbf{I} + 1.664\mathbf{J} - 5.699\mathbf{K}$ km/s, $\Delta\mathbf{v}_2 = -9.044\mathbf{I} + 1.366\mathbf{J} - 8.836\mathbf{K}$ km/s.)

11. The orbital elements of the planet Mars referred to the mean ecliptic plane are $a = 1.52372$ AU, $e = 0.09331$, $\Omega = 49.55°$, $\omega = 336.011°$, and $i = 1.8498°$. Assuming that Mars and Earth last passed perihelion exactly 400 and 60 mean solar days ago, respectively, design a transfer orbit for a spacecraft launched from Earth ($a = 1$ AU, $e = 0.01667$, $\omega = 103.059°$, $\Omega = 0$, $i = 0°$) such that it reaches Mars at the next Mars aphelion. Also calculate the two velocity impulses necessary to perform the flight. Neglect the effects of planetary gravitation on the spacecraft. (1 AU=149597870.691 km, $\mu = 1.32712440018 \times 10^{11}$ km^3/s^2 for the sun).

References

Battin RH 1999. *An Introduction to the Mathematics and Methods of Astrodynamics*. American Institute of Aeronautics and Astronautics (AIAA) Education Series, Reston, VA.

9

Orbital Perturbations

The two-body motion is an approximation of the actual flight of a spacecraft in the solar system, and the effects which have been neglected in deriving the Keplerian model, albeit small, are always present. The neglected dynamics can be regarded as perturbations to the Keplerian model, leading to deviations from the orbit calculated by the solution to the two-body problem. The perturbing accelerations are usually small in magnitude compared to the spherical gravity field of the central body, and therefore cause only a slow variation of the shape, size, and/or the plane of the orbit. Examples of such disturbances include a non-uniform gravitational field due to an aspherical mass distribution of the central body or caused by the presence of a distant third body, atmospheric drag, the solar radiation pressure, and the small but continuously applied thrust of an electric propulsion engine. Often the perturbation of the largest magnitude can be identified to prevail on the spacecraft, and the vehicle's motion can be corrected accordingly. For example, a satellite in a low orbit experiences the largest orbital perturbations due to the atmospheric drag and the non-spherical gravity field of a planet. In contrast, a spacecraft in a high orbit around a planet has a negligible atmospheric effect and a much smaller planetary gravitational anomaly when compared to the gravitational tug from the nearest moon as well as the sun. A spacecraft in deep space within the inner solar system, and far away from any planet, is perturbed appreciably by solar radiation pressure, especially if it is equipped with a large solar array. The trajectory of a spacecraft on an interplanetary voyage is affected by gravitational attraction of the closest planet, apart from solar gravitation. If the perturbations are not taken into account, the actual trajectory of the spacecraft would deviate wildly from the orbit calculated by solving the two-body problem. Hence, for navigational purposes, an adequate model of the perturbing acceleration and its effect on the spacecraft's motion are necessary. The objective of this chapter is to model the effects of small perturbations on a two-body orbit.

9.1 Perturbing Acceleration

The motion of a spacecraft around a spherical body perturbed by a continuous acceleration, $\mathbf{u}(t)$, is described by

$$\frac{\partial \mathbf{v}}{\partial t} + \mu \frac{\mathbf{r}}{r^3} = \mathbf{u} \tag{9.1}$$

Foundations of Space Dynamics, First Edition. Ashish Tewari.

and

$$\frac{\partial \mathbf{r}}{\partial t} = \mathbf{v}, \tag{9.2}$$

where $\mathbf{r}(t)$ is the position vector measured in an inertial reference frame with origin at the centre of the spherical body, and μ is the gravitational constant of the two-body system. The perturbing acceleration (also called the secondary acceleration) magnitude, u, is generally small in magnitude when compared to the primary acceleration due to gravity, μ/r^2. However, there is a significant departure of the trajectory from a two-body orbit ($u = 0$) even if u is very small (but non-zero).

The perturbing acceleration, $\mathbf{u}$, can arise out of two types of forces: (a) a conservative force, such as the gravity of an aspherical central body or a third body, and (b) a non-conservative force, such as the thrust produced by the engines, the atmospheric drag, the solar radiation pressure, etc. The perturbation produced by a conservative force can be expressed as the gradient of a scalar potential function which varies only with the position, $U(\mathbf{r})$, as follows:

$$\mathbf{u} = \nabla U(\mathbf{r}) = \left(\frac{\partial U}{\partial \mathbf{r}}\right)^T. \tag{9.3}$$

A conservative perturbation conserves the energy, ε, of a Keplerian orbit, but can cause the orbital plane to rotate due to the variation of angular momentum. A non-conservative perturbation by definition causes a variation in ε, and hence the shape and size of the orbit also vary with time, along with the possible rotation of the orbital plane.

9.2 Osculating Orbit

If the magnitude of the perturbing acceleration, u, is much smaller than that of the primary acceleration (i.e., $u \ll \mu/r^2$), the resulting trajectory is a slow departure from the Keplerian motion at any instant of time. This fact results in the useful approximation of an *osculating orbit*, defined to be the trajectory tangential to the perturbed orbit, which would be followed if the perturbing acceleration were *suddenly* to be removed. The osculating orbit is thus an imaginary two-body orbit satisfying Eq. (9.1) for $\mathbf{u}(t) = \mathbf{0}$, $t \geq \tau$, with the initial condition, $\mathbf{r}(\tau)$, $\mathbf{v}(\tau)$, and would be the future trajectory of the spacecraft if the perturbing acceleration, $\mathbf{u}(t)$, suddenly vanishes at $t = \tau$, and remains zero for $t > \tau$. The actual trajectory of the spacecraft can then be approximated by the osculating orbit beginning at any time, τ, called the *epoch*, where the initial condition, $\mathbf{r}(\tau)$, $\mathbf{v}(\tau)$, is applied. At a future time, $t_1 > \tau$, a correction can be applied to the osculating orbit by a suitable perturbation method considered in this chapter, such that the new Keplerian orbit with epoch, $t = t_1$, better approximates the actual trajectory. Since both the position and velocity vectors must match with those of the osculating orbit at the epoch, it follows that the osculating orbit is tangential to the actual trajectory at the epoch, τ, thereby explaining the adjective "osculating" (or "kissing"). If the perturbing acceleration is small at all times, $u(t) \ll \mu/r(t)^2$, $t \geq \tau$, then the proximity and tangency with the osculating orbit is maintained until a correction is applied. The concept of osculating orbit is depicted in Fig. 9.1.

If the perturbing acceleration, u, is large and were suddenly to be removed, there would be an impulsive change in the acceleration at the instant of removal, which translates into a step change in the velocity, $\mathbf{v}$. In such a case, the resulting orbit would *not* be tangential to the original (perturbed) orbit; hence there would be a "bump" in the trajectory at $t = \tau$. Therefore, the

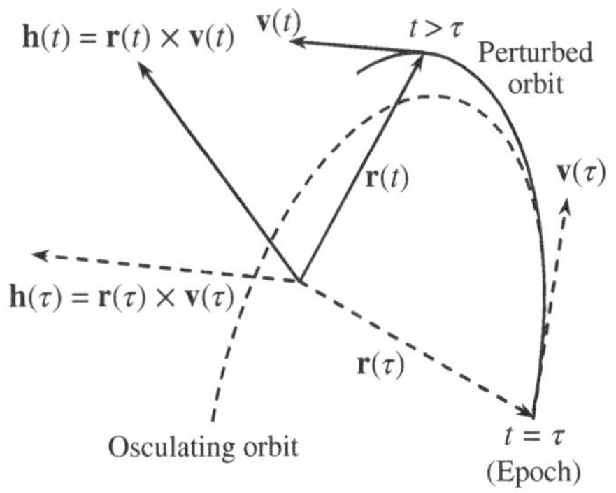

Figure 9.1 The osculating orbit as a reference for the trajectory in the presence of a small perturbing acceleration, **u**.

concept of an osculating orbit is valid only if the perturbation, $u(t)$, is always small, so that at any instant, a sudden vanishing of the perturbation would not cause an abrupt change in the velocity, thereby approximating the resulting orbit as being tangential (or osculating) to the original (perturbed) orbit at the epoch, $t = \tau$.

An osculating orbit can be represented by $\mathbf{h} = \mathbf{r}(\tau) \times \mathbf{v}(\tau) = $ const., whose constant direction relative to an inertial frame, $(\mathbf{I}, \mathbf{J}, \mathbf{K})$, is given by the direction cosine, $\cos\sigma = \mathbf{i_h} \cdot \mathbf{K}$, where $\mathbf{i_h} = \mathbf{h}/h$. Without a loss of generality, the osculating orbital plane can be taken normal to the $\mathbf{K}$-axis, i.e., $\sigma = 0$. The differential equations of perturbed motion, Eqs. (9.1) and (9.2), can then be resolved either in Cartesian or spherical coordinates relative to the osculating orbit, in terms of variations in $r, \dot{r}, h$, and σ caused by $\mathbf{u}$. For example, the use of the local-horizon frame, $(\mathbf{i_r}, \mathbf{i}_\theta, \mathbf{i_h})$, with $\mathbf{i_r} \times \mathbf{i}_\theta = \mathbf{i_h}$, where $\mathbf{i_h}$ is aligned with $\mathbf{K}$ at $t = \tau$, utilizes the spherical coordinates, (r, θ, σ), with $h = r^2\dot{\theta}$ to resolve the position and velocity in an orbit departing from the osculating orbit for $t \geq \tau$:

$$\ddot{r} - \frac{h^2}{r^3} + \frac{\mu}{r^2} = u_r\,, \tag{9.4}$$

$$\dot{h} = r u_\theta\,, \tag{9.5}$$

and

$$\dot{\sigma} = \frac{r}{h} u_n\,. \tag{9.6}$$

Here the overdot represents taking a partial derivative with respect to time, $\dot{(.)} = \partial(.)/\partial t$, and the perturbing acceleration is resolved into the radial, circumferential, and normal components, u_r, u_θ, and u_n, respectively:

$$\mathbf{u} = (u_r, u_\theta, u_n)^T\,. \tag{9.7}$$

The changing orientation of the local-horizon frame, $(\mathbf{i_r}, \mathbf{i}_\theta, \mathbf{i_h})$, relative to the inertial frame, $(\mathbf{I}, \mathbf{J}, \mathbf{K})$, is represented by the Euler angles, δ (declination), λ (right ascension), and σ (plane-change angle). While the angle σ satisfies $\cos\sigma = \mathbf{i_h} \cdot \mathbf{K}$, the angles δ and λ represent the direction of $\mathbf{i_r}$ as follows:

$$\mathbf{i_r} = \mathbf{I}\cos\delta\cos\lambda + \mathbf{J}\cos\delta\sin\lambda + \mathbf{K}\sin\delta\,. \tag{9.8}$$

The remaining two axes of the reference frame are updated at each time instant by

$$\mathbf{i_h} = \frac{\mathbf{r} \times \mathbf{v}}{|\,\mathbf{r} \times \mathbf{v}\,|} \tag{9.9}$$

and

$$\mathbf{i_e} = \mathbf{i_h} \times \mathbf{i_r} \,. \tag{9.10}$$

Equations (9.4)–(9.6) represent only four out of the required six motion variables, $(r, \dot{r}, h, \sigma)$. The remaining two variables are the angles δ and λ, whose first-order time variations, $\dot{\delta}$, $\dot{\lambda}$, are derived by differentiating Eq. (9.8) with time. Such a variation of the orientation of a reference frame with time is covered in Chap. 11 under *attitude kinematics*.

9.3 Variation of Parameters

Section 9.2 showed how the concept of an osculating orbit can be applied to approximate the perturbed equations of motion if the magnitude of the perturbing acceleration, u, is much smaller than that of the primary acceleration (i.e., $u \ll \mu/r^2$). An approach based upon the concept of the osculating orbit is the classical Lagrange method of *variation of parameters*. In this method, the nonlinear differential equations governing the perturbed motion, Eqs. (9.1) and (9.2), which *do not* possess any particular set of parameters for an arbitrary perturbing acceleration, u, are approximated by a set of slowly time-varying parameters representing the two-body, osculating orbit if $u \ll \mu/r^2$. The method of variation of parameters is especially useful when the small perturbing acceleration arises out of a conservative force, such as the gravity of an aspherical planet or that of a distant third body. In such a case, we have $\mathbf{u} = \nabla U(\mathbf{r}) = (\partial U/\partial \mathbf{r})^T$.

To describe the method of variation of parameters, the Cartesian coordinates are employed for simplicity of notation. In an inertial Cartesian reference frame, $(\mathbf{I}, \mathbf{J}, \mathbf{K})$, with its origin at the centre of the primary body of gravitational constant, μ, the position of the spacecraft is represented by the Cartesian coordinates, (X_1, X_2, X_3), as $\mathbf{r} = X_1\mathbf{I} + X_2\mathbf{J} + X_3\mathbf{K}$. When the perturbing acceleration is absent, the following equations of Keplerian motion are valid:

$$\ddot{X}_1 + \mu\frac{X_1}{r^3} = 0\,; \qquad \ddot{X}_2 + \mu\frac{X_2}{r^3} = 0\,; \qquad \ddot{X}_3 + \mu\frac{X_3}{r^3} = 0\,, \tag{9.11}$$

where $r = \sqrt{X_1^2 + X_2^2 + X_3^2}$. The solution to Eq. (9.11) is completely described by a set of six scalar constants, $(c_1, c_2, c_3, c_4, c_5, c_6)$, at any given time t. An example of such a set of constants is the classical orbital elements, $(a, e, t_0, \Omega, i, \omega)$. Thus we have the following solution to Eq. (9.11):

$$X_1(t) = f_1(c_1, c_2, c_3, c_4, c_5, c_6, t)\,, \tag{9.12}$$

$$X_2(t) = f_2(c_1, c_2, c_3, c_4, c_5, c_6, t)\,, \tag{9.13}$$

$$X_3(t) = f_3(c_1, c_2, c_3, c_4, c_5, c_6, t)\,, \tag{9.14}$$

$$\dot{X}_1(t) = g_1(c_1, c_2, c_3, c_4, c_5, c_6, t)\,, \tag{9.15}$$

$$\dot{X}_2(t) = g_2(c_1, c_2, c_3, c_4, c_5, c_6, t)\,, \tag{9.16}$$

$$\dot{X}_3(t) = g_3(c_1, c_2, c_3, c_4, c_5, c_6, t)\,, \tag{9.17}$$

where

$$g_i(c_1, c_2, c_3, c_4, c_5, c_6, t) = \frac{\partial f_i(c_1, c_2, c_3, c_4, c_5, c_6, t)}{\partial t}, (i = 1, 2, 3). \tag{9.18}$$

The solution vectors can be expressed in the following vector-matrix form:

$$\mathbf{r}(t) = \mathbf{r}(\boldsymbol{\alpha}, t) ; \qquad \mathbf{v}(t) = \mathbf{v}(\dot{\boldsymbol{\alpha}}, t), \tag{9.19}$$

where

$$\begin{aligned} \boldsymbol{\alpha} &= (c_1, c_2, c_3, c_4, c_5, c_6)^T \\ \dot{\boldsymbol{\alpha}} &= (\dot{c}_1, \dot{c}_2, \dot{c}_3, \dot{c}_4, \dot{c}_5, \dot{c}_6)^T . \end{aligned} \tag{9.20}$$

In the presence of a perturbing acceleration, $\mathbf{u} = (u_1, u_2, u_3)^T$, the perturbed equations of motion, Eqs. (9.1) and (9.2), are expressed in the Cartesian coordinates as follows:

$$\ddot{X}_1 + \mu \frac{X_1}{r^3} = u_1 ; \qquad \ddot{X}_2 + \mu \frac{X_2}{r^3} = u_2 ; \qquad \ddot{X}_3 + \mu \frac{X_3}{r^3} = u_3 . \tag{9.21}$$

The method of variation of parameters postulates that the perturbed motion is exactly described by the same set of six parameters as the unperturbed motion, $\boldsymbol{\alpha} = (c_1, c_2, c_3, c_4, c_5, c_6)^T$. However, these parameters are no longer constants, but vary slowly with time. It must be reiterated that such an approximation is valid only if the perturbing acceleration, $\mathbf{u}$, in Eq. (9.1) is small. Hence, the perturbed motion variables, $X_1, X_2, X_3, \dot{X}_1, \dot{X}_2, \dot{X}_3$, are also given by Eqs. (9.12)–(9.18), which were originally derived for Keplerian motion. Since these six parameters instantaneously describe a Keplerian orbit at any time, they are called *osculating elements*, and hence can be used to define an osculating orbit at a given time, t. To determine the solution, (f_i, g_i), $i = 1, 2, 3$, which satisfies Eq. (9.21), the first step is to determine the time variation of the osculating elements, $(c_1, c_2, c_3, c_4, c_5, c_6)$. The second step is to consider the time derivative of Eqs. (9.12)–(9.14) for the perturbed motion expressed as follows:

$$\dot{X}_1(t) = \frac{\partial f_1}{\partial t} + \sum_{k=1}^{6} \frac{\partial f_1}{\partial c_k} \dot{c}_k , \tag{9.22}$$

$$\dot{X}_2(t) = \frac{\partial f_2}{\partial t} + \sum_{k=1}^{6} \frac{\partial f_2}{\partial c_k} \dot{c}_k , \tag{9.23}$$

$$\dot{X}_3(t) = \frac{\partial f_3}{\partial t} + \sum_{k=1}^{6} \frac{\partial f_3}{\partial c_k} \dot{c}_k , \tag{9.24}$$

which are expressed in the vector-matrix form as follows:

$$\mathbf{v}(t) = \dot{\mathbf{r}} = \frac{\partial \mathbf{r}}{\partial t} + \frac{\partial \mathbf{r}}{\partial \boldsymbol{\alpha}} \dot{\boldsymbol{\alpha}} . \tag{9.25}$$

When compared with Eqs. (9.15)–(9.18), the validity of Eqs. (9.22)–(9.24) requires that the following conditions must hold:

$$\sum_{k=1}^{6} \frac{\partial f_1}{\partial c_k} \dot{c}_k = 0 , \tag{9.26}$$

$$\sum_{k=1}^{6} \frac{\partial f_2}{\partial c_k} \dot{c}_k = 0 \,, \tag{9.27}$$

and

$$\sum_{k=1}^{6} \frac{\partial f_3}{\partial c_k} \dot{c}_k = 0 \,. \tag{9.28}$$

Equations (9.26)–(9.28) are the differential equations to be satisfied by the parameters, c_k, $k = 1, 2, \ldots, 6$.

The second time derivatives of Eqs. (9.12)–(9.14) yield the following:

$$\ddot{X}_1(t) = \frac{\partial^2 f_1}{\partial t^2} + \sum_{k=1}^{6} \frac{\partial g_1}{\partial c_k} \dot{c}_k \,, \tag{9.29}$$

$$\ddot{X}_2(t) = \frac{\partial^2 f_2}{\partial t^2} + \sum_{k=1}^{6} \frac{\partial g_2}{\partial c_k} \dot{c}_k \,, \tag{9.30}$$

$$\ddot{X}_3(t) = \frac{\partial^2 f_3}{\partial t^2} + \sum_{k=1}^{6} \frac{\partial g_3}{\partial c_k} \dot{c}_k \,, \tag{9.31}$$

or

$$\dot{\mathbf{v}} = \frac{\partial \mathbf{v}}{\partial t} + \frac{\partial \mathbf{v}}{\partial \boldsymbol{\alpha}} \dot{\boldsymbol{\alpha}} \,. \tag{9.32}$$

When these are substituted into Eq. (9.21), the following differential equations are obtained:

$$\frac{\partial^2 f_1}{\partial t^2} + \mu \frac{f_1}{r^3} + \sum_{k=1}^{6} \frac{\partial g_1}{\partial c_k} \dot{c}_k = u_1 \,, \tag{9.33}$$

$$\frac{\partial^2 f_2}{\partial t^2} + \mu \frac{f_2}{r^3} + \sum_{k=1}^{6} \frac{\partial g_2}{\partial c_k} \dot{c}_k = u_2 \,, \tag{9.34}$$

$$\frac{\partial^2 f_3}{\partial t^2} + \mu \frac{f_3}{r^3} + \sum_{k=1}^{6} \frac{\partial g_3}{\partial c_k} \dot{c}_k = u_3 \,. \tag{9.35}$$

Since Eq. (9.11) is satisfied by the solution given by Eqs. (9.12)–(9.14), the first two terms on the left-hand side of each of the Eqs. (9.33)–(9.35) vanish for an osculating orbit at any given time instant t. Hence, a simplification of these equations yields the following additional conditions to be satisfied by the osculating elements, c_k, $k = 1, 2, \ldots, 6$:

$$\sum_{k=1}^{6} \frac{\partial g_1}{\partial c_k} \dot{c}_k = u_1 \,, \tag{9.36}$$

$$\sum_{k=1}^{6} \frac{\partial g_2}{\partial c_k} \dot{c}_k = u_2 \,, \tag{9.37}$$

$$\sum_{k=1}^{6} \frac{\partial g_3}{\partial c_k} \dot{c}_k = u_3 \,. \tag{9.38}$$

Equations (9.26)–(9.28) and (9.36)–(9.38) are the conditions to be satisfied by the first-order time variation of the osculating elements, $\dot{c}_k$, $k = 1, 2, \ldots, 6$. When the substitution, $X_i = f_i$, $\dot{X}_i = g_i$, $i = 1, 2, 3$, is made in these equations, they are expressed as follows:

$$\sum_{k=1}^{6} \frac{\partial X_1}{\partial c_k} \dot{c}_k = 0 , \tag{9.39}$$

$$\sum_{k=1}^{6} \frac{\partial X_2}{\partial c_k} \dot{c}_k = 0 , \tag{9.40}$$

$$\sum_{k=1}^{6} \frac{\partial X_3}{\partial c_k} \dot{c}_k = 0 , \tag{9.41}$$

$$\sum_{k=1}^{6} \frac{\partial \dot{X}_1}{\partial c_k} \dot{c}_k = u_1 , \tag{9.42}$$

$$\sum_{k=1}^{6} \frac{\partial \dot{X}_2}{\partial c_k} \dot{c}_k = u_2 , \tag{9.43}$$

$$\sum_{k=1}^{6} \frac{\partial \dot{X}_3}{\partial c_k} \dot{c}_k = u_3 \tag{9.44}$$

or in the following vector-matrix form:

$$\frac{\partial \mathbf{r}}{\partial \boldsymbol{\alpha}} \dot{\boldsymbol{\alpha}} = \mathbf{0} , \tag{9.45}$$

$$\frac{\partial \mathbf{v}}{\partial \boldsymbol{\alpha}} \dot{\boldsymbol{\alpha}} = \mathbf{u} . \tag{9.46}$$

9.3.1 Lagrange Brackets

Equations (9.39)–(9.44) must be solved to derive the first-order time variations of osculating elements. However, to do so the perturbing acceleration, $\mathbf{u} = (u_1, u_2, u_3)^T$, must be suitably modelled. When the perturbation arises out of a conservative force, such as the gravity of an aspherical planet, or that of a distant third body, the perturbing acceleration can be modelled to be the following gradient of a scalar potential function which varies only with the position, $\mathbf{r}$:

$$\mathbf{u} = \begin{Bmatrix} u_1 \\ u_2 \\ u_3 \end{Bmatrix} = \nabla U(\mathbf{r}) = \left(\frac{\partial U}{\partial \mathbf{r}} \right)^T = \begin{Bmatrix} \frac{\partial U}{\partial X_1} \\ \frac{\partial U}{\partial X_2} \\ \frac{\partial U}{\partial X_3} \end{Bmatrix} . \tag{9.47}$$

This is a crucial model, and leads to a great simplification in the variational equations. The substitution of Eq. (9.47) into Eqs. (9.39)–(9.44), followed by the multiplication of each successive equation of the set by $-\partial \dot{X}_1 / \partial c_j$, $-\partial \dot{X}_2 / \partial c_j$, $-\partial \dot{X}_3 / \partial c_j$, $-\partial X_1 / \partial c_j$, $-\partial X_2 / \partial c_j$, and $-\partial X_3 / \partial c_j$, then adding the resulting equations, produces the following:

$$\sum_{k=1}^{6} [c_j, \; c_k] \dot{c}_k = \frac{\partial U}{\partial c_j} , \qquad (j = 1, 2, \ldots, 6) , \tag{9.48}$$

where

$$[c_j,\ c_k] = \sum_{i=1}^{3} \left(\frac{\partial X_i}{\partial c_j} \frac{\partial \dot{X}_i}{\partial c_k} - \frac{\partial X_i}{\partial c_k} \frac{\partial \dot{X}_i}{\partial c_j} \right), \qquad (j = 1, \ldots, 6), (k = 1, \ldots, 6) \tag{9.49}$$

are termed the *Lagrange brackets*. This derivation is given in the vector-matrix form by the substitution of Eq. (9.47) into Eq. (9.46), the multiplication of Eq. (9.45) with $(\partial \mathbf{v}/\partial \boldsymbol{\alpha})^T$, followed by the multiplication of Eq. (9.46) with $(\partial \mathbf{r}/\partial \boldsymbol{\alpha})^T$ and the subtraction of the two equations, resulting in

$$\mathbf{L}\dot{\boldsymbol{\alpha}} = \left[\frac{\partial U(\mathbf{r})}{\partial \boldsymbol{\alpha}} \right]^T, \tag{9.50}$$

where

$$\mathbf{L} = \left(\frac{\partial \mathbf{r}}{\partial \boldsymbol{\alpha}} \right)^T \frac{\partial \mathbf{v}}{\partial \boldsymbol{\alpha}} - \left(\frac{\partial \mathbf{v}}{\partial \boldsymbol{\alpha}} \right)^T \frac{\partial \mathbf{r}}{\partial \boldsymbol{\alpha}}. \tag{9.51}$$

Here $\mathbf{L}$, which is of dimension 6×6, has its elements,

$$[c_i,\ c_j] = \frac{\partial \mathbf{r}^T}{\partial c_i} \frac{\partial \mathbf{v}}{\partial c_j} - \frac{\partial \mathbf{r}^T}{\partial c_j} \frac{\partial \mathbf{v}}{\partial c_i}, \qquad (j = 1, \ldots, 6), (k = 1, \ldots, 6), \tag{9.52}$$

as the Lagrange brackets. The Lagrange brackets have the properties $[c_j,\ c_k] = -[c_k,\ c_j]$ and $[c_j,\ c_j] = 0$, which imply that $\mathbf{L}$ is a skew-symmetric matrix, i.e., $\mathbf{L} = -\mathbf{L}^T$.

The Lagrange brackets were originally derived by Lagrange for orbital motion with a conservative perturbation, but as seen in Eq. (9.51), they are independent of the perturbation. Therefore, the variational equations, Eq. (9.50), can be generalized for an arbitrary perturbing acceleration, $\mathbf{u}$, as follows:

$$\mathbf{L}\dot{\boldsymbol{\alpha}} = \left(\frac{\partial \mathbf{r}}{\partial \boldsymbol{\alpha}} \right)^T \mathbf{u}. \tag{9.53}$$

Furthermore, if c_i, c_j are any two osculating elements, then the time derivative of the Lagrange bracket $[c_i, c_j]$, is given by

$$\frac{\mathrm{d}}{\mathrm{d}t}[c_i,\ c_j] = \frac{\partial}{\partial c_j} \left(\frac{\partial \mathbf{v}^T}{\partial t} \right) \frac{\partial \mathbf{r}}{\partial c_i} + \frac{\partial \mathbf{v}^T}{\partial c_j} \frac{\partial \mathbf{v}}{\partial c_i} - \frac{\partial}{\partial c_i} \left(\frac{\partial \mathbf{v}^T}{\partial t} \right) \frac{\partial \mathbf{r}}{\partial c_j} - \frac{\partial \mathbf{v}^T}{\partial c_i} \frac{\partial \mathbf{v}}{\partial c_j}, \tag{9.54}$$

where the second and the fourth terms on the right-hand side cancel each other. Also, the osculating elements c_i, c_j satisfy the Keplerian equations of motion, Eq. (9.11), which can be expressed as follows:

$$\frac{\partial^2 \mathbf{r}}{\partial t^2} = \left[\frac{\partial(\mu/r)}{\partial \mathbf{r}} \right]^T, \tag{9.55}$$

and imply that

$$\frac{\partial \mathbf{v}^T}{\partial t} = \frac{\partial(\mu/r)}{\partial \mathbf{r}}. \tag{9.56}$$

Thus we have

$$
\begin{aligned}
\frac{\mathrm{d}}{\mathrm{d}t}[c_i, c_j] &= \frac{\partial}{\partial c_j}\left[\frac{\partial(\mu/r)}{\partial \mathbf{r}}\right]\frac{\partial \mathbf{r}}{\partial c_i} - \frac{\partial}{\partial c_i}\left[\frac{\partial(\mu/r)}{\partial \mathbf{r}}\right]\frac{\partial \mathbf{r}}{\partial c_j} \\
&= \frac{\partial}{\partial \mathbf{r}}\left[\frac{\partial(\mu/r)}{\partial c_j}\right]\frac{\partial \mathbf{r}}{\partial c_i} - \frac{\partial}{\partial \mathbf{r}}\left[\frac{\partial(\mu/r)}{\partial c_i}\right]\frac{\partial \mathbf{r}}{\partial c_j} \\
&= \frac{\partial^2(\mu/r)}{\partial c_j \partial c_i} - \frac{\partial^2(\mu/r)}{\partial c_i \partial c_j} = 0 .
\end{aligned} \tag{9.57}
$$

Therefore, the Lagrange brackets, $L_{ij} = [c_i, c_j]$, are *time invariant*, which implies that they can be evaluated at any point in the orbit, and remain constant throughout time. This is an important property of the Lagrange brackets, quite useful in deriving the time variation of osculating elements.

9.4 Lagrange Planetary Equations

In this section, the Lagrange brackets are computed for an elliptic orbit ($0 \le e < 1$). To derive the Lagrange brackets for the perturbed elliptic motion, the traditional choice of osculating elements is that of the classical orbital elements; that is,

$$
\boldsymbol{\alpha} = (a, e, M_0, \Omega, i, \omega)^T , \tag{9.58}
$$

where $M_0 = -t_0\sqrt{\mu/a^3} = -nt_0$ is the mean anomaly at $t = 0$. Taking advantage of the fact that the Lagrange brackets are time invariant, they need to be computed only at a particular time. For simplicity, this time is selected to be the time of periapsis, $t = t_0$, for which we have $r = a(1 - e)$ and $\theta = E = M = 0$.

The computation of the Lagrange brackets requires expressing the position and velocity vectors in the celestial reference frame, $(\mathbf{I}, \mathbf{J}, \mathbf{K})$, using the classical orbital elements. For this purpose, the coordinate transformation between the perifocal frame, $(\mathbf{i}_\mathrm{e}, \mathbf{i}_\mathrm{p}, \mathbf{i}_\mathrm{h})$, and the celestial frame is necessary, which is given by

$$
\begin{Bmatrix} \mathbf{I} \\ \mathbf{J} \\ \mathbf{K} \end{Bmatrix} = \mathbf{C}^* \begin{Bmatrix} \mathbf{i}_\mathrm{e} \\ \mathbf{i}_\mathrm{p} \\ \mathbf{i}_\mathrm{h} \end{Bmatrix} . \tag{9.59}
$$

Here $\mathbf{C}^* = \mathbf{C_3}^T(\Omega)\mathbf{C_1}^T(i)\mathbf{C_3}^T(\omega)$ is the following rotation matrix (see Chap. 5):

$$
\mathbf{C}^* = \begin{pmatrix} c^*_{11} & c^*_{12} & c^*_{13} \\ c^*_{21} & c^*_{22} & c^*_{23} \\ c^*_{31} & c^*_{32} & c^*_{33} \end{pmatrix} , \tag{9.60}
$$

where

$$c_{11}^* = \cos\Omega\cos\omega - \sin\Omega\sin\omega\cos i \tag{9.61}$$

$$c_{12}^* = -\cos\Omega\sin\omega - \sin\Omega\cos\omega\cos i \tag{9.62}$$

$$c_{13}^* = \sin\Omega\sin i \tag{9.63}$$

$$c_{21}^* = \sin\Omega\cos\omega + \cos\Omega\sin\omega\cos i \tag{9.64}$$

$$c_{22}^* = -\sin\Omega\sin\omega + \cos\Omega\cos\omega\cos i \tag{9.65}$$

$$c_{23}^* = -\cos\Omega\sin i \tag{9.66}$$

$$c_{31}^* = \sin\omega\sin i \tag{9.67}$$

$$c_{32}^* = \cos\omega\sin i \tag{9.68}$$

$$c_{33}^* = \cos i\,. \tag{9.69}$$

The perifocal coordinates can be expressed either in terms of the true anomaly, θ, or the eccentric anomaly, E. Here we choose the latter for simplicity, as given by Eq. (4.23), and express the celestial position and velocity in the following Cartesian coordinates:

$$\mathbf{r} = \begin{Bmatrix} X_1 \\ X_2 \\ X_3 \end{Bmatrix} = \mathbf{C}^* \begin{Bmatrix} a(\cos E - e) \\ b\sin E \\ 0 \end{Bmatrix}, \tag{9.70}$$

$$\mathbf{v} = \begin{Bmatrix} \dot{X}_1 \\ \dot{X}_2 \\ \dot{X}_3 \end{Bmatrix} = \mathbf{C}^* \begin{Bmatrix} -\frac{an\sin E}{1-e\cos E} \\ \frac{bn\cos E}{1-e\cos E} \\ 0 \end{Bmatrix}, \tag{9.71}$$

where $b = a\sqrt{1-e^2}$ is the semi-minor axis of the ellipse, and E is the eccentric anomaly.

The time of periapsis, $t = t_0, E = 0$, is chosen for deriving the Lagrange brackets. To this end, we begin with the Lagrange brackets of the classical Euler angles:

$$\frac{\partial \mathbf{r}}{\partial i} = \begin{Bmatrix} \frac{\partial X_1}{\partial i} \\ \frac{\partial X_2}{\partial i} \\ \frac{\partial X_3}{\partial i} \end{Bmatrix} = a(1-e) \begin{Bmatrix} -\sin\Omega\sin\omega\sin i \\ -\cos\Omega\sin\omega\sin i \\ \sin\omega\cos i \end{Bmatrix} = a(1-e) \begin{Bmatrix} c_{13}^* \\ c_{23}^* \\ c_{33}^* \end{Bmatrix} \sin\omega\,. \tag{9.72}$$

$$\frac{\partial \mathbf{r}}{\partial \omega} = \begin{Bmatrix} \frac{\partial X_1}{\partial \omega} \\ \frac{\partial X_2}{\partial \omega} \\ \frac{\partial X_3}{\partial \omega} \end{Bmatrix} = a(1-e) \begin{Bmatrix} -\cos\Omega\sin\omega - \sin\Omega\cos\omega\cos i \\ -\sin\Omega\sin\omega + \cos\Omega\cos\omega\cos i \\ \cos\omega\sin i \end{Bmatrix} = a(1-e) \begin{Bmatrix} c_{12}^* \\ c_{22}^* \\ c_{32}^* \end{Bmatrix}. \tag{9.73}$$

$$\frac{\partial \mathbf{r}}{\partial \Omega} = \begin{Bmatrix} \frac{\partial X_1}{\partial \Omega} \\ \frac{\partial X_2}{\partial \Omega} \\ \frac{\partial X_3}{\partial \Omega} \end{Bmatrix} = a(1-e) \begin{Bmatrix} -\sin\Omega\cos\omega - \cos\Omega\sin\omega\cos i \\ \cos\Omega\cos\omega - \sin\Omega\sin\omega\cos i \\ 0 \end{Bmatrix} = a(1-e) \begin{Bmatrix} -c_{21}^* \\ c_{11}^* \\ 0 \end{Bmatrix}. \quad (9.74)$$

$$\frac{\partial \mathbf{v}}{\partial i} = \begin{Bmatrix} \frac{\partial X_1}{\partial i} \\ \frac{\partial X_2}{\partial i} \\ \frac{\partial X_3}{\partial i} \end{Bmatrix} = \frac{bn}{1-e} \begin{Bmatrix} \sin\Omega\sin i \\ -\cos\Omega\sin i \\ \cos i \end{Bmatrix} \cos\omega = \frac{bn}{1-e} \begin{Bmatrix} c_{13}^* \\ c_{23}^* \\ c_{33}^* \end{Bmatrix} \cos\omega. \quad (9.75)$$

$$\frac{\partial \mathbf{v}}{\partial \omega} = \begin{Bmatrix} \frac{\partial X_1}{\partial \omega} \\ \frac{\partial X_2}{\partial \omega} \\ \frac{\partial X_3}{\partial \omega} \end{Bmatrix} = \frac{bn}{1-e} \begin{Bmatrix} -\cos\Omega\cos\omega + \sin\Omega\sin\omega\cos i \\ -\sin\Omega\cos\omega - \cos\Omega\sin\omega\cos i \\ -\sin\omega\sin i \end{Bmatrix} = -\frac{bn}{1-e} \begin{Bmatrix} c_{11}^* \\ c_{21}^* \\ c_{31}^* \end{Bmatrix}. \quad (9.76)$$

$$\frac{\partial \mathbf{v}}{\partial \Omega} = \begin{Bmatrix} \frac{\partial X_1}{\partial \Omega} \\ \frac{\partial X_2}{\partial \Omega} \\ \frac{\partial X_3}{\partial \Omega} \end{Bmatrix} = \frac{bn}{1-e} \begin{Bmatrix} \sin\Omega\sin\omega - \cos\Omega\cos\omega\cos i \\ -\cos\Omega\sin\omega - \sin\Omega\cos\omega\cos i \\ 0 \end{Bmatrix} = \frac{bn}{1-e} \begin{Bmatrix} -c_{22}^* \\ c_{12}^* \\ 0 \end{Bmatrix}. \quad (9.77)$$

To evaluate the Lagrange brackets pertaining to the orbital shape, we consider the following derivatives of Kepler's equation, $E - e\sin E = nt + M_0$, with respect to a, e, and M_0 evaluated at $t = t_0$, $E = 0$:

$$\frac{\partial E}{\partial a} - e\cos E\frac{\partial E}{\partial a} = \frac{\partial n}{\partial a}t = -\frac{3n}{2a}t, \quad (9.78)$$

which at $E = 0$ yields

$$\frac{\partial E}{\partial a} = -\frac{3n}{2r}t = -\frac{3n}{2a(1-e)}t_0 = \frac{3M_0}{2a(1-e)}. \quad (9.79)$$

Similarly, we have

$$\frac{\partial E}{\partial e} - \sin E - e\cos E\frac{\partial E}{\partial e} = 0 \quad (9.80)$$

and

$$\frac{\partial E}{\partial M_0} - e\cos E\frac{\partial M_0}{\partial e} = 1, \quad (9.81)$$

which yield the following at $E = 0$:

$$\frac{\partial E}{\partial e} = 0 \quad (9.82)$$

and

$$\frac{\partial E}{\partial M_0} = \frac{1}{1-e}. \quad (9.83)$$

Thus we have

$$\frac{\partial}{\partial a}a(1-e\cos E)=1-e\,, \tag{9.84}$$

$$\begin{aligned}\frac{\partial}{\partial a}b\sin E &= \sqrt{1-e^2}\sin E + b\cos E\frac{\partial E}{\partial a}\\ &= \frac{3M_0 b}{2a(1-e)}\,,\end{aligned} \tag{9.85}$$

$$\begin{aligned}\frac{\partial}{\partial a}\frac{an\sin E}{1-e\cos E} &= -\frac{1}{2}a^{-3/2}\frac{\sqrt{\mu}\sin E}{1-e\cos E}\\ &= \frac{an[\cos E(1-e\cos E)-e\sin^2 E]}{(1-e\cos E)^2}\frac{\partial E}{\partial a}\\ &= \frac{an}{1-e}\frac{\partial E}{\partial a} = \frac{3}{2}\frac{nM_0}{(1-e)^2}\,,\end{aligned} \tag{9.86}$$

$$\begin{aligned}\frac{\partial}{\partial a}\frac{bn\cos E}{1-e\cos E} &= -\frac{1}{2}a^{-3/2}\sqrt{1-e^2}\frac{\sqrt{\mu}\cos E}{1-e\cos E}\\ &\quad + bn\frac{[-\sin E(1-e\cos E)-e\cos E\sin E]}{(1-e\cos E)^2}\frac{\partial E}{\partial a}\\ &= -\frac{1}{2}a^{-3/2}\sqrt{\mu}\sqrt{\frac{1+e}{1-e}} = -\frac{1}{2}n\sqrt{\frac{1+e}{1-e}}\,,\end{aligned} \tag{9.87}$$

$$\frac{\partial}{\partial e}a(1-e\cos E) = -a\cos E = -a\,, \tag{9.88}$$

$$\frac{\partial}{\partial e}b\sin E = -\frac{e}{\sqrt{1-e^2}}\sin E + b\cos E\frac{\partial E}{\partial e} = 0\,, \tag{9.89}$$

$$\frac{\partial}{\partial e}\frac{an\sin E}{1-e\cos E} = an\frac{\sin E\cos E}{(1-e\cos E)^2} + \frac{[an\cos E(1-e\cos E)-e\,\sin^2 E]}{(1-e\cos E)^2}\frac{\partial E}{\partial e} = 0\,, \tag{9.90}$$

$$\begin{aligned}\frac{\partial}{\partial e}\frac{bn\cos E}{1-e\cos E} &= -\frac{ane(1-e^2)^{-1/2}\cos E}{1-e\cos E} + \frac{bn\,\cos^2 E}{(1-e\cos E)^2}\\ &= \frac{an}{\sqrt{1-e^2}(1-e)^2}[-e(1-e)+1-e^2] = \frac{an}{\sqrt{1-e^2}(1-e)}\,,\end{aligned} \tag{9.91}$$

$$\frac{\partial}{\partial M_0}a(1-e\cos E) = ae\sin E\frac{\partial E}{\partial M_0} = 0\,, \tag{9.92}$$

$$\frac{\partial}{\partial M_0}b\sin E = b\cos E\frac{\partial E}{\partial M_0} = \frac{b}{1-e}\,, \tag{9.93}$$

$$\frac{\partial}{\partial M_0}\frac{an\sin E}{1-e\cos E} = an\frac{[\cos E(1-e\cos E)-e\sin^2 E]}{(1-e\cos E)^2}\frac{\partial E}{\partial M_0} = \frac{an}{(1-e)^2}\,, \tag{9.94}$$

$$\frac{\partial}{\partial M_0}\frac{bn\cos E}{1-e\cos E} = bn\frac{[-\sin E(1-e\cos E)-e\cos E\sin E]}{(1-e\cos E)^2}\frac{\partial E}{\partial M_0} = 0\,. \tag{9.95}$$

Thus taking the derivatives of Eqs. (9.70) and (9.71) results in the following:

$$\frac{\partial \mathbf{r}}{\partial a} = \begin{Bmatrix} \frac{\partial X_1}{\partial a} \\ \frac{\partial X_2}{\partial a} \\ \frac{\partial X_3}{\partial a} \end{Bmatrix} = (1-e)\begin{Bmatrix} c_{11}^* \\ c_{21}^* \\ c_{31}^* \end{Bmatrix} + \frac{3M_0 b}{2a(1-e)}\begin{Bmatrix} c_{12}^* \\ c_{22}^* \\ c_{32}^* \end{Bmatrix}. \tag{9.96}$$

$$\frac{\partial \mathbf{r}}{\partial e} = \begin{Bmatrix} \frac{\partial X_1}{\partial e} \\ \frac{\partial X_2}{\partial e} \\ \frac{\partial X_3}{\partial e} \end{Bmatrix} = -a\begin{Bmatrix} c_{11}^* \\ c_{21}^* \\ c_{31}^* \end{Bmatrix}. \tag{9.97}$$

$$\frac{\partial \mathbf{r}}{\partial M_0} = \begin{Bmatrix} \frac{\partial X_1}{\partial M_0} \\ \frac{\partial X_2}{\partial M_0} \\ \frac{\partial X_3}{\partial M_0} \end{Bmatrix} = \frac{b}{1-e}\begin{Bmatrix} c_{12}^* \\ c_{22}^* \\ c_{32}^* \end{Bmatrix}. \tag{9.98}$$

$$\frac{\partial \mathbf{v}}{\partial a} = \begin{Bmatrix} \frac{\partial X_1}{\partial a} \\ \frac{\partial X_2}{\partial a} \\ \frac{\partial X_3}{\partial a} \end{Bmatrix} = -\frac{3}{2}\frac{nM_0}{(1-e)^2}\begin{Bmatrix} c_{11}^* \\ c_{21}^* \\ c_{31}^* \end{Bmatrix} - \frac{1}{2}n\sqrt{\frac{1+e}{1-e}}\begin{Bmatrix} c_{12}^* \\ c_{22}^* \\ c_{32}^* \end{Bmatrix}. \tag{9.99}$$

$$\frac{\partial \mathbf{v}}{\partial e} = \begin{Bmatrix} \frac{\partial X_1}{\partial e} \\ \frac{\partial X_2}{\partial e} \\ \frac{\partial X_3}{\partial e} \end{Bmatrix} = \frac{an}{\sqrt{1-e^2}(1-e)}\begin{Bmatrix} c_{12}^* \\ c_{22}^* \\ c_{32}^* \end{Bmatrix}. \tag{9.100}$$

$$\frac{\partial \mathbf{v}}{\partial M_0} = \begin{Bmatrix} \frac{\partial X_1}{\partial M_0} \\ \frac{\partial X_2}{\partial M_0} \\ \frac{\partial X_3}{\partial M_0} \end{Bmatrix} = -\frac{an}{(1-e)^2}\begin{Bmatrix} c_{11}^* \\ c_{21}^* \\ c_{31}^* \end{Bmatrix}. \tag{9.101}$$

The Lagrange brackets are finally computed as follows:

$$\begin{aligned}[i,\ \Omega] &= \frac{\partial \mathbf{r}^T}{\partial i}\frac{\partial \mathbf{v}}{\partial \Omega} - \frac{\partial \mathbf{r}^T}{\partial \Omega}\frac{\partial \mathbf{v}}{\partial i} \\ &= a(1-e)\sin\omega(c_{13}^*,\ c_{23}^*,\ c_{33}^*)\frac{bn}{1-e}\begin{Bmatrix} -c_{22}^* \\ c_{12}^* \\ 0 \end{Bmatrix}\end{aligned}$$

$$
\begin{aligned}
&-a(1-e)(-c_{21}^*,\ c_{11}^*,\ 0)\frac{bn}{1-e}\begin{Bmatrix} c_{13}^* \\ c_{23}^* \\ c_{33}^* \end{Bmatrix}\cos\omega \\
&= abn\sin\omega(-c_{13}^*c_{22}^* + c_{23}^*c_{12}^*) - abn\cos\omega(-c_{21}^*c_{13}^* + c_{11}^*c_{23}^*) \\
&= abn\sin\omega[\sin\Omega\sin i(\sin\Omega\sin\omega - \cos\Omega\cos\omega\cos i) \\
&\quad - \cos\Omega\sin i(-\cos\Omega\sin\omega - \sin\Omega\cos\omega\cos i)] \\
&\quad -abn\cos\omega[-(\sin\Omega\cos\omega + \cos\Omega\sin\omega\cos i)\sin\Omega\sin i \\
&\quad -(\cos\Omega\cos\omega - \sin\Omega\sin\omega\cos i)\cos\Omega\sin i] \\
&= abn(\cos^2\omega + \sin^2\omega)\sin i = abn\sin i\,.
\end{aligned} \tag{9.102}
$$

$$
\begin{aligned}
[i,\ \omega] &= \frac{\partial \mathbf{r}^T}{\partial i}\frac{\partial \mathbf{v}}{\partial \omega} - \frac{\partial \mathbf{r}^T}{\partial \omega}\frac{\partial \mathbf{v}}{\partial i} \\
&= -a(1-e)\sin\omega(c_{13}^*,\ c_{23}^*,\ c_{33}^*)\frac{bn}{1-e}\begin{Bmatrix} c_{11}^* \\ c_{21}^* \\ c_{31}^* \end{Bmatrix} \\
&\quad -a(1-e)(c_{12}^*,\ c_{22}^*,\ c_{32}^*)\frac{bn}{1-e}\begin{Bmatrix} c_{13}^* \\ c_{23}^* \\ c_{33}^* \end{Bmatrix}\cos\omega \\
&= -abn[\sin\omega(c_{13}^*c_{11}^* + c_{23}^*c_{21}^* + c_{33}^*c_{31}^*) \\
&\quad + \cos\omega(c_{12}^*c_{13}^* + c_{22}^*c_{23}^* + c_{32}^*c_{33}^*)] \\
&= -abn(0-0) = 0\,.
\end{aligned} \tag{9.103}
$$

$$
\begin{aligned}
[\Omega,\ \omega] &= \frac{\partial \mathbf{r}^T}{\partial \Omega}\frac{\partial \mathbf{v}}{\partial \omega} - \frac{\partial \mathbf{r}^T}{\partial \omega}\frac{\partial \mathbf{v}}{\partial \Omega} \\
&= -abn(-c_{21}^*c_{11}^* + c_{11}^*c_{21}^*) - abn(-c_{12}^*c_{22}^* + c_{22}^*c_{12}^*) = 0\,.
\end{aligned} \tag{9.104}
$$

$$
\begin{aligned}
[a,\ i] &= \frac{\partial \mathbf{r}^T}{\partial a}\frac{\partial \mathbf{v}}{\partial i} - \frac{\partial \mathbf{r}^T}{\partial i}\frac{\partial \mathbf{v}}{\partial a} \\
&= bn\cos\omega(c_{11}^*c_{13}^* + c_{21}^*c_{23}^* + c_{31}^*c_{33}^*) \\
&\quad + \frac{3M_0b^2n}{2a}(c_{12}^*c_{13}^* + c_{22}^*c_{23}^* + c_{32}^*c_{33}^*) \\
&\quad + \frac{3}{2}\frac{anM_0}{(1-e)}\sin\omega(c_{13}^*c_{11}^* + c_{23}^*c_{21}^* + c_{33}^*c_{31}^*) \\
&\quad + \frac{1}{2}bn(c_{13}^*c_{12}^* + c_{23}^*c_{22}^* + c_{33}^*c_{32}^*) \\
&= 0+0+0+0 = 0\,.
\end{aligned} \tag{9.105}
$$

$$
\begin{aligned}
[a,\ \omega] &= \frac{\partial \mathbf{r}^T}{\partial a}\frac{\partial \mathbf{v}}{\partial \omega} - \frac{\partial \mathbf{r}^T}{\partial \omega}\frac{\partial \mathbf{v}}{\partial a} \\
&= -bn[(c_{11}^*)^2 + (c_{21}^*)^2 + (c_{31}^*)^2] - \frac{3nM_0 b^2}{2a(1-e)^2}(c_{12}^* c_{11}^* + c_{22}^* c_{21}^* + c_{32}^* c_{31}^*) \\
&\quad + \frac{3}{2}\frac{anM_0}{(1-e)}(c_{11}^* c_{12}^* + c_{21}^* c_{22}^* + c_{31}^* c_{32}^*) + \frac{1}{2}bn[(c_{12}^*)^2 + (c_{22}^*)^2 + (c_{32}^*)^2] \\
&= -bn + 0 + 0 + \frac{1}{2}bn = -\frac{1}{2}bn\,. \qquad (9.106)
\end{aligned}
$$

$$
\begin{aligned}
[a,\ \Omega] &= \frac{\partial \mathbf{r}^T}{\partial a}\frac{\partial \mathbf{v}}{\partial \Omega} - \frac{\partial \mathbf{r}^T}{\partial \Omega}\frac{\partial \mathbf{v}}{\partial a} \\
&= bn(-c_{22}^* c_{11}^* + c_{21}^* c_{12}^*) + \frac{3M_0 b^2 n}{2a(1-e)^2}(-c_{22}^* c_{12}^* + c_{22}^* c_{12}^*) \\
&\quad + \frac{3}{2}\frac{anM_0}{(1-e)}(-c_{21}^* c_{11}^* + c_{21}^* c_{11}^*) + \frac{1}{2}bn(-c_{21}^* c_{12}^* + c_{11}^* c_{22}^*) \\
&= \frac{1}{2}bn(-c_{22}^* c_{11}^* + c_{21}^* c_{12}^*) \\
&= \frac{1}{2}bn[-(-\sin\Omega\sin\omega + \cos\Omega\cos\omega\cos i)(\cos\Omega\cos\omega - \sin\Omega\sin\omega\cos i) \\
&\quad + (\sin\Omega\cos\omega + \cos\Omega\sin\omega\cos i)(-\cos\Omega\sin\omega - \sin\Omega\cos\omega\cos i)] \\
&= \frac{1}{2}bn[-\sin^2\omega(\sin^2\Omega + \cos^2\Omega) - \cos^2\omega(\sin^2\Omega + \cos^2\Omega)]\cos i \\
&= -\frac{1}{2}bn\cos i\,. \qquad (9.107)
\end{aligned}
$$

$$
\begin{aligned}
[a,\ e] &= \frac{\partial \mathbf{r}^T}{\partial a}\frac{\partial \mathbf{v}}{\partial e} - \frac{\partial \mathbf{r}^T}{\partial e}\frac{\partial \mathbf{v}}{\partial a} \\
&= \frac{an}{\sqrt{1-e^2}}(c_{11}^* c_{12}^* + c_{21}^* c_{22}^* + c_{31}^* c_{32}^*) + \frac{3nM_0 a}{2(1-e)^2}[(c_{12}^*)^2 + (c_{22}^*)^2 + (c_{32}^*)^2] \\
&\quad - \frac{3}{2}\frac{anM_0}{(1-e)^2}[(c_{11}^*)^2 + (c_{21}^*)^2 + (c_{31}^*)^2] - \frac{1}{2}an\sqrt{\frac{1+e}{1-e}}(c_{11}^* c_{12}^* + c_{21}^* c_{22}^* + c_{31}^* c_{32}^*) \\
&= 0 + 0 = 0\,. \qquad (9.108)
\end{aligned}
$$

$$
\begin{aligned}
[a,\ M_0] &= \frac{\partial \mathbf{r}^T}{\partial a}\frac{\partial \mathbf{v}}{\partial M_0} - \frac{\partial \mathbf{r}^T}{\partial M_0}\frac{\partial \mathbf{v}}{\partial a} \\
&= -\frac{an}{1-e}[(c_{11}^*)^2 + (c_{21}^*)^2 + (c_{31}^*)^2] - \frac{3M_0 n^2 b}{2(1-e)^2}(c_{12}^* c_{11}^* + c_{22}^* c_{21}^* + c_{32}^* c_{31}^*) \\
&\quad + \frac{3b}{2}\frac{nM_0}{(1-e)^3}(c_{11}^* c_{12}^* + c_{21}^* c_{22}^* + c_{31}^* c_{32}^*) + \frac{1}{2}\frac{an}{(1-e)}[(c_{12}^*)^2 + (c_{22}^*)^2 + (c_{32}^*)^2] \\
&= -\frac{an}{1-e} + \frac{1}{2}an\frac{(1-e^2)}{(1-e)^2} = -\frac{1}{2}an\,. \qquad (9.109)
\end{aligned}
$$

$$
\begin{aligned}
[e,\ i] &= \frac{\partial \mathbf{r}^T}{\partial e}\frac{\partial \mathbf{v}}{\partial i} - \frac{\partial \mathbf{r}^T}{\partial i}\frac{\partial \mathbf{v}}{\partial e} \\
&= -\frac{abn}{1-e}(c_{11}^* c_{13}^* + c_{21}^* c_{23}^* + c_{31}^* c_{33}^*) \\
&\quad -\frac{a^2 n}{\sqrt{1-e^2}}(c_{12}^* c_{13}^* + c_{22}^* c_{23}^* + c_{32}^* c_{33}^*) = 0 + 0 = 0\,.
\end{aligned} \tag{9.110}
$$

$$
\begin{aligned}
[e,\ \omega] &= \frac{\partial \mathbf{r}^T}{\partial e}\frac{\partial \mathbf{v}}{\partial \omega} - \frac{\partial \mathbf{r}^T}{\partial \omega}\frac{\partial \mathbf{v}}{\partial e} \\
&= \frac{abn}{1-e}[(c_{11}^*)^2 + (c_{21}^*)^2 + (c_{31}^*)^2] - \frac{a^2 n}{\sqrt{1-e^2}}[(c_{12}^*)^2 + (c_{22}^*)^2 + (c_{32}^*)^2] \\
&= \frac{abn}{1-e} - \frac{a^2 n}{\sqrt{1-e^2}} = \frac{a^2 ne}{\sqrt{1-e^2}}\,.
\end{aligned} \tag{9.111}
$$

$$
\begin{aligned}
[e,\ \Omega] &= \frac{\partial \mathbf{r}^T}{\partial e}\frac{\partial \mathbf{v}}{\partial \Omega} - \frac{\partial \mathbf{r}^T}{\partial \Omega}\frac{\partial \mathbf{v}}{\partial e} \\
&= -\frac{abn}{1-e}(-c_{22}^* c_{11}^* + c_{12}^* c_{21}^*) - \frac{a^2 n}{\sqrt{1-e^2}}(-c_{21}^* c_{12}^* + c_{22}^* c_{11}^*) \\
&= \left(\frac{abn}{1-e} - \frac{a^2 n}{\sqrt{1-e^2}}\right)(-c_{21}^* c_{12}^* + c_{22}^* c_{11}^*) \\
&= \frac{a^2 ne}{\sqrt{1-e^2}}[\sin^2\omega(\sin^2\Omega + \cos^2\Omega) + \cos^2\omega(\sin^2\Omega + \cos^2\Omega)]\cos i \\
&= \frac{a^2 ne}{\sqrt{1-e^2}}\cos i\,.
\end{aligned} \tag{9.112}
$$

$$
\begin{aligned}
[e,\ M_0] &= \frac{\partial \mathbf{r}^T}{\partial e}\frac{\partial \mathbf{v}}{\partial M_0} - \frac{\partial \mathbf{r}^T}{\partial M_0}\frac{\partial \mathbf{v}}{\partial e} \\
&= \frac{a^2 n}{(1-e)^2}[(c_{11}^*)^2 + (c_{21}^*)^2 + (c_{31}^*)^2] - \frac{a^2 n}{(1-e)^2}[(c_{12}^*)^2 + (c_{22}^*)^2 + (c_{32}^*)^2] \\
&= 0\,.
\end{aligned} \tag{9.113}
$$

$$
\begin{aligned}
[M_0,\ i] &= \frac{\partial \mathbf{r}^T}{\partial M_0}\frac{\partial \mathbf{v}}{\partial i} - \frac{\partial \mathbf{r}^T}{\partial i}\frac{\partial \mathbf{v}}{\partial M_0} \\
&= \frac{b^2 n}{(1-e)^2}(c_{12}^* c_{13}^* + c_{22}^* c_{23}^* + c_{32}^* c_{33}^*)\cos\omega \\
&\quad + \frac{a^2 n}{(1-e)}(c_{13}^* c_{11}^* + c_{23}^* c_{21}^* + c_{33}^* c_{31}^*)\sin\omega \\
&= 0 + 0 = 0\,.
\end{aligned} \tag{9.114}
$$

$$
\begin{aligned}
[M_0,\ \omega] &= \frac{\partial \mathbf{r}^T}{\partial M_0}\frac{\partial \mathbf{v}}{\partial \omega} - \frac{\partial \mathbf{r}^T}{\partial \omega}\frac{\partial \mathbf{v}}{\partial M_0} \\
&= -\frac{b^2 n}{(1-e)^2}(c_{12}^* c_{11}^* + c_{22}^* c_{21}^* + c_{32}^* c_{31}^*) \\
&\quad + \frac{a^2 n}{(1-e)}(c_{12}^* c_{11}^* + c_{22}^* c_{21}^* + c_{32}^* c_{31}^*) \\
&= 0 + 0 = 0 .
\end{aligned} \tag{9.115}
$$

$$
\begin{aligned}
[M_0,\ \Omega] &= \frac{\partial \mathbf{r}^T}{\partial M_0}\frac{\partial \mathbf{v}}{\partial \Omega} - \frac{\partial \mathbf{r}^T}{\partial \Omega}\frac{\partial \mathbf{v}}{\partial M_0} \\
&= \frac{b^2 n}{(1-e)^2}(-c_{22}^* c_{12}^* + c_{22}^* c_{12}^*) + \frac{a^2 n}{(1-e)}(-c_{21}^* c_{11}^* + c_{11}^* c_{21}^*) \\
&= 0 + 0 = 0 .
\end{aligned} \tag{9.116}
$$

Of the 15 Lagrange brackets, only six are non-zero, which are listed in Table 9.1, and complete the set required to determine the skew-symmetric matrix, **L**. The parametric equations for an elliptic orbit are now derived from Eq. (9.50) as follows:

$$
\begin{aligned}
-\frac{1}{2}bn\left(\cos i\frac{\mathrm{d}\Omega}{\mathrm{d}t} + \frac{\mathrm{d}\omega}{\mathrm{d}t}\right) - \frac{1}{2}an\frac{\mathrm{d}M_0}{\mathrm{d}t} &= \frac{\partial U}{\partial a} \\
\frac{a^2 ne}{\sqrt{1-e^2}}\left(\cos i\frac{\mathrm{d}\Omega}{\mathrm{d}t} + \frac{\mathrm{d}\omega}{\mathrm{d}t}\right) &= \frac{\partial U}{\partial e} \\
\frac{1}{2}an\frac{\mathrm{d}a}{\mathrm{d}t} &= \frac{\partial U}{\partial M_0} \\
-abn\sin i\frac{\mathrm{d}i}{\mathrm{d}t} + \frac{1}{2}bn\cos i\frac{\mathrm{d}a}{\mathrm{d}t} - \frac{a^2 ne}{\sqrt{1-e^2}}\cos i\frac{\mathrm{d}e}{\mathrm{d}t} &= \frac{\partial U}{\partial \Omega}
\end{aligned} \tag{9.117}
$$

Table 9.1 Non-zero Lagrange brackets for a perturbed elliptic orbit.

$[a,\ M_0]$	=	$-\frac{1}{2}an$
$[a,\ \Omega]$	=	$-\frac{1}{2}bn\cos i$
$[a,\ \omega]$	=	$-\frac{1}{2}bn$
$[e,\ \Omega]$	=	$\frac{a^2 ne}{\sqrt{1-e^2}}\cos i$
$[e,\ \omega]$	=	$\frac{a^2 ne}{\sqrt{1-e^2}}$
$[i,\ \Omega]$	=	$abn\sin i$

$$anb \sin i \frac{\mathrm{d}\Omega}{\mathrm{d}t} = \frac{\partial U}{\partial i}$$

$$\frac{1}{2} bn \frac{\mathrm{d}a}{\mathrm{d}t} - \frac{a^2 ne}{\sqrt{1-e^2}} \frac{\mathrm{d}e}{\mathrm{d}t} = \frac{\partial U}{\partial \omega}.$$

The solution to Eq. (9.117) is obtained by inverting **L**, which, by Eq. (9.50), yields the following:

$$\dot{\boldsymbol{\alpha}} = \mathbf{L}^{-1} \left[\frac{\partial U(\mathbf{r})}{\partial \boldsymbol{\alpha}} \right]^T, \tag{9.118}$$

or

$$\begin{aligned}
\frac{\mathrm{d}a}{\mathrm{d}t} &= \frac{2}{an} \frac{\partial U}{\partial M_0} \\
\frac{\mathrm{d}e}{\mathrm{d}t} &= -\frac{\sqrt{1-e^2}}{a^2 ne} \frac{\partial U}{\partial \omega} + \frac{(1-e^2)}{a^2 ne} \frac{\partial U}{\partial M_0} \\
\frac{\mathrm{d}M_0}{\mathrm{d}t} &= -\frac{2}{an} \frac{\partial U}{\partial a} - \frac{(1-e^2)}{a^2 ne} \frac{\partial U}{\partial e} \\
\frac{\mathrm{d}\Omega}{\mathrm{d}t} &= \frac{1}{a^2 n \sqrt{1-e^2} \sin i} \frac{\partial U}{\partial i} \\
\frac{\mathrm{d}i}{\mathrm{d}t} &= -\frac{1}{a^2 n \sqrt{1-e^2} \sin i} \left(\frac{\partial U}{\partial \Omega} - \cos i \frac{\partial U}{\partial \Omega} \right) \\
\frac{\mathrm{d}\omega}{\mathrm{d}t} &= -\frac{\cos i}{a^2 n \sqrt{1-e^2} \sin i} \frac{\partial U}{\partial i} + \frac{\sqrt{1-e^2}}{a^2 ne} \frac{\partial U}{\partial e}.
\end{aligned} \tag{9.119}$$

Equation (9.119), the set of first-order variational equations in the classical orbital elements due to a conservative perturbation, $U(\mathbf{r})$, is called *Lagrange planetary equations*. Lagrange planetary equations have been employed in the past to prepare the *Ephemeris charts* representing the motion of the planets using classical orbital elements. The variation of the classical orbital elements with time, beginning from a given time of *epoch* ($t = 0$), is taken into account, and the elements are listed as being constants between two epochs.

It is evident from Eq. (9.117) that the matrix **L** becomes singular for $e = 0$, which denotes a circular orbit, and for $i = 0, \pm\pi$, which are the singularities of the classical Euler angle representation. Furthermore, Eq. (9.119) indicates that some terms on the right-hand side, which multiply the perturbing acceleration components, become very large as e approaches unity. Therefore, for the cases where $e \simeq 0$, $e \simeq 1$, and $i = 0, \pm\pi$, the Lagrange planetary equations become ill conditioned, and yield inaccurate results.

Example 9.4.1 *An alternative parameter set, $\boldsymbol{\alpha}$ can be chosen to represent the Lagrange planetary equations. Such a set is the* Delaunay elements *(also called the* canonical variables*), which* *is given in terms of the classical orbital elements by*

$$\boldsymbol{\alpha} = \begin{Bmatrix} -\frac{1}{2}a^2n^2 \\ anb \\ anb\cos i \\ \frac{M_0}{n} \\ \omega \\ \Omega \end{Bmatrix} = \begin{Bmatrix} -\frac{\mu}{2a} \\ h \\ h\cos i \\ -t_0 \\ \omega \\ \Omega \end{Bmatrix} . \tag{9.120}$$

The Delaunay elements are grouped into two subsets, $\boldsymbol{\alpha} = (\boldsymbol{\alpha}_1^T, \boldsymbol{\alpha}_2^T)^T$, the first three elements being given by $\boldsymbol{\alpha}_1^T = (a_1, a_2, a_3) = [-\mu/(2a), h, h\cos i]$, and the last three elements constituting the other subset, $\boldsymbol{\alpha}_2^T = (a_4, a_5, a_6) = (-t_0, \omega, \Omega)$. The advantage of using the Delaunay elements is that they result in Lagrange brackets which are either zero or ± 1, and convert Lagrange planetary equations into the following canonical form*:*

$$\dot{\boldsymbol{\alpha}}_1 = \left(\frac{\partial U}{\partial \boldsymbol{\alpha}_2}\right)^T ; \qquad \dot{\boldsymbol{\alpha}}_2 = -\left(\frac{\partial U}{\partial \boldsymbol{\alpha}_1}\right)^T . \tag{9.121}$$

Hence the transformed variational model is much simpler to derive for the Delaunay elements than for the classical orbital elements.

The singularities of the classical orbital elements are retained if the perturbation potential derivatives in the Delaunay elements must be derived from those in the classical elements, which requires the inversion of the following set of equations:

$$\begin{aligned}
\frac{\partial U}{\partial a} &= \frac{an^2}{2}\frac{\partial U}{\partial a_1} + \frac{bn}{2}\left(\frac{\partial U}{\partial a_2} + \cos i\frac{\partial U}{\partial a_3}\right) + \frac{3M_0}{2an}\frac{\partial U}{\partial a_4} \\
\frac{\partial U}{\partial e} &= -\frac{a^2ne}{\sqrt{1-e^2}}\left(\frac{\partial U}{\partial a_2} + \cos i\frac{\partial U}{\partial a_3}\right) \\
\frac{\partial U}{\partial M_0} &= \frac{1}{n}\frac{\partial U}{\partial a_4} \\
\frac{\partial U}{\partial \Omega} &= \frac{\partial U}{\partial a_6} \\
\frac{\partial U}{\partial i} &= -anb\sin i\frac{\partial U}{\partial a_3} \\
\frac{\partial U}{\partial \omega} &= \frac{\partial U}{\partial a_5} .
\end{aligned} \tag{9.122}$$

However, if the perturbation potential, $U(\mathbf{r})$, is directly expressed in the Delaunay elements, then the problems associated with the classical orbital elements are removed.

9.5 Gauss Variational Model

Lagrange's variations-of-parameters model, applied to an elliptic orbit in Section 9.4,, can be generalized for a small but arbitrary (not necessarily conservative) perturbation, $\mathbf{u}$, applied to a general orbit with the use of Eq. (9.53) as follows:

$$\frac{\mathrm{d}\boldsymbol{\alpha}}{\mathrm{d}t} = \mathbf{L}^{-1}\left(\frac{\partial \mathbf{r}}{\partial \boldsymbol{\alpha}}\right)^T \mathbf{u}, \tag{9.123}$$

where the derivative $\partial \mathbf{r}/\partial \boldsymbol{\alpha}$ is evaluated at the current time, t. This provides an alternative derivation procedure, wherein the time variation of the osculating elements is analytically obtained from the perturbing acceleration, $\mathbf{u}$. The Gauss variational model is such a method where the small perturbing acceleration is expressed in spherical polar coordinates, $\mathbf{u} = (u_r, u_\theta, u_n)^T$, and the resulting variations in the two-body orbit are resolved into a set of first-order differential equations in time for the classical orbital elements, $a, e, t_0, i, \Omega, \omega$. Of course, it is expected that the Gauss variational equations are equivalent to the Lagrange planetary equations for an elliptic orbit and the given perturbation.

Consider the position vector, $\mathbf{r}$, resolved in the perifocal frame fixed to the osculating orbit, $(\mathbf{i_e}, \mathbf{i_p}, \mathbf{i_h})$. A local-horizon frame, $(\mathbf{i_r}, \mathbf{i}_\theta, \mathbf{i_h})$, can then be utilized to resolve the perturbing acceleration as follows:

$$\mathbf{u} = u_r \mathbf{i_r} + u_\theta \mathbf{i}_\theta + u_n \mathbf{i_h}, \tag{9.124}$$

such that $\mathbf{i_r} = \mathbf{r}/r$, $\mathbf{i_h} = \mathbf{h}/h$ and $\mathbf{i}_\theta = \mathbf{i_h} \times \mathbf{i_r}$. Finally, the celestial frame, $(\mathbf{I}, \mathbf{J}, \mathbf{K})$, is used as an inertial reference frame for the equations of perturbed motion. The coordinate transformations between these three coordinate frames are given by the following:

$$\begin{Bmatrix} \mathbf{I} \\ \mathbf{J} \\ \mathbf{K} \end{Bmatrix} = \mathbf{C}^* \begin{Bmatrix} \mathbf{i_e} \\ \mathbf{i_p} \\ \mathbf{i_h} \end{Bmatrix} = \begin{pmatrix} c_{11}^* & c_{12}^* & c_{13}^* \\ c_{21}^* & c_{22}^* & c_{23}^* \\ c_{31}^* & c_{32}^* & c_{33}^* \end{pmatrix} \begin{Bmatrix} \mathbf{i_e} \\ \mathbf{i_p} \\ \mathbf{i_h} \end{Bmatrix}, \tag{9.125}$$

$$\begin{Bmatrix} \mathbf{i_r} \\ \mathbf{i}_\theta \\ \mathbf{i_h} \end{Bmatrix} = \mathbf{C}_\theta \begin{Bmatrix} \mathbf{i_e} \\ \mathbf{i_p} \\ \mathbf{i_h} \end{Bmatrix}, \tag{9.126}$$

where $\mathbf{C}^*$ is the rotation matrix given by Eq. (9.61) in terms of the classical Euler angles, Ω, i, ω, representing the osculating orbital plane, and $\mathbf{C}_\theta$ gives the orientation of the local-horizon frame relative to the osculating orbital frame by

$$\mathbf{C}_\theta = \begin{pmatrix} \cos\theta & -\sin\theta & 0 \\ \sin\theta & \cos\theta & 0 \\ 0 & 0 & 1 \end{pmatrix} = \begin{pmatrix} \frac{a}{r}(\cos E - e) & -\frac{b}{r}\sin E & 0 \\ \frac{b}{r}\sin E & \frac{a}{r}(\cos E - e) & 0 \\ 0 & 0 & 1 \end{pmatrix}. \tag{9.127}$$

Here the angle θ would be the true anomaly, and E the eccentric anomaly, if the actual trajectory were a two-body orbit. If the vectors $\mathbf{r}^*$ and $\mathbf{u}^*$ denote the position and perturbing acceleration, respectively, resolved in the celestial reference frame, $(\mathbf{I}, \mathbf{J}, \mathbf{K})$, then we have

$$\mathbf{r}^* = \begin{Bmatrix} x \\ y \\ z \end{Bmatrix} = \mathbf{C}^*\mathbf{C}_\theta \begin{Bmatrix} r \\ 0 \\ 0 \end{Bmatrix} \tag{9.128}$$

$$\mathbf{u}^* = \begin{Bmatrix} u_x \\ u_y \\ u_z \end{Bmatrix} = \mathbf{C}^*\mathbf{C}_\theta \begin{Bmatrix} u_r \\ u_\theta \\ u_n \end{Bmatrix}. \tag{9.129}$$

The perturbation term on the right-hand side of Eq. (9.123),

$$\left(\frac{\partial \mathbf{r}}{\partial \boldsymbol{\alpha}}\right)^T \mathbf{u} = \mathbf{u}^T \frac{\partial \mathbf{r}}{\partial \boldsymbol{\alpha}}, \tag{9.130}$$

requires the inertial position and acceleration vectors given by Eqs. (9.128) and (9.129), respectively. Thus for a given element, α, of the set of osculating elements, $\boldsymbol{\alpha}$, we have

$$(\mathbf{u}^*)^T \frac{\partial \mathbf{r}^*}{\partial \alpha} = \mathbf{u}^T \mathbf{C}_\theta^T (\mathbf{C}^*)^T \frac{\partial \mathbf{C}^*}{\partial \alpha} \mathbf{C}_\theta \mathbf{r}. \tag{9.131}$$

The partial derivatives of the rotation matrix, $\mathbf{C}^*$, with respect to the classical Euler angles, Ω, i, ω, are the following:

$$\frac{\partial \mathbf{C}^*}{\partial \Omega} = \begin{pmatrix} -c_{21}^* & -c_{22}^* & -c_{23}^* \\ c_{11}^* & c_{12}^* & c_{13}^* \\ 0 & 0 & 0 \end{pmatrix}. \tag{9.132}$$

$$\frac{\partial \mathbf{C}^*}{\partial i} = \begin{pmatrix} c_{13}^* \sin\omega & c_{13}^* \cos\omega & \sin\Omega\cos i \\ c_{23}^* \sin\omega & c_{23}^* \cos\omega & -\cos\Omega\cos i \\ c_{33}^* \sin\omega & c_{33}^* \cos\omega & -\sin i \end{pmatrix}. \tag{9.133}$$

$$\frac{\partial \mathbf{C}^*}{\partial \omega} = \begin{pmatrix} c_{12}^* & -c_{11}^* & 0 \\ c_{22}^* & -c_{21}^* & 0 \\ c_{32}^* & -c_{31}^* & 0 \end{pmatrix}. \tag{9.134}$$

These results, substituted into Eq. (9.131) for each classical Euler angle, yield the following required expressions:

$$(\mathbf{u}^*)^T \frac{\partial \mathbf{r}^*}{\partial \Omega} = \mathbf{u}^T \begin{Bmatrix} 0 \\ r\cos i \\ -r\cos(\omega+\theta)\sin i \end{Bmatrix}. \tag{9.135}$$

$$(\mathbf{u}^*)^T \frac{\partial \mathbf{r}^*}{\partial i} = \mathbf{u}^T \begin{Bmatrix} 0 \\ 0 \\ r\sin(\omega+\theta) \end{Bmatrix}. \tag{9.136}$$

$$(\mathbf{u}^*)^T \frac{\partial \mathbf{r}^*}{\partial \omega} = \mathbf{u}^T \begin{Bmatrix} 0 \\ r \\ 0 \end{Bmatrix}. \tag{9.137}$$

The angle $(\omega + \theta)$ is the angle measured from the ascending node to the current radius along the osculating orbit, and is called the *argument of latitude*.

The perturbation terms for the variation of the other three osculating elements, a, e, M_0, are next derived as follows. Since the rotation matrix $\mathbf{C}^*$ is independent of these elements, the only contribution of a, e, M_0 to the perturbation terms in Eq. (9.131) arises through the rotation matrix $\mathbf{C}_\theta$ as follows:

$$\mathbf{C}_\theta \mathbf{r} = \mathbf{u}^T \begin{Bmatrix} a(\cos E - e) \\ b \sin E \\ 0 \end{Bmatrix}, \tag{9.138}$$

$$\frac{\partial}{\partial a}(\mathbf{C}_\theta \mathbf{r}) = \begin{Bmatrix} \cos E - e + \frac{3ant}{2r} \sin E \\ \frac{b}{a} \sin E - \frac{3ant}{2r} \cos E \\ 0 \end{Bmatrix} = \begin{Bmatrix} \frac{r}{a} \cos\theta + \frac{3ant}{2b} \sin\theta \\ \frac{r}{a} \sin\theta - \frac{3ant}{2b}(e + \cos\theta) \\ 0 \end{Bmatrix}, \tag{9.139}$$

$$\frac{\partial}{\partial e}(\mathbf{C}_\theta \mathbf{r}) = \begin{Bmatrix} -\frac{a^2}{r} \sin^2 E - a \\ -\frac{a^2 e}{b} \sin E + \frac{ab}{r} \sin E \cos E \\ 0 \end{Bmatrix} = \begin{Bmatrix} -\frac{a^2 r}{b^2} \sin^2\theta - a \\ \frac{a^2 r}{b^2} \sin\theta \cos\theta \\ 0 \end{Bmatrix}, \tag{9.140}$$

$$\frac{\partial}{\partial M_0}(\mathbf{C}_\theta \mathbf{r}) = \begin{Bmatrix} -\frac{a^2}{r} \sin E \\ \frac{ab}{r} \cos E \\ 0 \end{Bmatrix} = \begin{Bmatrix} -\frac{a^2}{b} \sin\theta \\ \frac{a^2}{b}(e + \cos\theta) \\ 0 \end{Bmatrix}. \tag{9.141}$$

The perturbation terms arising out of a, e, M_0 can be generically expressed as follows:

$$(\mathbf{u}^*)^T \frac{\partial \mathbf{r}^*}{\partial \alpha} = \mathbf{u}^T \mathbf{C}_\theta^T (\mathbf{C}^*)^T \mathbf{C}^* \frac{\partial}{\partial \alpha}(\mathbf{C}_\theta \mathbf{r}) = \mathbf{u}^T \mathbf{C}_\theta^T \frac{\partial}{\partial \alpha}(\mathbf{C}_\theta \mathbf{r}) \tag{9.142}$$

and result in

$$(\mathbf{u}^*)^T \frac{\partial \mathbf{r}^*}{\partial a} = \mathbf{u}^T \begin{Bmatrix} \frac{r}{a} - \frac{3ant}{2b} e \sin\theta \\ -\frac{3ant}{2b}(1 + e\cos\theta) \\ 0 \end{Bmatrix}. \tag{9.143}$$

$$(\mathbf{u}^*)^T \frac{\partial \mathbf{r}^*}{\partial e} = \mathbf{u}^T \left\{ \begin{array}{c} -a\cos\theta \\ \left(1+\frac{ar}{b^2}\right) a \sin\theta \\ 0 \end{array} \right\} . \tag{9.144}$$

$$(\mathbf{u}^*)^T \frac{\partial \mathbf{r}^*}{\partial M_0} = \mathbf{u}^T \left\{ \begin{array}{c} \frac{a^2}{b} e \sin\theta \\ \frac{a^2}{b}(1+e\sin\theta) \\ 0 \end{array} \right\} . \tag{9.145}$$

The presence of the two terms linearly varying with time in the expression for $(\mathbf{u}^*)^T \partial \mathbf{r}^*/\partial a$ (Eq. (9.143)) cause problems in the derivation of variational equations by matrix inversion, $\mathbf{L}^{-1}$. Such monotonically time-varying (aperiodic) terms are called *secular terms*, and in the present derivation, they arise due to $\mathrm{d}M_0/\mathrm{d}t$. Hence the secular terms can be eliminated by replacing M_0 with another element in the set of osculating elements. The current mean anomaly, $M = M_0 + nt$, is differentiated with time to yield

$$\frac{\mathrm{d}M}{\mathrm{d}t} = n - \frac{3nt}{2a}\frac{\mathrm{d}a}{\mathrm{d}t} + \frac{\mathrm{d}M_0}{\mathrm{d}t} . \tag{9.146}$$

To make the second term vanish on the right-hand side of Eq. (9.146) – which gives rise to the secular terms – recourse is made to the Lagrange planetary equations, Eq. (9.119), whose relevant terms are given by

$$\begin{aligned} \frac{\mathrm{d}a}{\mathrm{d}t} &= \frac{2}{an}(\mathbf{u}^*)^T \frac{\partial \mathbf{r}^*}{\partial M_0} \\ \frac{\mathrm{d}M_0}{\mathrm{d}t} &= -\frac{2}{an}(\mathbf{u}^*)^T \frac{\partial \mathbf{r}^*}{\partial a} - \frac{(1-e^2)}{a^2 ne}(\mathbf{u}^*)^T \frac{\partial \mathbf{r}^*}{\partial e} . \end{aligned} \tag{9.147}$$

A substitution of these into Eq. (9.146) results in the following:

$$\frac{\mathrm{d}M}{\mathrm{d}t} = n - \frac{2}{an}(\mathbf{u}^*)^T \left(\frac{3nt}{2a}\frac{\partial \mathbf{r}^*}{\partial M_0} + \frac{\partial \mathbf{r}^*}{\partial a} \right) - \frac{(1-e^2)}{a^2 ne}(\mathbf{u}^*)^T \frac{\partial \mathbf{r}^*}{\partial e} . \tag{9.148}$$

The substitution of Eqs. (9.143) and (9.145) into Eq. (9.148) removes the secular terms by cancellation within the parentheses. Therefore, a possible choice of the new osculating orbital element is obtained by writing Eq. (9.148) as follows:

$$\frac{\mathrm{d}M}{\mathrm{d}t} = n + \frac{\mathrm{d}\chi}{\mathrm{d}t} , \tag{9.149}$$

where the angle χ replaces M_0 as an orbital element, and is governed by the variation

$$\frac{\mathrm{d}\chi}{\mathrm{d}t} = -\frac{2}{an}(\mathbf{u}^*)^T \left(\frac{3nt}{2a}\frac{\partial \mathbf{r}^*}{\partial M_0} + \frac{\partial \mathbf{r}^*}{\partial a} \right) - \frac{(1-e^2)}{a^2 ne}(\mathbf{u}^*)^T \frac{\partial \mathbf{r}^*}{\partial e} . \tag{9.150}$$

Substituting Eqs. (9.135)–(9.137), (9.143), (9.144), and (9.149) into Eq. (9.123), and using the Lagrange brackets, **L**, which were computed in Section 9.5, solving for the remaining time derivatives constituting the vector, $\dot{\boldsymbol{\alpha}}$, we have

$$\frac{\mathrm{d}a}{\mathrm{d}t} = \frac{2a^2}{h}\left(eu_r \sin\theta + \frac{p}{r}u_\theta\right). \tag{9.151}$$

$$\frac{\mathrm{d}e}{\mathrm{d}t} = \frac{1}{h}\{pu_r \sin\theta + [(p+r)\cos\theta + re]u_\theta\}. \tag{9.152}$$

$$\frac{\mathrm{d}\chi}{\mathrm{d}t} = \frac{b}{aeh}[(p\cos\theta - 2re)u_r - (p+r)\sin\theta u_\theta]. \tag{9.153}$$

$$\frac{\mathrm{d}\Omega}{\mathrm{d}t} = \frac{r}{h\sin i}\sin(\theta+\omega)u_n. \tag{9.154}$$

$$\frac{\mathrm{d}i}{\mathrm{d}t} = \frac{r}{h}\cos(\theta+\omega)u_n. \tag{9.155}$$

$$\frac{\mathrm{d}\omega}{\mathrm{d}t} = \frac{1}{eh}[-p\cos\theta u_r + (p+r)\sin\theta u_\theta] - \frac{r\cos i\sin(\theta+\omega)}{h\sin i}u_n. \tag{9.156}$$

In these equations θ is the true anomaly obtained from solving Kepler's equation for the osculating orbit from the time of periapsis, t_0, to the present time, t, and $(a, e, \chi, \Omega, i, \omega)$ are the instantaneous orbital elements of the osculating orbit.

9.6 Variation of Vectors

In addition to Lagrange and Gauss variational models, a third method of deriving the variations of the osculating elements is by direct differentiation of the vectors constituting the Lagrange brackets. While the Lagrange planetary equations and the Gauss variational model have been derived for an elliptic orbit, the method of variation of vectors is valid for all orbital shapes, and for both conservative and non-conservative perturbations. Thus the method is more general than both Lagrange and Gauss models. However, the resulting equations for an elliptic orbit are equivalent to both of the previously discussed methods. Consider Eq. (9.123):

$$\frac{\mathrm{d}\boldsymbol{\alpha}}{\mathrm{d}t} = \mathbf{P}^T\left(\frac{\partial \mathbf{r}}{\partial \boldsymbol{\alpha}}\right)^T \mathbf{u}, \tag{9.157}$$

where

$$\mathbf{P}^T = \mathbf{L}^{-1} \tag{9.158}$$

is termed the *Poisson matrix*. Since the Lagrange matrix, **L**, is skew-symmetric, so is the Poisson matrix, and together they satisfy

$$\mathbf{PL} = \mathbf{LP} = -\mathbf{I}. \tag{9.159}$$

By substituting Eq. (9.51) into Eq. (9.158) and taking the transpose, we have

$$\mathbf{P} = \frac{\partial \boldsymbol{\alpha}}{\partial \mathbf{r}}\left(\frac{\partial \boldsymbol{\alpha}}{\partial \mathbf{v}}\right)^T - \frac{\partial \boldsymbol{\alpha}}{\partial \mathbf{v}}\left(\frac{\partial \boldsymbol{\alpha}}{\partial \mathbf{r}}\right)^T. \tag{9.160}$$

Substituting Eq. (9.160) into Eq. (9.157) yields the following:

$$\frac{\mathrm{d}\boldsymbol{\alpha}}{\mathrm{d}t} = \frac{\partial \boldsymbol{\alpha}}{\partial \mathbf{v}}\left(\frac{\partial \mathbf{r}}{\partial \mathbf{r}}\right)^T \mathbf{u} - \frac{\partial \boldsymbol{\alpha}}{\partial \mathbf{r}}\left(\frac{\partial \mathbf{r}}{\partial \mathbf{v}}\right)^T \mathbf{u}, \tag{9.161}$$

or

$$\frac{d\boldsymbol{\alpha}}{dt} = \frac{\partial \boldsymbol{\alpha}}{\partial \mathbf{v}}\mathbf{u}, \tag{9.162}$$

because $\partial \mathbf{r}/\partial \mathbf{r} = \mathbf{I}$ and $\partial \mathbf{r}/\partial \mathbf{v} = \mathbf{0}$.

Equation (9.162) provides the method of variation of parameters by taking the partial derivative of the vector of osculating elements with respect to the velocity vector, and multiplying it with the perturbing acceleration vector. This method is thus summarized as regarding the position vector, $\mathbf{r}$, as being fixed while varying the two-body orbital elements, $\boldsymbol{\alpha}$, and approximating the attendant time variation in the velocity, $\mathbf{v}$, by the perturbing acceleration, $\mathbf{u}$.

Beginning with the first element, a, of the set, $\boldsymbol{\alpha}$, from Eq. (9.162) we have

$$\frac{da}{dt} = \frac{\partial a}{\partial \mathbf{v}}\mathbf{u}, \tag{9.163}$$

where the derivative vector, $\partial a/\partial \mathbf{v}$, is calculated as follows by differentiating the energy integral, $\varepsilon = v^2/2 - \mu/r = -\mu/(2a)$:

$$\frac{\partial v^2}{\partial \mathbf{v}} = \frac{\partial \mathbf{v}^T\mathbf{v}}{\partial \mathbf{v}} = \mathbf{v}^T = \frac{\mu}{2a^2}\frac{\partial a}{\partial \mathbf{v}}, \tag{9.164}$$

which, substituted into Eq. (9.163), yields

$$\frac{da}{dt} = \frac{2a^2}{\mu}\mathbf{v}^T\mathbf{u}. \tag{9.165}$$

The next step is to consider the time derivative of the angular momentum vector, which can be used to derive the variations in e, i, Ω:

$$\frac{d\mathbf{h}}{dt} = \frac{\partial \mathbf{h}}{\partial \mathbf{v}}\mathbf{u}. \tag{9.166}$$

To do so, $\mathbf{h} = \mathbf{r} \times \mathbf{v}$ is resolved in the following matrix form:

$$\mathbf{h} = \mathbf{r} \times \mathbf{v} = \begin{Bmatrix} x \\ y \\ z \end{Bmatrix} \times \begin{Bmatrix} v_x \\ v_y \\ v_z \end{Bmatrix} = \begin{Bmatrix} yv_z - zv_y \\ zv_x - xv_z \\ xv_y - yv_x \end{Bmatrix} = \mathbf{S}(\mathbf{r})\mathbf{v}, \tag{9.167}$$

where

$$\mathbf{S}(\mathbf{r}) = \begin{pmatrix} 0 & -z & y \\ z & 0 & -x \\ -y & x & 0 \end{pmatrix} \tag{9.168}$$

represents the skew-symmetric matrix form of the vector product. This matrix product between any two vectors, $\mathbf{a}, \mathbf{b}$, can be defined as follows:

$$\mathbf{a} \times \mathbf{b} = \begin{Bmatrix} a_x \\ a_y \\ a_z \end{Bmatrix} \times \begin{Bmatrix} b_x \\ b_y \\ b_z \end{Bmatrix} = \begin{Bmatrix} a_yb_z - a_zb_y \\ a_zb_x - a_xb_z \\ a_xb_y - a_yb_x \end{Bmatrix}$$

$$= \mathbf{S}(\mathbf{a})\mathbf{b} = \begin{pmatrix} 0 & -a_z & a_y \\ a_z & 0 & -a_x \\ -a_y & a_x & 0 \end{pmatrix} \begin{Bmatrix} b_x \\ b_y \\ b_z \end{Bmatrix}$$

$$= -\mathbf{S}(\mathbf{b})\mathbf{a} = -\begin{pmatrix} 0 & -b_z & b_y \\ b_z & 0 & -b_x \\ -b_y & b_x & 0 \end{pmatrix} \begin{Bmatrix} a_x \\ a_y \\ a_z \end{Bmatrix} . \tag{9.169}$$

Thus we have

$$\frac{\partial \mathbf{h}}{\partial \mathbf{v}} = \mathbf{S}(\mathbf{r}) , \tag{9.170}$$

which, substituted into Eq. (9.166), yields the following:

$$\frac{\mathrm{d}\mathbf{h}}{\mathrm{d}t} = \mathbf{S}(\mathbf{r})\mathbf{u} = \mathbf{r} \times \mathbf{u} . \tag{9.171}$$

The variation of the magnitude, h, is derived as follows:

$$\frac{\partial h^2}{\partial \mathbf{v}} = 2h\frac{\partial h}{\partial \mathbf{v}} = \frac{\partial \mathbf{h}^T\mathbf{h}}{\partial \mathbf{v}} = 2\mathbf{h}^T\mathbf{S}(\mathbf{r}) , \tag{9.172}$$

or

$$\frac{\mathrm{d}h}{\mathrm{d}t} = \frac{\mathbf{h}^T}{h}\mathbf{S}(\mathbf{r})\mathbf{u} = \mathbf{i}_\mathrm{h} \cdot \mathbf{r} \times \mathbf{u} = \mathbf{i}_\mathrm{h} \times \mathbf{r} \cdot \mathbf{u} , \tag{9.173}$$

where $\mathbf{i}_\mathrm{h} = \mathbf{h}/h$. Resolving Eq. (9.173) in the local-horizon frame, $(\mathbf{i}_\mathrm{r}, \mathbf{i}_\theta, \mathbf{i}_\mathrm{h})$, we have

$$\frac{\mathrm{d}h}{\mathrm{d}t} = r\mathbf{i}_\theta \cdot \mathbf{u} . \tag{9.174}$$

Using the vector triple-product identity, we write

$$h^2 = (\mathbf{r} \times \mathbf{v}) \cdot (\mathbf{r} \times \mathbf{v}) = (\mathbf{r} \cdot \mathbf{r})(\mathbf{v} \cdot \mathbf{v}) - (\mathbf{r} \cdot \mathbf{v})(\mathbf{r} \cdot \mathbf{v}) , \tag{9.175}$$

from which the following alternative expression for $\dot{h}$ can be derived:

$$\frac{\mathrm{d}h}{\mathrm{d}t} = \frac{1}{h}\mathbf{r}^T(\mathbf{r}\mathbf{v}^T - \mathbf{v}\mathbf{r}^T)\mathbf{u}$$
$$= \frac{1}{h}[r^2\mathbf{v} \cdot \mathbf{u} - (\mathbf{r} \cdot \mathbf{v})(\mathbf{r} \cdot \mathbf{u})] . \tag{9.176}$$

In a similar manner, the variation of the eccentricity vector, $\mathbf{e}$, is derived by expressing the vector as follows:

$$\mathbf{e} = \frac{1}{\mu}(\mathbf{v} \times \mathbf{h}) - \mathbf{i}_\mathrm{r} = -\frac{1}{\mu}\mathbf{S}(\mathbf{h})\mathbf{v}) - \mathbf{i}_\mathrm{r}$$
$$= \frac{1}{\mu}\mathbf{S}(\mathbf{v})\mathbf{h}) - \mathbf{i}_\mathrm{r} , \tag{9.177}$$

where the skew-symmetric matrix operation representing the vector product, $\mathbf{S}(.)$, is given by Eq. (9.169). Then taking the partial derivative of $\mathbf{e}$ with respect to $\mathbf{v}$ according to the chain rule of differentiation and substituting Eq. (9.170), we have

$$\mu\frac{\partial \mathbf{e}}{\partial \mathbf{v}} = -\mathbf{S}(\mathbf{h})\frac{\partial \mathbf{v}}{\partial \mathbf{v}} + \mathbf{S}(\mathbf{v})\frac{\partial \mathbf{h}}{\partial \mathbf{v}} = -\mathbf{S}(\mathbf{h})\mathbf{I} + \mathbf{S}(\mathbf{v})\mathbf{S}(\mathbf{r}) . \tag{9.178}$$

Finally, we have the following time derivative of $\mathbf{e}$:

$$\begin{aligned}\frac{\mathrm{d}\mathbf{e}}{\mathrm{d}t} &= \frac{\partial \mathbf{e}}{\partial \mathbf{v}}\mathbf{u} = -\frac{1}{\mu}\mathbf{S}(\mathbf{h})\mathbf{u} + \frac{1}{\mu}\mathbf{S}(\mathbf{v})\mathbf{S}(\mathbf{r})\mathbf{u} \\ &= \frac{1}{\mu}\mathbf{u}\times\mathbf{h} + \frac{1}{\mu}\mathbf{v}\times(\mathbf{r}\times\mathbf{u}) \\ &= \frac{1}{\mu}\mathbf{u}\times(\mathbf{r}\times\mathbf{v}) + \frac{1}{\mu}\mathbf{v}\times(\mathbf{r}\times\mathbf{u}) \\ &= \frac{2}{\mu}(\mathbf{v}\cdot\mathbf{u})\mathbf{r} - \frac{1}{\mu}(\mathbf{r}\cdot\mathbf{u})\mathbf{v} - \frac{1}{\mu}(\mathbf{r}\cdot\mathbf{v})\mathbf{u} .\end{aligned} \tag{9.179}$$

The variation of orbital eccentricity can be derived by differentiating the osculating parameter, $p = h^2/\mu = a(1-e^2)$, as follows:

$$\frac{2h}{\mu}\frac{\mathrm{d}\mathbf{h}}{\mathrm{d}t} = (1-e^2)\frac{\mathrm{d}\mathbf{a}}{\mathrm{d}t} - 2ae\frac{\mathrm{d}\mathbf{e}}{\mathrm{d}t} . \tag{9.180}$$

The substitution of Eqs. (9.165) and (9.173) into Eq. (9.180) leads to the following:

$$\frac{\mathrm{d}\mathbf{e}}{\mathrm{d}t} = \frac{1}{\mu a e}[(\mathbf{r}\cdot\mathbf{v})(\mathbf{r}\cdot\mathbf{u}) - (r^2 - ap)(\mathbf{v}\cdot\mathbf{u})] . \tag{9.181}$$

The variation of the classical Euler angles, (Ω, i, ω), representing the osculating plane is derived next. Consider first the variation in the angular-momentum direction, $\mathbf{i}_\mathrm{h}$, caused by the first two elementary rotations of the Euler sequence $(\Omega_3, i_1, \omega_3)$, which are resolved along and perpendicular to the direction of the ascending node, $\mathbf{i}_\mathrm{n} = \mathbf{I}'$, (see Chap. 5). The angular velocity of $\mathbf{i}_\mathrm{h}$ is thus given by combining these two elementary rotations – the first about the axis $\mathbf{K}$ by angle Ω, followed by the rotation i about $\mathbf{i}_\mathrm{n} = \mathbf{I}'$ – as follows:

$$\dot{\Omega}\mathbf{K} + \frac{\mathrm{d}i}{\mathrm{d}t}\mathbf{i}_\mathrm{n} .$$

Consequently, the overall variation of the angular-momentum vector is given by

$$\begin{aligned}\frac{\mathrm{d}\mathbf{h}}{\mathrm{d}t} &= \dot{h}\mathbf{i}_\mathrm{h} + h\frac{\mathrm{d}\mathbf{i}_\mathrm{h}}{\mathrm{d}t} \\ &= \dot{h}\mathbf{i}_\mathrm{h} + h\left(\dot{\Omega}\mathbf{K} + \frac{\mathrm{d}i}{\mathrm{d}t}\mathbf{i}_\mathrm{n}\right)\times\mathbf{i}_\mathrm{h} \\ &= \dot{h}\mathbf{i}_\mathrm{h} + h\dot{\Omega}(\mathbf{K}\times\mathbf{i}_\mathrm{h}) + h\frac{\mathrm{d}i}{\mathrm{d}t}(\mathbf{i}_\mathrm{n}\times\mathbf{i}_\mathrm{h}) \\ &= \dot{h}\mathbf{i}_\mathrm{h} + h\sin i\dot{\Omega}\mathbf{i}_\mathrm{n} - h\frac{\mathrm{d}i}{\mathrm{d}t}\mathbf{J}'' ,\end{aligned} \tag{9.182}$$

where $\mathbf{J}'' = \mathbf{i}_h \times \mathbf{i}_n$ denotes the second axis of the reference frame after the two elementary rotations (Chap. 5). The scalar product of Eq. (9.171) with Eq. (9.182) yields the following variations:

$$\dot{\Omega} = \frac{\mathrm{d}\Omega}{\mathrm{d}t} = \frac{1}{h\sin i}(\mathbf{i}_n \times \mathbf{r}) \cdot \mathbf{u} = \frac{r\sin(\omega+\theta)}{h\sin i}\mathbf{i}_h \cdot \mathbf{u}, \tag{9.183}$$

$$\frac{\mathrm{d}i}{\mathrm{d}t} = -\frac{1}{h}(\mathbf{J}'' \times \mathbf{r}) \cdot \mathbf{u} = \frac{r\cos(\omega+\theta)}{h}\mathbf{i}_h \cdot \mathbf{u}, \tag{9.184}$$

where the *argument of latitude*, $(\omega+\theta) = \cos^{-1}(\mathbf{i}_r \cdot \mathbf{i}_n)$, locates $\mathbf{r}$ from the nodal vector, $\mathbf{i}_n$, in the osculating orbital plane.

The variation in the argument of periapsis, ω, is derived by considering the variation in the argument of latitude, $(\omega+\theta)$, while regarding the true anomaly, θ, as being fixed during the variation. The nodal vector resolved in the celestial frame is given by

$$\mathbf{i}_n = \mathbf{I}\cos\Omega + \mathbf{J}\sin\Omega, \tag{9.185}$$

from which it follows that

$$\cos(\omega+\theta) = \mathbf{i}_n \cdot \mathbf{i}_r = \mathbf{I}\cdot\mathbf{i}_r\cos\Omega + \mathbf{J}\cdot\mathbf{i}_r\sin\Omega. \tag{9.186}$$

Taking the partial derivative of Eq. (9.186) with respect to $\mathbf{v}$ while regarding $\mathbf{r}$ as being fixed, we have

$$-\sin(\omega+\theta)\frac{\partial(\omega+\theta)}{\partial\mathbf{v}} = [-\mathbf{I}\cdot\mathbf{i}_r\sin\Omega + \mathbf{J}\cdot\mathbf{i}_r\cos\Omega]\frac{\partial\Omega}{\partial\mathbf{v}}. \tag{9.187}$$

The direction cosines $\mathbf{I}\cdot\mathbf{i}_r$ and $\mathbf{J}\cdot\mathbf{i}_r$ are the following:

$$\begin{aligned} \mathbf{I}\cdot\mathbf{i}_r &= \cos\Omega\cos(\omega+\theta) - \sin\Omega\cos(\omega+\theta)\cos i \\ \mathbf{J}\cdot\mathbf{i}_r &= \sin\Omega\cos(\omega+\theta) + \cos\Omega\sin(\omega+\theta)\cos i, \end{aligned} \tag{9.188}$$

which, substituted into Eq. (9.187), produce

$$\frac{\partial(\omega+\theta)}{\partial\mathbf{v}} = -\cos i\frac{\partial\Omega}{\partial\mathbf{v}}, \tag{9.189}$$

or

$$\frac{\partial(\omega+\theta)}{\partial\mathbf{v}}\mathbf{u} = -\cos i\dot{\Omega}. \tag{9.190}$$

The net change in the argument of latitude, $(\omega+\theta)$, is caused by a variation of ω as well as the perturbation in the true anomaly, θ, caused by $\mathbf{u}$. Hence, one must subtract the latter variation from Eq. (9.190) to obtain the variation in ω as follows:

$$\frac{\mathrm{d}\omega}{\mathrm{d}t} = -\frac{\partial\theta}{\partial\mathbf{v}}\mathbf{u} - \cos i\dot{\Omega}. \tag{9.191}$$

To derive $\partial\theta/\partial\mathbf{v}$, the orbit equation, $r = p/(1+e\cos\theta)$, is differentiated with $\mathbf{v}$, resulting in

$$er\sin\theta\frac{\partial\theta}{\partial\mathbf{v}} = r\cos\theta\frac{\partial e}{\partial\mathbf{v}} - \frac{2h}{\mu}\frac{\partial h}{\partial\mathbf{v}}. \tag{9.192}$$

Furthermore, the radial velocity component is given by the following (Chap. 3):

$$\frac{\mu}{h} e r \sin\theta = \dot{r} = \mathbf{r} \cdot \mathbf{v}\,, \tag{9.193}$$

which on differentiation yields the following:

$$e r \cos\theta \frac{\partial \theta}{\partial \mathbf{v}} = -r \sin\theta \frac{\partial e}{\partial \mathbf{v}} + \frac{h}{\mu}\mathbf{r}^T + \frac{\mathbf{r}\cdot\mathbf{v}}{\mu}\frac{\partial h}{\partial \mathbf{v}}\,. \tag{9.194}$$

Elimination of the derivative, $\partial e/\partial \mathbf{v}$, from Eqs. (9.192) and (9.194) leads to

$$e r h \cos\theta \frac{\partial \theta}{\partial \mathbf{v}} = p \cos\theta \mathbf{r}^T - (r+p)\sin\theta \frac{\partial h}{\partial \mathbf{v}}\,, \tag{9.195}$$

and substitution of Eq. (9.173) yields the variation of the true anomaly with the velocity vector as follows:

$$\frac{\partial \theta}{\partial \mathbf{v}} = \frac{r}{e h^2}\left\{\left[\frac{h}{p}(e+\cos\theta) + \frac{eh}{r}\right]\mathbf{r}^T - (r+p)\sin\theta \mathbf{v}^T\right\}\,. \tag{9.196}$$

Finally, the variation in ω is obtained to be the following after substituting Eqs. (9.183) and (9.196) into Eq. (9.191):

$$\frac{\mathrm{d}\omega}{\mathrm{d}t} = -\frac{r}{e h^2}\left\{\left[\frac{h}{p}(e+\cos\theta) + \frac{eh}{r}\right]\mathbf{r}^T - (r+p)\sin\theta \mathbf{v}^T\right\}\mathbf{u} - \frac{r\sin(\omega+\theta)}{h\tan i}\mathbf{i}_\mathrm{h}\cdot\mathbf{u}\,. \tag{9.197}$$

9.7 Mean Orbital Perturbation

The variational models considered in Sections 9.3–9.6 provide the time rate of change of the osculating elements. However, it is much more useful to determine the variation of the osculating elements over a complete orbit, because such a variation is often periodic in nature. To do so, the variational equations – such as Eqs. (9.151)–(9.156) – are expressed in terms of the variation with respect to the eccentric anomaly, E, by dividing them by

$$\frac{\mathrm{d}E}{\mathrm{d}t} = \frac{n}{1 - e\cos E} = \frac{na}{r}\,,$$

where $n = \sqrt{\mu/a^3}$ is the orbital mean motion in an elliptic orbit. An averaging procedure is utilized to determine each element's mean rate of change by calculating its variation over a single revolution of the orbital period, $T = 2\pi/n$. The mean osculating vector is thus computed as follows:

$$\begin{aligned}\overline{\boldsymbol{\alpha}} &= \frac{1}{T}\int_{t_1}^{t_1+T}\frac{\mathrm{d}\boldsymbol{\alpha}}{\mathrm{d}t}\mathrm{d}t \\ &= \frac{1}{T}\int_{E_1}^{E_1+2\pi}\left(\frac{r}{na}\right)\frac{\mathrm{d}\boldsymbol{\alpha}}{\mathrm{d}t}\mathrm{d}E\,,\end{aligned} \tag{9.198}$$

where $\mathrm{d}\boldsymbol{\alpha}/\mathrm{d}t$ is calculated from Eqs. (9.151)–(9.156) with the terms on the right-hand side fixed at their mean values over a complete revolution. The integration limits of the averaging integral, $E_1, E_1 + 2\pi$, are conveniently selected based upon practical considerations (such as entering and exiting the central body's shadow). The same result can be obtained by first determining the mean perturbation, and substituting it into a variational model to find the mean variation per orbit, $\overline{\boldsymbol{\alpha}}$.

The concept of the mean perturbation (or variation) provides a rough approximation for determining the osculating orbital parameters at any instant. This method pertains to the following first-order approximation:

$$\boldsymbol{\alpha}(t) = \boldsymbol{\alpha}(0) + \frac{2k\pi}{n}\frac{\mathrm{d}\overline{\boldsymbol{\alpha}}}{\mathrm{d}t} + \delta\boldsymbol{\alpha}(t), \tag{9.199}$$

where k is an integer denoting the number of completed orbits since the epoch, $t = 0$, up to the present time; t, $\boldsymbol{\alpha}(0)$ is the parameter vector at epoch; $\delta\boldsymbol{\alpha}(t)$ is the short-period change computed by the variational equations:

$$\delta\boldsymbol{\alpha}(t) = \frac{\mathrm{d}\boldsymbol{\alpha}}{\mathrm{d}t}\left(t - \frac{2k\pi}{n}\right), \tag{9.200}$$

and $\mathrm{d}\overline{\boldsymbol{\alpha}}/\mathrm{d}t$ is the mean rate of variation per orbit given by

$$\frac{\mathrm{d}\overline{\boldsymbol{\alpha}}}{\mathrm{d}t} = \frac{1}{2\pi}\int_{M_1}^{M_1+2\pi}\frac{\mathrm{d}\boldsymbol{\alpha}}{\mathrm{d}t}\mathrm{d}M. \tag{9.201}$$

While this numerical averaging procedure can give quick results, it can suffer from inaccuracy if k is large. Such a computation is to be seen in the context of other numerical methods for the perturbed orbits (Cowell and Encke's methods), which are described later in this chapter.

9.8 Orbital Perturbation Due to Oblateness

The largest gravitational perturbation caused by the non-spherical shape of an axially symmetric central body on a two-body orbit is due to the first term in the spherical harmonics expansion for the gravitational potential, $U(\mathbf{r})$, derived in Chap. 2. This effect arises due to the spherical harmonic coefficient, J_2, which multiplies $(r_0/r)^2$ in the series expansion, and represents a uniform bulge around the equator (and a corresponding flattening at the poles); hence it is called the *oblateness effect*. Being by far the largest term in the series for most bodies, it can be used to approximate the perturbation due to the non-spherical shape of the central body by neglecting the terms of the higher powers in the expansion for $U(\mathbf{r})$ – that is, the terms containing $(r_0/r)^k$, $k = 3, 4, \ldots$. Such an approximation is given by

$$U(\mathbf{r}) = -\frac{\mu}{r}J_2\left(\frac{r_0}{r}\right)^2 P_2(\cos\varphi), \tag{9.202}$$

where r_0 is the equatorial radius of the central body of gravitational constant μ, $P_k(.)$ denotes the Legendre polynomial of degree k, J_2 is the first Jeffery's constant (called the *oblateness parameter*; Chap. 2), and $\varphi = \pi/2 - \delta$ is the complementary of the declination (latitude), called the co-latitude, of the orbiting body at radius r. Since J_2 is usually quite small (Table 9.2 lists the value of J_2 for some major bodies), it causes a small but continuous perturbation which can produce an appreciable change in the orbit only over a long period of time. This is possible only if the orbit is periodic (i.e., elliptic). Hence, the discussion in this section is confined to elliptic orbits.

Consider the position, $\mathbf{r}$, in the orbit around the oblate central body resolved in a celestial reference frame, $(\mathbf{I}, \mathbf{J}, \mathbf{K})$, with the its origin O located at the centroid of the body, such that

Table 9.2 The oblateness parameter, J_2, of some major bodies.

Central Body	J_2
Mercury	0.0000503
Venus	0.0000045
Earth	0.00108263
Mars	0.00196045
Jupiter	0.014696
Saturn	0.016291
Uranus	0.003511
Neptune	0.003408
Moon	0.0002034

the plane (**I**, **J**), is the equatorial plane and **K** is the polar axis, as shown in Fig. 9.2. Since the co-latitude, φ, is the angle between the radius vector, **r**, and the **K**-axis of the celestial frame (Fig. 9.2), we write

$$\cos\varphi = \mathbf{i}_r \cdot \mathbf{K} = \sin(\omega + \theta)\sin i\,, \tag{9.203}$$

where the argument of latitude, $(\omega + \theta)$, locates **r** from the nodal vector, $\mathbf{i}_n$, along the orbit (θ being the true anomaly), and i is the orbital inclination given by the angle between the angular-momentum vector, **h**, and the **K**-axis.

The derivation steps for Eq. (9.203) are the following:

$$\mathbf{i}_h = \frac{\mathbf{i}_n \times \mathbf{i}_r}{|\,\mathbf{i}_n \times \mathbf{i}_r\,|} = \frac{\mathbf{i}_n \times \mathbf{i}_r}{\sin(\omega + \theta)}\,,$$

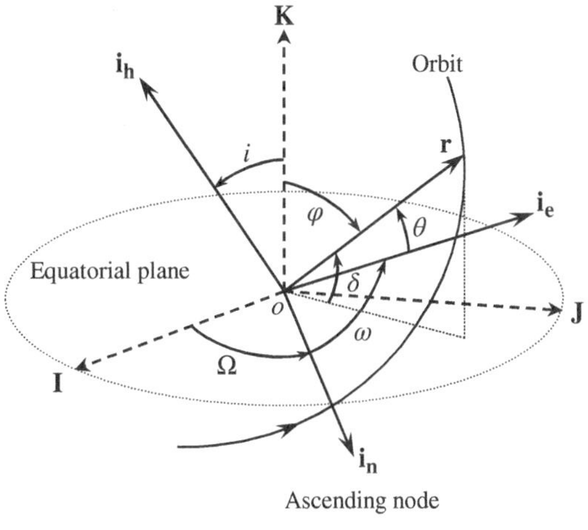

Figure 9.2 Orbital position around an oblate central body.

where

$$\mathbf{i_n} = \frac{\mathbf{K} \times \mathbf{i_h}}{|\,\mathbf{K} \times \mathbf{i_h}\,|} = \frac{\mathbf{K} \times \mathbf{i_h}}{\sin i},$$

from which it follows that

$$\mathbf{i_h} \sin(\omega + \theta) \sin i = (\mathbf{K} \times \mathbf{i_h}) \times \mathbf{i_r} = (\mathbf{K} \cdot \mathbf{i_r})\mathbf{i_h} - (\mathbf{i_h} \cdot \mathbf{i_r})\mathbf{K} = (\mathbf{K} \cdot \mathbf{i_r})\mathbf{i_h} \,.$$

Recalling from Chap. 2 that the Legendre polynomial of second degree is given by

$$P_2(v) = \frac{1}{2}(3v^2 - 1)\,,$$

we have the following from Eq. (9.203):

$$P_2(\cos\varphi) = \frac{1}{2}(3\cos^2\varphi - 1) = \frac{1}{2}[3\sin^2(\omega + \theta)\sin^2 i - 1]\,. \tag{9.204}$$

Hence the approximate perturbation potential given by Eq. (9.202) is expressed as follows:

$$U(\mathbf{r}) = -\frac{\mu J_2 r_0^2}{2p^3}(1 + e\cos\theta)^3[3\sin^2(\omega + \theta)\sin^2 i - 1]\,, \tag{9.205}$$

where the orbit equation, $r = p/(1 + e\cos\theta)$, has been substituted.

It is necessary to derive the effects of planetary oblateness over a complete orbit. This is carried out by first evaluating the mean perturbation potential, $\bar{U}$, per revolution as follows:

$$\begin{aligned}
\overline{U} &= \frac{1}{T}\int_0^{2\pi}\left(\frac{r}{na}\right)U(\mathbf{r})\mathrm{d}E \\
&= \frac{1}{2\pi}\int_0^{2\pi}\frac{r}{a}U(\mathbf{r})\mathrm{d}E \\
&= \frac{1}{2\pi}\int_0^{2\pi}\frac{nr^2}{h}U(\mathbf{r})\mathrm{d}\theta \\
&= -\frac{\mu J_2 r_0^2 n}{4\pi ph}\int_0^{2\pi}(1 + e\cos\theta)[3\sin^2(\omega + \theta)\sin^2 i - 1]\mathrm{d}\theta \\
&= -\frac{n^2 J_2 r_0^2}{4\pi(1 - e^2)^{3/2}}\left[3\sin^2 i\int_0^{2\pi}(1 + e\cos\theta)\sin^2(\omega + \theta)\mathrm{d}\theta - \int_0^{2\pi}(1 + e\cos\theta)\mathrm{d}\theta\right] \\
&= -\frac{n^2 J_2 r_0^2}{4\pi(1 - e^2)^{3/2}}(3\pi\sin^2 i - 2\pi) \\
&= \frac{n^2 J_2 r_0^2}{4(1 - e^2)^{3/2}}(2 - 3\sin^2 i)\,.
\end{aligned} \tag{9.206}$$

The following identities are employed in the derivation given in Eq. (9.206):

$$\frac{r}{a}\mathrm{d}E = n\mathrm{d}t = \mathrm{d}M = \frac{nr^2}{h}\mathrm{d}\theta\,.$$

$$\int_0^{2\pi} (1 + e\cos\theta)\mathrm{d}\theta = (\theta + e\sin\theta)_0^{2\pi} = 2\pi .$$

$$\begin{aligned}
\int_0^{2\pi} (1 + e\cos\theta)\sin^2(\omega + \theta)\mathrm{d}\theta &= \int_0^{2\pi} [1 + e\cos(\tilde{\theta} - \omega)]\sin^2\tilde{\theta}\mathrm{d}\tilde{\theta} \\
&= \int_0^{2\pi} \sin^2\tilde{\theta}\mathrm{d}\tilde{\theta} + e\cos\omega \int_0^{2\pi} \sin^2\tilde{\theta}\cos\tilde{\theta}\mathrm{d}\tilde{\theta} \\
&\quad + e\sin\omega \int_0^{2\pi} \sin^3\tilde{\theta}\mathrm{d}\tilde{\theta} \\
&= \pi + 0 + 0 = \pi .
\end{aligned}$$

The mean value of the perturbation potential, $\overline{U}$, given by Eq. (9.206) is independent of Ω, ω, and M_0, and its partial derivatives with respect to the orbital elements are as follows:

$$\begin{aligned}
\frac{\partial\overline{U}}{\partial a} &= \frac{-3n^2J_2r_0^2}{4a(1-e^2)^{3/2}}(2 - 3\sin^2 i) \\
\frac{\partial\overline{U}}{\partial e} &= \frac{3n^2eJ_2r_0^2}{4(1-e^2)^{5/2}}(2 - 3\sin^2 i) \\
\frac{\partial\overline{U}}{\partial\Omega} &= 0 \\
\frac{\partial\overline{U}}{\partial i} &= \frac{-3n^2J_2r_0^2}{2(1-e^2)^{3/2}}\sin i\cos i \\
\frac{\partial\overline{U}}{\partial\omega} &= 0 \\
\frac{\partial\overline{U}}{\partial M_0} &= 0 .
\end{aligned} \tag{9.207}$$

When these partial derivatives of $\bar{U}$ are substituted into the Lagrange's planetary equations, Eq. (9.119), we have the following variations of the classical orbital elements per orbit:

$$\begin{aligned}
\frac{\mathrm{d}\bar{a}}{\mathrm{d}t} &= 0 \\
\frac{\mathrm{d}\bar{e}}{\mathrm{d}t} &= 0 \\
\frac{\mathrm{d}\overline{M}_0}{\mathrm{d}t} &= \frac{3nJ_2r_0^2}{4a^2(1-e^2)^{3/2}}(2 - 3\sin^2 i) \\
\frac{\mathrm{d}\overline{\Omega}}{\mathrm{d}t} &= \frac{-3}{2}nJ_2\left(\frac{r_0}{p}\right)^2\cos i \\
\frac{\mathrm{d}\bar{\zeta}}{\mathrm{d}t} &= 0
\end{aligned} \tag{9.208}$$

$$\frac{d\overline{\omega}}{dt} = \frac{3}{4} n J_2 \left(\frac{r_0}{p}\right)^2 (5\cos^2 i - 1) .$$

The overbar in the quantities in Eq. (9.208) represents the average rate of change of the particular orbital element taken over a complete orbit. It is evident from Eq. (9.208) that oblateness has no effect on the semi-major axis, the eccentricity, and the orbital inclination, while the eccentricity vector, **e**, and the nodal vector, $\mathbf{i}_n$, both rotate at fixed rates per orbit. The rotation also causes a change in the time of periapsis, $t_0 = -M_0/n$, at the average rate of $-(d\overline{M}_0/dt)/n$ per orbit. Such a rotation of the orbital plane while maintaining a constant shape and inclination is termed *precession*, which is a term applied to the rotation of a rigid body, and is discussed in Chap. 11. As will be seen in Chap. 11, the precession of an axisymmetric rigid body is caused by the application of a torque normal to the spin axis. This concept can be extended to the orbital plane, which preserves its shape and size (and hence acts like a rigid body), and experiences an out-of-plane torque caused by the accretion of mass near the equator of the central body. Thus the orbital plane precesses like a spinning top about the polar axis, **K**, at an average frequency, $d\overline{\Omega}/dt$, thereby causing a change in the direction of the ascending node per orbit. Furthermore, the orbital plane rotates about the angular momentum vector such that the line joining the periapsis and apoapsis (the *line of apsides*) varies with the time at the constant rate, $d\overline{\omega}/dt$. The variation in the time of periapsis, $t_0 = -M_0/n$, can be physically understood as being caused by a speeding up of the orbiting body as it approaches the equatorial bulge, and a slowing down as it moves away from it. This shifts both the time (t_0) and the location (ω) of the periapsis.

For a direct orbit ($0 < i < \pi/2$), the ascending node direction, $\mathbf{i}_n$, rotates *backwards* (or westwards), such that the value of Ω constantly decreases with the time, as depicted in Fig. 9.3. This behavior is called the *regression of nodes*. The movement of $\mathbf{i}_n$ is in the forward (eastward) direction (called progression) for a retrograde orbit ($\pi/2 < i < \pi$).

The rotation of the line of apsides is called the *apsidal rotation*. There is a critical value of the orbital inclination, $i = i_c = \cos^{-1}\frac{1}{\sqrt{5}}$, which determines the direction of the variation in the argument of periapsis, ω. If $i > i_c$, the line of apsides regresses; that is, the angle ω decreases with time. For $i < i_c$, the argument of periapsis, ω, increases with the time after every orbit (Fig. 9.3). Both nodal and apsidal rotation rates diminish as the orbit size, a, increases (which

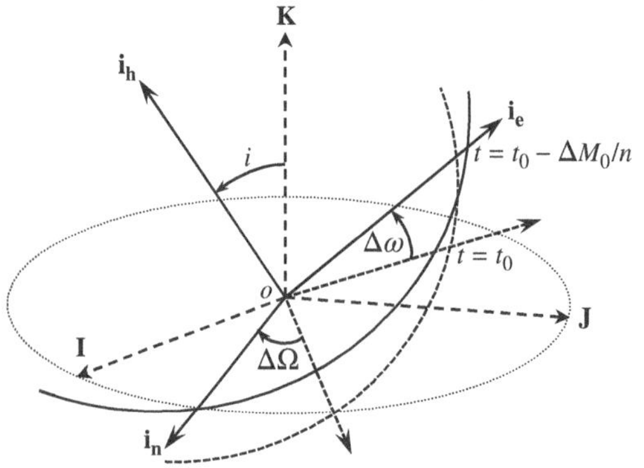

Figure 9.3 Rotation of apsides, $\Delta\omega$, and regression of nodes, $\Delta\Omega$, due to an oblate central body.

results in a larger p, and a smaller n), and are directly proportional to the oblateness parameter, J_2.

The variation per orbit in the orbital elements due to oblateness effect can be calculated from the average rates as follows:

$$\Delta M_0 = \frac{2\pi}{n}\frac{d\overline{M}_0}{dt} = \frac{3\pi J_2 r_0^2}{2a^2(1-e^2)^{3/2}}(2 - 3\sin^2 i)$$
$$\Delta\Omega = \frac{2\pi}{n}\frac{d\overline{\Omega}}{dt} = -3\pi J_2\left(\frac{r_0}{p}\right)^2 \cos i \tag{9.209}$$
$$\Delta\omega = \frac{2\pi}{n}\frac{d\overline{\omega}}{dt} = \frac{3\pi}{2}J_2\left(\frac{r_0}{p}\right)^2 (5\cos^2 i - 1)\,.$$

9.8.1 *Sun-Synchronous Orbits*

A beneficial effect of nodal regression due to planetary oblateness is the *Sun-synchronous orbit*, which is effectively utilized in all photographic reconnaissance and observation satellites. In general, the normal to a satellite's orbital plane continuously departs from the line joining the sun and the planet. A Sun-synchronous orbit is a special orbit in which the orbital plane maintains a constant angle relative to the sun (called the *solar inclination*), which is highly desirable for taking photographs of the surface features of a planet. When photographs are taken with the light source (the sun) at a fixed angle relative to the camera, the shadows are always of the same length, regardless of when the pictures are taken. This fact is useful in comparing the photographs taken at different times, thereby determining any changes which may have taken place at a target location.

The sun synchronization is achieved by matching the rate of departure of the satellite's orbital plane from the sun's direction, $360°/365.242 = 0.98565°/\text{day}$, with an equal and opposite rate of nodal regression achieved due to the planetary oblateness, $d\overline{\omega}/dt$, thereby rendering the solar inclination time invariant. By definition, a sun-synchronous orbit must cross a given latitude at a fixed solar (local) time; thus the photographs of the ground locations passing below the spacecraft can be taken in approximately the same lighting conditions. Since J_2 is a positive number, a positive nodal regression rate can be achieved only by having $\cos i < 0$, implying $i > \pi/2$. Hence, a sun-synchronous orbit is always a retrograde orbit. For planets such as Earth and Mars, the low Sun-synchronous orbits used in photography missions are nearly polar, which gives the additional advantage of covering the whole planet in a relatively small number of orbits.

Example 9.8.1 *Calculate the orbital inclination of a Sun-synchronous Earth satellite of* $a =$ 6800 *km and* $e = 0.001$.

The parameter and the mean motion of the satellite are determined as follows:

$$p = a(1-e^2) = 6799.9932\ km, \quad n = \sqrt{\frac{\mu}{a^3}} = 0.001125915\ \text{rad./s}\,.$$

Table 9.3 The inclination angle for Sun-synchronous spacecraft in circular orbit around Earth.

a	i
6500	96.067°
7500	100.044°
8000	102.6275°
8500	105.681°
9000	109.2776°
10000	118.5139°
11000	131.7894°
12000	154.6396°
12352.5	179.887°

For earth, $r_0 = 6378.14$ *km,* $J_2 = 0.00108263$*, which results in the following calculation of the orbital inclination by the fourth equation in Eq. (9.208):*

$$\begin{aligned}\cos i &= -\frac{2\left(\frac{6799.9932}{6378.14}\right)^2}{3 \times 0.001125915 \times 0.00108263}\frac{2\pi}{(365.242)(24)(3600)} \\ &= -0.123776\end{aligned} \tag{9.210}$$

or

$$i = \cos^{-1}(-0.123776) = 97.11^\circ .$$

Thus the satellite is in a nearly polar, retrograde orbit.

As the size, a, of the orbit increases, the inclination angle required for sun synchronization also increases. This is shown in Table 9.3, which lists the required inclination for various possible circular ($e = 0$) Sun-synchronous orbits around Earth. The largest possible value of semi-major axis for such an orbit is $a = 12352.506862$ km, which has $i = 180^\circ$.

9.8.2 Molniya Orbits

A class of special orbits based on oblateness effects are the *Molniya* orbits used by the erstwhile Soviet Union and modern Russia for telecommunications relay at high latitudes. The necessity of remaining at high latitudes for long periods requires a Molniya orbit to be highly elliptic ($e = 0.73$), of approximately a 12-hour period and with an apogee near the geographic north pole. To maintain the apogee at high latitudes requires that the perigee must be constant with $\omega \simeq 3\pi/2$; hence there should be no rotation of the apsides due to the oblateness effect. This requires the following from the last of the set in Eq. (9.208):

$$\frac{d\bar{\omega}}{dt} = \frac{3}{4}nJ_2\left(\frac{r_0}{p}\right)^2(5\cos^2 i - 1) = 0\,, \tag{9.211}$$

which implies that $\cos i = \sqrt{1/5}$, which can be achieved by either $i = 63.435°$ or $i = 296.565°$. Since Russia has a launch site located at Plesetsk (latitude 62.8°) it is practical to launch a spacecraft into a Molniya orbit of the critical inclination $i = 63.435°$. While a Molniya spacecraft has zero rotation of apsides, it experiences a regression of nodes at the average rate, $d\bar{\omega}/dt = -0.0024°/\text{day}$. With a period of nearly 12 hours, two Molniya satellites can cover the telecommunications requirements of Russia per 24-hour period. However, being highly eccentric orbits, Molniya satellites experience a fair amount of atmospheric drag at every perigee, which limits their life in orbit (as explored in Section 9.9). Consequently, Molniya satellites must be replaced in orbit every few months.

9.9 Effects of Atmospheric Drag

A spacecraft in a low orbit experiences a significant atmospheric drag, whose magnitude is directly proportional to the product of the atmospheric density, ρ, and the square of the velocity, v^2, relative to the atmosphere. Since the drag opposes the orbital motion, the perturbed equations of motion of a spacecraft of mass, m, can be expressed as follows, assuming the atmosphere is at rest:

$$\frac{d\mathbf{v}}{dt} + \frac{\mu}{r^3}\mathbf{r} = -qv\mathbf{v}$$

$$\frac{d\mathbf{r}}{dt} = \mathbf{v}, \tag{9.212}$$

where

$$q = \frac{1}{2}\rho\frac{C_D A}{m} \tag{9.213}$$

and C_D is the *drag coefficient* of the spacecraft based upon a reference area, A. Generally, the *free-molecular flow* assumption (Tewari, 2006) is valid at the orbital altitudes, which yields a drag coefficient, $C_D \simeq 2$, for a sphere. Since the drag coefficient is roughly invariant with the velocity, the ratio, $\beta = C_D A/m$, is a constant parameter of the spacecraft, and is called the *ballistic coefficient*.

The dependence of the atmospheric drag on the velocity makes it a non-conservative force, which causes a decline of the orbital energy, $\varepsilon = -\mu/(2a)$, and therefore the semi-major axis, a. Taking the scalar product of the deceleration due to drag with the relative velocity, we have the following rate of change of the orbital energy:

$$\dot{\varepsilon} = \frac{\mu}{2a^2}\dot{a} = -qv\mathbf{v}\cdot\mathbf{v} = -qv^3 = -\frac{1}{2}\beta\rho v^3, \tag{9.214}$$

or

$$\dot{a} = -\frac{2a^2}{\mu}qv^3 = -\frac{\beta a^2}{\mu}\rho v^3. \tag{9.215}$$

Thus for a given $q = \frac{1}{2}\rho\beta$, the rate of decline of the orbit, $-\dot{a}$, increases proportionally with v^3, as well as with the square of the orbit size, a^2.

The atmospheric density at orbital altitudes is approximated to be an exponentially decaying function of the altitude, $z = r - R_e$:

$$\rho = \rho_0 \exp\left(-\frac{z}{H}\right), \tag{9.216}$$

where the parameters, ρ_0 (the *base density*) and H (the *scale height*), depend upon the prevailing thermodynamic properties of the atmosphere, and can be considered to be constants in a specific range of altitudes. The substitution of Eq. (9.216) into Eq. (9.215) results in the following:

$$\dot{a} = -\frac{\beta\rho_0 a^2}{\mu}\exp\left(-\frac{z}{H}\right)v^3 . \tag{9.217}$$

9.9.1 Life of a Satellite in a Low Circular Orbit

The rate of decline of the orbit due to atmospheric drag given by Eq. (9.217) is used to determine the life of the satellite in a low circular orbit. Let the initial radius of the satellite's orbit be $a_0 = r_0 + z_0$ at $t = 0$, where r_0 is the planetary surface radius. The initial speed in the circular orbit is given by $v_0 = \sqrt{\mu/a_0}$. Since the drag causes a small deceleration compared to the acceleration due to gravity, and acts tangentially to the originally circular orbit, it is a good approximation to assume that a nearly circular shape of the orbit is maintained, which implies that

$$v \simeq \sqrt{\frac{\mu}{a}}, \tag{9.218}$$

which, substituted into Eq. (9.217), yields

$$\dot{a} = -\beta\rho_0 \exp\left[-\frac{(a-r_0)}{H}\right]\sqrt{\mu a} . \tag{9.219}$$

An integration of Eq. (9.219) from a_0 to $a(t)$ results in the following:

$$\int_{a_0}^{a}\frac{\exp\left(\frac{a-r_0}{H}\right)}{\sqrt{a}}\mathrm{d}a = -\beta\rho_0\sqrt{\mu}t , \tag{9.220}$$

or

$$\int_{z_0}^{z}\frac{\exp(z/H)}{\sqrt{r_0+z}}\mathrm{d}z = -\beta\rho_0\sqrt{\mu}t . \tag{9.221}$$

Due to the fact that the satellite is initially in a low orbit, $z \ll r_0$, the following approximation is valid:

$$\frac{1}{\sqrt{r_0+z}} \simeq \frac{1}{\sqrt{r_0}}, \tag{9.222}$$

which, substituted into Eq. (9.221), produces

$$\frac{H}{\sqrt{r_0}}[\exp(z/H) - \exp(z_0/H)] \simeq -\beta\rho_0\sqrt{\mu}t , \tag{9.223}$$

or

$$z(t) = H\ln\left[\exp(z_0/H) - \frac{\beta\rho_0\sqrt{\mu r_0}}{H}t\right] . \tag{9.224}$$

Equation (9.224) is used to predict the life of a satellite in a low circular orbit of initial altitude, z_0, as follows. For small values of the time, t, the exponential term in the square brackets on the right-hand side is predominant, which implies that for some time, a nearly constant altitude, $z \simeq z_0$, is maintained. As the time becomes large, the second term in the square brackets on the right-hand side becomes appreciable in magnitude, thereby subtracting from the exponential term, which results in a rapid decay of the altitude with time. The approximate life of the satellite, t_d, is therefore obtained as the time taken to reach the surface, and is given by

$$0 = H \ln \left[\exp(z_0/H) - \frac{\beta \rho_0 \sqrt{\mu r_0}}{H} t_d \right], \tag{9.225}$$

or

$$t_d = \frac{H}{\beta \rho_0 \sqrt{\mu r_0}} [\exp(z_0/H) - 1]. \tag{9.226}$$

It is evident from Eq. (9.226) that the life of the satellite, t_d, increases approximately by a factor of three when the initial altitude, z_0, is increased by one scale height, H.

Example 9.9.1 *Estimate the life of a satellite of frontal cross-sectional area, $A = 50$ m^2; drag coefficient based on the frontal area, $C_D = 2.2$; mass, $m = 200$ kg., initially placed in a circular orbit of altitude 250 km around Earth ($\mu = 398600.4$ km^3/s^2, $r_0 = 6378.14$ km), assuming an exponential atmosphere with $\rho_0 = 1.752$ kg/m^3 and $H = 6.7$ km.*

The ballistic coefficient and the constant, $\sqrt{\mu r_0}$, are calculated to be the following:

$$\beta = \frac{C_D A}{m} = 0.55 \; ; \qquad \sqrt{\mu r_0} = 50421514805.25 \text{ m}^2/\text{s} ,$$

which is substituted into Eq. (9.226) as follows:

$$t_d = \frac{6700 \, [\exp(250/6.7) - 1]}{1.752 \times 0.55 \times 50421514805.25} = 2210956020.612 \text{ s. } (70.06 \text{ yr.}) .$$

Thus the life of this satellite is estimated to be approximately 70 years. A plot of Eq. (9.224) in Fig. 9.4 shows the approximate variation of the altitude with time, indicating a steep fall at $t = 70$ years.

The accurate estimate of the orbital life requires a knowledge of the actual atmospheric properties at high altitudes prevailing over long periods of time (years and decades), which is seldom possible due to random external disturbances caused by the interaction between solar radiation and the planet's magnetic field.

9.9.2 *Effect on Orbital Angular Momentum*

Taking the vector product of the deceleration due to drag with the relative velocity, we have the following rate of change of angular momentum:

$$\dot{\mathbf{h}} = -\mathbf{r} \times qv\mathbf{v} = -qv\mathbf{h} . \tag{9.227}$$

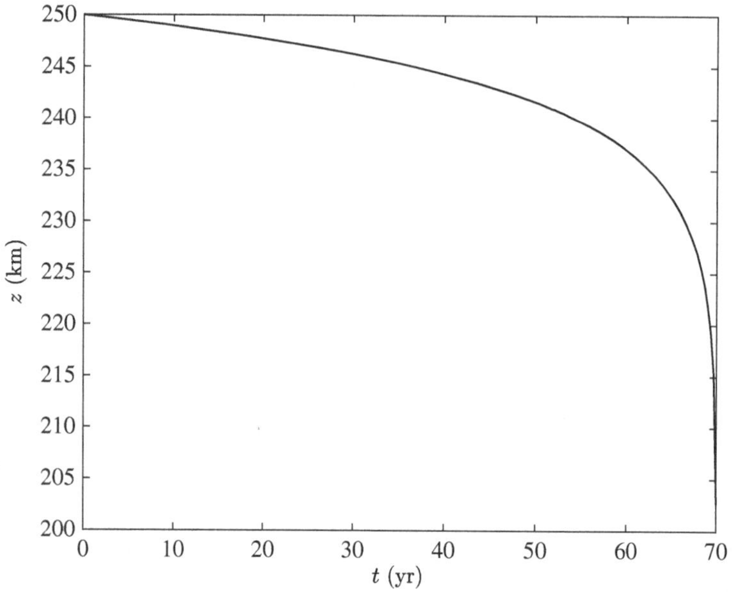

Figure 9.4 The variation of altitude with time of a satellite initially placed in a circular orbit of altitude 250 km around Earth.

Taking the time derivative of the equation $h^2 = \mathbf{h} \cdot \mathbf{h}$ results in the following:

$$h\dot{h} = \mathbf{h} \cdot \dot{\mathbf{h}} . \tag{9.228}$$

Substitution of Eq. (9.228) into Eq. (9.227) yields

$$h\dot{h} = -qv\mathbf{h} \cdot \mathbf{h} = -qvh^2 , \tag{9.229}$$

or

$$\dot{h} = -qvh . \tag{9.230}$$

Since $\mathbf{h} = h\mathbf{i}_h$, we have

$$\dot{\mathbf{h}} = \dot{h}\mathbf{i}_h + h\frac{\mathrm{d}\mathbf{i}_h}{\mathrm{d}t} . \tag{9.231}$$

The substitution of Eqs. (9.227) and (9.230) into Eq. (9.231) produces

$$h\frac{\mathrm{d}\mathbf{i}_h}{\mathrm{d}t} = qvh\mathbf{i}_h - qv\mathbf{h} = \mathbf{0} , \tag{9.232}$$

which implies that

$$\frac{\mathrm{d}\mathbf{i}_h}{\mathrm{d}t} = \mathbf{0} . \tag{9.233}$$

Therefore, the orbital plane is unaffected by the atmospheric drag; hence the Euler angles, i, Ω, defining the orientation of the orbit normal, $\mathbf{i}_h$, remain invariant with time.

Due to an almost exponential increase in the atmospheric density with a decrease in the altitude, the largest drag occurs at the periapsis, and the smallest at the apoapsis. Such a variation in the drag with the true anomaly can cause a change in the orbital eccentricity, as well as in the time and the location of the periapsis, which are explored in Section 9.9.3.

9.9.3 *Effect on Orbital Eccentricity and Periapsis*

As seen in Section 9.9.2, the orbital plane, $\mathbf{i_h} = \mathbf{h}/h$, is unchanged by the atmospheric drag. This implies that the equations of the perturbed Keplerian motion due to drag can be resolved in the radial and circumferential directions, $\mathbf{i_r}$ and $\mathbf{i}_\theta$, respectively [Eqs. (6.50) and (6.51)]:

$$\ddot{r} - r\dot{\theta}^2 + \frac{\mu}{r^2} = \ddot{r} - \frac{h^2}{r^3} + \frac{\mu}{r^2} = u_r \,, \tag{9.234}$$

$$r\ddot{\theta} + 2\dot{r}\dot{\theta} = \frac{1}{r}\dot{h} = u_\theta \,, \tag{9.235}$$

where

$$\mathbf{u} = u_r\mathbf{i_r} + u_\theta\mathbf{i}_\theta = -qv\mathbf{v} \,, \tag{9.236}$$

and $q = \frac{1}{2}\rho\beta$.

Since the drag always acts in the tangential direction, the perturbed equations are expressed in the tangential and velocity-normal directions, $\mathbf{i_v} = \mathbf{v}/v$ and $\mathbf{i_n} = \mathbf{i_v} \times \mathbf{i_h}$, respectively, as follows:

$$\frac{\mathrm{d}\mathbf{r}}{\mathrm{d}t} = \mathbf{v}$$

$$\frac{\mathrm{d}\mathbf{v}}{\mathrm{d}t} + \frac{\mu}{r^3}\mathbf{r} = -qv^2\mathbf{i_v} \,. \tag{9.237}$$

The coordinate transformation between the two right-handed coplanar frames, $(\mathbf{i_r}, \mathbf{i}_\theta, \mathbf{i_h})$ and $(\mathbf{i_n}, \mathbf{i_v}, \mathbf{i_h})$, can be derived as follows:

$$\begin{aligned}\frac{\mathrm{d}\mathbf{v}}{\mathrm{d}t} &= \dot{v}\mathbf{i_v} + \boldsymbol{\omega} \times \mathbf{v} \\ &= \dot{v}\mathbf{i_v} + (\dot{\theta} - \dot{\phi})\mathbf{i_h} \times \mathbf{i_v} = \dot{v}\mathbf{i_v} - v\dot{\theta}\mathbf{i_n} \,,\end{aligned} \tag{9.238}$$

where θ is the true anomaly along the osculating orbit, and the radius and velocity vectors are resolved in the velocity-normal frame as follows in terms of the flight-path angle, ϕ (Fig. 6.8):

$$\mathbf{v} = v\mathbf{i_v}$$

$$\mathbf{r} = r\sin\phi\mathbf{i_v} + r\cos\phi\mathbf{i_n} \,. \tag{9.239}$$

Thus we have

$$\begin{Bmatrix} \mathbf{i_r} \\ \mathbf{i}_\theta \end{Bmatrix} = \begin{pmatrix} \sin\phi & \cos\phi \\ \cos\phi & -\sin\phi \end{pmatrix} \begin{Bmatrix} \mathbf{i_v} \\ \mathbf{i_n} \end{Bmatrix} \tag{9.240}$$

or

$$\mathbf{i_v} = \mathbf{i_r}\sin\phi + \mathbf{i}_\theta\cos\phi \,, \tag{9.241}$$

which, substituted into Eq. (9.236), yields the radial and circumferential components of perturbing acceleration to be the following:

$$u_r = -qv^2 \sin\phi\,; \qquad u_\theta = -qv^2 \cos\phi\,. \tag{9.242}$$

The flight-path angle is given in terms of the true anomaly along the osculating orbit by Eqs. (3.88) and (3.88) as follows:

$$\cos\phi = \frac{\mu(1+e\cos\theta)}{hv}\,; \qquad \sin\phi = \frac{\mu e\sin\theta}{hv}\,. \tag{9.243}$$

Hence the acceleration components are the following:

$$u_r = -\frac{\mu qv}{h} e\sin\theta\,; \qquad u_\theta = -\frac{\mu qv}{h}(1+e\cos\theta)\,. \tag{9.244}$$

The substitution of Eq. (9.240) into the second of Eqs. (9.237) yields the following scalar components:

$$\dot{v} + \frac{\mu}{r^2}\sin\phi = -qv^2 \tag{9.245}$$

and

$$-v(\dot{\theta} - \dot{\phi}) + \frac{\mu}{r^2}\cos\phi = 0\,. \tag{9.246}$$

The Eqs. (9.234), (9.235) pair and the Eqs. (9.245), (9.246) pair offer two alternative routes for deriving the variational equations for the elements of the osculating orbit. Here, the former set is employed. Substitution of Eq. (9.244) into the Gauss variational model, Eqs. (9.151)–(9.156), yields the following variational equations for an orbit perturbed by atmospheric drag:

$$\begin{aligned}
\frac{\mathrm{d}a}{\mathrm{d}t} &= \frac{2a^2}{h}\left(eu_r\sin\theta + \frac{p}{r}u_\theta\right) \\
&= -\frac{2a^2\mu qv}{h^2}\left[e^2\sin^2\theta + \frac{p}{r}(1+e\cos\theta)\right] \\
&= -\frac{2a^2\mu qv}{h^2}[e^2\sin^2\theta + (1+e\cos\theta)^2] \\
&= -\frac{2a^2\mu qv}{h^2}\left(\frac{h^2v^2}{\mu^2}\right)(\sin^2\phi + \cos^2\phi) \\
&= -\frac{2a^2}{\mu}qv^3\,.
\end{aligned} \tag{9.247}$$

$$\begin{aligned}
\frac{\mathrm{d}e}{\mathrm{d}t} &= \frac{1}{h}\{pu_r\sin\theta + [(p+r)\cos\theta + re]u_\theta\} \\
&= -qv\left\{e\,\sin^2\theta + \left[\left(1+\frac{r}{p}\right)\cos\theta + \frac{r}{p}e\right](1+e\cos\theta)\right\} \\
&= -2qv(e+\cos\theta)\,.
\end{aligned} \tag{9.248}$$

$$\frac{d\beta}{dt} = \frac{b}{aeh}[(p\cos\theta - 2re)u_r - (p+r)\sin\theta u_\theta]$$
$$= -\frac{b\mu q v}{aeh^2}[(p\cos\theta - 2re)e\sin\theta - (p+r)\sin\theta(1+e\cos\theta)]$$
$$= -\frac{2bqv}{ae}\left[\frac{1+e(e+\cos\theta)}{1+e\cos\theta}\right]\sin\theta\,. \tag{9.249}$$

$$\frac{d\Omega}{dt} = \frac{r}{h\sin i}\sin(\theta+\omega)u_n = 0\,. \tag{9.250}$$

$$\frac{di}{dt} = \frac{r}{h}\cos(\theta+\omega)u_n = 0\,. \tag{9.251}$$

$$\frac{d\omega}{dt} = \frac{1}{eh}[-p\cos\theta u_r + (p+r)\sin\theta u_\theta] - \frac{r\cos i\sin(\theta+\omega)}{h\sin i}u_n$$
$$= -\frac{\mu q v}{eh^2}[-ep\cos\theta\sin\theta + (p+r)\sin\theta(1+e\cos\theta)]$$
$$= -\frac{2qv}{e}\sin\theta\,. \tag{9.252}$$

Of these, Eqs. (9.247), (9.250), and (9.251) have already been accounted for earlier in this section by considering the variation in the orbital energy, ε, and the direction of the angular momentum vector, $\mathbf{i}_\mathbf{h}$, while Eqs. (9.248), (9.249), and (9.252) specify the variation in the eccentricity, e, the time of periapsis, t_0, and the argument of the periapsis, ω, respectively.

For $0 \le e \le 1$, the averaged variations over a complete orbit are calculated by expressing the velocity, v, and the atmospheric density, ρ, in terms of the true anomaly, θ, as follows:

$$v = \sqrt{\frac{2\mu}{r} - \frac{\mu}{a}} = \sqrt{\frac{\mu}{p}(1+2e\cos\theta+e^2)} = \sqrt{\left(\frac{\mu}{a}\right)\frac{1+e\cos E}{1-e\cos E}}\,. \tag{9.253}$$

$$\rho = \rho_0\exp[-(r-r_0)/H] = \rho_0\exp(r_0/H)\exp[-p/(H+eH\cos\theta)]$$
$$= \rho_0\exp[(r_0-a)/H]\exp(ae\cos E/H)\,. \tag{9.254}$$

These are substituted into Eqs. (9.248), (9.249), and (9.252), and integrated per orbit as follows:

$$\frac{d\bar{e}}{dt} = \frac{1}{2\pi}\int_0^{2\pi}\frac{nr^2}{h}\frac{de}{dt}d\theta = \frac{1}{2\pi}\int_0^{2\pi}\frac{r}{a}\frac{de}{dt}dE$$
$$= -\frac{np\rho_0\exp(r_0/H)\beta}{2\pi}\int_0^{2\pi}\frac{e+\cos\theta}{(1+e\cos\theta)^2}\sqrt{2(1+e\cos\theta)^2+e^2-1}$$
$$\times\exp[-p/(H+eH\cos\theta)]d\theta \tag{9.255}$$
$$= -\frac{np\beta\rho_0\exp[(r_0-a)/H]}{2\pi}\int_0^{2\pi}\cos E\sqrt{\frac{1+e\cos E}{1-e\cos E}}\exp(ae\cos E/H)dE\,.$$

$$\begin{aligned}\frac{\mathrm{d}\overline{\beta}}{\mathrm{d}t} &= \frac{1}{2\pi}\int_0^{2\pi}\frac{r}{a}\frac{\mathrm{d}\beta}{\mathrm{d}t}\mathrm{d}E \\ &= -\frac{\beta\rho_0\exp[(r_0-a)/H]}{2\pi e}\int_0^{2\pi}\sin E(1-e^3\cos E)\sqrt{\frac{1+e\cos E}{1-e\cos E}}\exp(ae\cos E/H)\mathrm{d}E \\ &= 0.\end{aligned} \tag{9.256}$$

$$\begin{aligned}\frac{\mathrm{d}\overline{\omega}}{\mathrm{d}t} &= \frac{1}{2\pi}\int_0^{2\pi}\frac{r}{a}\frac{\mathrm{d}\omega}{\mathrm{d}t}\mathrm{d}E \\ &= -\frac{b\beta\rho_0\exp[(r_0-a)/H]}{2\pi ae}\int_0^{2\pi}\sin E\sqrt{\frac{1+e\cos E}{1-e\cos E}}\exp(ae\cos E/H)\mathrm{d}E \\ &= 0.\end{aligned} \tag{9.257}$$

Hence, while the orbital eccentricity decreases per orbit, there is no variation over a complete orbit in the argument and the time of periapsis.

The evaluation of the integral in Eq. (9.255) is carried out by expanding the first two factors of the integrand in the following cosine Fourier series:

$$F(E) = \cos E\sqrt{\frac{1+e\cos E}{1-e\cos E}} = \sum_{k=0}^{\infty} A_k \cos kE\,, \tag{9.258}$$

where $\cos E$ is first expanded in a power series, and then series manipulations are employed for $F(E)$ to evaluate the coefficients, A_k, as follows (Battin, 1999):

$$\begin{aligned}A_0 &= \frac{1}{2\pi}\int_{-\pi}^{\pi}F(E)\mathrm{d}E = \frac{1}{2}e\left(1+\frac{3}{8}e^2\right)+\cdots \\ A_1 &= \frac{1}{\pi}\int_{-\pi}^{\pi}F(E)\cos E\mathrm{d}E = 1+\frac{3}{8}e^2+\frac{15}{64}e^4+\cdots \\ A_2 &= \frac{1}{\pi}\int_{-\pi}^{\pi}F(E)\cos 2E\mathrm{d}E = \frac{1}{2}e\left(1+\frac{1}{2}e^2\right)+\cdots \\ A_3 &= \frac{1}{\pi}\int_{-\pi}^{\pi}F(E)\cos 3E\mathrm{d}E = \frac{1}{8}e^2\left(1+\frac{15}{16}e^2\right)+\cdots \\ A_4 &= \frac{1}{\pi}\int_{-\pi}^{\pi}F(E)\cos 4E\mathrm{d}E = \frac{1}{16}e^3+\cdots \\ A_5 &= \frac{1}{\pi}\int_{-\pi}^{\pi}F(E)\cos 5E\mathrm{d}E = \frac{3}{128}e^4+\cdots.\end{aligned} \tag{9.259}$$

The substitution of Eq. (9.258) into Eq. (9.255), results in the following:

$$\frac{\mathrm{d}\bar{e}}{\mathrm{d}t} = -\frac{np\beta\rho_0 \exp[(r_0 - a)/H]}{2\pi} \int_0^{2\pi} \cos E \sqrt{\frac{1 + e\cos E}{1 - e\cos E}} \exp(ae\cos E/H)\mathrm{d}E$$

$$= -np\beta\rho_0 \exp[(r_0 - a)/H] \sum_{k=0}^{\infty} I_k(\nu), \tag{9.260}$$

where $\nu = ae/H$, and

$$I_k(\nu) = \frac{1}{\pi} \int_0^{\pi} \exp(\nu \cos E) \cos kE\mathrm{d}E \tag{9.261}$$

is the *modified Bessel function* of the first kind and order k (Abramowitz and Stegun 1974), which is numerically evaluated by Bessel functions with an imaginary argument, $J_k(i\nu)$, with $i = \sqrt{-1}$, as follows:

$$I_k(\nu) = \frac{1}{i^k} J_k(i\nu) = \sum_{j=0}^{\infty} \frac{\left(\frac{1}{2}\nu\right)^{k+2j}}{j!(k+j)!}. \tag{9.262}$$

The above-demonstrated invariance of the orbital plane, as well as the time and the argument of periapsis, in the presence of drag can be advantageously utilized in effectively modifying the shape and size of an orbit by passing through the planetary atmosphere, and is termed an *aeroassited orbital transfer*. The reduction in the orbital eccentricity and the semi-major axis by making successive passes through a fixed periapsis inside the atmosphere is called *aerobraking*. While Eqs. (9.247) and (9.255) mathematically describe the concept of aerobraking, it can be physically understood as the application of a negative, tangential velocity impulse during each passage of a nearly fixed periapsis, $r_p = a(1 - e)$, thereby causing a reduction in both the orbital energy, $\varepsilon = -\mu/(2a)$, and the orbital angular momentum magnitude, $h = \sqrt{\mu a(1 - e^2)} = \sqrt{\mu(1 + e)r_p}$. This is continued until the orbit becomes circular, i.e., $r_p = a$, $e = 0$, and all points in the orbit have the same constant. Aerobraking has been successfully employed in several missions to circularize the orbit around the target planet, such as the *Magellan* Venus mission, the *Mars Global Surveyor* (MGS), and the *Mars Odyssey* mission.

Eqs. (9.247) and (9.248) also indicate that it is possible to change a parabolic or hyperbolic trajectory of a spacecraft as it arrives at a planet to an elliptic orbit around the planet, by making an initial pass through the atmosphere. This method of capturing a spacecraft in a bound orbit is called *aerocapture*, and offers a significant saving in the propellant mass (hence the mission cost) by avoiding a propulsive manoeuvre, which would be otherwise required for this purpose (called an *orbit insertion burn*).

9.10 Third-Body Perturbation

The third-body gravitational perturbation on a two-body orbit is modelled in a manner similar to the non-spherical gravity of the primary body. Consider the mutual orbit of the spherical masses, m_1 and m_2, perturbed by the gravity of the third spherical mass, m_3, as shown in Fig. 9.4. The equations of motion for the two orbiting bodies in an inertial reference frame are the following:

$$\begin{aligned}
\ddot{\mathbf{R}}_1 - \frac{Gm_1}{r_{21}^3}\mathbf{r_{21}} - \frac{Gm_3}{r_{31}^3}\mathbf{r_{31}} &= \mathbf{0} \\
\ddot{\mathbf{R}}_2 + \frac{Gm_2}{r_{21}^3}\mathbf{r_{21}} - \frac{Gm_3}{r_{32}^3}\mathbf{r_{32}} &= \mathbf{0}\,,
\end{aligned} \tag{9.263}$$

where $\mathbf{R_i}$, $i = 1, 2, 3$, denote the respective inertial position of the three bodies, and $\mathbf{r_{ji}} = \mathbf{R_j} - \mathbf{R_i}$, $i \neq j$, are the relative positions of the masses. A subtraction of the two equations from each other results in the following equation of relative motion between m_1 and m_2:

$$\ddot{\mathbf{r}} + \frac{\mu}{r^3}\mathbf{r} = Gm_3\left(\frac{\mathbf{r_{32}}}{r_{32}^3} - \frac{\mathbf{r_{31}}}{r_{31}^3}\right), \tag{9.264}$$

where $\mathbf{r} = \mathbf{r_{21}}$ and $\mu = G(m_1 + m_2)$ represent the two-body problem of m_2 relative to m_1. Let a disturbance potential function, U, be defined by

$$U = Gm_3\left(\frac{1}{r_{32}} - \frac{\mathbf{r}^T\mathbf{r_{31}}}{r_{31}^3}\right). \tag{9.265}$$

Since we have $\mathbf{r_{32}} = \mathbf{r_{31}} - \mathbf{r}$, the potential function, U, can be regarded to be the function of only two vector variables: $\mathbf{r_{31}}$ and $\mathbf{r}$. The gradient of the disturbance potential function with respect to $\mathbf{r}$ is therefore given by

$$\begin{aligned}
\frac{\partial U}{\partial \mathbf{r}} &= Gm_3\left[-\frac{1}{r_{32}^2}\frac{\partial r_{32}}{\partial \mathbf{r}} - \frac{\partial}{\partial \mathbf{r}}\left(\frac{\mathbf{r}^T\mathbf{r_{31}}}{r_{31}^3}\right)\right] \\
&= Gm_3\left(-\frac{1}{r_{32}^2}\frac{\partial r_{32}}{\partial \mathbf{r_{32}}}\frac{\partial \mathbf{r_{32}}}{\partial \mathbf{r}} - \frac{\mathbf{r_{31}}^T}{r_{31}^3}\right) \\
&= Gm_3\left(\frac{\mathbf{r_{32}}^T}{r_{32}^3} - \frac{\mathbf{r_{31}}^T}{r_{31}^3}\right),
\end{aligned} \tag{9.266}$$

where the following identities are utilized:

$$\begin{aligned}
\frac{\partial r_{32}}{\partial \mathbf{r_{32}}} &= \mathbf{r_{32}}^T \\
\frac{\partial \mathbf{r_{32}}}{\partial \mathbf{r}} &= \frac{\partial(\mathbf{r_{31}} - \mathbf{r})}{\partial \mathbf{r}} = -\mathbf{I}\,.
\end{aligned}$$

The substitution of Eq. (9.266) into Eq. (9.264) results in the following form of the perturbed orbital equation:

$$\ddot{\mathbf{r}} + \frac{\mu}{r^3}\mathbf{r} = \left(\frac{\partial U}{\partial \mathbf{r}}\right)^T, \tag{9.267}$$

which has the same form of the conservative perturbation as given by Eq. (9.3).

To consider the disturbance caused by more bodies, one only has to add the gradients of disturbance potentials of the additional bodies on the right hand-side of Eq. (9.267). Thus, no qualitative insight is available by additional perturbation potentials, and only the third-body disturbance needs to be considered. This is usually sufficient in most spacecraft missions, where at any given time the spacecraft experiences the gravitational attraction of no more than two primary bodies.

Since the third body, m_3, is at a much larger distance than the separation between the orbiting bodies ($r_{31} \gg r$, $r_{32} \gg r$), the perturbing acceleration given by Eq. (9.266) requires the computation of the difference between two nearly equal vectors. To alleviate this difficulty, the perturbing potential is expressed as follows:

$$U = \frac{Gm_3}{r_{31}} \left(\frac{r_{31}}{r_{32}} - \nu x \right) , \tag{9.268}$$

where $x = r/r_{31}$ and $\nu = \cos\gamma$, with γ being the angle between $\mathbf{r_{31}}$ and $\mathbf{r}$, as shown in Fig. 9.5. Define a number q such that

$$q = \left(\frac{r_{32}}{r_{31}} \right)^2 - 1 . \tag{9.269}$$

The application of the cosine law to the triangle of Fig. 9.5 yields

$$r_{32}^2 = r_{31}^2 + r^2 - 2rr_{31}\cos\gamma , \tag{9.270}$$

or

$$q = x^2 - 2\nu x . \tag{9.271}$$

The substitution of Eq. (9.271) into Eq. (9.268) results in the following:

$$U = \frac{Gm_3}{r_{31}} \left(\frac{1}{\sqrt{1+q}} - \nu x \right) = \frac{Gm_3}{r_{31}} \left(\frac{1}{\sqrt{1+x^2-2\nu x}} - \nu x \right) . \tag{9.272}$$

Using binomial theorem to expand the term in the square root, we have

$$\frac{1}{\sqrt{1+q}} = \sum_{i=0}^{\infty} \frac{(-1)^i(2i)!}{(2^i i!)^2} q^i \tag{9.273}$$

and

$$q^i = (x^2 - 2\nu x)^i = x^i \sum_{j=0}^{i} \frac{i!}{j!(i-j)!} x^j (-2\nu)^{i-j} . \tag{9.274}$$

Thus we have

$$\frac{1}{\sqrt{1+x^2-2\nu x}} = \sum_{i=0}^{\infty} \sum_{j=0}^{i} \frac{(-1)^j(2i)!}{2^{i+j} i! j! (i-j)!} x^{i+j} \nu^{i-j} . \tag{9.275}$$

Equation (9.275) with $k = i + j$ is expressed as follows:

$$\frac{1}{\sqrt{1+x^2-2\nu x}} = \sum_{k=0}^{\infty} P_k(\nu) x^k , \tag{9.276}$$

where $P_k(\nu)$ denotes the Legendre polynomial in ν of degree k (Chap. 2).

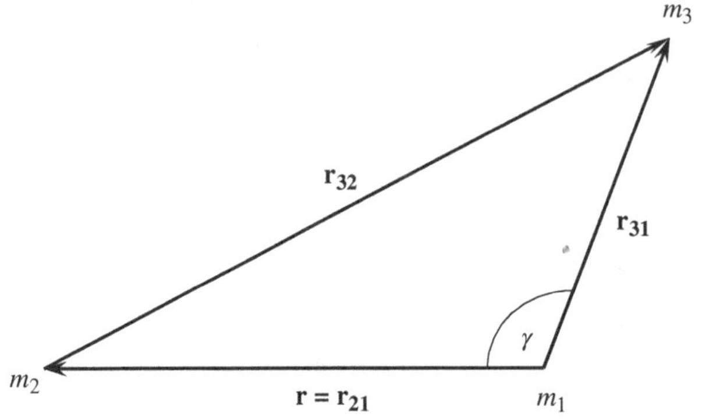

Figure 9.5 The orbit of m_2 relative to m_1, perturbed by m_3.

The substitution of Eq. (9.276) into Eq. (9.272) results in the following expression for the perturbation potential in terms of the Legendre polynomials:

$$U = \frac{Gm_3}{r_{31}}\left[1 + \sum_{k=2}^{\infty}\left(\frac{r}{r_{31}}\right)^k P_k(\cos\gamma)\right] . \tag{9.277}$$

The series in Eq. (9.277) is absolutely convergent for $r < r_{31}$, and converges rapidly for $r \ll r_{31}$. Hence only a small number of terms are necessary to approximate the series in Eq. (9.277). The main advantage of Eq. (9.277) is in the independence of the disturbance potential, U, from $\mathbf{r}_{32}$, which is continuously changing in the orbit of m_2 relative to m_1. Therefore, the gradient of the potential is directly calculated, and using the properties of the Legendre polynomials given in Chap. 2, we have the following perturbed equation of motion:

$$\ddot{\mathbf{r}} + \frac{\mu}{r^3}\mathbf{r} = G\frac{m_3}{r_{31}^2}\sum_{k=1}^{\infty}\left(\frac{r}{r_{31}}\right)^k\left[P'_{k+1}(\cos\gamma)\frac{\mathbf{r}_{31}}{r_{31}} - P'_k(\cos\gamma)\frac{\mathbf{r}}{r}\right] , \tag{9.278}$$

where the prime indicates the derivative of a Legendre polynomial with respect to its argument, and the following identities are utilized:

$$\frac{\partial}{\partial\mathbf{r}}\left(\frac{r}{r_{31}}\right)^k = k\left(\frac{r}{r_{31}}\right)^k\frac{\mathbf{r}^T}{r}$$

$$\frac{\partial\cos\gamma}{\partial\mathbf{r}} = \frac{\partial}{\partial\mathbf{r}}\left(\frac{\mathbf{r}^T\mathbf{r}_{31}}{rr_{31}}\right) = \frac{1}{r}\left(\frac{{\mathbf{r}_{31}}^T}{r_{31}} - \frac{\mathbf{r}^T}{r}\cos\gamma\right)$$

$$kP'_k(v) = vP'_k(v) - P'_{k-1}(v) .$$

9.10.1 *Lunar and Solar Perturbations on an Earth Satellite*

A satellite in an Earth's orbit experiences significant orbital perturbations caused by the gravity of the moon and the sun, called the *luni-solar attraction*. These perturbations can be individually calculated according to the perturbation potential derived earlier in this section, substituted into the Lagrange planetary equations. Let m_3 represent the mass of either the moon or the sun as the perturbing body. Since $r \ll r_{31}$ in either case, only the first two terms of the series in Eq. (9.277) are sufficient to estimate the perturbing potential, U, and the corresponding acceleration, $(\partial U/\partial \mathbf{r})^T$. Thus we have

$$\begin{aligned} U &= \frac{Gm_3}{r_{31}}\left[1+\sum_{k=2}^{3}\left(\frac{r}{r_{31}}\right)^k P_k(\cos\gamma)\right] \\ &= \frac{Gm_3}{r_{31}}\left[1+\left(\frac{r}{r_{31}}\right)^2 P_2(\cos\gamma)+\left(\frac{r}{r_{31}}\right)^3 P_3(\cos\gamma)\right] \\ &= \frac{Gm_3}{r_{31}}\left[1+\frac{1}{2}\left(\frac{r}{r_{31}}\right)^2(3\cos^2\gamma-1)+\frac{1}{2}\left(\frac{r}{r_{31}}\right)^3(5\cos^3\gamma-3\cos\gamma)\right] \end{aligned} \tag{9.279}$$

and

$$\begin{aligned} \left(\frac{\partial U}{\partial \mathbf{r}}\right)^T &\simeq \frac{Gm_3}{r_{31}^2}\sum_{k=1}^{2}\left(\frac{r}{r_{31}}\right)^k\left[P'_{k+1}(\cos\gamma)\frac{\mathbf{r_{31}}}{r_{31}}-P'_k(\cos\gamma)\frac{\mathbf{r}}{r}\right] \\ &= \frac{Gm_3}{r_{31}^2}\left\{\frac{3r}{r_{31}}\left[\cos\gamma+\frac{1}{2}(5\cos\gamma-1)\frac{r}{r_{31}}\right]\frac{\mathbf{r_{31}}}{r_{31}}\right. \\ &\quad \left.-\frac{r}{r_{31}}\left[1+3\left(\frac{r}{r_{31}}\right)\cos\gamma\right]\frac{\mathbf{r}}{r}\right\}. \end{aligned} \tag{9.280}$$

The effects of the third-body gravity on a satellite's orbit are similar to that due to planetary oblateness, wherein the inclined orbital plane of the satellite with respect to the apparent orbital plane of the perturbing body experiences an out-of-plane torque, causing a precession of the orbital plane. The variations in the orbital plane per orbit consist of the regression of nodes, $\Delta\Omega$, and the rotation of apsides, $\Delta\omega$, and can be estimated by applying Eqs. (9.154) and (9.156) to the perturbing acceleration, $\mathbf{u} = (\partial U/\partial \mathbf{r})^T$, given by Eq. (9.280). Consider the case of Earth satellite in an elliptic orbit of mean motion, n, semi-major axis, a, eccentricity, e, and inclination, i, relative to the equatorial plane. The third body of mass, m_3, is assumed to be in an apparent circular orbit of radius r_{13} around m_1, with an inclination, i_3, and mean motion, $n_3 = \sqrt{Gm_3/r_{13}^3}$. The inclination of the satellite's orbital plane, $\mathbf{i_h} = \mathbf{h}/h$, relative to the apparent orbital plane, $\mathbf{K}$, of the third body is $i - i_3$, with the ascending node denoted by $\mathbf{I}$, and the argument of periapsis, ω, measured from the ascending node, as shown in Fig. 9.6. A third unit vector, $\mathbf{J} = \mathbf{K}\times\mathbf{I}$, completes the right-handed frame, $(\mathbf{I}, \mathbf{J}, \mathbf{K})$, which can be used to resolve the radius vectors, $\mathbf{r}$ and $\mathbf{r_{31}}$, as follows:

$$\begin{aligned} \mathbf{r_{31}} &= r_{31}[\cos(n_3 t)\mathbf{I}+\sin(n_3 t)\mathbf{J}] \\ \mathbf{r} &= r[\cos(\theta+\omega)\mathbf{I}+\sin(\theta+\omega)\cos(i-i_3)\mathbf{J}+\sin(i-i_3)\mathbf{K}], \end{aligned} \tag{9.281}$$

where the time, t, is measured from the time of m_3 crossing the ascending node, when $t = 0$ and $\gamma(0) = 2\pi - [\theta(0) + \omega(0)]$.

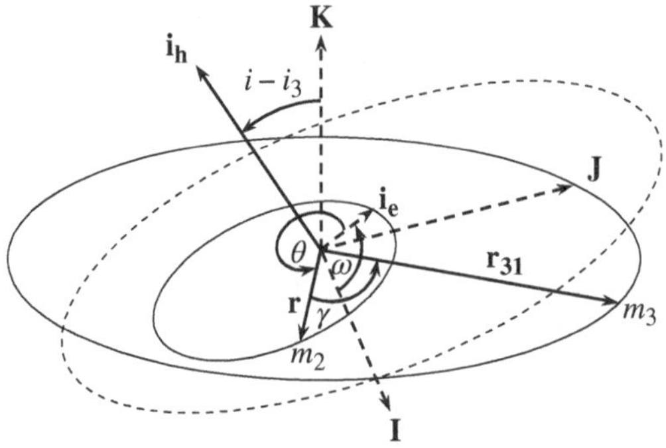

Figure 9.6 The perturbation caused by the apparent orbit of the moon or the sun, m_3, on a satellite, m_2, in an elliptic orbit around Earth.

The normal component of the perturbing acceleration is expressed as follows:

$$\begin{aligned} u_n &= \left(\frac{\partial U}{\partial \mathbf{r}}\right)^T \cdot \mathbf{i}_\mathrm{h} \\ &= \frac{3Gm_3}{r_{31}^3} r \left[\cos\gamma + \frac{1}{2}(5\cos\gamma - 1)\frac{r}{r_{31}}\right] \frac{\mathbf{r}_{31}}{r_{31}} \cdot \mathbf{i}_\mathrm{h} \\ &= 3n_3^2 \cos(i - i_3) r \left[\cos\gamma + \frac{1}{2}(5\cos\gamma - 1)\frac{r}{r_{31}}\right] . \end{aligned} \tag{9.282}$$

The regression of nodes over a complete orbit of m_2 around m_1 is then given by the following average change per revolution [Eq. (9.154)]:

$$\begin{aligned} \frac{\partial \overline{\Omega}}{\partial t} &= \frac{1}{2\pi} \int_0^{2\pi} \frac{nr^2}{h} \frac{\partial \Omega}{\partial t} \mathrm{d}\theta \\ &= \frac{3n_3^2 \cos(i - i_3)}{2\pi n a^4 (1 - e^2) \sin i} \int_0^{2\pi} r^4 \sin(\theta + \omega) \left[\cos\gamma + \frac{1}{2}(5\cos\gamma - 1)\frac{r}{r_{31}}\right] \mathrm{d}\theta . \end{aligned} \tag{9.283}$$

The rotation of apsides (Eq. (9.156)) requires, in addition to u_n, the following perturbing acceleration components resolved along the radial and the circumferential directions:

$$\begin{aligned} u_r &= \left(\frac{\partial U}{\partial \mathbf{r}}\right)^T \cdot \frac{\mathbf{r}}{r} \\ &= -\frac{Gm_3}{r_{31}^3} r \left[1 + 3\left(\frac{r}{r_{31}}\right)\cos\gamma\right] \\ &= -n_3^2 r \left[1 + 3\left(\frac{r}{r_{31}}\right)\cos\gamma\right] , \end{aligned} \tag{9.284}$$

$$\begin{aligned} u_\theta &= \left(\frac{\partial U}{\partial \mathbf{r}}\right)^T \cdot \left(\mathbf{i}_h \times \frac{\mathbf{r}}{r}\right) \\ &= \frac{3Gm_3}{r_{31}^3} r \left[\cos\gamma + \frac{1}{2}(5\cos\gamma - 1)\frac{r}{r_{31}}\right] \frac{\mathbf{r}_{31}}{r_{31}} \cdot \left(\mathbf{i}_h \times \frac{\mathbf{r}}{r}\right) \\ &= 3n_3^2 r \left[\cos\gamma + \frac{1}{2}(5\cos\gamma - 1)\frac{r}{r_{31}}\right] \mathbf{i}_h \cdot \left(\frac{\mathbf{r}}{r} \times \frac{\mathbf{r}_{31}}{r_{31}}\right), \end{aligned} \tag{9.285}$$

where a vector triple-product identity has been utilized. These substituted into Eq. (9.156), and averaged over a complete orbit, yield

$$\begin{aligned} \frac{\partial \bar{\omega}}{\partial t} &= \frac{1}{2\pi}\int_0^{2\pi} \frac{nr^2}{h}\frac{\partial \omega}{\partial t} d\theta \\ &= -\frac{3nn_3^2 \cos(i - i_3)\cos i}{2\pi h^2 \sin i} \int_0^{2\pi} r^4 \sin(\theta + \omega)\left[\cos\gamma + \frac{1}{2}(5\cos\gamma - 1)\frac{r}{r_{31}}\right] d\theta \\ &+ \frac{nn_3^2 p}{2\pi e h^2}\int_0^{2\pi} r^3 \cos\theta \left[1 + 3\left(\frac{r}{r_{31}}\right)\cos\gamma\right] d\theta \\ &+ \frac{3nn_3^2}{2\pi e h^2}\int_0^{2\pi} r^3 (r + p)\sin\theta \left[\cos\gamma + \frac{1}{2}(5\cos\gamma - 1)\frac{r}{r_{31}}\right] \mathbf{i}_h \cdot \left(\frac{\mathbf{r}}{r} \times \frac{\mathbf{r}_{31}}{r_{31}}\right) d\theta . \end{aligned} \tag{9.286}$$

The evaluation of the integrals in Eqs. (9.283) and (9.286) requires expressing the integrands as functions of θ. The orbit equation yields

$$r = \frac{a(1 - e^2)}{1 + e\cos\theta},$$

the following expression is employed for $\cos\gamma$:

$$\begin{aligned} \cos\gamma &= \frac{\mathbf{r} \cdot \mathbf{r}_{31}}{rr_{31}} \\ &= \cos(\theta + \omega)\cos(n_3 t) + \sin(\theta + \omega)\sin(n_3 t)\cos(i - i_3), \end{aligned} \tag{9.287}$$

and the vector triple product which appears in u_θ is given by

$$\begin{aligned} \mathbf{i}_h \cdot \left(\frac{\mathbf{r}}{r} \times \frac{\mathbf{r}_{31}}{r_{31}}\right) &= [\cos(\theta + \omega)\sin(n_3 t) - \sin(\theta + \omega)\cos(n_3 t)\cos(i - i_3)]\cos(i - i_3) \\ &\quad -\sin^2(i - i_3)\cos(n_3 t). \end{aligned} \tag{9.288}$$

The time, t, occurring explicitly in the integrands, can be eliminated via Kepler's equation, $E - e\sin E = n(t - t_0)$.

The evaluation of the integrals in Eqs. (9.283) and (9.286) is a formidable task, requiring numerical computation. However, for the case where $r \ll r_{31}$, the terms involving the squares (and higher powers) of the ratio, r/r_{31}, appearing in the perturbing acceleration can be neglected.

The application of such an approximation results in the following expressions for the regression of nodes and the rotation of apsides averaged per orbit:

$$\frac{\partial \overline{\Omega}}{\partial t} = -\frac{3}{8}\frac{n_3^2}{n}\frac{\left(1+\frac{3}{2}e^2\right)}{\sqrt{1-e^2}}\cos(i-i_3)(3\cos^2 i_3 - 1)$$

$$\frac{\partial \overline{\omega}}{\partial t} = \frac{3}{4}\frac{n_3^2}{n}\frac{\left(1-\frac{3}{2}\sin^2 i_3\right)}{\sqrt{1-e^2}}\left[2-\frac{5}{2}\sin^2(i-i_3)+\frac{e^2}{2}\right]. \tag{9.289}$$

Example 9.10.1 *Calculate the regression of nodes and the rotation of apsides on a GEO satellite caused by the gravity of (a) the Moon, and (b) the sun.*

For a GEO spacecraft, we have $e = 0$, $i = 0$, *and* $n = 7.292116 \times 10^{-5}$ *rad./s.*

(a) The Moon's orbit around Earth is inclined at $i_3 = 28.582°$ *relative to Earth's equatorial plane, with a mean motion of*

$$n_3 = 2\pi/(27.3 \times 24 \times 3600) = 2.6638 \times 10^{-6} \text{ rad/s} .$$

Equation (9.289) yields the following average values for the perturbations caused by the moon on a GEO satellite:

$$\frac{\partial \overline{\Omega}}{\partial t} = -\frac{3}{8}\frac{n_3^2}{n}\frac{\left(1+\frac{3}{2}e^2\right)}{\sqrt{1-e^2}}\cos(i-i_3)(3\cos^2 i_3 - 1) = -4.2085 \times 10^{-8} \text{ rad./s } (-0.20834°/day)$$

$$\frac{\partial \overline{\omega}}{\partial t} = \frac{3}{4}\frac{n_3^2}{n}\frac{\left(1-\frac{3}{2}\sin^2 i_3\right)}{\sqrt{1-e^2}}\left[2-\frac{5}{2}\sin^2(i-i_3)+\frac{e^2}{2}\right] = 6.8428 \times 10^{-8} \text{ rad./s } (0.3387°/day) .$$

(b) For the sun's apparent orbit around Earth (called the solar ring*) we have* $i_3 = 23.442°$ *(the obliquity of the ecliptic) and*

$$n_3 = 2\pi/(365.25636 \times 24 \times 3600) = 1.99099 \times 10^{-7} \text{ rad./s} ,$$

which yields

$$\frac{\partial \overline{\Omega}}{\partial t} = -\frac{3}{8}\frac{n_3^2}{n}\frac{\left(1+\frac{3}{2}e^2\right)}{\sqrt{1-e^2}}\cos(i-i_3)(3\cos^2 i_3 - 1)$$
$$= -2.85256 \times 10^{-10} \text{ rad./s } (-0.001412°/day)$$

$$\frac{\partial \overline{\omega}}{\partial t} = \frac{3}{4}\frac{n_3^2}{n}\frac{\left(1-\frac{3}{2}\sin^2 i_3\right)}{\sqrt{1-e^2}}\left[2-\frac{5}{2}\sin^2(i-i_3)+\frac{e^2}{2}\right]$$
$$= 4.9882 \times 10^{-10} \text{ rad./s } (0.002469°/day) .$$

Hence the effect of the sun's gravity on the GEO orbit is about 100 times smaller than that caused by the Moon's gravity. If uncorrected over a period of a few weeks, these perturbations can cause the satellite to deviate significantly from its assigned geostationary position, thereby jeopardizing its mission. Hence periodic orbital corrections must be applied to the GEO satellite's orbit by propulsive means, which causes a depletion of onboard propellants and limits the useful life of the satellite in orbit.

9.10.2 Sphere of Influence and Conic Patching

The concept of a *sphere of influence* allows the approximation of a multi-body trajectory by a two-body problem, and is useful in quickly designing a lunar or an interplanetary mission. Laplace introduced this approximation when studying the trajectory of a comet having passed close to Jupiter, by assigning a spherical region to the planet such that the planet's gravity dominates within the sphere, while outside it the sun's influence predominates the trajectory.

The region of influence of m_1 is developed by considering the motion of m_2 around m_1 perturbed by the gravity of m_3, such that either of the following two equations of motion is valid for m_2:

$$\ddot{\mathbf{r}} + \frac{G(m_1+m_2)}{r^3}\mathbf{r} = -Gm_3\left(\frac{\mathbf{r_{23}}}{r_{23}^3} + \frac{\mathbf{r_{31}}}{r_{31}^3}\right)$$

$$\ddot{\mathbf{r}}_{\mathbf{23}} + \frac{G(m_2+m_3)}{r_{23}^3}\mathbf{r_{23}} = -Gm_1\left(\frac{\mathbf{r}}{r^3} - \frac{\mathbf{r_{31}}}{r_{31}^3}\right), \tag{9.290}$$

where $\mathbf{r_{23}} = -\mathbf{r_{32}}$ locates m_2 from m_3. Of the two equations comprising Eq. (9.290) the one which has a smaller magnitude of the ratio of the perturbation, $\mathbf{u}$, to the primary two-body acceleration is considered to be the better model. This ratio for the motion of m_2 around m_1 is given by

$$\begin{aligned}
&\frac{m_3}{G(m_1+m_2)}r^2\left\{\left(\frac{\mathbf{r_{23}}}{r_{23}^3} + \frac{\mathbf{r_{31}}}{r_{31}^3}\right)^T\left(\frac{\mathbf{r_{23}}}{r_{23}^3} + \frac{\mathbf{r_{31}}}{r_{31}^3}\right)\right\}^{1/2} \\
&= \frac{m_3}{m_1+m_2}r^2\left(\frac{\mathbf{r_{32}}^T\mathbf{r_{32}}}{r_{32}^6} + \frac{\mathbf{r_{31}}^T\mathbf{r_{31}}}{r_{31}^6} - 2\frac{\mathbf{r_{32}}^T\mathbf{r_{31}}}{r_{32}^3 r_{31}^3}\right)^{1/2} \\
&= \frac{m_3}{m_1+m_2}r^2\left(\frac{1}{r_{32}^4} + \frac{1}{r_{31}^4} - \frac{2\cos\lambda}{r_{32}^2 r_{31}^2}\right)^{1/2} \\
&= \frac{m_3}{m_1+m_2}\frac{r^2}{r_{31}^2}\left(\frac{r_{31}^4}{r_{32}^4} + 1 - \frac{2\cos\lambda r_{31}^2}{r_{32}^2}\right)^{1/2} \\
&= \frac{m_3}{m_1+m_2}x^2\sqrt{1 + \frac{1}{(1+x^2-2\nu x)^2} - \frac{2}{(1+x^2-2\nu x)}\sqrt{1-\frac{x^2(1-\nu^2)}{1+x^2-2\nu x}}},
\end{aligned} \tag{9.291}$$

where $x = r/r_{31}$, $\nu = \cos\gamma = \mathbf{r}\cdot\mathbf{r_{31}}/(rr_{31})$ (Fig. 9.5), and the following identities are employed:

$$\mathbf{r_{23}}^T\mathbf{r_{31}} = r_{32}r_{31}\cos\lambda, \tag{9.292}$$

with

$$\sin\lambda = \frac{r}{r_{32}}\sqrt{1-\nu^2} \tag{9.293}$$

and

$$\frac{r}{r_{32}} = \frac{x}{\sqrt{1+x^2-2\nu x}}\,. \tag{9.294}$$

Similarly, the ratio of the perturbing acceleration to the primary acceleration of the orbit of m_2 around m_3 is given by

$$\begin{aligned}
&\frac{m_1}{m_2+m_3} r_{32}^2 \left(\frac{\mathbf{r}}{r^3} - \frac{\mathbf{r_{31}}}{r_{31}^3}\right)^T \left(\frac{\mathbf{r}}{r^3} - \frac{\mathbf{r_{31}}}{r_{31}^3}\right)^{1/2} \\
&= \frac{m_1}{m_2+m_3} r_{32}^2 \left(\frac{\mathbf{r}^T\mathbf{r}}{r^6} + \frac{\mathbf{r_{31}}^T\mathbf{r_{31}}}{r_{31}^6} - 2\frac{\mathbf{r}^T\mathbf{r_{31}}}{r^3 r_{31}^3}\right)^{1/2} \\
&= \frac{m_1}{m_2+m_3} \left(\frac{r_{32}^4}{r^4} + \frac{r_{32}^4}{r_{31}^4} - \frac{2r_{32}^4\nu}{r^2 r_{31}^2}\right)^{1/2} \\
&= \frac{m_1}{m_2+m_3} \left[(1+x^2-2\nu x)^2 + \frac{1}{x^4}(1+x^2-2\nu x)^2 - \frac{2\nu}{x^2}(1+x^2-2\nu x)^2\right]^{1/2} \\
&= \frac{m_1}{m_2+m_3} \frac{(1+x^2-2\nu x)}{x^2}\sqrt{1+x^4-2\nu x^2}\,.
\end{aligned} \tag{9.295}$$

The region of influence is derived for the mass, m_1, by comparing the two ratios. If the ratio given by Eq. (9.291) is smaller than that given by Eq. (9.295), then m_2 is within the region of influence of m_1; otherwise, it is in an orbit of m_3. The boundary of the region of influence of m_1 is obtained by equating the two ratios, which yields an equation for the variation of the non-dimensional radius, $x = x_s$, with ν. For the limiting case of $r \ll r_{31}$ (i.e., $x \ll 1$), the following approximations are employed:

$$\begin{aligned}
&\frac{m_3}{m_1+m_2} x^2 \sqrt{1 + \frac{1}{(1+x^2-2\nu x)^2} - \frac{2}{(1+x^2-2\nu x)}\sqrt{1-\frac{x^2(1-\nu^2)}{1+x^2-2\nu x}}} \\
&\simeq \frac{m_3}{m_1+m_2} x^3 \sqrt{1+3\nu^2}
\end{aligned} \tag{9.296}$$

and

$$\begin{aligned}
&\frac{m_1}{m_2+m_3} \frac{(1+x^2-2\nu x)}{x^2}\sqrt{1+x^4-2\nu x^2} \\
&\simeq \frac{m_1}{m_2+m_3}\frac{1}{x^2}\,.
\end{aligned} \tag{9.297}$$

Equating Eqs. (9.296) and (9.297) yields the following approximate expression for the region of influence:

$$x_s = \frac{r_s}{r_{31}} \simeq \left[\frac{m_1(m_1+m_2)}{m_3(m_2+m_3)}\right]^{1/5} (1+3\nu^2)^{-1/10}\,. \tag{9.298}$$

Table 9.4 Mass ratio to the solar mass, average distance from the sun, and the radius of the sphere of influence of the planets in the solar system.

Planet	m_1/m_3	r_{31} (AU)	r_s (km)
Mercury	1.66005×10^{-7}	0.387098	112407.248
Venus	2.4471×10^{-6}	0.723332	616201.317
Earth	3.00255×10^{-6}	1.0	924530.498
Mars	3.22613×10^{-7}	1.523679	577153.807
Jupiter	0.000954301	5.2044	48213650.533
Saturn	0.0002857272	9.5826	54801091.558
Uranus	4.36494×10^{-5}	19.2184	51836265.225
Neptune	5.14871×10^{-5}	30.11	86759162.416

With the further approximation $m_3 \gg m_1 \gg m_2$, the region of influence of m_1 reduces to a spherical region centred at m_1, whose radius, r_s, is approximated by

$$x_s = \frac{r_s}{r_{31}} \approx \left(\frac{m_1}{m_3}\right)^{\frac{2}{5}} . \tag{9.299}$$

Here the variation with v is neglected by considering that the factor $(1 + 3v^2)^{-1/10}$ ranges between 0.87055 and unity, with a mean value of 0.9353. Hence, $(1 + 3v^2)^{-1/10} \approx 1$.

For all the planets in the solar system, the sphere of influence is a good approximation. A large planet orbiting far away from the sun has a large sphere of influence, whereas a small planet closer to the sun has a smaller sphere. Table 9.4 lists the spheres of influence of the various planets. It is interesting to note that the largest sphere of influence is for Neptune, a massive planet farthest from the sun, while the smallest is for Mercury. The Earth-Moon system can be used to estimate the sphere of influence of the Moon, which has a radius of $r_s = 66182.656$ km centred at the Moon. However, the sphere is not a good approximation for the region of influence of the moon due to the mass ratio of the moon relative to Earth not being small enough. The actual shape is determined by solving Eq. (9.296) for r_s corresponding to various values of v.

The concept of sphere of influence enables a quick estimation of an interplanetary trajectory by connecting the two-body trajectories calculated by taking either m_1 (planet), or m_3 (the sun) as the primary body, depending upon whether the spacecraft (m_2) is inside or outside the region of influence of m_1. Such an approach is called the *patched-conic approximation*, and consists of smoothly patching the two-body solutions at the boundary of the sphere of influence of m_1. This is essential because the actual trajectory does not encounter any discontinuity in the flight-path angle at that purely mathematical boundary. The design of an interplanetary mission involves aiming the spacecraft from an initial heliocentric position, $\mathbf{r}_{23}(0)$, at time $t = 0$, in the vicinity of the first planet to a final heliocentric position, $\mathbf{r}_{23}(\mathbf{t_f})$, close to another planet, in a given flight time, t_f. The spacecraft transits between the spheres of influence of the two planets, through the intervening solar gravitation region outside the two spheres. To design such a mission by the patched-conic approach, the following steps are employed: (a) Given the initial position and velocity relative to the first planet, calculate the heliocentric position, $\mathbf{r}_{23}(0) = \mathbf{r_i} - \mathbf{r}_{31}(0)$, and the required velocity of the spacecraft along the departure hyperbola at the sphere of influence. This step requires the solution of the two-body problem with the first planet as the primary mass, m_1. (b) Select a point of arrival, $\mathbf{r}_{23}(t_f) = \mathbf{r_f} - \mathbf{r}_{31}(t_f)$, at the sphere of influence of the target planet, and solve the Lambert's problem associated with $\mathbf{r_{23}(0)}$, $\mathbf{r_{23}(t_f)}$, and flight time, $\mathbf{t_f}$, using the sun ($\mathbf{m_3}$) as the primary body. (c) From $\mathbf{r_{23}(t_f)}$ and the arrival velocity, estimate the hyperbolic trajectory relative to the target planet, and check whether it passes through the desired final relative position, $\mathbf{r_f}$. (d) Iterate for $\mathbf{r_{23}(0)}$ and $\mathbf{r_{23}(t_f)}$ until the velocities from the relative hyperbolae match with those of the heliocentric transfer ellipse at the respective boundaries of the departure and arrival spheres.

The patched-conic procedure is a practical method for the preliminary design and analysis of interplanetary missions. However, it is not sufficiently accurate to be used in the navigation of spacecraft, where even a small error in the heliocentric position results in the spacecraft completely missing the target planet. When designing lunar missions by the patched-conic method, advantage is taken of the fact that the Moon's orbit lies within Earth's sphere of influence ($\mathbf{r_s} = 924530.498$ km). Therefore, solar gravity would equally perturb Earth, the Moon, and the spacecraft during such a mission, and can be effectively removed from the calculations. However, since the Moon's region of influence cannot be regarded as being spherical, the patching of the conic solutions becomes complicated on the non-spherical boundary. An alternative solution procedure is to determine a trans-lunar trajectory by the numerical solution of the restricted three-body problem (Chap. 10).

9.11 Numerical Methods for Perturbed Keplerian Motion

A numerical procedure can be employed to integrate the nonlinear, ordinary differential equations, Eqs., (9.1) and (9.2), in time, using a known perturbation model for $\mathbf{u}(t)$, and with a given initial condition, $\mathbf{r}(0), \mathbf{v}(0)$. In this section, we will consider two broad types of numerical methods used for solving the perturbed two-body problem.

9.11.1 Cowell's Method

A procedure where a Runge-Kutta – type method (see Appendix A) is directly applied to integrate Eqs., (9.1) and (9.2) in time is called *Cowell's method.* Solving the coupled, non-linear, ordinary differential equations, Eqs., (9.1) and (9.2), by Cowell's method is often difficult due to convergence issues, especially when the magnitude of the perturbation, $u(t) = \mid \mathbf{u}(t) \mid$, is very small, which requires a large number of time steps for an accurate solution (Appendix A). However, Cowell's method is the only available option when $u(t)$ is either larger than or of a comparable magnitude to the primary acceleration, μ/r^2.

9.11.2 Encke's Method

An indirect numerical procedure which utilizes the concept of the osculating orbit is *Encke's method*, which avoids the numerical issues of Cowell's method when $u \ll \mu/r^2$. Encke's method is based upon calculating the deviation, $\bar{\mathbf{r}}(t), \bar{\mathbf{v}}(t)$, from the osculating orbit, projected forward in time, $t \geq \tau$, from the current position and velocity, $\mathbf{r}(\tau), \mathbf{v}(\tau)$. The instantaneous conic solution comprising the osculating orbit satisfies

$$\dot{\bar{\mathbf{v}}} + \mu \frac{\bar{\mathbf{r}}}{\bar{r}^3} = \mathbf{0}$$
$$\dot{\bar{\mathbf{r}}} = \bar{\mathbf{v}}\,, \tag{9.300}$$

and the classical orbital elements of the osculating orbit are calculated at time τ, from the current position and velocity at the given instant, such that

$$\bar{\mathbf{r}}(\tau) = \mathbf{r}(\tau)$$
$$\bar{\mathbf{v}}(\tau) = \mathbf{v}(\tau)\,. \tag{9.301}$$

At a slightly later time, $t = \tau + \Delta t$, we have

$$\mathbf{r}(t) = \bar{\mathbf{r}}(t) + \boldsymbol{\sigma}(t)$$
$$\mathbf{v}(t) = \bar{\mathbf{v}}(t) + \boldsymbol{\xi}(t)\,. \tag{9.302}$$

Substituting Eqs. (9.302) and (9.300) into Eq. (9.1), we have

$$\frac{\mathrm{d}^2\boldsymbol{\sigma}}{\mathrm{d}t^2} + \frac{\mu}{\bar{r}^3}\boldsymbol{\sigma} = \frac{\mu}{\bar{r}^3}\left(1 - \frac{\bar{r}^3}{r^3}\right)\mathbf{r} + \mathbf{u}\,, \tag{9.303}$$

which is subject to the initial condition

$$\boldsymbol{\sigma}(\tau) = \mathbf{0}$$
$$\dot{\boldsymbol{\sigma}}(\tau) = \boldsymbol{\xi}(\tau) = \mathbf{0}\,. \tag{9.304}$$

Since r and $\bar{r}$ are nearly equal, the term in the brackets on the right-hand side of Eq. (9.303) may present numerical difficulties associated with machine round-off error. In order to avoid such a problem, this term is expressed as follows:

$$1 - \frac{\bar{r}^3}{r^3} = -x\frac{3 + 3x + x^2}{1 + (1 + x)^{\frac{3}{2}}}\,, \tag{9.305}$$

where

$$x = \frac{\boldsymbol{\sigma}\cdot(\boldsymbol{\sigma} - 2\mathbf{r})}{r^2}\,. \tag{9.306}$$

The numerical implementation of Encke's method can be expressed by the following steps:

(a) Compute $\bar{\mathbf{r}}(t), \bar{\mathbf{v}}(t)$ using the osculating orbit obtained at $t = \tau$ from Eq. (9.301). For this purpose, one can apply Lagrange's coefficients (Chap. 2) for the osculating orbit.
(b) Numerically integrate Eq. (9.303), and compute the true position and velocity by Eq. (9.302).
(c) If at any time, the magnitudes of $\boldsymbol{\sigma}, \boldsymbol{\xi}$ exceed specified tolerances, rectify the osculating orbit by making $\tau = t$ and going back to step (a).

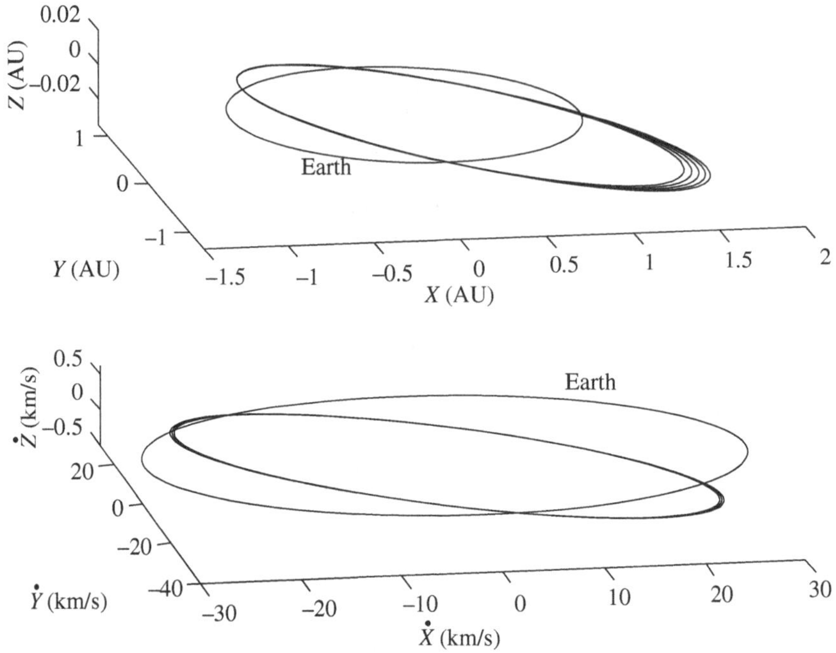

Figure 9.7 The heliocentric position and velocity components of a spacecraft perturbed by Earth's gravity computed by Cowell's method for 2500 mean solar days.

Due to the in-built rectification procedure, Encke's method never faces the likelihood of numerical instability. One can use a sufficiently small time interval, Δt, to ensure an essentially forward-marching solution, such as that with a finite-difference scheme.

Example 9.11.1 *A spacecraft is launched from Earth with the following initial heliocentric position and velocity referred to the ecliptic synodic frame at $t = 0$:*

$$\mathbf{r}(0) = \begin{pmatrix} 50 \\ -200 \\ 0.01 \end{pmatrix} \times 10^6 \; km, \quad \mathbf{v}(0) = \begin{pmatrix} 25 \\ -1 \\ -0.5 \end{pmatrix} ; \text{km/s} .$$

The Earth's current orbital elements computed from the J2000 epoch are the following:

$$a = 149598023 \; km$$

$$e = 0.0167086$$

$$t_0 = -10 \text{ mean solar days} .$$

Plot the heliocentric position and velocity of the spacecraft for the next 2500 mean solar days.

In this example, the spacecraft's initial position is well within the earth's sphere of influence. However, its initial velocity is quite large, which implies that the sphere of influence would be crossed in a few days, resulting in a diminishing influence of Earth's gravity. While a fourth-order Runge-Kutta algorithm (Appendix A) is chosen to directly integrate the equations

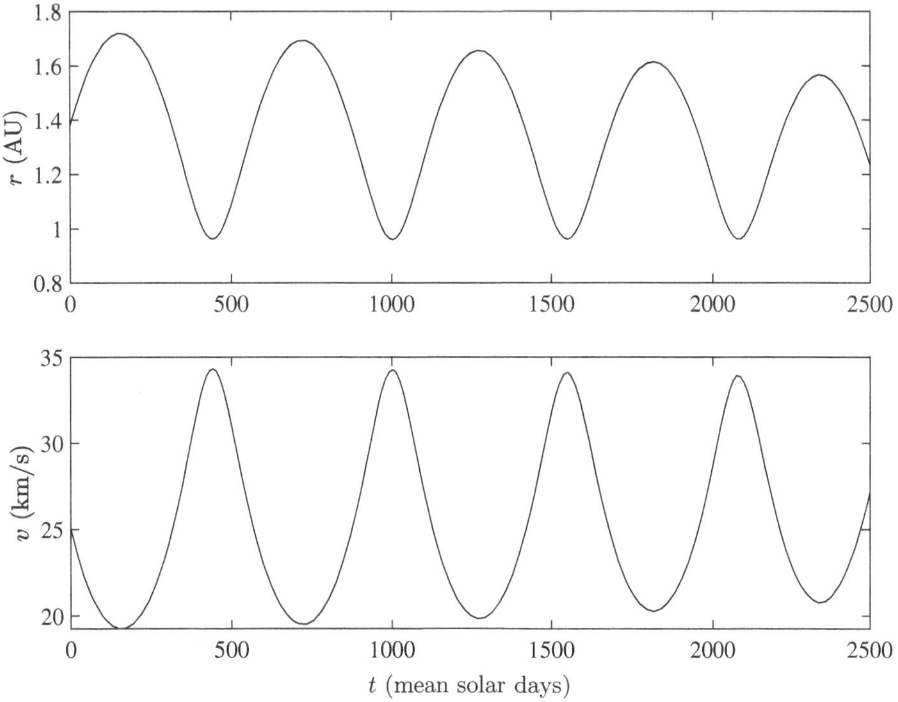

Figure 9.8 The heliocentric radius and inertial speed of a spacecraft perturbed by Earth's gravity computed by Cowell's method for 2500 mean solar days.

of perturbed motion with Cowell's method, a rather crude forward-difference approximation is applied for the time derivatives in the time-marching solution by Encke's method, with an update in the osculating orbit after each day. The direct numerical integration by Cowell's method produces a trajectory which is plotted for 2500 mean solar days in Fig. 9.7, with the corresponding heliocentric radius and speed plotted in Fig. 9.8. The decaying elliptic solar orbit due to Earth's gravitational perturbation is evident in both the figures, while the orbital inclination is seen to be constant. The final position predicted by Cowell's method at $t = 2500$ *m.s.d. is the following:*

$$X(2500) = 0.3139035\ AU;\ Y(2500) = 1.19178029\ AU;\ Z(2500) = -0.01242055\ AU$$

when its velocity components are

$$\dot{X}(2500) = -27.16132 \text{ km/s};\ \dot{Y}(2500) = 0.4766378 \text{ km/s};\ \dot{Z}(2500) = 0.5463395 \text{ km/s}\ .$$

Next, the computation is repeated by using Encke's method, and the results are plotted in Figs. 9.9 and 9.10. The heliocentric position predicted by Encke's method at $t = 2500$ *m.s.d. is the following:*

$$X(2500) = 1.143382\ \overline{AU};\ Y(2500) = 0.929732\ \overline{AU};\ Z(2500) = -0.02784292\ \overline{AU}$$

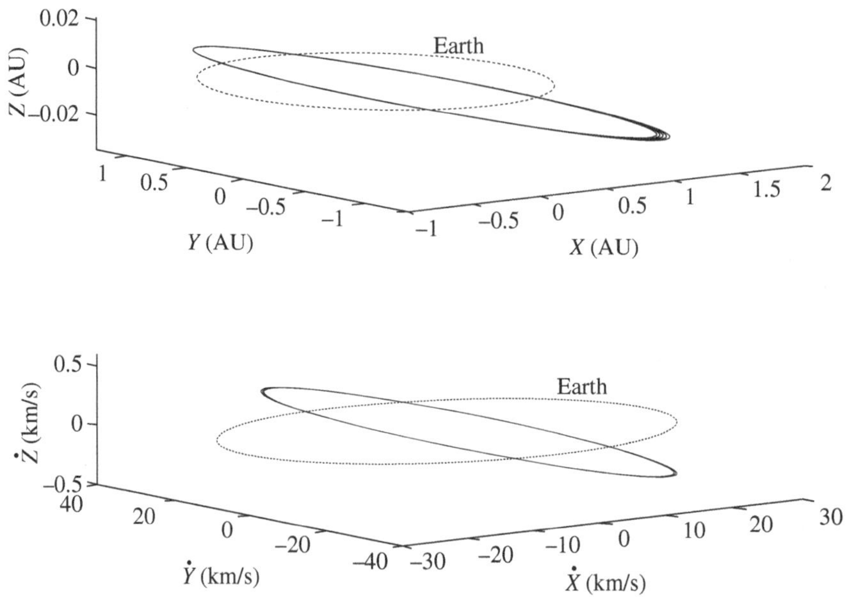

Figure 9.9 The heliocentric position and velocity components of a spacecraft perturbed by Earth's gravity computed by Encke's method for 2500 mean solar days.

and the velocity components are

$$\dot{X}(2500) = -17.8356 \text{ km/s}; \ \dot{Y}(2500) = 14.4263 \text{ km/s}; \ \dot{Z}(2500) = 0.286774 \text{ km/s} .$$

There is clearly a large difference in the final position and velocity predicted by the two methods, which is due to an accumulation of the truncation error in Cowell's method.

The geocentric position and velocity, $\mathbf{r}_{23}$, *and* $\dot{\mathbf{r}}_{23}$ *of the spacecraft computed by Encke's method for 2500 mean solar days are shown in Fig. 9.11, which indicates the near-Earth nature of the spacecraft's solar orbit.*

The radial distance and speed relative to Earth are plotted in Fig. 9.12, showing a gradual increase in the speed and a decrease in the radius after every orbit around the sun.

Exercises

1. For a spacecraft powered by an ion engine which applies a constant radial acceleration,

$$\mathbf{u} = u\mathbf{r}/r ,$$

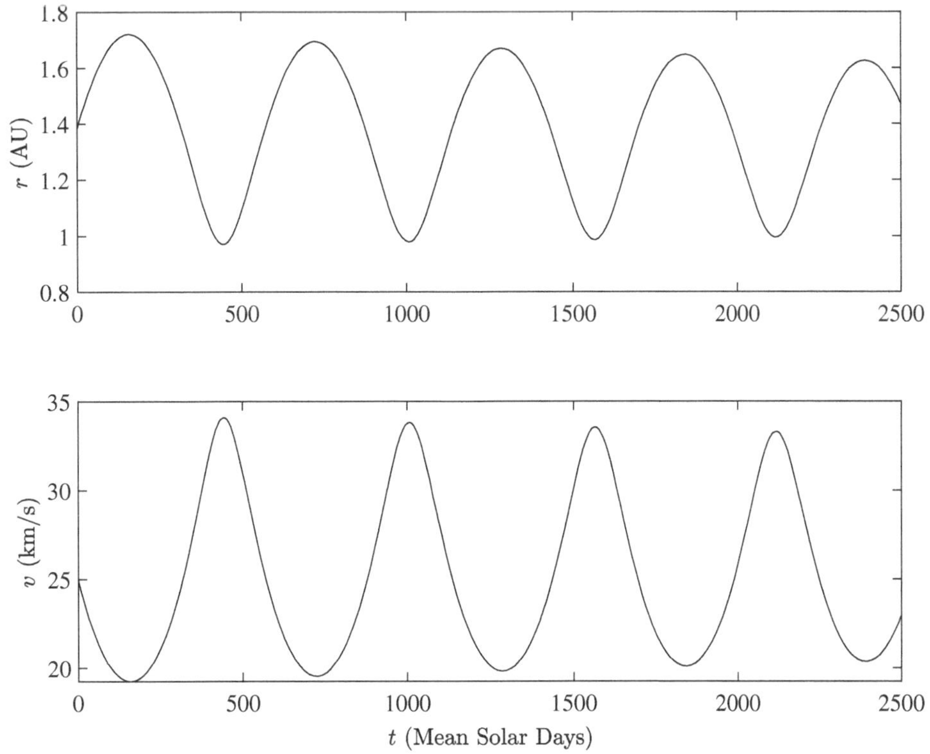

Figure 9.10 The heliocentric radius and inertial speed of a spacecraft perturbed by Earth's gravity computed by Encke's method for 2500 mean solar days.

with $u \ll \mu/r^2$, derive the perturbation potential, U, and using the Lagrange planetary equations, formulate the equations governing the time variation of the classical orbital elements.

2. Show that the Lagrange brackets for a perturbed elliptic orbit in terms of the osculating elements, $(M_0, \Omega, \omega, L, h, H)$, where

$$L = \sqrt{\mu a} = h/\sqrt{1-e^2}\ ; \qquad h = L\sqrt{1-e^2}\ ; \qquad H = h\cos i\,,$$

are the following:

$$[M_0,\ L] = 1\ ; \qquad [\Omega,\ H] = 1\ ; \qquad [\omega,\ h] = 1\,,$$

and all the remaining brackets are zero.

3. Show that the set of osculating elements given in Exercise 2 constitutes a set of canonical variables (see Example 9.4.1), which satisfy the following canonical form of the Lagrange

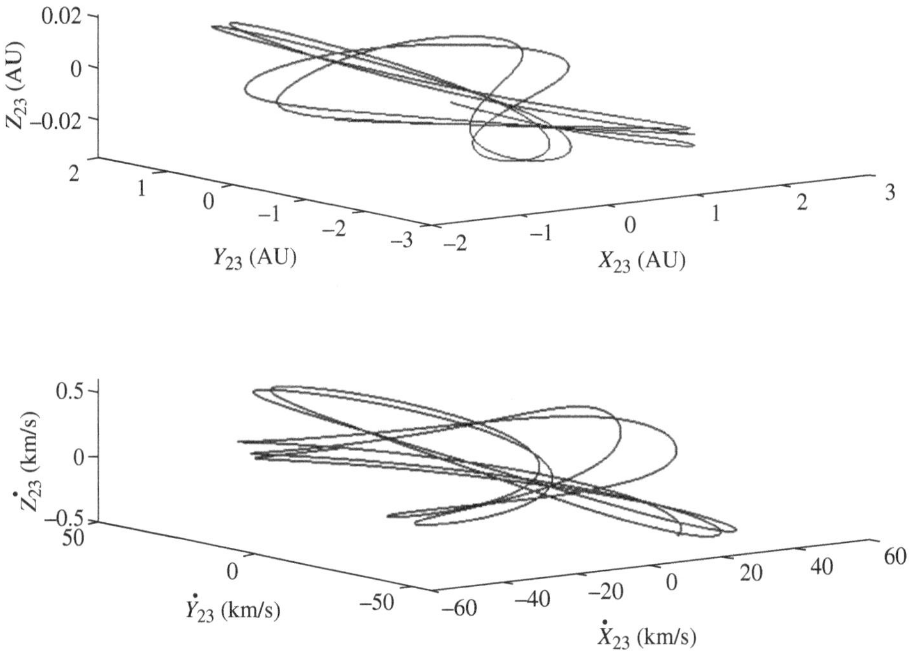

Figure 9.11 The geocentric position and velocity components of the spacecraft computed by Encke's method.

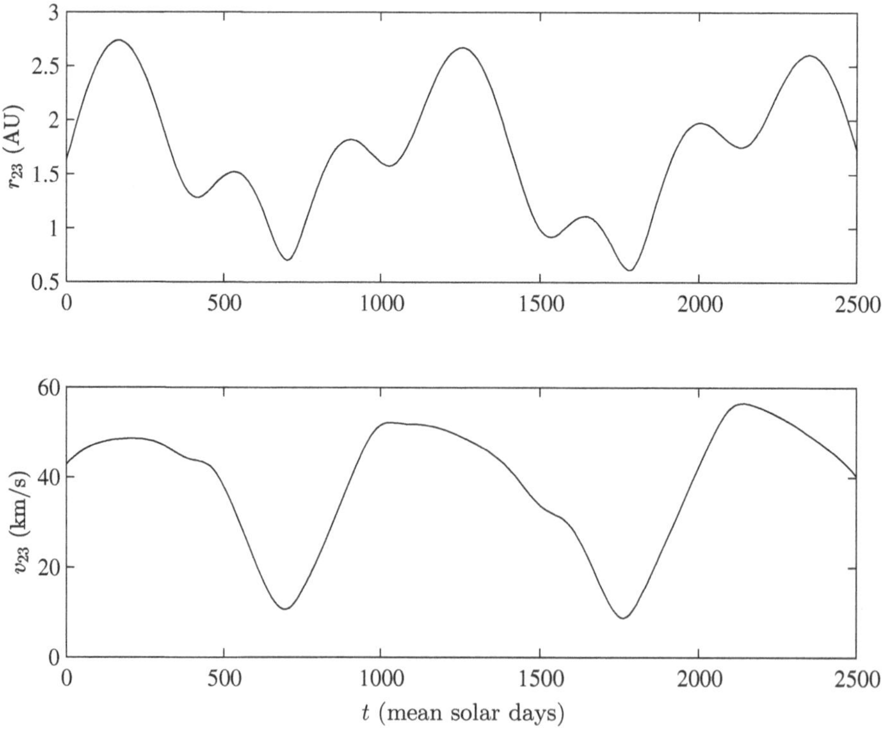

Figure 9.12 The geocentric radius and speed of the spacecraft computed by Encke's method.

planetary equations:

$$\frac{dL}{dt} = \frac{\partial U}{\partial M_0}; \qquad \frac{dM_0}{dt} = -\frac{\partial U}{\partial L}$$
$$\frac{dh}{dt} = \frac{\partial U}{\partial \omega}; \qquad \frac{d\omega}{dt} = -\frac{\partial U}{\partial h}$$
$$\frac{dH}{dt} = \frac{\partial U}{\partial \Omega}; \qquad \frac{d\Omega}{dt} = -\frac{\partial U}{\partial H}.$$

4. In order to avoid the secular term in the canonical equations derived in Exercise 3, the mean anomaly at $t = 0$, M_0, is replaced by the current mean anomaly, M, ω by g, and Ω by f, which transforms the equations to the following:

$$\frac{dL}{dt} = \frac{\partial F}{\partial M}; \qquad \frac{dM}{dt} = -\frac{\partial F}{\partial L}$$
$$\frac{dh}{dt} = \frac{\partial F}{\partial g}; \qquad \frac{dg}{dt} = -\frac{\partial F}{\partial h}$$
$$\frac{dH}{dt} = \frac{\partial F}{\partial f}; \qquad \frac{df}{dt} = -\frac{\partial F}{\partial H},$$

where F is the following transformed perturbation potential:

$$F = U + \frac{\mu^2}{2L^2}.$$

Find the expressions for the canonical variables, f and g.

5. Derive the Lagrange planetary equations for a perturbed elliptic orbit in terms of the osculating elements, $(a, e, \lambda, \Omega, i, \tilde{\omega})$, where

$$\lambda = M + \tilde{\omega}; \qquad \tilde{\omega} = \Omega + \omega.$$

6. Derive the Lagrange planetary equations for a perturbed elliptic orbit in terms of the osculating elements, $(e \sin\tilde{\omega}, e\cos\tilde{\omega}, \sin i \sin\Omega, \sin i \cos\Omega)$, and show that they are free of the singularity at $i = 0, \pi$ and $e = 0$.

7. Let the gravitational perturbation potential due to an oblate body [Eq. (9.202)] of equatorial radius, r_0, and gravitational constant, μ, be expressed as follows:

$$U(\mathbf{r}) = -\frac{\mu}{r} J_2 \left(\frac{r_0}{r}\right)^2 \left(-\frac{1}{2} + \frac{3}{2}\sin^2\delta\right),$$

where

$$\sin\delta = \sin(\omega + \theta)\sin i,$$

with δ being the latitude, ω the argument of the periapsis, θ the true anomaly, and i the orbital inclination. Show that the variational equations caused by the oblateness can be expressed as the following set of canonical equations:

$$\frac{\mathrm{d}L}{\mathrm{d}t} = \frac{\partial F}{\partial M}\,; \qquad \frac{\mathrm{d}M}{\mathrm{d}t} = -\frac{\partial F}{\partial L}$$
$$\frac{\mathrm{d}h}{\mathrm{d}t} = \frac{\partial F}{\partial \omega}\,; \qquad \frac{\mathrm{d}\omega}{\mathrm{d}t} = -\frac{\partial F}{\partial h}$$
$$\frac{\mathrm{d}H}{\mathrm{d}t} = 0\,,$$

where

$$F = U(r, \beta) + \frac{\mu^2}{2L^2}$$

and

$$L = \sqrt{\mu a} = h/\sqrt{1 - e^2}\,; \qquad H = h \cos i\,.$$

8. The orbit of a spacecraft around Earth has the elements $i = 28.8°$, $e = 0.732$, and period $T =$ 10.6 hours. Find the rotation of apsides and regression of nodes due to Earth's oblateness.
9. The mean motion of Mars in its orbit around the sun is 0.5240°/day. Design a photographing orbit for Mars at the altitude 300 km. (For Mars: $\mu = 42828.3$ km^3/s^2, $r_0 = 3397$ km.)
10. A satellite of frontal area 35 m^2 and mass 200 kg is placed in a 150 km high circular Earth orbit, where its drag coefficient is 2.2. Assuming an exponential atmosphere with a base density of 1.752 kg/m^3 and a scale height of 6.7 km, estimate the life of the satellite in orbit. (*Ans.* $t_d = 1041.058$ s.)
11. For the ratio of the perturbing acceleration caused by a third body to the primary acceleration given by Eq. (9.291), use the approximation $x \ll 1$ to derive the approximate expression of Eq. (9.296) by neglecting the terms of order x^4 in the series expansion of the radical.

References

Abramowitz M and Stegun IA 1974. *Handbook of Mathematical Functions*. Dover, New York.

Battin RH 1999. *An Introduction to the Mathematics and Methods of Astrodynamics*. American Institute of Aeronautics and Astronautics (AIAA) Education Series, Reston, VA.

Tewari A 2006. *Atmospheric and Space Flight Dynamics*. Birkhäuser, Boston.

10

Three-Body Problem

The *three-body problem* refers to the motion of three spherical masses under mutual gravitational attraction. The Moon's orbit influenced by the combined gravitation of Earth and the sun is modelled as a three-body problem. The sun, being a very massive body compared to Earth, but also much farther away, exerts an influence of a comparable magnitude on the moon when compared to that exerted by a much smaller Earth. Hence, the approximation of a small perturbation to a two-body orbit considered in Chapter 9 becomes invalid when applied to the Moon's orbit around Earth. Similarly, the flight of a spacecraft in the Moon's vicinity is accurately modelled as a three-body problem with Earth, the Moon, and the spacecraft as the three bodies. However, since the mass of the spacecraft is negligible in comparison with the masses of Earth and the Moon, the concerned problem is said to be a *restricted three-body problem*. The restricted problem refers to the approximation when the relative motion of the other two bodies (called the *primary* bodies, or just the *primaries*) is unaffected by the gravity of the third body, which can be regarded to be a test mass placed in the gravitational field created by the primaries. Other examples of a restricted three-body problem are the interplanetary voyage of a spacecraft, and the flight of a comet under the combined gravity of the sun and Jupiter. In general, the restricted three-body problem is a much better approximation of the actual motion of a spacecraft at a point sufficiently far away from a given body, when compared to the two-body problem of the previous chapters. An important application of the restricted three-body problem is the design of circumlunar and interplanetary trajectories where a spacecraft crosses the gravitational fields generated by various pairs of primaries. A special case of the restricted problem occurs when the two primaries revolve around the common centre of mass in circular orbits, and is called the *circular restricted three-body problem*. This is a useful approximation in most cases, and possesses important analytical features. Another important case of the restricted problem is that of the *elliptic restricted three-body problem*, in which the primaries orbit their common centre of mass in elliptic orbits. In either case, the relative motion of the primaries is obtained by the solution of the two-body problem, while that of the negligible third mass relative to the primaries is determined by solving the restricted three-body problem.

Foundations of Space Dynamics, First Edition. Ashish Tewari.

10.1 Equations of Motion

The three-body problem has attracted the attention of mathematicians and physicists over the past three centuries, including Euler, Lagrange, Jacobi, Poincare, Birkhoff, and Hill. For a historical development of the topic, the reader can refer to Szebehely (1967). The first systematic study of the three-body problem was by Lagrange, who derived the particular solutions of the coplanar problem, and investigated the stability of the equilibrium solutions. The time in this chapter is denoted by t', which is done in order to save the notation t for the non-dimensional time in the later parts of the chapter. The equations of motion for the three-body problem are derived as a special case of the n-body problem (Chapter 2) as follows:

$$\frac{\mathrm{d}^2\mathbf{R_i}}{\mathrm{d}(t')^2} = G\sum_{j\neq i}^{3}\frac{m_j}{r_{ji}^3}\mathbf{r_{ji}}\,; \quad (i = 1, 2, 3)\,, \tag{10.1}$$

where G is the universal gravitational constant, $\mathbf{R_i}$ denotes the position of the centre of the ith body in an inertial reference frame, and $\mathbf{r_{ji}} = \mathbf{R_j} - \mathbf{R_i}$ denotes the relative position of the centres of the bodies, i,j $(i \neq j)$. The position of the centre of mass (called the *barycentre*) is given by

$$\mathbf{R_c} = \frac{\sum_{i=1}^{3} m_i\mathbf{R_i}}{\sum_{i=1}^{3} m_i}\,. \tag{10.2}$$

Being a special case of the n-body problem with $n = 3$, the three-body problem possesses the same constants of motion as derived in Chapter 2; that is, the total energy, the linear momentum of the centre of mass, and the net angular momentum about the centre of mass. The potential energy of the three-body system is given by

$$\begin{aligned} V &= \frac{1}{2}G\sum_{i=1}^{3} m_i\sum_{j\neq i}^{3}\frac{m_j}{r_{ji}} \\ &= -G\left(\frac{m_1m_2}{r_{21}} + \frac{m_2m_3}{r_{32}} + \frac{m_1m_3}{r_{31}}\right)\,, \end{aligned} \tag{10.3}$$

while its kinetic energy is the following:

$$T = \frac{1}{2}\sum_{i=1}^{3}\sum_{j\neq i}^{3} m_i\left(\frac{\mathrm{d}r_{ji}}{\mathrm{d}t'}\right)^2\,. \tag{10.4}$$

Since no external force acts upon the system, the total energy, $E = T + V$, is conserved, and represents a scalar constant. Another six scalar constants are obtained by considering the motion of the centre of mass, $\mathbf{R_c}$, which follows a straight line at constant velocity, $\mathbf{v_{c0}}$, beginning from a constant initial position, $\mathbf{R_{c0}}$, by Newton's first law of motion:

$$\mathbf{R_c}(t') = \frac{\sum_{i=1}^{3} m_i\mathbf{R_i}}{\sum_{i=1}^{3} m_i} = \mathbf{v_{c0}}t' + \mathbf{R_{c0}}\,. \tag{10.5}$$

Three more scalar constants arise out of the conserved net angular momentum vector, **H**, about the barycentre:

$$\mathbf{H} = \sum_{i=1}^{3} m_i \dot{\mathbf{R}}_i \times \mathbf{R}_i = \text{const.}, \tag{10.6}$$

where the overdot represents taking the time derivative, d/dt'. The constants, $\mathbf{R}_{c0}$, $\mathbf{v}_{c0}$, E, and **H**, constitute ten scalar constants of the three-body problem, whereas $2n \times 3 = 18$ are required for a general solution. Hence eight more constants are needed to solve the three-body problem. Lagrange showed that certain particular solutions of the problem exist when the motion of the three bodies is confined to a plane. Such a coplanar motion of bodies is a common occurrence in the universe.

10.2 Particular Solutions by Lagrange

Before addressing Lagrange's particular solutions, the equations of motion are expressed in the following form:

$$\mathbf{f}_i = Gm_i \sum_{j \neq i}^{3} \frac{m_j}{r_{ji}^3} \mathbf{r}_{ji} \; ; \quad (i = 1, 2, 3), \tag{10.7}$$

where $\mathbf{f}_i$ is the net force experienced by the mass m_i due to the combined gravity of the other two masses. Lagrange showed that certain particular solutions of the three-body problem exist when the motion of the three bodies is always confined to a single plane. These are discussed next.

Equilibrium Solutions in a Rotating Frame

Consider the case when the three bodies describe concentric, coplanar circles about the origin of an inertial reference frame. Such a motion has a constant separation of the three bodies, and the straight line joining any two bodies rotates at a constant angular speed, $\omega = \dot{\theta}$, due to the angular-momentum conservation. The equations of motion of the coplanar three-body problem are the following:

$$\mathbf{f}_i = -m_i \omega^2 \mathbf{R}_i \; ; \quad (i = 1, 2, 3), \tag{10.8}$$

where

$$\mathbf{f}_i = G \sum_{j=1}^{3} \frac{m_i m_j}{r_{ji}^3} \mathbf{r}_{ji} \; ; \qquad (i = 1, 2, \ldots, n, j \neq i). \tag{10.9}$$

A summation applied to Eq. (10.8) yields

$$\sum_{i=1}^{3} \mathbf{f}_i = \mathbf{0} = -\omega^2 \sum_{i=1}^{3} m_i \mathbf{R}_i , \tag{10.10}$$

which implies that the rotation must take place about an axis passing through the barycentre. Hence, the origin of the inertial frame coincides with the barycentre, and we have $\mathbf{R}_c = \mathbf{0}$. Consider a right-handed coordinate frame (called a *synodic frame*), $(\mathbf{i}, \mathbf{j}, \mathbf{k})$, with its origin at the barycentre such that the three bodies always move on the plane $(\mathbf{i}, \mathbf{j})$, with $\boldsymbol{\omega} = \omega \mathbf{k}$ being the constant angular velocity of the frame. Taking the time derivative of the position of the ith mass, we have

$$\frac{d\mathbf{R_i}}{dt'} = \omega \times \mathbf{R_i} \; ; \quad (i = 1, 2, 3) \,, \tag{10.11}$$

which means that the vector $\mathbf{R_i}$ does not change in magnitude, and hence appears to be a constant vector in the synodic frame. Therefore, the solution given by the concentric circles represents fixed locations, $\mathbf{R_i} = \mathbf{R_{i_e}}$, $i = 1, 2, 3$, of the masses in the rotating frame, called the *equilibrium points* (or the *Lagrangian points*).

The substitution of Eq. (10.10) into Eq. (10.8) yields the following equations of motion to be satisfied by the equilibrium points:

$$\begin{aligned}
\left(\frac{\omega^2}{G} - \frac{m_2}{r_{21}^3} - \frac{m_3}{r_{31}^3}\right)\mathbf{R_1} + \frac{m_2}{r_{21}^3}\mathbf{R_2} + \frac{m_3}{r_{31}^3}\mathbf{R_3} &= \mathbf{0} \\
\frac{m_1}{r_{21}^3}\mathbf{R_1} + \left(\frac{\omega^2}{G} - \frac{m_1}{r_{21}^3} - \frac{m_3}{r_{32}^3}\right)\mathbf{R_2} + \frac{m_3}{r_{32}^3}\mathbf{R_3} &= \mathbf{0} \\
m_1\mathbf{R_1} + m_2\mathbf{R_2} + m_3\mathbf{R_3} &= \mathbf{0} \,,
\end{aligned} \tag{10.12}$$

where the equation corresponding to $i = 3$ in Eq. (10.8) is replaced by Eq. (10.10).

Particular equilibrium solutions of Eq. (10.12) include the *equilateral triangle* configurations of the three masses, wherein the masses are at the same constant distance from the barycentre ($r_{21} = r_{31} = r_{32} = \rho$); hence the angular speed is given by

$$\omega = \sqrt{\frac{G(m_1 + m_2 + m_3)}{\rho^3}} \,. \tag{10.13}$$

These are termed the *equilateral Lagrangian points* (or the *triangular Lagrangian points*).

Another set of equilibrium points are the *collinear Lagrangian points*, where the three masses fall on a straight line in the synodic frame. Let the axis $\mathbf{i}$ of the synodic frame represent the straight line of the collinear points, such that the axial locations of the masses, m_1, m_2, m_3, are given by x_1, x_2, x_3, respectively, and the convention $x_1 < x_2 < x_3$ is adopted. Then it follows that

$$\begin{aligned}
\mathbf{R_1} &= x_1\mathbf{i} \\
\mathbf{R_2} &= x_2\mathbf{i} = (x_1 + r_{21})\mathbf{i} \\
\mathbf{R_3} &= x_3\mathbf{i} = (x_1 + r_{21} + r_{32})\mathbf{i} \,.
\end{aligned} \tag{10.14}$$

A substitution of these into Eq. (10.12) produces the following for the collinear points:

$$\begin{aligned}
r_{21}^3 \frac{\omega^2}{G}\frac{x_1}{r_{21}} + m_2 + \frac{1}{(1+\alpha)^2}m_3 &= 0 \\
r_{21}^3 \frac{\omega^2}{G}\left(1 + \frac{x_1}{r_{21}}\right) - m_1 + \frac{1}{\alpha^2}m_3 &= 0 \\
(m_1 + m_2 + m_3)\frac{x_1}{r_{21}} + m_2 + (1+\alpha)m_3 &= 0 \,,
\end{aligned} \tag{10.15}$$

where $\alpha = r_{32}/r_{21}$ is the non-dimensional distance ratio. The first and the third equations of the set Eq. (10.15) can be solved for x_1 and ω to be the following:

$$x_1 = -r_{21}\frac{m_2 + (1+\alpha)m_3}{m_1 + m_2 + m_3}$$

$$\omega^2 = \frac{G(m_1 + m_2 + m_3)}{r_{21}^3(1+\alpha)^2}\frac{m_2(1+\alpha)^2 + m_3}{m_2 + (1+\alpha)m_3}, \tag{10.16}$$

which, substituted into the second of Eq. (10.15), yields the following *quintic equation of Lagrange* for the non-dimensional distance ratio, α:

$$\begin{aligned}(m_1 + m_2)\alpha^5 + (3m_1 + 2m_2)\alpha^4 + (3m_1 + m_2)\alpha^3 \\ -(m_2 + 3m_3)\alpha^2 - (2m_2 + 3m_3)\alpha - (m_2 + m_3) = 0\,.\end{aligned} \tag{10.17}$$

Equation (10.17) has only one positive root, since the coefficients change sign only once. However, there are three values possible for the positive root, α, obtained by cycling through the nomenclature of the three masses.

For example, the values of the distance ratio, α, can be numerically determined for the collinear Earth, Moon, spacecraft system by taking the ratio of the Moon's mass to Earth's mass is $1/81.3$ and neglecting the spacecraft's mass in comparison with the masses of heavenly bodies. These are $\alpha = 0.1678, 0.1778$, or 1.0071, depending upon whether the Moon, the spacecraft, or Earth, respectively, falls between the other two bodies.

Conic Section Solutions

Consider a right-handed synodic frame, $(\mathbf{i}, \mathbf{j}, \mathbf{k})$, with the origin at the barycentre such that the three bodies always appear to be moving radially on the plane $(\mathbf{i}, \mathbf{j})$. Let the rate of rotation of the frame about $\mathbf{k}$ be $\omega = \dot{\theta}$. Furthermore, let $(\mathbf{i}', \mathbf{j}', \mathbf{k})$, be a right-handed, inertial reference frame with origin at the barycentre such that $\mathbf{i}' = \mathbf{i}$ and $\mathbf{j}' = \mathbf{j}$ at the time $t' = 0$. Such an inertial reference frame is called a *sidereal frame*. The angle made by the axes, $\mathbf{i}$ and $\mathbf{j}$, with the axes, $\mathbf{i}'$ and $\mathbf{j}'$, respectively, is $\theta(t') = \omega t'$, as shown in Fig. 10.1. The coordinate transformation between the sidereal and synodic frames at the current time, t', is given by

$$\begin{Bmatrix} \mathbf{i}' \\ \mathbf{j}' \end{Bmatrix} = \mathbf{C}(t') \begin{Bmatrix} \mathbf{i} \\ \mathbf{j} \end{Bmatrix}, \tag{10.18}$$

where the rotation matrix, $\mathbf{C}(t')$, is the following (Chapter 5):

$$\mathbf{C}(t') = \begin{bmatrix} \cos\theta(t') & -\sin\theta(t') \\ \sin\theta(t') & \cos\theta(t') \end{bmatrix}. \tag{10.19}$$

The location of the mass m_i at time t' in the rotating frame is given by

$$\mathbf{R}_i(t') = a(t')\mathbf{R}_i(0)\,, \tag{10.20}$$

where $\mathbf{R}_i(0)$ is the initial location of the mass at $t' = 0$. At any given instant, t', the three bodies share the same values of $\theta(t')$ and $a(t')$. However, due to the radial movement of the bodies,

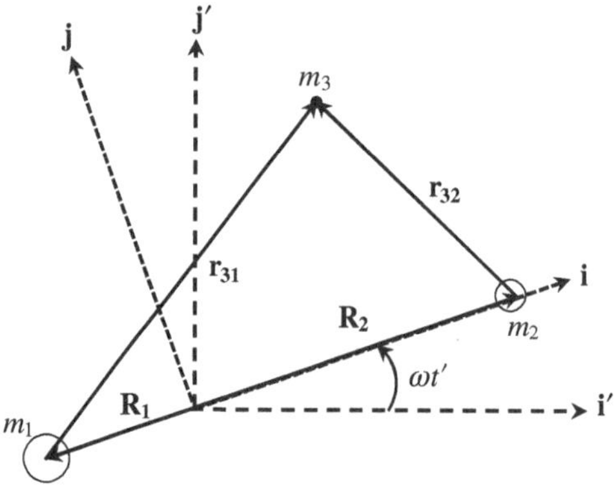

Figure 10.1 The synodic frame, $(\mathbf{i}, \mathbf{j}, \mathbf{k})$, and the sidereal frame, $(\mathbf{i}', \mathbf{j}', \mathbf{k})$, with common origin at the barycentre, with the axis $\mathbf{k} = \mathbf{i} \times \mathbf{k}$ normal to the plane of the primaries, m_1, and m_2.

the angular speed, $\dot{\theta}$, of the frame keeps changing with time due to conservation of angular momentum. By substituting Eq. (10.20) into Eq. (10.7), the external force acting on the mass, m_i, is expressed as follows:

$$\mathbf{f}_\mathrm{i}(t') = \frac{\mathbf{f}_\mathrm{i}(0)}{a^2}, \tag{10.21}$$

and the skew-symmetric matrix, $\mathbf{S}(\boldsymbol{\omega})$, representing the vector product with the frame's angular velocity, $\boldsymbol{\omega} = \dot{\theta}\mathbf{k}$, is the following (Chapter 10):

$$\mathbf{S}(\boldsymbol{\omega}) = \mathbf{C}^T\dot{\mathbf{C}} = \begin{pmatrix} 0 & -1 \\ 1 & 0 \end{pmatrix} \dot{\theta}. \tag{10.22}$$

The acceleration of m_i is resolved in the synodic frame as follows:

$$\frac{\mathrm{d}^2\mathbf{R}_\mathrm{i}}{\mathrm{d}(t')^2} = \mathbf{C}\left(\frac{\partial^2\mathbf{R}_\mathrm{i}}{\partial(t')^2} + 2\boldsymbol{\omega} \times \frac{\partial \mathbf{R}_\mathrm{i}}{\partial t'} + \frac{\mathrm{d}\boldsymbol{\omega}}{\mathrm{d}t'}\mathbf{R}_\mathrm{i} + \boldsymbol{\omega} \times [\boldsymbol{\omega} \times \mathbf{R}_\mathrm{i}]\right), \tag{10.23}$$

which is also expressed by

$$\frac{\mathrm{d}^2\mathbf{R}_\mathrm{i}}{\mathrm{d}(t')^2} = \mathbf{C}\left(\frac{\partial^2\mathbf{R}_\mathrm{i}}{\partial(t')^2} + 2\mathbf{S}(\boldsymbol{\omega})\frac{\partial \mathbf{R}_\mathrm{i}}{\partial t'} + \dot{\mathbf{S}}(\boldsymbol{\omega})\mathbf{R}_\mathrm{i} + \mathbf{S}^2(\boldsymbol{\omega})\mathbf{R}_\mathrm{i}\right). \tag{10.24}$$

Substituting Eqs. (10.20) and (10.21) into Eq. (10.24), we have

$$m_i\left[\left(\frac{\mathrm{d}^2 a}{\mathrm{d}(t')^2} - a\dot{\theta}^2\right)\mathbf{I} + \frac{1}{a}\frac{\mathrm{d}(a^2\dot{\theta})}{\mathrm{d}t'}\mathbf{J}\right]\mathbf{R}_\mathrm{i}(0) = \frac{\mathbf{f}_\mathrm{i}(0)}{a^2}, \tag{10.25}$$

where $\mathbf{I}$ is the identity matrix and J is the following skew-symmetric matrix (which can be regarded as the imaginary matrix, because $\mathbf{J}^2 = -\mathbf{I}$):

$$\mathbf{J} = \begin{pmatrix} 0 & -1 \\ 1 & 0 \end{pmatrix} . \tag{10.26}$$

Since the net external force experienced by m_i is towards the barycentre, causing the radial and rotary motion of the mass, it must be a centripetal force given by

$$\mathbf{f_i}(t') = -m_i b^2 \mathbf{R_i}(t') , \tag{10.27}$$

where b is a constant. In a planar motion with radial acceleration, we have

$$R_i^2 \dot{\theta} = \text{constant} , \tag{10.28}$$

or

$$\frac{\mathrm{d}(a^2\dot{\theta})}{\mathrm{d}t'} = 0 . \tag{10.29}$$

Therefore, it follows from Eq. (10.25) that

$$\frac{\mathrm{d}^2 a}{\mathrm{d}(t')^2} - a\dot{\theta}^2 = -\frac{b^2}{a^2} . \tag{10.30}$$

Equations (8.30) and (8.31) represent a conic section (Chapter 3) in polar coordinates, which is the equation governing the relative motion of two bodies. Hence, each mass in the coplanar three-body problem traces a conic section about the barycentre.

10.3 Circular Restricted Three-Body Problem

When one of the three masses – say, m_3 – is negligible in comparison with the other two, m_1, and m_2 (called the *primaries*), those latter two of which are in coplanar circular orbits about the barycentre, a major simplification occurs in the coplanar three-body equations (Eq. (10.20)). Since $m_3 \ll m_1$ and $m_3 \ll m_2$, the gravitational pull of m_3 on both m_1 and m_2 is negligible, and the natural motion of m_3 relative to the primaries is referred to as the *circular restricted three-body problem* (or CR3BP). The barycentre in a CR3BP is approximated to be the common centre of mass of the two primaries, i.e., $m_1 R_1 \simeq m_2 R_2$. Since the motion of the primaries is known, the solution to the CR3BP determines the motion of the negligible mass, m_3, relative to the primaries.

10.3.1 Equations of Motion in the Inertial Frame

The equations of motion for the CR3BP are relatively easily derived in the inertial frame. Since the primaries, m_1 and m_2, traverse coplanar circles centred at the origin of the inertial frame (also the barycentre), they are described by Eqs. (10.8) and (10.9), expressed for the two masses as follows:

$$\mathbf{f_1} = G\frac{m_1 m_2}{r_{21}^3}\mathbf{r}_{21} = -m_1\omega^2\mathbf{R_1} \; ; \qquad \mathbf{f_2} = -\mathbf{f_1} = -m_2\omega^2\mathbf{R_2} , \tag{10.31}$$

where $\mathbf{r}_{21} = \mathbf{R}_2 - \mathbf{R}_1$. As seen earlier, the two primaries always maintain a constant separation, r_{21}, and a straight line joining them rotates at a constant rate, ω. Since the axis of rotation passes through the barycentre, we have $m_1\mathbf{R}_1 = -m_2\mathbf{R}_2$ (which is also implied by Eq. (10.31)); hence it follows that

$$m_1R_1 = m_2R_2\ ; \qquad R_2 = \frac{m_1r_{21}}{m_1+m_2}\ ; \qquad R_1 = \frac{m_2r_{21}}{m_1+m_2}\ . \tag{10.32}$$

Equation (10.31) further implies that

$$Gm_1 = \omega^2R_2r_{21}^2\ ; \qquad Gm_2 = \omega^2R_1r_{21}^2\ ; \qquad G(m_1+m_2) = \omega^2r_{21}^2\ . \tag{10.33}$$

Let $\mathbf{i}$ denote the unit vector along the straight line joining the primaries, m_1 and m_2, such that $\mathbf{R}_1 = R_1\mathbf{i}$ and $\mathbf{R}_2 = -R_2\mathbf{i}$. Let the position vectors of the masses, $\mathbf{R}_1$, $\mathbf{R}_2$, $\mathbf{R}_3$, be resolved as follows in an inertial (sidereal) reference frame, $(\mathbf{i}', \mathbf{j}', \mathbf{k})$, with origin at the barycentre, such that $\mathbf{i}' = \mathbf{i}$ at the time, $t' = 0$ (Fig. 10.1):

$$\begin{aligned}\mathbf{R}_1 &= X_1\mathbf{i}' + Y_1\mathbf{j}'\\ \mathbf{R}_2 &= X_2\mathbf{i}' + Y_2\mathbf{j}'\\ \mathbf{R}_3 &= X\mathbf{i}' + Y\mathbf{j}' + Z\mathbf{k}\,.\end{aligned} \tag{10.34}$$

The angle made by the unit vector, $\mathbf{i}$, with the axis, $\mathbf{i}'$, is $\omega t'$ (Fig. 10.1). This implies that $X_1(0) = R_1$, $X_2(0) = -R_2$, and $Y_1(0) = Y_2(0) = 0$. Since $\mathbf{i}$ rotates with a constant frequency, ω, given by Eq. (10.33), about the $\mathbf{k}$ axis, we have the following Cartesian coordinates for the position of the primaries at a time, t':

$$\begin{aligned}X_1 &= R_1\cos\omega t'\ ; \qquad Y_1 = R_1\sin\omega t'\\ X_2 &= -R_2\cos\omega t'\ ; \qquad Y_2 = -R_2\sin\omega t'\,.\end{aligned}$$

The inertial position of the mass, m_3, relative to the primaries is then described by

$$\begin{aligned}\mathbf{r}_{31} &= \mathbf{R}_3 - \mathbf{R}_1 = (X - R_1\cos\omega t')\mathbf{i}' + (Y - R_1\sin\omega t')\mathbf{j}' + Z\mathbf{k}\\ \mathbf{r}_{32} &= \mathbf{R}_3 - \mathbf{R}_2 = (X + R_2\cos\omega t')\mathbf{i}' + (Y + R_2\sin\omega t')\mathbf{j}' + Z\mathbf{k}\,.\end{aligned}$$

The equation of motion of m_3 is given by Eq. (10.1) with $i = 3$, and is expressed as follows:

$$\frac{\mathrm{d}^2\mathbf{R}_3}{\mathrm{d}(t')^2} = G\sum_{j=1}^{2}\frac{m_j}{r_{j3}^3}\mathbf{r}_{j3} = -G\sum_{j=1}^{2}\frac{m_j}{r_{3j}^3}\mathbf{r}_{3j} = \frac{\partial U'}{\partial\mathbf{R}_3}\,, \tag{10.35}$$

where U' is the net gravitational potential seen by m_3, and is given by

$$U' = G\sum_{j=1}^{2}\frac{m_j}{r_{3j}} = G\left(\frac{m_1}{r_{31}} + \frac{m_2}{r_{32}}\right)\,. \tag{10.36}$$

Substituting Eq. (10.35) into Eq. (10.35) and collecting terms, we have

$$\begin{aligned}
\frac{d^2X}{d(t')^2} &= -G\left[\frac{m_1(X - R_1\cos\omega t')}{r_{31}^3} + \frac{m_2(X + R_2\cos\omega t')}{r_{32}^3}\right] \\
\frac{d^2Y}{d(t')^2} &= -G\left[\frac{m_1(Y - R_1\sin\omega t')}{r_{31}^3} + \frac{m_2(Y + R_2\sin\omega t')}{r_{32}^3}\right] \\
\frac{d^2Z}{d(t')^2} &= -GZ\left(\frac{m_1}{r_{31}^3} + \frac{m_2}{r_{32}^3}\right),
\end{aligned} \tag{10.37}$$

where

$$\begin{aligned}
r_{31} &= \sqrt{(X - R_1\cos\omega t')^2 + (Y - R_1\sin\omega t')^2 + Z^2} \\
r_{32} &= \sqrt{(X + R_2\cos\omega t')^2 + (Y + R_2\sin\omega t')^2 + Z^2}\,.
\end{aligned} \tag{10.38}$$

10.4 Non-dimensional Equations in the Synodic Frame

For convenience, the CR3BP equations of motion are rendered non-dimensional by dividing each of the masses by the total mass of the primaries, $m_1 + m_2$, and dividing all the distances by the constant separation between the primaries, $r_{21} = |\,\mathbf{R_2} - \mathbf{R_1}\,|$. Furthermore, the notation is simplified by dropping the subscript 3 used in denoting the relative position of m_3. Thus the following non-dimensional quantities are introduced:

$$\begin{aligned}
\mu &= \frac{m_2}{m_1 + m_2} \\
r_1 &= \frac{r_{13}}{r_{21}} \\
r_2 &= \frac{r_{23}}{r_{21}}\,.
\end{aligned} \tag{10.39}$$

By convention, $m_1 \geq m_2$ is taken, which results in $0 \leq \mu \leq 1/2$. Hence, the normalized masses of the primaries, m_1 and m_2, are given by $(1 - \mu)$ and μ, respectively. The non-dimensional distances of the two primaries from the barycentre are $R_1 = \mu$ and $R_2 = (1 - \mu)$, respectively. The gravitational constant, G, is also rendered non-dimensional as follows, such that the angular velocity, ω, of the primaries is unity:

$$\omega = \sqrt{G(m_1 + m_2)/r_{21}^3} = \sqrt{\mu + (1 - \mu)} = 1 \text{ rad/s}\,, \tag{10.40}$$

which produces the following non-dimensional time:

$$t = \omega t' = t'\sqrt{\frac{r_{21}^3}{G(m_1 + m_2)}}\,, \tag{10.41}$$

such that the orbital period of the primaries on the non-dimensional time scale equals 2π.

If the inertial position of the mass, m_3, is described by the non-dimensional Cartesian coordinates, (ξ, η, ζ), then Eqs. (10.37) and (10.38) are rendered non-dimensional as follows:

$$\begin{aligned}\ddot{\xi} &= -\frac{(1-\mu)(\xi-\xi_1)}{r_1^3} - \frac{\mu(\xi-\xi_2)}{r_2^3} \\ \ddot{\eta} &= -\frac{(1-\mu)(\eta-\eta_1)}{r_1^3} - \frac{\mu(\eta-\eta_2)}{r_2^3} \\ \ddot{\zeta} &= -\frac{(1-\mu)\zeta}{r_1^3} - \frac{\mu\zeta}{r_2^3},\end{aligned} \tag{10.42}$$

where the overdot represents the derivative with respect to the non-dimensional time, t, the Cartesian coordinates of the primaries, m_1 and m_2, at time t are $(\xi_1, \eta_1, 0)$ and $(\xi_2, \eta_2, 0)$, respectively, and

$$\begin{aligned} r_1 &= \sqrt{(\xi-\xi_1)^2 + (\eta-\eta_1)^2 + \zeta^2} \\ r_2 &= \sqrt{(\xi-\xi_2)^2 + (\eta-\eta_2)^2 + \zeta^2}. \end{aligned} \tag{10.43}$$

Clearly, we have

$$\begin{aligned} \xi_1 &= \mu \cos t\,; & \xi_2 &= -(1-\mu)\cos t \\ \eta_1 &= \mu \sin t\,; & \eta_2 &= -(1-\mu)\sin t. \end{aligned} \tag{10.44}$$

In terms of the following non-dimensional gravitational potential:

$$U = \frac{(1-\mu)}{r_1} + \frac{\mu}{r_2}, \tag{10.45}$$

Eq. (10.42) is expressed as follows:

$$\begin{aligned} \ddot{\xi} &= \frac{\partial U}{\partial \xi} \\ \ddot{\eta} &= \frac{\partial U}{\partial \eta} \\ \ddot{\zeta} &= \frac{\partial U}{\partial \zeta}. \end{aligned} \tag{10.46}$$

The equations of motion are now resolved in a synodic frame, $(\mathbf{i}, \mathbf{j}, \mathbf{k})$, with the origin at the barycentre, and rotating with a non-dimensional angular velocity, $\mathbf{k}$. The axis, $\mathbf{i}$, points from m_1 towards m_2, as shown in Fig. 10.2. The axis, $\mathbf{j}$, being normal to both $\mathbf{i}$ and $\mathbf{k}$, completes the right-handed triad, $\mathbf{i} \times \mathbf{j} = \mathbf{k}$. Let the non-dimensional position of the mass, m_3, in the synodic frame be given by

$$\mathbf{r} = x\mathbf{i} + y\mathbf{j} + z\mathbf{k}. \tag{10.47}$$

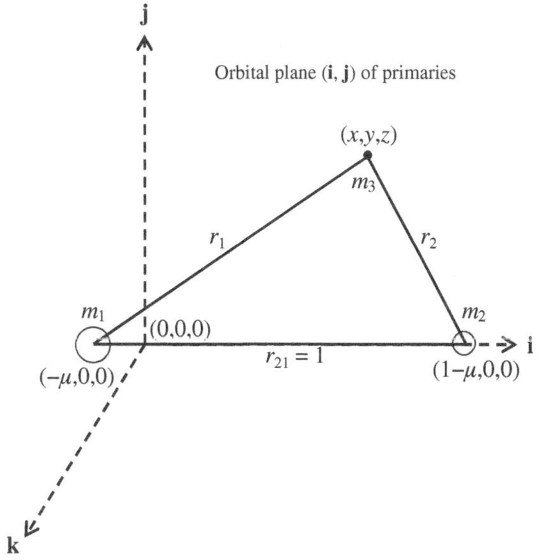

Figure 10.2 Geometry of the circular restricted three-body problem.

The coordinate transformation between the inertial (sidereal) and the rotating (synodic) frames is the following:

$$\begin{Bmatrix} \xi \\ \eta \\ \zeta \end{Bmatrix} = \begin{pmatrix} \cos t & -\sin t & 0 \\ \sin t & \cos t & 0 \\ 0 & 0 & 1 \end{pmatrix} \begin{Bmatrix} x \\ y \\ z \end{Bmatrix}. \tag{10.48}$$

The substitution of Eq. (10.48) into Eq. (10.42) and the elimination of $\sin t$ and $\cos t$ yield the following:

$$\begin{aligned} \ddot{x} - 2\dot{y} - x &= -\frac{(1-\mu)(x-x_1)}{r_1^3} - \frac{\mu(x-x_2)}{r_2^3} \\ \ddot{y} + 2\dot{x} - y &= -\frac{(1-\mu)(y-y_1)}{r_1^3} - \frac{\mu(y-y_2)}{r_2^3} \\ \ddot{z} &= -\frac{(1-\mu)z}{r_1^3} - \frac{\mu z}{r_2^3}, \end{aligned} \tag{10.49}$$

where $x_1 = -\mu$, $x_2 = 1 - \mu$, $y_1 = y_2 = 0$, are the fixed positions of the primaries in the synodic frame. Therefore, we have the following equations of CR3BP motion in terms of the non-dimensional, synodic, Cartesian coordinates, (x, y, z) (Fig. 10.2):

$$\ddot{x} - 2\dot{y} - x = -\frac{(1-\mu)(x+\mu)}{r_1^3} - \frac{\mu(x-1+\mu)}{r_2^3}$$

$$\ddot{y} + 2\dot{x} - y = -\frac{(1-\mu)y}{r_1^3} - \frac{\mu y}{r_2^3} \tag{10.50}$$

$$\ddot{z} = -\frac{(1-\mu)z}{r_1^3} - \frac{\mu z}{r_2^3},$$

where

$$r_1 = \sqrt{(x+\mu)^2 + y^2 + z^2}$$

$$r_2 = \sqrt{(x-1+\mu)^2 + y^2 + z^2}\,. \tag{10.51}$$

The set in Eq. (10.50) is expressed as follows in terms of the non-dimensional gravitational potential, $U = (1-\mu)/r_1 + \mu/r_2$, resolved in the synodic frame:

$$\begin{aligned}\ddot{x} - 2\dot{y} - x &= \frac{\partial U}{\partial x}\\ \ddot{y} + 2\dot{x} - y &= \frac{\partial U}{\partial y}\\ \ddot{z} &= \frac{\partial U}{\partial z}\,.\end{aligned} \tag{10.52}$$

A comparison with the inertial non-dimensional equations, Eq. (10.46), shows the presence of additional terms on the left-hand side of Eqs. (10.50) and (10.52). These arise when the acceleration, $\ddot{\mathbf{r}} = \mathrm{d}^2\mathbf{r}/\mathrm{d}t^2$, is resolved in the synodic frame, $(\mathbf{i}, \mathbf{j}, \mathbf{k})$, which is rotating with the non-dimensional angular velocity, $\mathbf{k}$. A step-by-step derivation of the acceleration terms is given as follows:

$$\begin{aligned}\dot{\mathbf{r}} &= \dot{x}\mathbf{i} + \dot{y}\mathbf{j} + \dot{z}\mathbf{k} + \mathbf{k}\times\mathbf{r}\\ &= \dot{x}\mathbf{i} + \dot{y}\mathbf{j} + \dot{z}\mathbf{k} + x\mathbf{j} - y\mathbf{i}\\ &= (\dot{x} - y)\mathbf{i} + (\dot{y} + x)\mathbf{j} + \dot{z}\mathbf{k}\,.\end{aligned} \tag{10.53}$$

$$\begin{aligned}\ddot{\mathbf{r}} &= (\ddot{x} - \dot{y})\mathbf{i} + (\ddot{y} + \dot{x})\mathbf{j} + \ddot{z}\mathbf{k} + \mathbf{k}\times\dot{\mathbf{r}}\\ &= (\ddot{x} - \dot{y})\mathbf{i} + (\ddot{y} + \dot{x})\mathbf{j} + \ddot{z}\mathbf{k} + (\dot{x} - y)\mathbf{j} - (\dot{y} + x)\mathbf{i}\\ &= (\ddot{x} - 2\dot{y} - x)\mathbf{i} + (\ddot{y} + 2\dot{x} - y)\mathbf{j} + \ddot{z}\mathbf{k}\,.\end{aligned} \tag{10.54}$$

The non-dimensional equations, Eq. (10.50), are alternatively expressed in terms of a *pseudo-potential*, Ω, given by

$$\Omega = U + \frac{1}{2}(x^2 + y^2)\,, \tag{10.55}$$

as follows:

$$\begin{aligned}\ddot{x} - 2\dot{y} &= \frac{\partial \Omega}{\partial x}\\ \ddot{y} + 2\dot{x} &= \frac{\partial \Omega}{\partial y}\\ \ddot{z} &= \frac{\partial \Omega}{\partial z}\,.\end{aligned} \tag{10.56}$$

The pseudo-potential plays an important role in deriving an integral constant of the CR3BP, as discussed later.

The set of coupled, non-linear, ordinary differential equations, Eq. (10.50), representing the CR3BP in the synodic frame cannot be integrated in a closed form. However, certain important analytical insights into the problem can be derived without actually solving the problem. An important aspect of the CR3BP equations is the symmetrical nature of its solutions. If $x(t), y(t), z(t)$ is a solution to the equations, then it is easily verified that $x(t), y(t), -z(t)$ is also a solution, thereby implying a reflection across the (x, y) plane. Similarly, $x(-t), -y(-t), z(-t)$ is also a solution, which implies a reflection across the (x, z) plane with time reversal. This property is due to the time-invariant nature of the gravitational potential, U, when resolved in the synodic frame, which will be further utilized later. The third symmetry is related to the parameter, μ. If $x(t), y(t), z(t)$ satisfies the equations with parameter μ, then $-x(t), -y(t), z(t)$ also satisfies the equations with parameter $1 - \mu$. This means that one needs to study the CR3BP only for $\mu \in [0, 1/2]$.

10.5 Lagrangian Points and Stability

As in the case of the general three-body problem, coplanar equilibrium solutions are available in the synodic frame for the CR3BP represented by Eq. (10.50). These equilibrium points are called the *Lagrangian points* (also *libration points*), and can be derived for the CR3BP model by equating the time derivatives in Eq. (10.50) to zero, resulting in the following algebraic equations:

$$\begin{aligned} x &= \frac{(1-\mu)(x+\mu)}{r_1^3} + \frac{\mu(x-1+\mu)}{r_2^3} \\ y &= \frac{(1-\mu)y}{r_1^3} + \frac{\mu y}{r_2^3} \\ 0 &= \frac{(1-\mu)z}{r_1^3} + \frac{\mu z}{r_2^3}\,. \end{aligned} \tag{10.57}$$

From the last of Eq. (10.57), we have the result $z = 0$ for all the Lagrangian points, which implies $r_1 = \sqrt{(x+\mu)^2 + y^2}$ and $r_2 = \sqrt{(x+\mu-1)^2 + y^2}$. Hence all the equilibrium points are on the plane of the primaries. These coplanar equilibrium points consist of the three possible collinear positions of m_3, called the *collinear Lagrangian points*, L_1, L_2, L_3, as well as the two equilateral triangle positions with the primaries, called *triangular Lagrangian points*, L_4, L_5, discussed earlier.

The collinear Lagrangian points, L_1, L_2, L_3, are derived from the positive real solutions to the quintic equation of Lagrange [the last of Eq. (10.17)], which for $m_1 = 1 - \mu$, $m_2 = \mu$, $m_3 = 0$, and $\alpha = r_{23}/r_{12}$ is written as follows:

$$\alpha^5 + (3-\mu)\alpha^4 + (3-2\mu)\alpha^3 - \mu\alpha^2 - 2\mu\alpha - \mu = 0\,. \tag{10.58}$$

As discussed earlier, there is only one real, positive root of the polynomial in Eq. (10.58). Thus the three collinear points are obtained by cycling through the three possible values of μ for a given three-body system. For example, with $\mu = 1/82.3$ for the Earth-Moon system, Eq. (10.58)

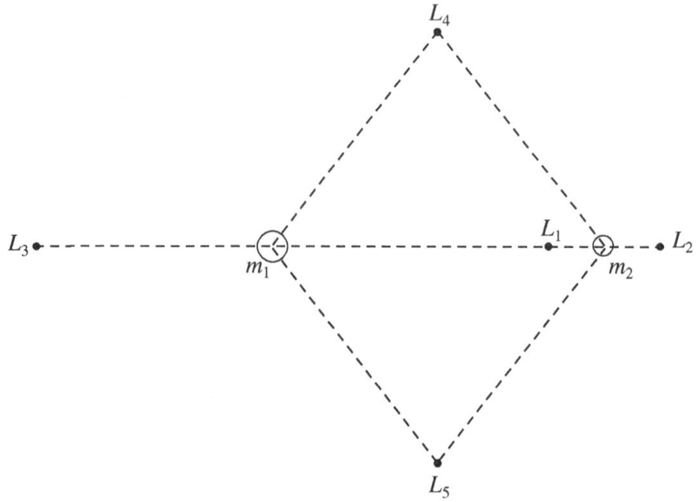

Figure 10.3 Lagrangian points of the circular restricted three-body problem.

yields $\alpha = 0.167833$, or $x = 1.15568$ for the Lagrangian point, L_2. With $m_1 = \mu$ (Moon), $m_2 = 1 - \mu$ (Earth), and $m_3 = 0$ (spacecraft), the quintic equation becomes the following:

$$\alpha^5 + (2+\mu)\alpha^4 + (1+2\mu)\alpha^3 - (1-\mu)\alpha^2 - 2(1-\mu)\alpha - (1-\mu) = 0\,, \tag{10.59}$$

whose real root for the Earth-Moon system is $\alpha = 0.992912$, thereby yielding $x = -\alpha - \mu = -1.00506$ for the Lagrangian point, L_3. The Lagrangian point, L_1, is obtained by putting $m_2 = 0$ (spacecraft), $m_1 = \mu$ (Moon), and $m_3 = 1 - \mu$ (Earth) into the quintic equation, which becomes

$$\mu\alpha^5 + 3\mu\alpha^4 + 3\mu\alpha^3 - 3(1-\mu)\alpha^2 - 3(1-\mu)\alpha - (1-\mu) = 0\,, \tag{10.60}$$

whose positive root for the Earth-Moon system is $\alpha = r_{23}/r_{12} = 5.62538553$ with $r_{12} + r_{23} = 1$, thereby yielding

$$x = r_{23} - \mu = \frac{1}{1 + 1/\alpha} - \mu = 0.836914719\,.$$

The triangular Lagrangian points, L_4, L_5, correspond to $r_1 = r_2 = 1$, and are given by the coordinates, $x = \frac{1}{2} - \mu$, $y = \pm\sqrt{3}/2$, and $z = 0$. Figure 10.3 depicts the positions of the Lagrangian points relative to the primaries in the general case, $m_1 > m_2$.

10.5.1 Stability Analysis

To investigate the stability of a Lagrangian point, consider infinitesimal displacements, $(\delta x, \delta y, \delta z)$, from the equilibrium position, $(x_0, y_0, 0)$. If the displacements remain small, the equilibrium point is said to be stable; otherwise, unstable. Such an analysis easily reveals that the out-of-plane motion represented by δz is unconditionally stable about all the equilibrium points. However, the stability of the coplanar motion described by $(\delta x, \delta y)$ crucially depends upon the location (x_0, y_0) of the given equilibrium point.

For the small displacement, coplanar motion about a triangular Lagrangian point, say, L_4, the following linearized model can be applied:

$$\delta\ddot{x} - 2\delta\dot{y} - \frac{3}{4}\delta x - \frac{3\sqrt{3}(\mu - \frac{1}{2})}{2}\delta y = 0$$

$$\delta\ddot{y} + 2\delta\dot{x} - \frac{3\sqrt{3}(\mu - \frac{1}{2})}{2}\delta x - \frac{9}{4}\delta y = 0\,, \tag{10.61}$$

whose general solution to initial displacement $\delta x_0, \delta y_0$, can be expressed as follows (see Chapter 7):

$$\delta \mathbf{r} = e^{\mathbf{A}t}\delta \mathbf{r_0}\,, \tag{10.62}$$

where $\delta \mathbf{r} = (\delta x, \delta y)^T$ is the state vector of linearized dynamics, $\delta \mathbf{r_0} = (\delta x_0, \delta y_0)^T$ is the initial state, and $\mathbf{A}$ is the following Jacobian matrix:

$$\mathbf{A} = \begin{pmatrix} 0 & 0 & 1 & 0 \\ 0 & 0 & 0 & 1 \\ \frac{3}{4} & \frac{3\sqrt{3}(\mu-\frac{1}{2})}{2} & 0 & 2 \\ \frac{3\sqrt{3}(\mu-\frac{1}{2})}{2} & \frac{9}{4} & -2 & 0 \end{pmatrix}. \tag{10.63}$$

The characteristic equation for the eigenvalue, λ, of $\mathbf{A}$, is thus the following:

$$\det(\lambda \mathbf{I} - \mathbf{A}) = \lambda^4 + \lambda^2 + \frac{27}{4}\mu(1-\mu) = 0\,, \tag{10.64}$$

whose roots are

$$\lambda^2 = \frac{1}{2}[\pm\sqrt{1 - 27\mu(1-\mu)} - 1]\,. \tag{10.65}$$

For stability, the values of λ must be purely imaginary, representing a simple harmonic oscillation about the equilibrium point. If the quantity in the square root is negative, there is at least one value of λ with a positive real part and represents an unstable system. Therefore, the critical values of μ representing the boundary between stable and unstable behavior of L_4 are those that correspond to

$$1 - 27\mu(1-\mu) = 0\,. \tag{10.66}$$

Hence, for the stability of the triangular Lagrangian points, we require either $\mu \leq 0.0385209$ or $\mu \geq 0.9614791$, which is always satisfied by any two CR3BP primaries in the solar system. Therefore, we expect that the triangular Lagrangian points would provide stable locations for smaller bodies in the solar system. The existence of such bodies for the Sun-Jupiter system ($\mu = 0.00095369$) have been verified in the form of Trojan asteroids. For the Earth-Moon system, $\mu = 0.01215$ is within the stable region; hence one could expect the triangular points to be populated. However, the Earth-Moon system is not a good example of the restricted three-body problem, because of an appreciable influence of sun's gravity (a fourth body), which renders the triangular points of Earth-Moon system unstable. Hence, L_4, L_5 for Earth-Moon system are empty regions in space.

For a given CR3BP system, one can determine the actual small displacement dynamics about the stable triangular Lagrangian points, in terms of the eigenvalues, λ. For example, the Earth-Moon system has non-dimensional natural frequencies 0.2982 and 0.9545, which indicate *long-period* and *short-period* modes of the small displacement motion.

For the stability of the collinear Lagrangian points, we employ $y_0 = 0$, and the resulting equations of small displacement can be written as follows:

$$\delta\ddot{x} - 2\delta\dot{y} - \left[\frac{2(1-\mu)}{(x_0-\mu)^3} + \frac{2\mu}{(x_0+1-\mu)^3} + 1\right]\delta x$$
$$\delta\ddot{y} + 2\delta\dot{x} + \left[\frac{1-\mu}{(x_0-\mu)^3} + \frac{\mu}{(x_0+1-\mu)^3} - 1\right]\delta y\,. \tag{10.67}$$

The eigenvalues of this system are the eigenvalues of the following square matrix:

$$\mathbf{A} = \begin{pmatrix} 0 & 0 & 1 & 0 \\ 0 & 0 & 0 & 1 \\ \alpha & 0 & 0 & 2 \\ 0 & -\beta & -2 & 0 \end{pmatrix}, \tag{10.68}$$

where

$$\alpha = \frac{2(1-\mu)}{(x_0-\mu)^3} + \frac{2\mu}{(x_0+1-\mu)^3} + 1$$
$$\beta = \frac{1-\mu}{(x_0-\mu)^3} + \frac{\mu}{(x_0+1-\mu)^3} - 1\,. \tag{10.69}$$

The resulting characteristic equation for the eigenvalues, λ, is the following:

$$\lambda^4 + (4+\beta-\alpha)\lambda^2 - \alpha\beta = 0\,. \tag{10.70}$$

For the collinear Lagrangian points, we have $\alpha > 0$ and $\beta < 0$, which implies that the constant term in the characteristic equation is always positive. Therefore, there is always at least one eigenvalue with a positive real part; hence the system is unconditionally unstable. Thus, the collinear Lagrangian points of any restricted three-body system are always unstable. However, as discussed in Section 10.6, a spacecraft can successfully orbit a collinear point with small energy expenditure. Such an orbit around a Lagrangian point is termed a *halo orbit*, and is useful in practical space missions, such as stationing a probe around the Sun-Earth L_1 and L_2 points for monitoring the solar and interplanetary zone ahead of and behind the earth.

10.6 Orbital Energy and Jacobi's Integral

The total orbital energy of the three masses is *not* conserved in the restricted three-body problem. This is due to the fact that the motion of the primaries, m_1 and m_2, is governed by the two-body problem, which is unaffected by the gravity of the negligible mass, m_3. Therefore, while the primaries have their net orbital energy conserved, the motion of the mass, m_3, has its energy varying with time. This can be seen by multiplying the corresponding equations of Eqs. (10.42) with the non-dimensional inertial velocity components, $\dot{\xi}$, $\dot{\eta}$, and $\dot{\zeta}$, summing the results, and then integrating in non-dimensional time to yield the following:

$$\frac{1}{2}(\dot{\xi}^2+\dot{\eta}^2+\dot{\zeta}^2)=(v')^2/2=\int_0^t\left(\frac{\partial U}{\partial \xi}\dot{\xi}+\frac{\partial U}{\partial \eta}\dot{\eta}+\frac{\partial U}{\partial \zeta}\dot{\zeta}\right)\mathrm{d}t$$

$$\int_0^t\left(\mathrm{d}U-\frac{\partial U}{\partial t}\right)\mathrm{d}t \tag{10.71}$$

$$=U-\int_0^t\frac{\partial U}{\partial t}\mathrm{d}t\,,$$

where $v'=\sqrt{\dot{\xi}^2+\dot{\eta}^2+\dot{\zeta}^2}$ is the inertial speed of the mass, m_3. Thus we have

$$\frac{1}{2}(v')^2-U=-\int_0^t\frac{\partial U}{\partial t}\mathrm{d}t\,,$$

which implies that the net orbital energy of the mass, m_3, is not constant, but varies with time. This non-conservation of the total energy, E, is a violation of the principle of energy conservation of the exact three-body problem, and only arises due to the approximation involved in the restricted problem.

A similar analysis reveals that the net angular momentum, $\mathbf{H}$, is also not conserved for the restricted three-body problem, and the barycentre approximated to be the centre of mass of the two primaries does not yield the constant vectors, $\mathbf{R}_{c0}$, $\mathbf{v}_{c0}$, of the actual problem. Therefore, it may appear that the ten integral constants of the actual three-body problem are lost by resorting to the approximation, $m_3 \ll m_1$, and $m_3 \ll m_2$. However, this is not so, and one (and the only) integral of the restricted problem can be retrieved as follows.

Consider the multiplication of the equations constituting Eq. (10.56) with the non-dimensional inertial velocity components, $\dot{x}$, $\dot{y}$, and $\dot{z}$, respectively, summing the results, and then integrating in non-dimensional time to yield the following:

$$\frac{1}{2}(\dot{x}^2+\dot{y}^2+\dot{z}^2)=v^2/2=\int_0^t\left(\frac{\partial \Omega}{\partial x}\dot{x}+\frac{\partial \Omega}{\partial y}\dot{y}+\frac{\partial \Omega}{\partial z}\dot{z}\right)\mathrm{d}t$$

$$\int_0^t\left(\mathrm{d}\Omega-\frac{\partial \Omega}{\partial t}\right)\mathrm{d}t \tag{10.72}$$

$$=\Omega-\int_0^t\frac{\partial \Omega}{\partial t}\mathrm{d}t\,,$$

where, the pseudo-potential is given by Eq. (10.55), and $v=\sqrt{\dot{x}^2+\dot{y}^2+\dot{z}^2}$ is the apparent speed of the mass, m_3, in the synodic frame. Since the gravitational potential, U, is time-invariant in the synodic frame, we have

$$\frac{\partial \Omega}{\partial t}=0\,,$$

which, substituted into Eq. (10.72), results in the following:

$$\frac{1}{2}v^2-\Omega=\frac{1}{2}v^2-\frac{1}{2}(x^2+y^2)-\frac{(1-\mu)}{r_1}-\frac{\mu}{r_2}=-\frac{1}{2}C=\text{const.} \tag{10.73}$$

The constant, $C = 2\Omega - v^2$, is the only known integral of the restricted problem, and is called *Jacobi's integral*. Jacobi's integral can be thought to represent a pseudo-energy of the mass m_3, which is a sum of the relative kinetic energy, gravitational potential energy, and an additional pseudo-potential energy, $(x^2 + y^2)$. The value of C at any point is thus a measure of the relative energy of m_3.

10.6.1 Zero-Relative-Speed Contours

The most useful interpretation of Jacobi's integral consists of the contours of zero relative speed, which are surfaces in the (x, y, z) space corresponding to a constant Jacobi's integral, $C = C_0$, and $v = 0$:

$$C_0 = 2\Omega = (x^2 + y^2) + \frac{2(1-\mu)}{r_1} + \frac{2\mu}{r_2} = \text{const.} \tag{10.74}$$

Cross sections of such contours can be plotted for various constant values of the out-of-plane coordinate, z, producing curves of the (x, y) coordinates. Of particular interest are the curves obtained on the plane of the primaries, $z = 0$, which have

$$r_1 = \sqrt{(x+\mu)^2 + y^2}$$
$$r_2 = \sqrt{(x-1+\mu)^2 + y^2}\,, \tag{10.75}$$

or

$$x^2 + y^2 = (1-\mu)r_1^2 + \mu r_2^2 - \mu(1-\mu)\,. \tag{10.76}$$

The substitution of Eq. (10.76) into Eq. (10.74) yields the family of curves in the (x, y) space, representing the zero-relative-speed contours for $z = 0$. A convenient expression of Jacobi's integral for the $z = 0$ curves is the following:

$$C_0^* = 2\Omega^* = (1-\mu)\left(r_1^2 + \frac{2}{r_1}\right) + \mu\left(r_2^2 + \frac{2}{r_2}\right) = \text{const.}\,, \tag{10.77}$$

where the star represents the special case, $z = 0$, for which Ω^* is the following pseudo-potential modified by adding the constant, $\mu(1-\mu)/2$, to Ω:

$$\Omega^* = (\Omega)_{z=0} + \frac{1}{2}\mu(1-\mu) = (U)_{z=0} + \frac{1}{2}(x^2 + y^2) + \frac{1}{2}\mu(1-\mu)\,. \tag{10.78}$$

A zero-relative-speed contour represents the boundary between the motion towards the higher value of the pseudo-potential, Ω, or the return trajectory to the lower pseudo-potential. On such a boundary, the mass, m_3, must stop and turn back. Hence, the zero-relative-speed contours demarcate the regions accessible to a spacecraft, and cannot be crossed for the given value of $C = C_0$.

Construction of the zero-relative-speed contours for the coplanar problem ($z = 0$) reveals them to be curves symmetric about the axis, $\mathbf{i}$, for $C_0^* > 3$. There is no real solution, (x, y), of Eq. (10.77) for $C_0^* \leq 3$. For the value of C_0^* slightly greater than 3, two closed curves appear around the triangular Lagrangian points, L_4 and L_5, as shown in Fig. 10.4 for $C_0^* = 3.01$. As the value of Jacobi's integral is increased, this pair of contours increases in size and merge into a single closed contour curving around the larger primary, m_1, and enclosing the Lagrangian

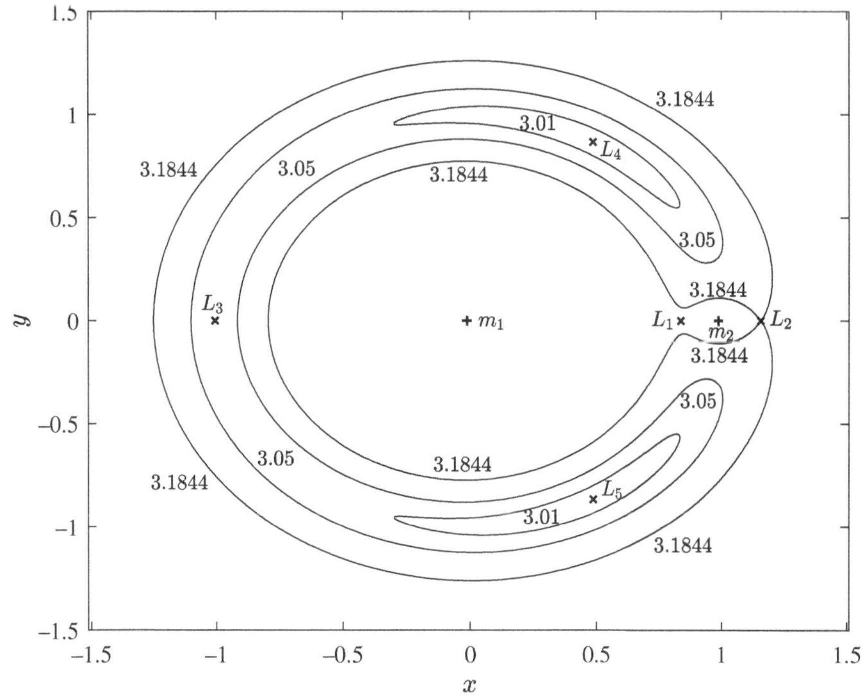

Figure 10.4 zero-relative-speed contours for the Earth-Moon system for $C_0^* = 3.01,\ 3.05$, and 3.1844.

points, L_3, L_4, and L_5, exemplified by the contour corresponding to $C_0^* = 3.05$ in Fig. 10.4. As C_0^* is increased to $C_0^* = 3.1844$, the two arms of the contour meet at the collinear Lagrangian point, L_2, leading to two completely enclosed regions as shown in Fig. 10.4. The outer region of the contour encloses the Lagrangian points, L_3, L_4, and L_5, while the inner region encompasses the two primaries and the Lagrangian point, L_1. This indicates the possibility of travel between the primaries inside the inner region.

A further increase in C_0^* causes the single closed contour to split into two different closed contours. The outer contour encircles all the Lagrangian points, while the inner contour passes through the inner collinear Lagrangian point, L_1, wrapping around the smaller primary, m_2, and encircling m_1 in the opposite sense. This situation is shown in Fig. 10.5 for $C_0^* = 3.2$. As C_0^* is increased beyond this critical value to $C_0^* = 3.3$, the inner contour splits into two separate contours, each individually enclosing the primaries, and the outer contour enclosing all the Lagrangian points increases in size, as shown in Fig. 10.5. The inner contour enclosing m_2 is of a much smaller size than that enclosing m_1. As the value of Jacobi's integral is increased, there is no further change in the qualitative nature of the zero-relative-speed contours, but the inner contours individually encircling the primaries decrease in size, while the outer contour becomes larger. Figure 10.6 shows the complete qualitative set of the zero-relative-speed contours for the Earth-Moon system up to the value $C_0^* = 3.5$.

The closed $v = 0$ contours enveloping any one of the primaries indicate that a flight from one primary to the other is impossible at the given energy level. Furthermore, for the relatively small

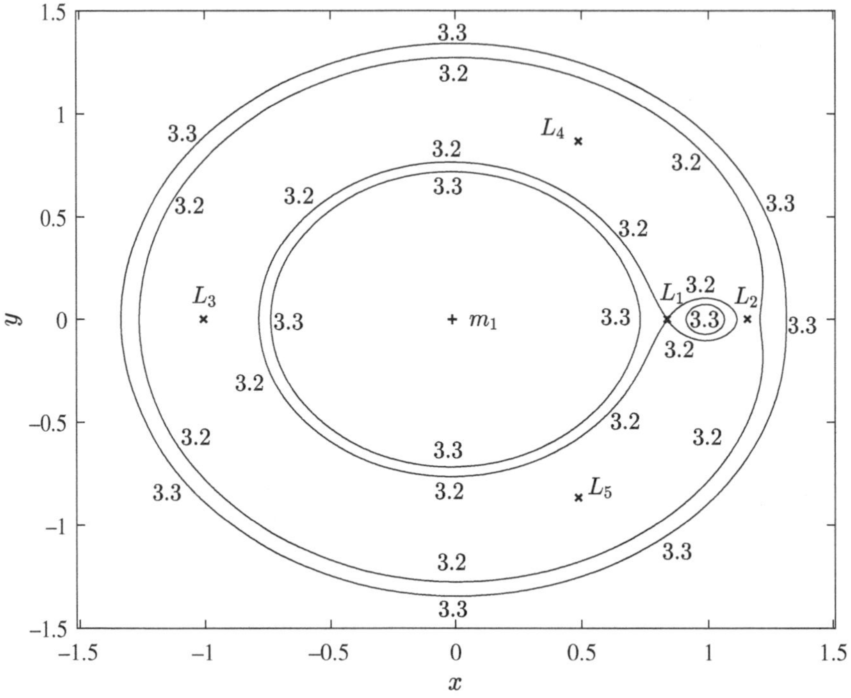

Figure 10.5 zero-relative-speed contours for the Earth-Moon system for $C_0^* = 3.01,\ 3.05$, and 3.1844.

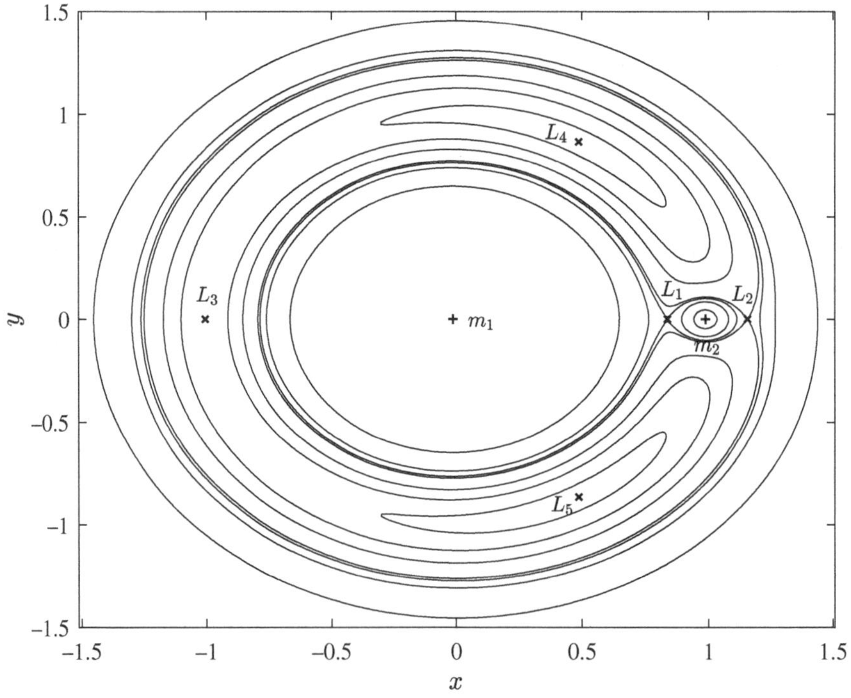

Figure 10.6 Complete qualitative set of zero-relative-speed contours for the Earth-Moon system in the range, $3.01 \le C_0^* \le 3.5$.

values of C_0^*, zero-relative-speed contours enclose the triangular Lagrangian points, indicating that a flight to reach them is impossible from any one of the primaries. As the value of C_0^* is increased, the forbidden regions around L_4, L_5, shrink but never actually vanish, thereby denoting that the stable triangular Lagrangian points can be approached (but never reached) only with a very high initial energy.

10.6.2 *Tisserand's Criterion*

As noted earlier, Jacobi's integral is the only constant of the restricted three-body problem. It can be used to derive an important relationship for the special case when the negligible mass, m_3, makes a close pass of the smaller primary, m_2. In such a case, the two-body orbit of m_3 relative to the larger primary, m_1, undergoes an abrupt change, which can be approximately expressed by this relationship, called *Tisserand's criterion*. Such an approximation of a three-body trajectory by an impulsively perturbed two-body (Keplerian) orbit of m_3 around m_1 is valid when the mass of the second primary is very small in comparison with the first, i.e., $m_2 \ll m_1$, which implies $\mu \simeq 0$. In the solar system, this approximate analysis is applied to a comet having passed close to Jupiter, because $\mu = 0.00095369$ for the Sun-Jupiter system.

Consider the inertial velocity and position of m_3 given by $\mathbf{v}'$ and $\mathbf{r}$, respectively, relative to the barycentre. The inertial velocity is resolved as follows in the synodic frame, $(\mathbf{i}, \mathbf{j}, \mathbf{k})$:

$$\mathbf{v}' = \mathbf{v} + \boldsymbol{\omega} \times \mathbf{r}\,, \tag{10.79}$$

where $\boldsymbol{\omega} = \mathbf{k}$ rad./s is the angular velocity of the synodic frame, and $\mathbf{r} = x\mathbf{i} + y\mathbf{j} + z\mathbf{k}$ is the non-dimensional inertial position vector resolved in the synodic frame. Thus we have

$$\begin{aligned} v^2 &= \mathbf{v} \cdot \mathbf{v} = (\mathbf{v}' - \boldsymbol{\omega} \times \mathbf{r}) \cdot (\mathbf{v}' - \boldsymbol{\omega} \times \mathbf{r}) \\ x^2 + y^2 &= -\mathbf{r} \cdot \boldsymbol{\omega} \times (\boldsymbol{\omega} \times \mathbf{r})\,. \end{aligned} \tag{10.80}$$

These substituted into Eq. (10.73) yield the following expression for the Jacobi's integral:

$$\begin{aligned} C = 2\Omega - v^2 &= \frac{2(1-\mu)}{r_1} + \frac{2\mu}{r_2} - [(\mathbf{v}' - \boldsymbol{\omega} \times \mathbf{r}) \cdot (\mathbf{v}' - \boldsymbol{\omega} \times \mathbf{r})] \\ &\quad -\mathbf{r} \cdot \boldsymbol{\omega} \times (\boldsymbol{\omega} \times \mathbf{r}) \\ &= \frac{2(1-\mu)}{r_1} + \frac{2\mu}{r_2} - (v')^2 + 2\boldsymbol{\omega} \cdot (\mathbf{r} \times \mathbf{v}')\,, \end{aligned} \tag{10.81}$$

where the triple-product identities have been applied. Making the approximation $m_2 \ll m_1$ results in the barycentre moving to the centre of the primary, m_1, which implies that $\mathbf{r} \simeq \mathbf{r}_1$. Consequently, v' and $\mathbf{h} \simeq \mathbf{r} \times \mathbf{v}'$ are the approximate two-body orbital speed and angular momentum, respectively, of m_3 relative to m_1. Furthermore, this approximation results in $\mu \simeq 0$, which results in the following approximate Jacobi's integral:

$$C = \frac{2}{r} - (v')^2 + 2\boldsymbol{\omega} \cdot \mathbf{h}\,. \tag{10.82}$$

Since $\boldsymbol{\omega}$ has units of rad./s, C has the same units. In terms of the non-dimensional Keplerian orbital elements of m_3 relative to m_1, $(\bar{a}, e, i)$, which correspond to $Gm_1 = 1, \omega = 1$, we have the following:

$$C = -2\varepsilon + 2\boldsymbol{\omega} \cdot \mathbf{h} = \frac{1}{\bar{a}} + 2h\cos i = \frac{1}{\bar{a}} + 2\sqrt{\bar{a}(1-e^2)}\cos i\,, \tag{10.83}$$

where $\bar{a} = a/r_{21} = a/(Gm_1)^{1/3}$ is the non-dimensional semi-major axis, e the orbital eccentricity, and i the orbital inclination. Thus we have the following criterion for the variation in the two-body orbital elements of m_3 relative to m_1, after making a close pass of m_2:

$$\frac{Gm_1}{2a} + \sqrt{a(1-e^2)(Gm_1)}\cos i = C(Gm_1)^{2/3} = \text{const.}\,, \tag{10.84}$$

which is Tisserand's criterion. The units of the constant $C(Gm_1)^{2/3}$ in Eq. (10.84) are km^2/s^2.

Example 10.6.1 *The comet* C/1980 E1 *was observed on 11 February, 1980, to have a nearly parabolic trajectory of period approximately 7.1 million years relative to the sun* ($Gm_1 = 1.32712 \times 10^{11}$ km^3/s^2)*, with a perihelion of 3.3639 astronomical units (AU), an aphelion of* 74300 *AU, and inclination from the ecliptic of* 1.6617°*. Assuming no change in its orbital inclination, what was the semi-major axis, a, of the comet after it passed within* 0.228 *AU of Jupiter on 9 December 1980, when its orbital eccentricity was seen to increase to* $e = 1.057$*? (1 AU =* 1.495978×10^8 km*).*

We begin by putting the values of a and e before the encounter with Jupiter to yield the constant in Eq. (10.84) as follows:

$$a = \frac{74300 + 3.3639}{2} = 37151.68195\ AU\ (5557809886019.71\ \text{km})$$

$$e = \frac{74300 - 3.3639}{74300 + 3.3639} = 0.9999095$$

$$\frac{Gm_1}{2a} + \sqrt{a(1-e^2)(Gm_1)}\cos i = C(Gm_1)^{2/3} = 11552121085.1686\ \text{km}^2/\text{s}^2\,.$$

Next, the new value of the semi-major axis is computed by putting $e = 1.057$ *and* $i = 1.6617°$ *and solving the following cubic equation for* $-a$*:*

$$(-a)^{3/2}\sqrt{(e^2-1)(Gm_1)}\cos i - (-a)C(Gm_1)^{2/3} - Gm_1/2 = 0\,,$$

whose only real root is $a = -8583599988.319$ km *(*-57.377849 *AU). Thus we have the semi-major axis of the hyperbolic orbit after the close pass of Jupiter to be* $a = -57.377849$ *AU. The new value of the perihelion is* $a(1-e) = 3.27054$ *AU.*

10.7 Canonical Formulation

An analytical insight into the CR3BP equations of motion is possible through their derivation by the Lagrangian method (Chapter 2) in terms of the canonical coordinates vector, $\mathbf{q} = (x, y, z)^T$, as follows:

$$\ddot{\mathbf{q}} - 2\mathbf{K}\dot{\mathbf{q}} = \nabla\Omega\,, \tag{10.85}$$

where $\nabla = (\partial/\partial x, \partial/\partial y, \partial/\partial z)^T$ is the gradient operator

$$\mathbf{K} = \begin{pmatrix} 0 & 1 & 0 \\ -1 & 0 & 0 \\ 0 & 0 & 0 \end{pmatrix} \tag{10.86}$$

and

$$\Omega = -\frac{1}{2}\mathbf{q}^T\mathbf{K}^2\mathbf{q} + \frac{\mu}{\mid \mathbf{q} - (1-\mu)\mathbf{i} \mid} + \frac{(1-\mu)}{\mid \mathbf{q} + \mu\mathbf{i} \mid} = \frac{1}{2}(x^2 + y^2) + \frac{(1-\mu)}{r_1} + \frac{\mu}{r_2} \tag{10.87}$$

is the pseudo-potential function with $\mathbf{i} = (1, 0, 0)^T$ and r_1, r_2 given by Eq. (10.51). The scalar components of Eq. (10.85) are the following:

$$\begin{aligned} \ddot{x} - 2\dot{y} &= \frac{\partial \Omega}{\partial x} \\ \ddot{y} + 2\dot{x} &= \frac{\partial \Omega}{\partial y} \\ \ddot{z} &= \frac{\partial \Omega}{\partial z}. \end{aligned} \tag{10.88}$$

The Hamiltonian associated with Eq. (10.85) is the following:

$$H(\mathbf{q}, \dot{\mathbf{q}}) = -\frac{1}{2}\dot{\mathbf{q}}^T(\dot{\mathbf{q}} - \mathbf{K}\mathbf{q}) - L(\mathbf{q}, \dot{\mathbf{q}}), \tag{10.89}$$

with the Lagrangian given by

$$L(\mathbf{q}, \dot{\mathbf{q}}) = \frac{1}{2}(\dot{\mathbf{q}} - \mathbf{K}\mathbf{q})^T(\dot{\mathbf{q}} - \mathbf{K}\mathbf{q}) - U(\mathbf{q}) \tag{10.90}$$

and the gravitational potential by

$$U(\mathbf{q}) = -\Omega - \frac{1}{2}\mathbf{q}^T\mathbf{K}^2\mathbf{q} = -\frac{(1-\mu)}{r_1} - \frac{\mu}{r_2}. \tag{10.91}$$

By defining the generalized momentum vector as follows:

$$\mathbf{p} = \frac{\partial L}{\partial \dot{\mathbf{q}}} = \dot{\mathbf{q}} - \mathbf{K}\mathbf{q} \tag{10.92}$$

the Hamiltonian is expressed as

$$H(\mathbf{q}, \mathbf{p}) = \frac{1}{2}\mathbf{p}^T\mathbf{p} + \mathbf{p}^T\mathbf{K}\mathbf{q} + U(\mathbf{q}). \tag{10.93}$$

This yields the canonical equations of motion, Eq. (10.85):

$$\begin{aligned} \dot{\mathbf{q}} &= \frac{\partial H}{\partial \mathbf{p}} \\ \dot{\mathbf{p}} &= -\frac{\partial H}{\partial \mathbf{q}}, \end{aligned} \tag{10.94}$$

which are compactly expressed in terms of the state vector, $\mathbf{X} = (\mathbf{q}^T, \mathbf{p}^T)^T$, as follows:

$$\dot{\mathbf{X}} = \mathbf{J}\nabla_X H(\mathbf{X}), \tag{10.95}$$

where ∇_X is the gradient operator relative to $\mathbf{X}$, and

$$\mathbf{J} = \begin{pmatrix} \mathbf{0}_{3\times3} & \mathbf{I}_{3\times3} \\ -\mathbf{I}_{3\times3} & \mathbf{0}_{3\times3} \end{pmatrix}. \tag{10.96}$$

Jacobi's integral can be derived by taking the scalar product of Eq. (10.85) with $\dot{\mathbf{q}}$:

$$\ddot{\mathbf{q}} \cdot \dot{\mathbf{q}} - 2(\mathbf{K}\dot{\mathbf{q}}) \cdot \dot{\mathbf{q}} = \frac{\partial\Omega}{\partial\mathbf{q}}^T \cdot \dot{\mathbf{q}}, \tag{10.97}$$

or

$$\frac{1}{2}\frac{\mathrm{d}\dot{q}^2}{\mathrm{d}t} = \frac{\mathrm{d}\Omega}{\mathrm{d}t}, \tag{10.98}$$

which is an exact differential, and can be integrated to obtain

$$\Omega = \frac{1}{2}(\dot{q}^2 + C). \tag{10.99}$$

Here, the integration constant, C, is Jacobi's integral expressed as

$$C = 2\Omega - \dot{q}^2 = (x^2 + y^2) - 2\frac{1-\mu}{r_1} - 2\frac{\mu}{r_2} - (\dot{x}^2 + \dot{y}^2 + \dot{z}^2). \tag{10.100}$$

The Hamiltonian, H, is related to Jacobi's integral as follows:

$$\begin{aligned} H &= \frac{1}{2}\mathbf{p}^T\mathbf{p} + \mathbf{p}^T\mathbf{K}\mathbf{q} + U(\mathbf{q}) \\ &= \frac{1}{2}\dot{q}^2 + \frac{1}{2}\mathbf{q}^T\mathbf{K}^2\mathbf{q} + U(\mathbf{q}) \\ &= \frac{1}{2}\dot{q}^2 - \frac{1}{2}(x^2 + y^2) + U = -C/2. \end{aligned} \tag{10.101}$$

10.8 Special Three-Body Trajectories

A closed-form solution to the circular restricted three-body problem (CR3BP) is unavailable, as first demonstrated by Poincaré using phase-space surfaces. The CR3BP belongs to the group of *chaotic systems*, wherein a small change in the initial condition results in an arbitrarily large change in the trajectory. Numerical integration of the CR3BP equations of motion to certain initial conditions by an appropriate technique, such as the Runge-Kutta method (Appendix A), reveals trajectories that are possible flight paths around the primaries and the Lagrangian points. Some trajectories, which can be derived within one time period of the rotation of the primaries, are investigated in this section.

10.8.1 Perturbed Orbits About a Primary

When a spacecraft is orbiting one of the primaries, its trajectory is perturbed by the presence of the more distant primary. The precession of the orbital plane comprising an apsidal rotation and a nodal regression can be approximated by the Lagrange planetary equations in Chapter 9 for a close orbit of a primary. When the spacecraft's orbit is coplanar with the plane of the primaries, such a perturbation in the orbit caused by the other primary is investigated here by solving the CR3BP problem, and plotting the results in the synodic frame. Consider for example the perturbation in a spacecraft's orbit around Earth due to the Moon's gravity, as plotted in Fig. 10.7, when the three bodies are always in the same plane. The value of the non-dimensional mass parameter for the Earth-Moon system is $\mu = 0.01215$, which implies Earth's location at $(-0.01215, 0)$ marked by the + symbol in Fig. 10.7. The trajectory is simulated for $t = 1$ (about 4.8 mean solar days in the real time, t'). The initial condition for this plot is $x(0) = 0.1$, $y(0) = 0$, $\dot{x} = 0$, and $\dot{y} = 0.25$.

Figure 10.8 shows the orbit of a spacecraft around the moon perturbed by Earth's gravity. The initial condition for the given coplanar trajectory plotted for $t = 1$, is $x(0) = 1.0$, $y(0) = 0$, $\dot{x} = 0$, and $\dot{y} = 0.25$.

10.8.2 Free-Return Trajectories

An interesting periodic solution of the CR3BP is the free-return trajectory, which passes close to both the primaries and can be used for an essentially fuel-free transfer from one primary to

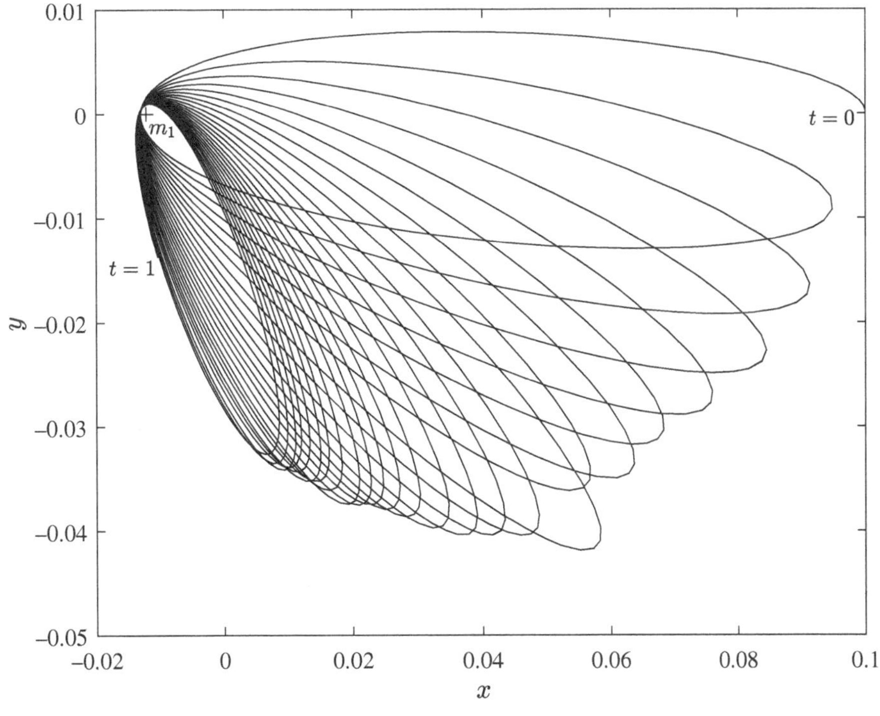

Figure 10.7 Spacecraft's elliptical orbit around Earth perturbed by the Moon, simulated for $t = 1$ (4.8 days).

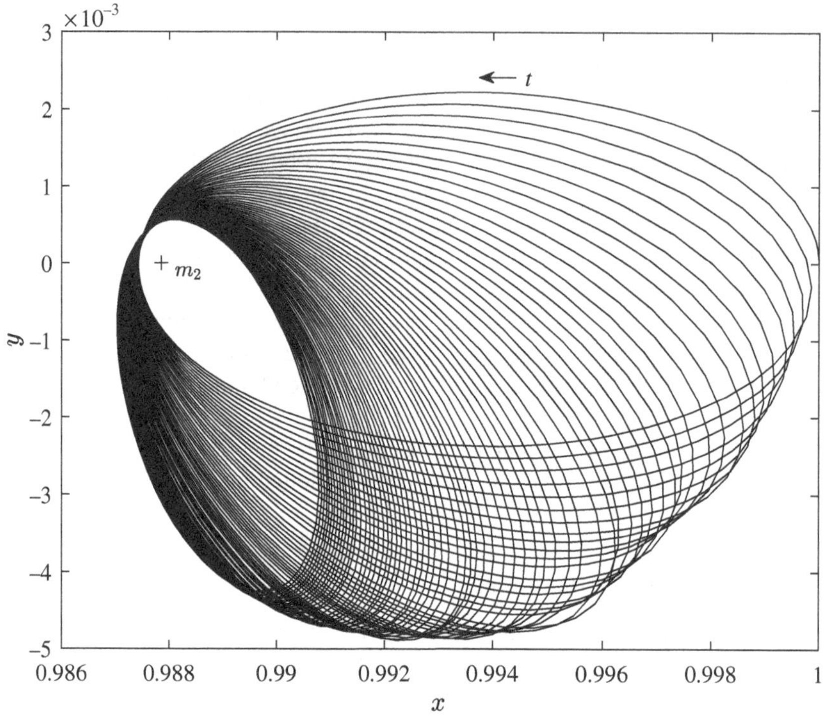

Figure 10.8 Spacecraft's elliptical orbit around the moon perturbed by Earth, simulated for $t = 1$ (4.8 days).

the other. The free-return concept was utilized by the manned lunar exploration through the *Apollo* lunar missions, wherein the initial insertion was performed at a given location with the desired velocity, both on the outward and the return journeys. Being a natural solution, it is a fuel-optimal path of the planar CR3BP. An example of the free-return Earth-Moon trajectory is plotted in Fig. 10.9 for $t = 3.57$, which extends from a radius of 33770 km from Earth, m_1, to a point close to the Moon, m_2, and encloses both m_1 and the Lagrangian point, L_1. The value of Jacobi's integral for this trajectory is $C = 1.2988$.

Another free-return trajectory is depicted in Fig. 10.10, seen to be passing between the Moon, m_2, and the Lagrangian point, L_2. The total flight time is reduced significantly in this case to about $t = 2.387$, and the value of Jacobi's integral for this trajectory is $C = 2.1995$, which indicates that a smaller energy expenditure is required than that for the trajectory shown in Fig. 10.9. The much smaller energy expenditure of the trajectory of Fig. 10.10 when compared to that of Fig. 10.9 can be explained by the swing around the moon seen in Fig. 10.10. The pass around the moon boosts the orbital kinetic energy of the spacecraft relative to Earth, without any fuel expenditure. Such an increase in the Keplerian orbital energy derived by passing around a third body in an orbit around the central body is called a *gravity-assist manoeuvre* (or a *swing-by* pass).

An extreme case of a lunar gravity-assist manoeuvre is depicted in Fig. 10.11, where the spacecraft escapes the Earth-Moon system by swinging by very close to the Moon.

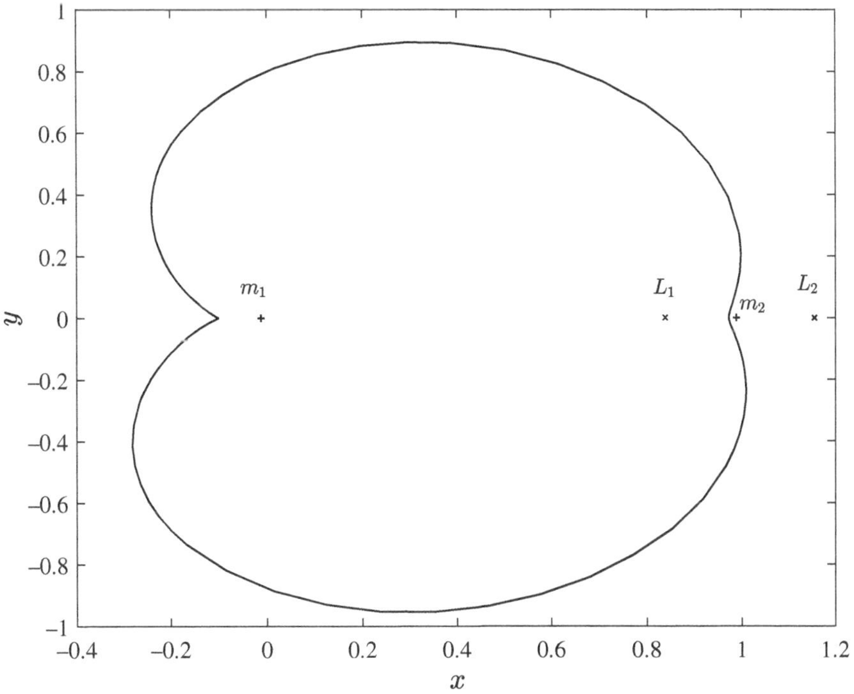

Figure 10.9 Free-return trajectory passing close to Earth and the moon in $t = 3.5$ (16.7 days).

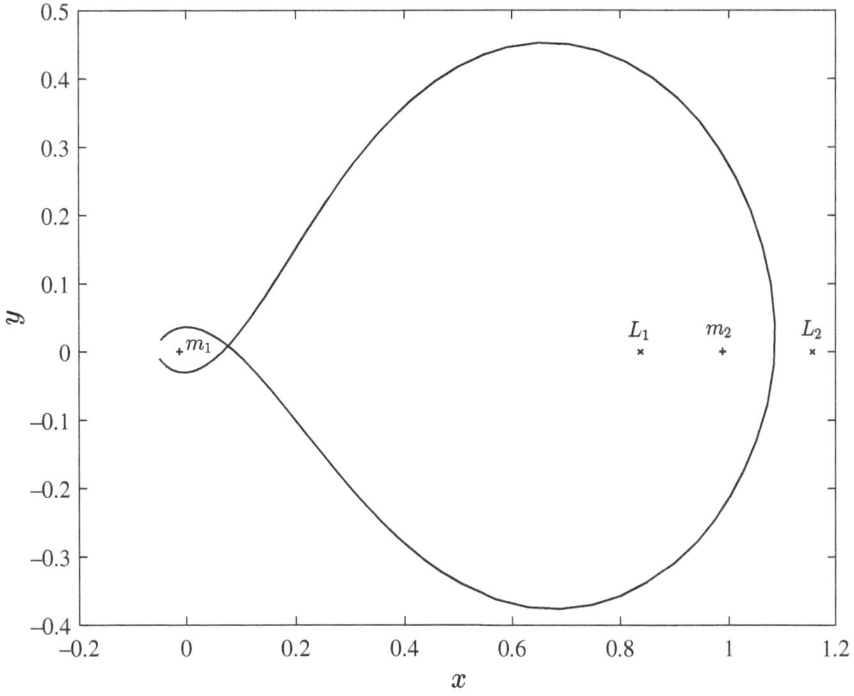

Figure 10.10 Lunar swing-by free-return trajectory passing close to Earth and the moon in $t = 2.387$ (11.4 days).

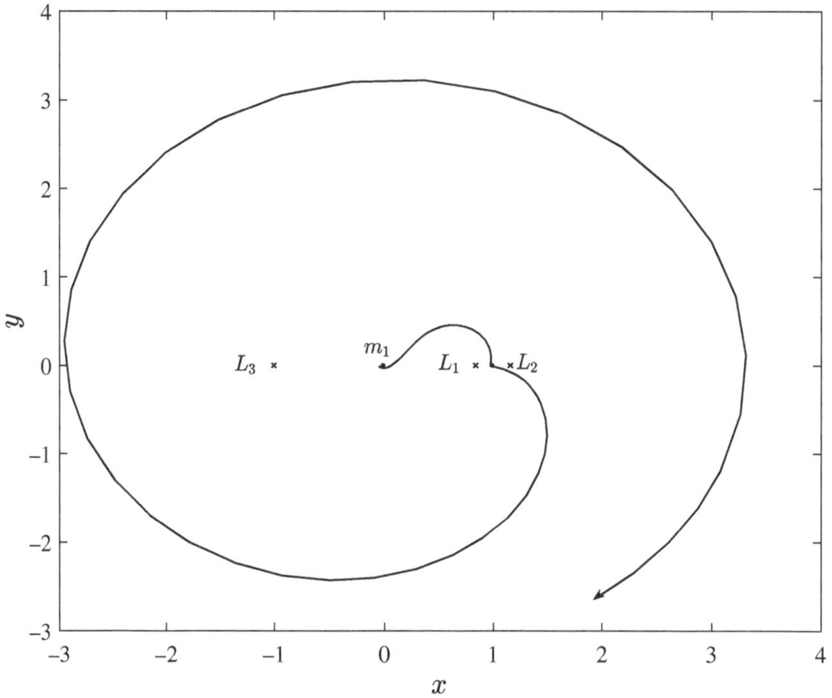

Figure 10.11 Lunar swing-by escape trajectory simulated for $t = 10$ (47.76 days).

The initial velocity from the same initial location for the escape trajectory of Fig. 10.11 is *smaller* ($v = 6.935, C = 2.3982$) than that of the free-return trajectory of Fig. 10.10, which has ($v = 6.9493, C = 2.1995$). Lunar swing-by trajectories have been practically employed in launching several spacecraft to the Sun-Earth Lagrangian points, such as the *ISEE-3, MAP*, and *ACE* missions of NASA. Multiple planetary swing-bys have been useful in reducing the mission cost of travel to distant planets around the sun, such as the *Voyager* spacecraft of NASA. Figure 10.12 shows a trajectory beginning near the Moon, and orbiting all five Lagrangian points of the Earth-Moon system in $t = 28.5$.

Exercises

1. Calculate the locations of the Lagrangian points of the Sun-Jupiter system ($\mu = 0.00095369$).
2. Use the linearized dynamics of the circular restricted three-body problem described by Eqs. (10.61) and (10.62), to express the solution, $\delta\mathbf{r} = (\delta x, \delta y)^T$, representing a small-displacement oscillation about the triangular Lagrangian points, L_4, L_5, with a given small initial condition, $\delta\mathbf{r_0} = (\delta x_0, \delta y_0)^T$, when the value of μ falls in the region of stability. Show that the oscillation is a linear combination of two modes:

$$\lambda_1 = \pm i\omega_s \, ; \qquad \lambda_1 = \pm i\omega_p \, ,$$

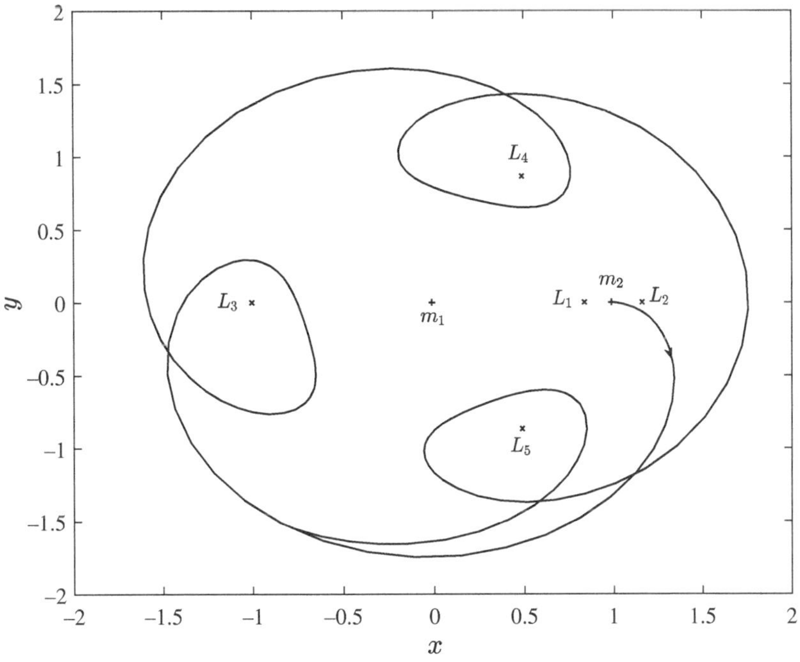

Figure 10.12 Free trajectory orbiting all five Lagrangian points of the Earth-Moon system in $t = 28.5$ (136 days).

where $i = \sqrt{-1}$ and ω_s and ω_p are the natural frequencies of the *short-period mode* and the *long-period mode*, respectively. Calculate these natural frequencies for the Sun-Jupiter system ($\mu = 0.00095369$).

3. For the restricted two-body problem consisting of the motion of the test mass, m_3, around a spherical body of mass, m_1, use Jacobi's integral with $\mu = 0$ to plot zero-relative-speed contours corresponding to $C_0 = 4$. What are the permissible regions for the motion of m_3 if the initial condition is given by the following?
   (a) $r_1(0) = 2$, $v(0) = 1$.
   (b) $r_1(0) = 1/2$, $v(0) = 1/2$.

Reference

Szebehely V 1967. *Theory of Orbits – The Restricted Problem of Three Bodies*. Academic Press, New York.

11

Attitude Dynamics

Thus far, the discussion has been confined to the orbital motion of spacecraft, which is represented by the translation of the centre of mass in a celestial reference frame. However, the rotational motion of the spacecraft about its centre of mass, called the *attitude dynamics*, is equally important as it involves the changing attitude (orientation) of the spacecraft. The attitude dynamics is crucial in accurately pointing a spacecraft's payload (cameras and other imaging devices and sensors) in specific directions for performing its basic mission. Furthermore, various antennae must be pointed towards Earth for communication, and solar arrays should be maintained nearly normal to the sun for generating the electrical power necessary for charging the onboard batteries. When orbital manoeuvres have to be performed, it is critical that the spacecraft is accurately manoeuvred to the correct attitude, and maintains that attitude while firing the rocket engine. Stability of the torque-free rotational dynamics is an important objective in any spacecraft, because it allows it to maintain an equilibrium attitude in the presence of small external disturbances. A spacecraft which has unstable rotational dynamics cannot maintain a constant attitude required for performing any task, and hence becomes useless in its mission. Therefore, it is important that the attitude dynamics should be stable, either in the open loop (that is, without applying any control inputs), or at least in the closed loop by the help of rotors, attitude thrusters, and/or magnetic torquers. This chapter addresses the attitude dynamics of rigid spacecraft. Attitude dynamics consists of two parts: (a) *attitude kinematics*, which describes a changing orientation of a coordinate frame rigidly fixed to the body, and which can be represented by the attitude kinematics models introduced in Chapter 5, and (b) *attitude kinetics*, which describes the variation of the angular velocity of the body. Both of these aspects of attitude dynamics are discussed in this chapter. The stability analyses presented in this chapter range from torque-free motion of both axisymmetric and asymmetric spacecraft to gravity-gradient satellites.

Chapter 5 introduced the attitude description of an orbital coordinate frame relative to a celestial reference frame. However, when the attitude of a coordinate frame varies with time, the instantaneous attitude is described by a set of ordinary differential equations in time. Such a description of the changing attitude is termed the *rotational kinematics* (or *attitude kinematics*), and requires a set of kinematic parameters, which change with time in a particular manner for a given angular velocity of the frame. If the spacecraft is assumed to be a rigid body, then a

Foundations of Space Dynamics, First Edition. Ashish Tewari.

reference frame rigidly attached to the vehicle is used to represent the vehicle's attitude. Such a frame is called a *body-fixed frame*. However, the angular velocity of the body-fixed frame cannot be arbitrary, but must satisfy the laws of *rotational kinetics* (or *attitude kinetics*) which take into account the mass distribution of the vehicle, as well as the external torques acting about the centre of mass.

11.1 Euler's Equations of Attitude Kinetics

The rotational kinetics of a body taken to be a collection of a large number of particles of elemental mass, dm is described by the following equation of motion derived from Newton's second law of motion (Chapter 2):

$$\mathbf{M} = \frac{\mathrm{d}\mathbf{H}}{\mathrm{d}t}, \tag{11.1}$$

where

$$\mathbf{H} = \int \mathbf{r} \times \mathbf{v} \mathrm{d}m \tag{11.2}$$

is the net *angular momentum* of the body about its centre of mass, o. Here $\mathbf{v}$ is the inertial velocity of the particle, $\mathbf{r}$ is its position from the centre of mass, o, and $\mathbf{M}$ is the net *external torque* acting on the body about o. In the derivation of Eq. (11.1) it has been assumed (Chapter 2) that all internal torques cancel each other by virtue of Newton's third law. This is due to the fact that the internal forces between any two particles constituting the body act along the line joining the particles. Most forces of interaction among particles obey this principle, with the exception of the magnetic force.

The total (inertial) velocity of an arbitrary point on the rigid body located at $\mathbf{r}$ relative to o is the following:

$$\mathbf{v} = \mathbf{v}_0 + \frac{\mathrm{d}\mathbf{r}}{\mathrm{d}t}, \tag{11.3}$$

where $\mathbf{v}_0$ denotes the velocity of the centre of mass, o.

If the body is *rigid*, the distance between any two points constituting the body does not change with time. To derive the basic equations of attitude kinetics, it is assumed that the spacecraft is a rigid body; hence the magnitude of the relative position vector, $r = \mid \mathbf{r} \mid$, is invariant with time. Therefore, we have for a point located at $\mathbf{r}$ relative to o on the rigid body

$$\frac{\mathrm{d}\mathbf{r}}{\mathrm{d}t} = \dot{r}\frac{\mathbf{r}}{r} + \boldsymbol{\omega} \times \mathbf{r} = \boldsymbol{\omega} \times \mathbf{r}, \tag{11.4}$$

where $\boldsymbol{\omega}$ is the angular velocity of $\mathbf{r}$ about o. Since the distance between any two points on the rigid body is fixed, all straight lines contained in the body remain straight, and rotate with the same angular velocity, $\boldsymbol{\omega}$, about o. A reference frame, $oxyz$, can thus be chosen with its origin, o, at the centre of mass (see Fig. 2.3) to describe the rotational motion of the body. Such a frame is said to be a *body frame*. A substitution of Eq. (11.4) into Eq. (11.3) gives the velocity of an arbitrary point on the rigid body to be the following:

$$\mathbf{v} = \mathbf{v}_0 + \boldsymbol{\omega} \times \mathbf{r}, \tag{11.5}$$

and a substitution of Eq. (11.5) into Eq. (11.2) yields the following expression for the net angular momentum of the rigid body about o:

$$\mathbf{H} = \int \mathbf{r} \times \mathbf{v}_o \mathrm{d}m + \int \mathbf{r} \times (\boldsymbol{\omega} \times \mathbf{r}) \mathrm{d}m . \tag{11.6}$$

The first term on the right-hand side of Eq. (11.6) is expressed as follows:

$$\int \mathbf{r} \times \mathbf{v}_o \mathrm{d}m = \left(\int \mathbf{r} \mathrm{d}m \right) \times \mathbf{v}_o , \tag{11.7}$$

which vanishes by virtue of o being the centre of mass ($\int \mathbf{r}\mathrm{d}m = 0$). Therefore, it follows that

$$\mathbf{H} = \int \mathbf{r} \times (\boldsymbol{\omega} \times \mathbf{r}) \mathrm{d}m . \tag{11.8}$$

Let all the vectors describing the rotational kinetics be resolved in the right-handed body frame, with axes ox, oy, oz along the unit vectors, $\mathbf{i}_1, \mathbf{i}_2, \mathbf{i}_3$, respectively, as follows:

$$\mathbf{r} = x\mathbf{i}_1 + y\mathbf{i}_2 + z\mathbf{i}_3 \tag{11.9}$$

$$\boldsymbol{\omega} = \omega_x\mathbf{i}_1 + \omega_y\mathbf{i}_2 + \omega_z\mathbf{i}_3 \tag{11.10}$$

$$\mathbf{H} = H_x\mathbf{i}_1 + H_y\mathbf{i}_2 + H_z\mathbf{i}_3 \tag{11.11}$$

$$\mathbf{M} = M_x\mathbf{i}_1 + M_y\mathbf{i}_2 + M_z\mathbf{i}_3 . \tag{11.12}$$

By substituting these vector components into Eq. (11.8) and simplifying the result, we arrive at the following expression for the angular-momentum vector:

$$\mathbf{H} = \mathbf{J}\boldsymbol{\omega} , \tag{11.13}$$

where $\mathbf{J}$ is the *inertia tensor*, given by

$$\mathbf{J} = \begin{pmatrix} \int (y^2 + z^2)\mathrm{d}m & -\int xy\mathrm{d}m & -\int xz\mathrm{d}m \\ -\int xy\mathrm{d}m & \int (x^2 + z^2)\mathrm{d}m & -\int yz\mathrm{d}m \\ -\int xz\mathrm{d}m & -\int yz\mathrm{d}m & \int (x^2 + y^2)\mathrm{d}m \end{pmatrix} . \tag{11.14}$$

It is evident from the elements of $\mathbf{J}$ that it is a symmetric matrix, $\mathbf{J}^T = \mathbf{J}$, and hence can be expressed as follows:

$$\mathbf{J} = \begin{pmatrix} J_{xx} & -J_{xy} & -J_{xz} \\ -J_{xy} & J_{yy} & -J_{yz} \\ -J_{xz} & -J_{yz} & J_{zz} \end{pmatrix} . \tag{11.15}$$

The elements of the inertia tensor are grouped into the *moments of inertia*, J_{xx}, J_{yy}, J_{zz}, and the *products of inertia*, J_{xy}, J_{yz}, J_{xz}. Since $\boldsymbol{\omega}$ is the angular velocity of the body frame, $oxyz$, if the axes of the frame are *not* fixed to the rigid body, the angular velocity of the body would be different from $\boldsymbol{\omega}$. In such a case, the moments and products of inertia would be time-varying. Since the main purpose of writing the angular momentum in the form of Eq. (11.13) is the introduction of an inertia tensor whose elements describe the constant mass distribution of the rigid body, it is desirable to have a constant inertia tensor. Hence, the axes of the body frame, *oxyz*, are deliberately chosen to be rigidly attached to the rigid body, and therefore rotating with the same angular velocity, $\boldsymbol{\omega}$, as that of the body. Such a choice of the body frame with its axes tied rigidly to the body is called a *body-fixed frame*, and results in a time-invariant inertia tensor, **J**. From this point on, the body frame (*oxyz*) will be taken to be a body-fixed frame. Hence, $\boldsymbol{\omega}$ in Eq. (11.13) is the angular velocity of the rigid body, and **J** is a constant matrix.

The equations of rotational kinetics of the rigid body are resolved in the body-fixed frame by substituting Eq. (11.13) into Eq. (11.1), and applying the chain rule of taking the time derivative of a vector (Chapter 2):

$$\begin{aligned}\mathbf{M} &= \frac{d\mathbf{H}}{dt} = \frac{\partial \mathbf{H}}{\partial t} + \boldsymbol{\omega} \times \mathbf{H} \\ &= \mathbf{J}\frac{\partial \boldsymbol{\omega}}{\partial t} + \boldsymbol{\omega} \times (\mathbf{J}\boldsymbol{\omega}),\end{aligned} \tag{11.16}$$

where $\partial/\partial t$ represents the time derivative taken with reference to the body-fixed frame, such as

$$\frac{\partial \boldsymbol{\omega}}{\partial t} = \begin{Bmatrix} \frac{d\omega_x}{dt} \\ \frac{d\omega_y}{dt} \\ \frac{d\omega_z}{dt} \end{Bmatrix} = \begin{Bmatrix} \dot{\omega}_x \\ \dot{\omega}_y \\ \dot{\omega}_z \end{Bmatrix}. \tag{11.17}$$

By replacing the vector product in Eq. (11.16) with a matrix product (Chapter 10), we have

$$\mathbf{M} = \mathbf{J}\frac{\partial \boldsymbol{\omega}}{\partial t} + \mathbf{S}(\boldsymbol{\omega})\mathbf{J}\boldsymbol{\omega}, \tag{11.18}$$

where

$$\mathbf{S}(\boldsymbol{\omega}) = \begin{pmatrix} 0 & -\omega_z & \omega_y \\ \omega_z & 0 & -\omega_x \\ -\omega_y & \omega_x & 0 \end{pmatrix}. \tag{11.19}$$

Equation (11.18) represents three scalar, coupled, first-order non-linear ordinary differential equations in $\boldsymbol{\omega}(t)$, and are called *Euler's equations*. These are the governing equations for the rotational kinetics of a rigid body, and are solved for $\boldsymbol{\omega}(t)$ for a given initial condition, $\boldsymbol{\omega}(0)$, and a known external torque, $\mathbf{M}(t)$, for $t \geq 0$.

11.2 Attitude Kinematics

To complete the description of rotational dynamics of the spacecraft regarded as a rigid body, an additional set of differential equations is required which represents the time-varying attitude of the body-fixed frame, $(\mathbf{i}_1, \mathbf{i}_2, \mathbf{i}_3)$, relative to a fixed (inertial) reference frame, $(\mathbf{I}_1, \mathbf{I}_2, \mathbf{I}_3)$. Such a set of first-order differential equations is termed *attitude kinematics*, and can be derived by differentiating with time the following coordinate transformation described by the rotation matrix, **C** (Chapter 5):

$$\begin{Bmatrix} \mathbf{i}_1 \\ \mathbf{i}_2 \\ \mathbf{i}_3 \end{Bmatrix} = \mathbf{C} \begin{Bmatrix} \mathbf{I}_1 \\ \mathbf{I}_2 \\ \mathbf{I}_3 \end{Bmatrix}. \tag{11.20}$$

Differentiating Eq. (11.20) with time results in the following:

$$\frac{d}{dt}\begin{Bmatrix} \mathbf{i}_1 \\ \mathbf{i}_2 \\ \mathbf{i}_3 \end{Bmatrix} = \begin{Bmatrix} \frac{d\mathbf{i}_1}{dt} \\ \frac{d\mathbf{i}_2}{dt} \\ \frac{d\mathbf{i}_3}{dt} \end{Bmatrix} = \begin{Bmatrix} \boldsymbol{\omega} \times \mathbf{i}_1 \\ \boldsymbol{\omega} \times \mathbf{i}_2 \\ \boldsymbol{\omega} \times \mathbf{i}_3 \end{Bmatrix} = \frac{d\mathbf{C}}{dt}\begin{Bmatrix} \mathbf{I}_1 \\ \mathbf{I}_2 \\ \mathbf{I}_3 \end{Bmatrix}. \tag{11.21}$$

The time derivative of the rotation matrix, $d\mathbf{C}/dt$, can be evaluated once a set of kinematical parameters is chosen to represent the rotation matrix. Chapter 5 presented the rotation matrix in terms of various alternative attitude representations, such as the Euler angles, the Euler-axis/principal angle combination, the quaternion, and the modified Rodrigues parameters. Once a set of kinematical parameters is selected, it is used to calculate the time derivative of the rotation matrix, $d\mathbf{C}/dt$, which along with Euler's equations of rotational dynamics completes the set of differential equations needed to describe the changing attitude of a rigid body under the influence of an external torque vector, $\mathbf{M}(t)$. The variables of the rotational dynamics are thus the kinematical parameters representing the instantaneous attitude of the body-fixed frame, $(\mathbf{i}_1, \mathbf{i}_2, \mathbf{i}_3)$, and the angular velocity, $\boldsymbol{\omega}(t)$, of the rigid body obtained as the solution to Euler equations. Recall from Chapter 5 that a set of three kinematical parameters is sufficient to describe a general attitude, and is termed a minimal set. Along with the three components of the angular velocity, $(\omega_x, \omega_y, \omega_z)$, a minimal set of kinematical parameters forms a set of six scalar variables required to completely describe the rotational dynamics of a rigid spacecraft.

A commonly employed set of minimal attitude kinematical parameters is the 3-1-3 Euler-angle representation, $(\psi)_3, (\theta)_1, (\phi)_3$. The 3-1-3 sequence was employed in Chapter 5 for the orientation of an orbital plane, $(\Omega)_3, (i)_1, (\omega)_3$, and is expressed here as follows:

$$\begin{Bmatrix} \mathbf{i}_1 \\ \mathbf{i}_2 \\ \mathbf{i}_3 \end{Bmatrix} = \mathbf{C}_3(\phi)\mathbf{C}_1(\theta)\mathbf{C}_3(\psi)\begin{Bmatrix} \mathbf{I}_1 \\ \mathbf{I}_2 \\ \mathbf{I}_3 \end{Bmatrix}. \tag{11.22}$$

The rotation matrix representing the orientation of the body-fixed frame, $(\mathbf{i}_1, \mathbf{i}_2, \mathbf{i}_3)$, relative to an inertial reference frame, $(\mathbf{I}_1, \mathbf{I}_2, \mathbf{I}_3)$, via the $(\psi)_3, (\theta)_1, (\phi)_3$ representation is the following:

$$\mathbf{C} = \mathbf{C}_3(\phi)\mathbf{C}_1(\theta)\mathbf{C}_3(\psi) = \begin{pmatrix} c_{11} & c_{12} & c_{13} \\ c_{21} & c_{22} & c_{23} \\ c_{31} & c_{32} & c_{33} \end{pmatrix}, \tag{11.23}$$

where the elements, c_{ij}, are the following:

$$c_{11} = \cos\psi\cos\phi - \sin\psi\sin\phi\cos\theta \tag{11.24}$$

$$c_{12} = \sin\psi\cos\phi + \cos\psi\sin\phi\cos\theta \tag{11.25}$$

$$c_{13} = \sin\phi\sin\theta \tag{11.26}$$

$$c_{21} = -\cos\psi\sin\phi - \sin\psi\cos\phi\cos\theta \tag{11.27}$$

$$c_{22} = -\sin\psi\sin\phi + \cos\psi\cos\phi\cos\theta \tag{11.28}$$

$$c_{23} = \cos\phi\sin\theta \tag{11.29}$$

$$c_{31} = \sin\psi \sin\theta \tag{11.30}$$

$$c_{32} = -\cos\psi \sin\theta \tag{11.31}$$

$$c_{33} = \cos\theta\,. \tag{11.32}$$

The angular velocity vector, $\boldsymbol{\omega}(t)$, can be expressed as follows in terms of the rates of the Euler angles, (ψ, θ, ϕ), by using the respective axes of the three elementary rotations of the 3-1-3 sequence:

$$\begin{aligned}\boldsymbol{\omega} &= \dot{\psi}\mathbf{I}_3 + \dot{\theta}\mathbf{I}_1{}' + \dot{\phi}\mathbf{i}_3 \\ &= \omega_x\mathbf{i}_1 + \omega_y\mathbf{i}_2 + \omega_z\mathbf{i}_3\,,\end{aligned} \tag{11.33}$$

where $\mathbf{I}_1{}' = \mathbf{I}_1{}''$ is the first axis of the reference frame following the first and second elementary rotations, and is given by

$$\mathbf{I}_1{}' = \mathbf{I}_1{}'' = \mathbf{i}_1 \cos\phi - \mathbf{i}_2 \sin\phi\,. \tag{11.34}$$

The third axis of the reference frame resolved in the body-fixed frame is the following:

$$\mathbf{I}_3 = \mathbf{i}_1 \sin\phi \sin\theta + \mathbf{i}_2 \cos\phi \sin\theta + \mathbf{i}_3 \cos\theta\,. \tag{11.35}$$

By substituting Eqs. (11.34) and (11.35) into Eq. (11.33), and comparing the components, we arrive at the following expression for the angular velocity components:

$$\begin{Bmatrix}\omega_x \\ \omega_y \\ \omega_z\end{Bmatrix} = \begin{pmatrix}\sin\phi\sin\theta & \cos\phi & 0 \\ \cos\phi\sin\theta & -\sin\phi & 0 \\ \cos\theta & 0 & 1\end{pmatrix}\begin{Bmatrix}\dot{\psi} \\ \dot{\theta} \\ \dot{\phi}\end{Bmatrix}. \tag{11.36}$$

The inversion of the matrix on the right-hand side of Eq. (11.36) yields the following expression for the time-rate of change of the Euler angles:

$$\begin{Bmatrix}\dot{\psi} \\ \dot{\theta} \\ \dot{\phi}\end{Bmatrix} = \frac{1}{\sin\theta}\begin{pmatrix}\sin\phi & \cos\phi & 0 \\ \cos\phi\sin\theta & -\sin\phi\sin\theta & 0 \\ -\sin\phi\cos\theta & -\cos\phi\cos\theta & \sin\theta\end{pmatrix}\begin{Bmatrix}\omega_x \\ \omega_y \\ \omega_z\end{Bmatrix}. \tag{11.37}$$

Such an inversion would not be possible if $\theta = 0, \pm\pi$, which are the singularities of the Euler-angle representation, $(\psi)_3, (\theta)_1, (\phi)_3$, and cause the matrix in Eq. (11.36) to become singular.

11.3 Rotational Kinetic Energy

The net kinetic energy of a body is derived by the summation of the individual kinetic energies of all elemental particles of elemental mass, dm, constituting the body as follows:

$$T = \frac{1}{2}\int \mathbf{v}^T\mathbf{v}\mathrm{d}m = \frac{1}{2}mv_o^2 + \int \frac{\mathrm{d}\mathbf{r}^T}{\mathrm{d}t}\frac{\mathrm{d}\mathbf{r}}{\mathrm{d}t}\mathrm{d}m\,, \tag{11.38}$$

where $\mathbf{r}$ locates the particle relative to the centre of mass, o, which has the velocity, $\mathbf{v}_0$, $m = \int dm$ is the total mass of the body, and $\mathbf{r}$, $\mathbf{v}$, and $\mathbf{v}_0$ are related by Eq. (11.3). For a rigid body, we have $d\mathbf{r}/dt = \boldsymbol{\omega} \times \mathbf{r}$ by Eq. (11.3); hence Eq. (11.38) becomes the following:

$$T = \frac{1}{2}mv_o^2 + \frac{1}{2}\int (\boldsymbol{\omega} \times \mathbf{r}) \cdot (\boldsymbol{\omega} \times \mathbf{r}) dm . \tag{11.39}$$

The first term on the right-hand side of Eq. (11.39) represents the kinetic energy due to the translation of the centre of mass, while the second term denotes the kinetic energy of rotation of the rigid body about the centre of mass. The expression for the rotational kinetic energy of the rigid body, T_r, can be simplified by substituting Eq. (11.13) for the the angular momentum, leading to

$$T_r = \frac{1}{2}\int (\boldsymbol{\omega} \times \mathbf{r}) \cdot (\boldsymbol{\omega} \times \mathbf{r}) dm = \frac{1}{2}\boldsymbol{\omega} \cdot \mathbf{H} = \frac{1}{2}\boldsymbol{\omega}^T \mathbf{J} \boldsymbol{\omega} . \tag{11.40}$$

This expression for the rotational kinetic energy of a rigid body is useful in simplifying Euler's equations.

An alternative expression for T_r is the following in terms of the angular-momentum vector:

$$T_r = \frac{1}{2}\mathbf{H}^T \mathbf{J}^{-1} \mathbf{H} . \tag{11.41}$$

For motion in which the rotational kinetic energy is conserved, (T_r = const.), Eq. (11.41) can be expressed as follows:

$$\frac{H_x^2}{2J_{xx}T_r} + \frac{H_y^2}{2J_{yy}T_r} + \frac{H_z^2}{2J_{zz}T_r} = 1 , \tag{11.42}$$

which represents the equation of an *ellipsoid* in the angular-momentum space, i.e., the space spanned by the axes, oH_x, oH_y, and oH_z. The semi-axis lengths of this energy ellipsoid are $\sqrt{2J_{xx}T_r}$, $\sqrt{2J_{yy}T_r}$, and $\sqrt{2J_{zz}T_r}$.

The rotational kinetic energy is conserved if there is no external torque applied to the rigid body. This fact is evident by taking the time derivative of Eq. (11.40), and substituting Euler's equations, Eq. (11.16), with $\mathbf{M} = \mathbf{0}$:

$$\frac{dT_r}{dt} = \frac{1}{2}\frac{d\boldsymbol{\omega}}{dt} \cdot \mathbf{H} + \frac{1}{2}\boldsymbol{\omega} \cdot \frac{d\mathbf{H}}{dt} = 0 . \tag{11.43}$$

Since $\mathbf{M} = \mathbf{0}$, the second term on the right-hand side of Eq. (11.43) vanishes due to Eq. (11.1), while the first term on the right-hand side vanishes by virtue of Eq. (11.16), which produces

$$\frac{d\boldsymbol{\omega}}{dt} \cdot \mathbf{H} = -\boldsymbol{\omega} \cdot (\boldsymbol{\omega} \times \mathbf{H}) = \mathbf{0} .$$

In the absence of an external torque, the conservation of both the rotational kinetic energy and the angular momentum can be effectively utilized in obtaining the analytical relationships between the angular velocity and the inertia tensor.

11.4 Principal Axes

As seen in Chapter 2, the translation of a coordinate frame is trivially handled by shifting the origin of the coordinate frame to a new location. A translation of the origin (centre of mass of the rigid body) involves no rotation of the frame's axes, and therefore produces a new inertia tensor which is derived merely by shifting the limits of integration in Eq. (11.14). The transformation in the inertia tensor caused by a shift of the origin o of the reference frame, $oxyz$, yields the *parallel axes theorem*. However, a rotation of the body-fixed frame about the origin is a non-trivial transformation, as it profoundly changes the inertia tensor. The body-fixed frame, $oxyz$, used in deriving Euler's equations has an arbitrary orientation relative to the rigid body. There are infinitely many ways in which the mutually orthogonal axes of the frame can be fixed to a given rigid body at the centre of mass, o. However, a major simplification in Euler's equations is possible by choosing a specific orientation, $(ox_py_pz_p)$, of the body-fixed frame relative to the rigid body, such that the products of inertia, J_{xy}, J_{yz}, J_{xz}, vanish. Such a frame is called the *principal body-fixed frame* (or in short, the *principal axes*). The inertia tensor resolved in the principal axes is a diagonal matrix, $\mathbf{J_p}$.

To derive the coordinate transformation that produces the principal axes, $(\mathbf{i_{1p}}, \mathbf{i_{2p}}, \mathbf{i_{3p}})$, from an arbitrary body-fixed frame, $(\mathbf{i_1}, \mathbf{i_2}, \mathbf{i_3})$, consider a rotation matrix, $\mathbf{C_p}$, defined by

$$\begin{Bmatrix} \mathbf{i_1} \\ \mathbf{i_2} \\ \mathbf{i_3} \end{Bmatrix} = \mathbf{C_p} \begin{Bmatrix} \mathbf{i_{1p}} \\ \mathbf{i_{2p}} \\ \mathbf{i_{3p}} \end{Bmatrix} . \tag{11.44}$$

Let the subscript p be used to denote a vector resolved in the principal frame. The components of a vector, $\mathbf{a} = a_x\mathbf{i_1} + a_y\mathbf{i_2} + a_z\mathbf{i_3}$, resolved in the arbitrary body-fixed frame, are related to the same vector in the principal axes, $\mathbf{a_p} = a_{xp}\mathbf{i_{1p}} + a_{yp}\mathbf{i_{2p}} + a_{zp}\mathbf{i_{3p}}$, via multiplication with the rotation matrix, $\mathbf{C_p}$, as follows:

$$\mathbf{a} = \begin{Bmatrix} a_x \\ a_y \\ a_z \end{Bmatrix} = \mathbf{C_p} \begin{Bmatrix} a_{xp} \\ a_{yp} \\ a_{zp} \end{Bmatrix} = \mathbf{C_p a_p} . \tag{11.45}$$

Similarly, we write $\boldsymbol{\omega} = \mathbf{C_p}\boldsymbol{\omega}_p$. Since there is no change in the rotational kinetic energy, T_r, caused by the coordinate transformation, $\mathbf{C_p}$, we have

$$T_r = \frac{1}{2}\boldsymbol{\omega}^T\mathbf{J}\boldsymbol{\omega} = \frac{1}{2}\boldsymbol{\omega}_p^T\mathbf{J_p}\boldsymbol{\omega}_p . \tag{11.46}$$

The substitution of $\boldsymbol{\omega} = \mathbf{C_p}\boldsymbol{\omega}_p$ into Eq. (11.46) and the comparison of the terms on both the sides of the resulting equation yields the following:

$$\boldsymbol{\omega}_p^T\mathbf{J_p}\boldsymbol{\omega}_p = \boldsymbol{\omega}^T\mathbf{J}\boldsymbol{\omega} = \boldsymbol{\omega}_p^T\mathbf{C_p}^T\mathbf{J}\mathbf{C_p}\boldsymbol{\omega}_p , \tag{11.47}$$

which, on applying the orthogonality property of a rotation matrix (Chapter 5), produces

$$\mathbf{J_p} = \mathbf{C_p}^T\mathbf{J}\mathbf{C_p} = \text{diag}(J_1, J_2, J_3) . \tag{11.48}$$

Determination of the rotation matrix $\mathbf{C_p}$ from Eq. (11.48) is tantamount to solving an eigenvalue problem. Since $\mathbf{J_p} = \text{diag}(J_1, J_2, J_3)$ is a diagonal matrix, it follows that the diagonal elements, J_i, $i = 1, 2, 3$, of $\mathbf{J_p}$ are the eigenvalues of $\mathbf{J}$, while $\mathbf{C_p}$ has the eigenvectors, $\mathbf{v}_i$, $i = 1, 2, 3$, of $\mathbf{J}$ as its columns:

$$J_i \mathbf{v}_i = \mathbf{J}\mathbf{v}_i \qquad (i = 1, 2, 3) \tag{11.49}$$

$$\mathbf{C_p} = [\mathbf{v}_1, \mathbf{v}_2, \mathbf{v}_3] \, . \tag{11.50}$$

Thus Eq. (11.48) is the formula for deriving the inertia tensor in the principal frame, and the coordinate transformation matrix, $\mathbf{C_p}$, from the eigenvalue analysis of $\mathbf{J}$. A comparison of Eqs. (11.44) and (11.50) indicates that the eigenvectors, $\mathbf{v}_i$, relate the unit vectors representing the principal axes, $(ox_p y_p z_p)$, to those representing the arbitrary frame, $oxyz$. Hence, it is necessary that the eigenvectors are normalized; that is, $|\ \mathbf{v}_i\ | = 1$, $i = 1, 2, 3$. Otherwise the unit vectors will not be preserved in the magnitude during the transformation.

However, the eigenvectors are linearly independent, if and only if the eigenvalues are distinct, i.e. $(J_i \neq J_k,\ i \neq k)$. In case the eigenvalues of $\mathbf{J}$ are repeated, which happens when the mass distribution has at least one axis of symmetry, the rotation matrix $\mathbf{C_p}$ cannot be determined by Eq. (11.50).

Example 11.4.1 *A rigid body has the following inertia tensor:*

$$\mathbf{J} = \begin{pmatrix} 1000 & -100 & 135 \\ -100 & 1200 & -50 \\ 135 & -50 & 760 \end{pmatrix} \text{kg.m}^2 \, .$$

Find the inertia tensor in the principal axes, and the coordinate transformation matrix, $\mathbf{C_p}$.

The eigenvalues of the inertia tensor, $\mathbf{J}$, *are first determined by solving the following characteristic equation:*

$$\det(\lambda \mathbf{I} - \mathbf{J}) = 0,$$

which yields the eigenvalues $\lambda_1 = J_1 = 699.3298$ kg.m^2, $\lambda_2 = J_2 = 998.6671$ kg.m^2, *and* $\lambda_3 = J_3 = 1262.0031$ kg.m^2.

The eigenvectors, $\mathbf{v}_i$, $i = 1, 2, 3$, *corresponding to the eigenvalues,* λ_i, $i = 1, 2, 3$, *respectively, are next determined by solving the eigenvalue problem:*

$$\lambda_i \mathbf{v}_i = \mathbf{J}\mathbf{v}_i \qquad (i = 1, 2, 3) \, .$$

The normalized eigenvectors, $|\ \mathbf{v}_i\ | = 1$, *are computed to be the following:*

$$\mathbf{v}_1 = \begin{Bmatrix} -0.4068224 \\ 0.0099674 \\ 0.9134529 \end{Bmatrix} ; \ \mathbf{v}_2 = \begin{Bmatrix} -0.8008139 \\ -0.4850157 \\ -0.3513643 \end{Bmatrix} ; \ \mathbf{v}_3 = \begin{Bmatrix} 0.4395368 \\ -0.8744487 \\ 0.2052973 \end{Bmatrix} .$$

Hence, the rotation matrix of the transformation and the principal inertia tensor are the following:

$$\mathbf{C_p} = \begin{pmatrix} -0.4068224 & -0.8008139 & 0.4395368 \\ 0.0099674 & -0.4850157 & -0.8744487 \\ 0.9134529 & -0.3513643 & 0.2052973 \end{pmatrix} ,$$

$$\mathbf{J_p} = \begin{pmatrix} 699.3298 & 0 & 0 \\ 0 & 998.6671 & 0 \\ 0 & 0 & 1262.0031 \end{pmatrix} \text{kg.m}^2 .$$

In this computation, the eigenvalues of $\mathbf{J}$ *have been arranged in the order of increasing magnitude,* $J_3 > J_2 > J_1$*, although this order is arbitrary. Similarly, the computed eigenvectors can be scaled by* ± 1*, and are thus not unique. However, the eigenvectors should be normalized,* $|\mathbf{v}_i| = 1$*, which results in a proper rotation matrix,* $\det(\mathbf{C_p}) = 1$*.*

The computation of eigenvalues and eigenvectors can be performed by any standard numerical algorithm, such as the MATLAB® function eig.

The eigenvalues of $\mathbf{J}$ must be distinct in order to compute the principal axes by the method of diagonalization presented above. If a body is axisymmetric, two of its principal moments of inertia are identical. However, in such a case eigenvalue analysis is unnecessary for deriving the principal axes, because the axis of symmetry being one of the principal axes, is identified from the shape of the body. The other two principal axes in such a case are any pair of orthogonal axes normal to the axis of symmetry. Since an arbitrary body-fixed frame, *oxyz*, can be transformed into the principal frame, all further discussion in this chapter assumes that *oxyz* denotes the principal axes of the spacecraft, without explicitly carrying the subscript, *p*. Furthermore, for convenience and unless otherwise stated, it is assumed that *oz* is the *major axis* and *ox* is the *minor axis*, that is, the principal axes with the largest and the smallest moments of inertia, respectively. Therefore, $J_{zz} > J_{yy} > J_{xx}$, and $J_{xy} = J_{xz} = J_{yz} = 0$.

11.5 Torque-Free Rotation of Spacecraft

Torque-free motion ($\mathbf{M} = \mathbf{0}$) is an essential characteristic of a spacecraft's attitude dynamics described by Eqs. (11.17) and (11.21). Stability in torque-free rotation is necessary in order to maintain an equilibrium state in the presence of small torque perturbations arising out of the neglected dynamics when deriving Euler's equations. To derive the torque-free dynamics, substitute $\mathbf{M} = \mathbf{0}$ into Eq. (11.17) to yield the following equations of kinetics in the principal axes (*oxyz*):

$$\begin{aligned} J_{xx}\dot{\omega}_x + \omega_y\omega_z(J_{zz} - J_{yy}) &= 0 \\ J_{yy}\dot{\omega}_y + \omega_x\omega_z(J_{xx} - J_{zz}) &= 0 \\ J_{zz}\dot{\omega}_z + \omega_x\omega_y(J_{yy} - J_{xx}) &= 0 , \end{aligned} \tag{11.51}$$

where the overdot represents the time derivative, $\mathrm{d}/\mathrm{d}t$. An equilibrium kinetic state of the rigid spacecraft is the constant solution, $\boldsymbol{\omega}(t) = \boldsymbol{\omega}(0) = \text{const.}$, to Eq. (11.51), for which at least two of the angular velocity components, $\omega_x, \omega_y, \omega_z$, must vanish. This implies that a rotation about a single principal axis which remains fixed in space is an equilibrium state of the spacecraft's rotational kinetics. There are therefore three possible equilibrium states – that is, ($\omega_x = \omega_y = 0$), ($\omega_x = \omega_z = 0$), and ($\omega_y = \omega_z = 0$), which correspond to states of pure spin about the *oz*, *oy*, and *ox*, axes, respectively. It is yet to be determined which of the three possible equilibrium states are *stable*.

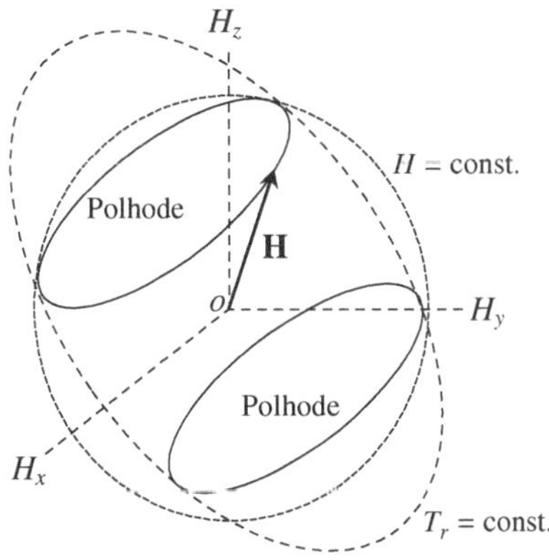

Figure 11.1 Torque-free rotation of a rigid spacecraft in the angular-momentum space.

An alternative description of torque-free kinetics is in the angular-momentum space via energy ellipsoid. As seen earlier, the constant energy state, T_r = const., results in an energy ellipsoid in the angular-momentum space, $(oH_xH_yH_z)$. While $\mathbf{H}$ is a constant vector in the inertial space, it rotates in the body-fixed principal axes, $(oxyz)$, with a constant magnitude, $H = |\mathbf{H}|$. Hence the tip of the vector $\mathbf{H}$ describes a curve on the surface of the angular momentum sphere of radius H centred at the origin, o. The trajectory of a torque-free rotation, $\boldsymbol{\omega}(t) = \mathbf{J}^{-1}\mathbf{H}(t)$, follows the intersection of the energy ellipsoid,

$$\frac{H_x^2}{2J_{xx}T_r} + \frac{H_y^2}{2J_{yy}T_r} + \frac{H_z^2}{2J_{zz}T_r} = 1\,,$$

and the angular momentum sphere, given by

$$H_x^2 + H_y^2 + H_z^2 = H^2\,. \tag{11.52}$$

The intersection of an energy ellipsoid with the angular-momentum sphere results in two closed curves called *polhodes*, as shown in Fig. 11.1. The trajectory $\boldsymbol{\omega}(t) = \mathbf{J}^{-1}\mathbf{H}(t)$ follows any one of the polhodes, depending upon the initial condition, $\boldsymbol{\omega}(0) = (\omega_{x0}, \omega_{y0}, \omega_{z0})^T$. Since the polhodes are closed curves, they denote the fact that the angular velocity magnitude, $\omega(t) = |\boldsymbol{\omega}(t)|$, is always bounded. Furthermore, since the polhodes for a constant T_r and H are non-intersecting, it is not possible for a trajectory to jump from one polhode to another. In this manner, a graphical analysis of the torque-free motion can be carried out for a given combination of the inertia parameters, J_{xx}, J_{yy}, J_{zz}, and for a given initial state, $\boldsymbol{\omega}(0) = (\omega_{x0}, \omega_{y0}, \omega_{z0})^T$, which determine the constants, H and T_r.

11.5.1 Stability of Rotational States

Stability is a property of an equilibrium state of a system. A stable equilibrium is the one from which a small disturbance does not produce a large departure. The disturbance can be regarded as the initial condition applied to the system of torque-free dynamics, expressed as an initial deviation of the motion variables, $(\omega_x, \omega_y, \omega_z)$, from the equilibrium state. A more precise definition of a stable equilibrium is the ability to make a trajectory, $\boldsymbol{\omega}(t)$, $t \geq 0$, remain arbitrarily close to the equilibrium state, $\boldsymbol{\omega}_e = \text{const.}$, by selecting an appropriate initial condition, $\boldsymbol{\omega}(0)$, at $t = 0$.

In a stability analysis, it is usually sufficient to study the response to a small initial deviation, because stability – being an inherent property of the equilibrium – is independent of the magnitude of the initial condition. Furthermore, since a rigid body does not have any energy dissipation mechanism, the rotational kinetic energy, T_r, is conserved in the torque-free motion of a rigid body. This implies that any deviation in the angular velocity is bounded in magnitude, because $T_r = \frac{1}{2}\boldsymbol{\omega}^T\mathbf{J}\boldsymbol{\omega} = \text{const.}$, and $H = \text{const.}$ To verify this fact, consider, for example, an equilibrium state given by $\boldsymbol{\omega}_e = n\mathbf{i}_3 = \text{const.}$ Let a small initial deviation be applied at $t = 0$, denoted by $\delta\boldsymbol{\omega}(0) = (\omega_{x0}, \omega_{y0}, 0)^T$, such that

$$\boldsymbol{\omega}(0) = n\mathbf{i}_3 + \delta\boldsymbol{\omega}(0) = \omega_{x0}\mathbf{i}_1 + \omega_{y0}\mathbf{i}_2 + n\mathbf{i}_3\,. \tag{11.53}$$

Conservation of the angular momentum implies that

$$\begin{aligned}\mathbf{H} = \mathbf{J}\boldsymbol{\omega} = \mathbf{J}\boldsymbol{\omega}(0) &= J_{xx}\omega_{x0}\mathbf{i}_1 + J_{yy}\omega_{y0}\mathbf{i}_2 + J_{zz}n\mathbf{i}_3 \\ &= J_{xx}\omega_x\mathbf{i}_1 + J_{yy}\omega_y\mathbf{i}_2 + J_{zz}\omega_z\mathbf{i}_3 = \text{const.}\,,\end{aligned} \tag{11.54}$$

with

$$H^2 = J_{xx}^2\omega_x^2 + J_{yy}^2\omega_y^2 + J_{zz}^2\omega_z^2 = J_{xx}^2\omega_{x0}^2 + J_{yy}^2\omega_{y0}^2 + J_{zz}^2n^2 = \text{const.}$$

The conservation of the rotational kinetic energy requires that

$$\begin{aligned}T_r &= \frac{1}{2}(J_{xx}\omega_{x0}^2 + J_{yy}\omega_{y0}^2 + J_{zz}n^2) \\ &= \frac{1}{2}(J_{xx}\omega_x^2 + J_{yy}\omega_y^2 + J_{zz}\omega_z^2) = \text{const.}\,,\end{aligned} \tag{11.55}$$

where the angular velocity components satisfy Eq. (11.51), and the principal axes, $(\mathbf{i}_1, \mathbf{i}_2, \mathbf{i}_3)$, satisfy the attitude kinematics equations, Eq. (11.21). The conservation of H and T_r implies that

$$\omega = |\,\boldsymbol{\omega}\,| = \sqrt{\omega_x^2 + \omega_y^2 + \omega_z^2} \leq c\,,$$

where $c > 0$ is a constant. The same information is derived from the polhodes in the angular-momentum space (Fig. 11.1) which are closed curves. Thus it is evident that while the perturbed motion involves a variation in the angular velocity components, $[\omega_x(t), \omega_y(t), \omega_z(t)]$, the magnitude of the angular velocity is bounded by a constant. However, while the perturbations remain bounded in magnitude, the rotation beginning from one equilibrium state could shift to another equilibrium state, which implies an instability of the original equilibrium. The stability analysis thus involves solving for the individual angular velocity components in a perturbed equilibrium, and investigating whether each component remains close to the particular equilibrium state. While in general Eq. (11.51) does not possess an easily expressible closed-form solution, for the special case of an axisymmetric body, such as solution is readily obtained. However, since stability analysis requires a small perturbation from an equilibrium state, an approximate first-order model of the non-linear dynamics could be employed to yield analytical results.

To investigate the stability of the state of pure spin of rate n about the principal axis oz, prior to the time, $t = 0$, when a small disturbance, ω_{x0}, ω_{y0}, with $\omega_{x0}^2 + \omega_{y0}^2 \ll n^2$ is applied, consider the following expression of the perturbed angular velocity at a subsequent time t:

$$\boldsymbol{\omega}(t) = \omega_x(t)\mathbf{i}_1 + \omega_y(t)\mathbf{i}_2 + [n + \epsilon(t)]\mathbf{i}_3 , \tag{11.56}$$

where the z-component perturbation, $\epsilon(t)$, which was initially zero ($\epsilon(0) = 0$), should remain small for stability, i.e., $| \epsilon(t) | \ll n$ for $t > 0$. If $| \epsilon(t) |$ grows with time, it indicates that the equilibrium state, $\boldsymbol{\omega}_e = n\mathbf{i}_3$, is unstable. The criterion for stability can be derived from the assumption of a small deviation from equilibrium wherein the second-order terms involving $\epsilon, \omega_x, \omega_y$ are neglected. This approximation results in the following linearized Euler's equations:

$$\begin{aligned} J_{xx}\dot{\omega}_x + n\omega_y(J_{zz} - J_{yy}) &\approx 0 \\ J_{yy}\dot{\omega}_y + n\omega_x(J_{xx} - J_{zz}) &\approx 0 \\ J_{zz}\dot{\epsilon} &\approx 0 . \end{aligned} \tag{11.57}$$

We note that the linearized motion about the z-axis is decoupled from that about the x and y-axes. However, the exact z-axis dynamics is governed by the last of Eq. (11.51), which shows that if either ω_x or ω_y grows to be large, so does ϵ, thereby indicating an unstable equilibrium. Therefore, a stability criterion can be derived from the first two equations of Eq. (11.57), which are expressed in the following vector matrix form:

$$\begin{Bmatrix} \dot{\omega}_x \\ \dot{\omega}_y \end{Bmatrix} = \begin{pmatrix} 0 & -k_1 \\ k_2 & 0 \end{pmatrix} \begin{Bmatrix} \omega_x \\ \omega_y \end{Bmatrix} , \tag{11.58}$$

where $k_1 = n\frac{(J_{zz}-J_{yy})}{J_{xx}}$, and $k_2 = n\frac{(J_{zz}-J_{xx})}{J_{yy}}$. Being in a linear, time-invariant state-space form, these approximate equations are solved using the matrix exponential as follows:

$$\begin{Bmatrix} \omega_x(t) \\ \omega_y(t) \end{Bmatrix} = e^{\mathbf{A}t} \begin{Bmatrix} \omega_{x0} \\ \omega_{y0} \end{Bmatrix} , \tag{11.59}$$

where $e^{\mathbf{A}t}$ is the matrix exponential denoting the state transition matrix (Chapter 8), and

$$\mathbf{A} = \begin{pmatrix} 0 & -k_1 \\ k_2 & 0 \end{pmatrix} . \tag{11.60}$$

The eigenvalues of **A** determine whether the ensuing motion will be bounded, and thus denote stability or instability. These are obtained as follows:

$$\det(s\mathbf{I} - \mathbf{K}) = s^2 + k_1k_2 = 0 , \tag{11.61}$$

or

$$s_{1,2} = \pm\sqrt{-k_1k_2} . \tag{11.62}$$

From the eigenvalues of **A**, it is clear that two possibilities exist: (a) $k_1k_2 < 0$, for which one eigenvalue has a positive real part, indicating an exponentially growing (unbounded) motion in both (ω_x, ω_y), or (b) $k_1k_2 > 0$, for which both the eigenvalues are imaginary, and the motion is a constant amplitude (bounded) oscillation about the equilibrium, $\omega_x = \omega_y = 0$. Therefore, for stability we must have $k_1 k_2 > 0$, which implies that either ($J_{zz} > J_{xx}$, $J_{zz} > J_{yy}$), or ($J_{zz} < J_{xx}$, $J_{zz} < J_{yy}$). This implies that the z-axis should be either the major axis or the minor axis of the spacecraft. Hence, a rigid spacecraft can be *spin stabilized* (i.e., rendered stable) by spinning it at a constant rate n about its minor or major axis.

11.6 Precession and Nutation

In summary of the discussion in Section 11.5, a stable rotation of a spacecraft about the axis oz, ($\omega_z = n$, $\omega_x = \omega_y = 0$), caused by a small perturbation from the equilibrium state due to the initial condition, $\boldsymbol{\omega}(0) = (\omega_{x0}, \omega_{y0}, n)^T$, with $\omega_{x0}^2 + \omega_{y0}^2 \ll n^2$, consists of a small, constant amplitude oscillation in the angular velocity components normal to the spin axis, $\omega_x(t), \omega_y(t)$, as well as a constant amplitude perturbation, $\epsilon(t)$, in the angular velocity about the spin axis, $\omega_z(t) = n + \epsilon(t)$. When the attitude kinematics of the principal frame, $oxyz$, is taken into account, the stable perturbed motion of the spacecraft consists of *precession* and *nutation*. Precession is defined to be the rotation of the angular velocity, $\boldsymbol{\omega}$, about the spin axis, oz, and is therefore described by the variation of the angular velocity components normal to the spin axis, $\omega_x(t), \omega_y(t)$. Nutation is the small amplitude oscillation of the spin axis with time in the inertial space caused by $\epsilon(t)$. In terms of the 3-1-3 Euler angles, $(\psi)_3, (\theta)_1, (\phi)_3$ (depicted in Fig. 11.2), the attitude kinematics is described by Eq. (11.37), which for the perturbed stable rotation about oz are expressed as follows:

$$\begin{Bmatrix} \dot{\psi} \\ \dot{\theta} \\ \dot{\phi} \end{Bmatrix} = \frac{1}{\sin\theta} \begin{pmatrix} \sin\phi & \cos\phi & 0 \\ \cos\phi\sin\theta & -\sin\phi\sin\theta & 0 \\ -\sin\phi\cos\theta & -\cos\phi\cos\theta & \sin\theta \end{pmatrix} \begin{Bmatrix} \omega_x \\ \omega_y \\ n+\epsilon \end{Bmatrix}. \qquad (11.63)$$

Let the inertial space, $(\mathbf{I}_1, \mathbf{I}_2, \mathbf{I}_3)$, be fixed by the constant angular momentum vector such that $\mathbf{I}_3 = \mathbf{H}/H$. Thus the angle θ is the angle between the spin axis, $\mathbf{i}_3$, and the angular-momentum

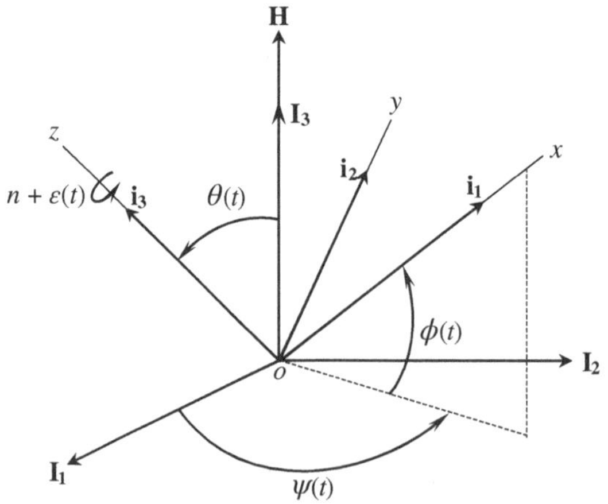

Figure 11.2 Attitude kinematics of a rigid spacecraft in a torque-free state in terms of the Euler-angle representation, $(\psi)_3, (\theta)_1, (\phi)_3$.

vector (Fig. 11.2), given by

$$\cos\theta = \mathbf{i}_3 \cdot \frac{\mathbf{H}}{H} = \frac{H_z}{H} = \frac{J_{zz}(n+\epsilon)}{\sqrt{J_{xx}^2\omega_{x0}^2 + J_{yy}^2\omega_{y0}^2 + J_{zz}^2 n^2}}, \tag{11.64}$$

and the variation of θ with time caused by the stable perturbation, $\epsilon(t)$, can be derived by differentiating Eq. (11.64), resulting in

$$-\dot{\theta}\sin\theta = \frac{\dot{H}_z}{H} = \frac{J_{zz}\dot{\epsilon}}{\sqrt{J_{xx}^2\omega_{x0}^2 + J_{yy}^2\omega_{y0}^2 + J_{zz}^2 n^2}}, \tag{11.65}$$

or

$$\dot{\theta} = \frac{-\dot{H}_z}{H\sin\theta} = \frac{-J_{zz}\dot{\epsilon}}{\sqrt{H^2 - H_z^2}} = \frac{-J_{zz}\dot{\epsilon}}{\sqrt{J_{xx}^2\omega_{x0}^2 + J_{yy}^2\omega_{y0}^2 + J_{zz}^2[n^2 - (n+\epsilon)^2]}}. \tag{11.66}$$

The solution for the *nutation rate*, $\dot{\theta}$, requires an integration of Eq. (11.66), where $\epsilon(t) = \omega_z(t) - n$ is obtained from the integration of Euler's equations, Eq. (11.51). The term "nutation" is derived from the Latin *nutare*, which means *to nod*, and describes the nodding motion, $\dot{\theta}(t)$, of the spin axis during a stable rotation.

Precession of the rigid spacecraft in a stable rotation is described by the variation of the Euler angles, $\psi(t)$, $\phi(t)$ (Fig. 11.2), and the rate of precession is given by

$$\omega_{xy} = \sqrt{\omega_x^2 + \omega_y^2} = \sqrt{\dot{\theta}^2 + \dot{\psi}^2\sin^2\theta}, \tag{11.67}$$

where the 3-1-3 rotational kinematics, Eq. (11.36), has been substituted. The rate $\dot{\psi}$, is given by the first of Eq. (11.63) to be the following:

$$\dot{\psi} = \frac{1}{\sin\theta}(\omega_x \sin\phi + \omega_y\cos\phi). \tag{11.68}$$

An alternative expression for $\dot{\psi}$ is the following, which can be used even in the case of a singularity ($\sin\theta = 0$) of the 3-1-3 Euler angles:

$$\dot{\psi} = H\frac{J_{xx}\omega_x^2 + J_{yy}\omega_y^2}{J_{xx}^2\omega_x^2 + J_{yy}^2\omega_y^2}. \tag{11.69}$$

Deriving the solution for precession and nutation rates requires the time integration of the coupled, non-linear ordinary differential equations, Eqs. (11.51) and (11.63). The procedures for deriving such a solution will be described in Section 11.8.

11.7 Semi-Rigid Spacecraft

The stability analysis detailed in Section 11.6 has revealed an important result, namely that it is possible to have a stable rotational state of a rigid spacecraft by spinning it about either the major axis or the minor axis. This conclusion held until the middle of the twentieth century, when the first US satellite – the *Explorer 1* – was launched into a low-Earth orbit. This spacecraft was spin stabilized by spinning it about the minor axis. However, soon it was discovered that the satellite had switched axes by first tumbling into a multi-axis spin, and then settling into a spin about the major axis! This was an entirely unexpected result, and led to a loss of communication with the satellite because its antennae were now spinning wildly rather than maintaining a desired orientation towards the ground.

To explain the *Explorer 1* behaviour, consider the fact that a spacecraft is actually a semi-rigid body. Apart from the presence of flexible structural components, such as communications antennae and solar arrays, a spacecraft can have liquid propellants inside the fuel tanks, and can be equipped with rotors for attitude stabilization and control (to be discussed later). All of these make an actual spacecraft only a semi-rigid body, wherein some parts can have relative motion with respect to one another, creating friction and consequently a dissipation of the rotational kinetic energy. Such energy dissipation mechanisms come into effect when the lowest possible rotational energy state is not reached. If such a state can be determined, then that would be the true equilibrium state of a spinning spacecraft.

Consider a spacecraft with $J_{zz} > J_{yy} > J_{xx}$. Let it be spinning initially about the minor axis (x-axis) with a constant rate n. The rotational kinetic energy in this state is given by $T_x = \frac{1}{2}J_{xx}n^2$, whereas the constant angular momentum is $H = J_{xx}n$. Comparing this state with that of a spin about the major axis, with the same angular momentum, H, but of rotational kinetic energy, $T_z = \frac{1}{2}J_{zz}\omega_z^2$, we have

$$\omega_z = \frac{J_{xx}n}{J_{zz}}\ ; \quad T_z = \frac{1}{2}J_{zz}\left(\frac{J_{xx}n}{J_{zz}}\right)^2 = \frac{1}{2}J_{xx}n^2\left(\frac{J_{xx}}{J_{zz}}\right) < T_x\,.$$

Therefore, for a given angular momentum, a spin about the major axis has a smaller rotational kinetic energy than that about the minor axis. A similar analysis for the intermediate axis, y, reveals the following:

$$\omega_y = \frac{J_{xx}n}{J_{yy}}\ ; \quad T_y = \frac{1}{2}J_{yy}\left(\frac{J_{xx}n}{J_{yy}}\right)^2 = \frac{1}{2}J_{xx}n^2\left(\frac{J_{xx}}{J_{yy}}\right) < T_x\,,$$

$$\omega_z = \frac{J_{yy}n}{J_{zz}}\ ; \quad T_z = \frac{1}{2}J_{zz}\left(\frac{J_{yy}n}{J_{zz}}\right)^2 = \frac{1}{2}J_{yy}n^2\left(\frac{J_{yy}}{J_{zz}}\right) < T_y\,.$$

Therefore we have $T_x > T_y > T_z$, or the spin about the major axis has the lowest possible rotational kinetic energy, for a given angular momentum. This analysis reveals that a spin beginning from either the minor axis or the intermediate axis is ultimately transformed into a spin about the major axis, because it has the smallest possible rotational kinetic energy.

The torque-free motion of a semi-rigid spacecraft can be graphically depicted by a family of polhodes which are the intersection of a time-varying energy ellipsoid with the unit angular-momentum sphere, until T_r reaches its minimum possible value corresponding to a pure spin about the major axis. The continuous variation of the polhodes on the $\mathbf{H}$ = const. sphere ends with the boundary polhode, which is a closed curve corresponding to $\omega_z = H/J_{zz}$ and $T_r = \frac{1}{2}H^2/J_{zz}$.

11.7.1 Dual-Spin Stability

In summary of the foregoing discussion, a spin about the major axis is the only stable equilibrium for a semi-rigid body. This is validated by the natural bodies such as the planets and their moons having the largest principal moment of inertia, J_{zz}, about the spin axis *oz*. If the spin axis, *oz*, with the largest moment of inertia, J_{zz}, is also the axis of symmetry, then the body is said to be *oblate*. If the axis of symmetry, *oz*, has the smallest moment of inertia, J_{zz}, i.e., the axis of symmetry is the minor axis, then the body is said to be *prolate*. A prolate, semi-rigid spacecraft which is spinning about its axis of symmtery is in an unstable equilibrium, and soon the equilibrium will shift to spin about a major axis normal to the axis of symmetry (called a *lateral axis*) due to the internal energy dissipation.

Many spacecraft, such as telecommunications satellites, are designed to be prolate in shape, often because a prolate spacecraft fits neatly into the aerodynamically efficient payload bay of the launch vehicle. To stabilize such a spacecraft about its axis of symmetry, a rotor is used in a *dual-spin* configuration. Consider a prolate spacecraft with a large rotor about its axis of symmetry, *oz* (called the *longitudinal axis*), and a platform on which a communications payload is mounted (Fig. 11.3). The platform is required to spin at a very low rate, which is dictated by the rate of rotation of the ground relative to the orbit, ω_p, such that the communications antennae are always pointed towards the receiving station. The net angular momentum of the dual-spin configuration in the presence of a lateral disturbance, (ω_x, ω_y), is the following:

$$\mathbf{H} = [J_p\omega_p + J_r(\omega_p + \omega_r)]\mathbf{i}_3 + J_{xy}(\omega_x\mathbf{i}_1 + \omega_y\mathbf{i}_2)\,, \tag{11.70}$$

where J_p is the moment of inertia of the platform about the spin axis, J_r is the moment of inertia of the rotor about the spin axis, and J_{xy} is the moment of inertia of the total system (platform and rotor) about a lateral (major) axis. The net rotational kinetic energy of the system is given by

$$T = \frac{1}{2}(J_p + J_r)\omega_p^2 + \frac{1}{2}J_r\omega_r^2 + J_r\omega_r\omega_p + \frac{1}{2}J_{xy}\omega_{xy}^2\,, \tag{11.71}$$

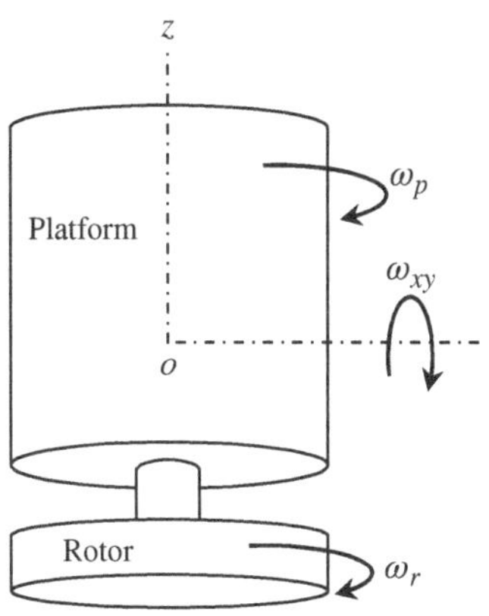

Figure 11.3 A dual-spin spacecraft.

where $\omega_{xy}^2 = \omega_x^2 + \omega_y^2$. Although the net angular momentum is conserved, the net rotational kinetic energy varies with time due to the internal energy dissipation caused by the friction of the bearing between the platform and the rotor, and perhaps also due to other mechanisms such as the sloshing of propellants and the flexing of appendages. The dissipation of rotational kinetic energy for the platform is different from that of the rotor; hence each body must be modelled as a separate rigid body with different frictional torques acting upon it. The rate of change of total rotational kinetic energy is given by

$$\dot{T} = J_p\omega_p\dot{\omega}_p + J_r(\omega_p + \omega_r)(\dot{\omega}_p + \dot{\omega}_r) + J_{xy}\omega_{xy}\dot{\omega}_{xy} . \tag{11.72}$$

Since the angular momentum is constant, we have the following from Eq. (11.70):

$$H\dot{H} = [J_p\omega_p + J_r(\omega_p + \omega_r)][J_p\dot{\omega}_p + J_r(\dot{\omega}_p + \dot{\omega}_r)] + J_{xy}^2\omega_{xy}\dot{\omega}_{xy} = 0 , \tag{11.73}$$

from which the term pertaining to the rate of change of kinetic energy by precession can be calculated as

$$J_{xy}\omega_{xy}\dot{\omega}_{xy} = -\frac{1}{J_{xy}}[J_p\omega_p + J_r(\omega_p + \omega_r)][J_p\dot{\omega}_p + J_r(\dot{\omega}_p + \dot{\omega}_r)] . \tag{11.74}$$

By substituting Eq. (11.74) into Eq. (11.72) we have

$$\dot{T} = \dot{T}_p + \dot{T}_r , \tag{11.75}$$

where $\dot{T}_p$ and $\dot{T}_r$ represent the rates of change of rotational kinetic energy of the platform and rotor, respectively, given by

$$\dot{T}_p = J_p[\omega_p - \frac{1}{J_{xy}}\{J_p\omega_p + J_r(\omega_p + \omega_r)\}]\dot{\omega}_p , \tag{11.76}$$

and

$$\dot{T}_r = J_r[\omega_r + \omega_p - \frac{1}{J_{xy}}\{J_p\omega_p + J_r(\omega_p + \omega_r)\}](\dot{\omega}_p + \dot{\omega}_r) . \tag{11.77}$$

Both $\dot{T}_p$ and $\dot{T}_r$ are negative due to internal energy dissipation. However, the stability of the motion depends upon the relative magnitudes of these dissipation rates, in order that the kinetic energy of precession is reduced to zero. Therefore, for stability it is crucial that the rotor provides an *energy sink* for the precessional motion, ω_{xy}, which requires that

$$J_{xy}\omega_{xy}\dot{\omega}_{xy} = (\dot{T}_p - J_p\omega_p\dot{\omega}_p) + [\dot{T}_r - J_r(\omega_p + \omega_r)(\dot{\omega}_p + \dot{\omega}_r)] < 0 , \tag{11.78}$$

or

$$-J_{xy}\omega_{xy}\dot{\omega}_{xy} = [J_p\omega_p + J_r(\omega_p + \omega_r)]\left[\frac{J_p}{J_{xy}}\dot{\omega}_p + \frac{J_r}{J_{xy}}(\dot{\omega}_p + \dot{\omega}_r)\right] > 0 , \tag{11.79}$$

which leads to the following requirement:

$$J_p\dot{\omega}_p + J_r(\dot{\omega}_p + \dot{\omega}_r) > 0 \tag{11.80}$$

because $\omega_p > 0$ and $\omega_r > 0$. Since ω_p is small in magnitude, the second-order terms involving it and its time derivative are neglected, leading to the approximations

$$\dot{T}_p \approx -\frac{J_p J_r}{J_{xy}}(\omega_p + \omega_r)\dot{\omega}_p$$

$$\dot{T}_r \approx J_r\left(1 - \frac{J_r}{J_{xy}}\right)(\omega_p + \omega_r)(\dot{\omega}_p + \dot{\omega}_r)\,. \tag{11.81}$$

Equation (11.81) must be satisfied by negative values of both $\dot{T}_p$ and $\dot{T}_r$. Therefore, if the rotor is an oblate body ($J_{xy} < J_r$), it follows from Eq. (11.81) that $\dot{\omega}_p > 0$ and $\dot{\omega}_r > 0$. For a prolate rotor ($J_{xy} > J_r$), and $\dot{\omega}_r < 0$. Hence, the platform and an oblate rotor speed up, while a prolate rotor slows down in the presence of the lateral disturbance, ω_{xy}. Hence the stability requirement of Eq. (11.80) is unconditionally satisfied by an oblate rotor. However, in a practical case, the rotor is usually prolate, for which stability requires that

$$(J_p + J_r)\dot{\omega}_p > -J_r\dot{\omega}_r\,. \tag{11.82}$$

In terms of the energy dissipation terms, the stability requirement for a prolate rotor is obtained by eliminating $\dot{\omega}_p$ and $\dot{\omega}_r$ from Eqs. (11.81) and (11.82), and making the approximation $\omega_p << \omega_r$:

$$-\dot{T}_p > -\dot{T}_r\frac{J_r}{J_{xy} - J_r}\,. \tag{11.83}$$

Therefore, for a stable configuration of a prolate spacecraft with a small spin rate coupled with a prolate rotor, the platform must lose energy at a greater rate than the rotor. Due to the friction between the rotor and the platform, the rotor's spin rate decreases, and the platform speeds up, even in the absence of a lateral disturbance. If this situation is uncorrected, both the rotor and the platform will eventually be spinning at the same rate, which leads to an unstable configuration (prolate semi-rigid body spinning about its minor axis). To prevent this from happening, a motor is used to continually apply a small torque to the rotor bearing.

In summary, a prolate spacecraft is unconditionally stabilized about its minor spin axis by an oblate rotor. However, if a prolate rotor is to be used for the same purpose, the spacecraft must lose its rotational kinetic energy at a greater rate than that of the rotor.

Most communications satellites employ a dual-spin configuration for rotational stability. A recent interesting application of the dual-spin stabilization was by the *Galileo* interplanetary spacecraft of NASA. This spacecraft had an inertial (non-spinning) platform for carrying out communications with Earth during its six-year voyage to Jupiter, while its rotor, on which several navigational and scientific sensors were mounted, rotated at three revolutions per minute.

11.8 Solution to Torque-Free Euler's Equations

Solving the Euler's equations for the torque-free condition requires a numerical solution procedure due to their non-linear, coupled character. This can be derived either in a closed form as given by Jacobi, or by a direct numerical integration in time by a Runge-Kutta method. However, before proceeding to solve the general, three-axis Euler's equations of an asymmetric spacecraft, let us first consider the exact analytical solution for the case of an axisymmetric spacecraft.

11.8.1 Axisymmetric Spacecraft

When the spacecraft possesses an axis of symmetry, the principal moments of inertia about two axes become identical, say, $J_{xx} = J_{yy}$, and the Euler's equations are simplified as follows:

$$\begin{aligned} J_{xx}\dot{\omega}_x + \omega_y\omega_z(J_{zz} - J_{xx}) &= 0 \\ J_{xx}\dot{\omega}_y + \omega_x\omega_z(J_{xx} - J_{zz}) &= 0 \\ J_{zz}\dot{\omega}_z &= 0\,, \end{aligned} \tag{11.84}$$

which denotes a rotation about the axis of symmetry, *oz*, called the *longitudinal axis*. It is clear from Eq. (11.84) that the spacecraft is in a state of equilibrium whenever $\omega_x = \omega_y = 0$, and $\omega_z = n =$ const., called a state of *pure spin* about the axis of symmetry. It is also evident from the last of Eq. (11.84) that we have $\dot{\omega}_z = 0$, or $\omega_z = n$, irrespective of the magnitudes of ω_x, ω_y. Assume that the spacecraft is in a state of pure spin, when a disturbance, ω_{x0}, ω_{y0}, is applied at the time $t = 0$. Even after the perturbation is applied, the spin rate about the axis of symmetry remains constant, $\omega_z(t) = n$, $t \geq 0$, which, compared to the dynamics of an asymmetric spacecraft, implies that $\epsilon(t) = 0$. Hence the angular momentum about the spin axis, $H_z = J_{zz}\omega_z = J_z zn$, is conserved. Because the net angular momentum magnitude, H, is also conserved in the torque-free motion, it implies that the net angular momentum of rotation about the other two axes must also be conserved.

The effect of the perturbation is thus described by the solution of the first two equations of Eq. (11.84), which are written in the following vector matrix form:

$$\begin{Bmatrix} \dot{\omega}_x \\ \dot{\omega}_y \end{Bmatrix} = \begin{pmatrix} 0 & -k \\ k & 0 \end{pmatrix} \begin{Bmatrix} \omega_x \\ \omega_y \end{Bmatrix}, \tag{11.85}$$

where

$$k = n\frac{(J_{zz} - J_{xx})}{J_{xx}}\,.$$

The magnitude, $|\, k \,|$, is the natural frequency of precession.

The equations in Eq. (11.85) are linear, time-invariant state-equations, whose solution subject to the initial condition, ω_{x0}, ω_{y0}, at $t = 0$, is expressed as follows:

$$\begin{Bmatrix} \omega_x(t) \\ \omega_y(t) \end{Bmatrix} = e^{\mathbf{K}t} \begin{Bmatrix} \omega_{x0} \\ \omega_{y0} \end{Bmatrix}, \tag{11.86}$$

where $e^{\mathbf{K}t}$ is the matrix exponential (Chapter 7), and

$$\mathbf{K} = \begin{pmatrix} 0 & -k \\ k & 0 \end{pmatrix}. \tag{11.87}$$

Using one of the methods given in Chapter 7, the matrix exponential is calculated by the inverse Laplace transform as follows:

$$e^{\mathbf{K}t} = \mathscr{L}^{-1}(s\mathbf{I} - \mathbf{K})^{-1} = \begin{bmatrix} \cos(kt) & -\sin(kt) \\ \sin(kt) & \cos(kt) \end{bmatrix}. \tag{11.88}$$

Therefore, the solution is given by

$$\omega_x(t) = \omega_{x0}\cos(kt) - \omega_{y0}\sin(kt)$$
$$\omega_y(t) = \omega_{x0}\sin(kt) + \omega_{y0}\cos(kt)\,. \qquad (11.89)$$

The set in Eq. (11.89) implies that the rotational motion of an axisymmetric, rigid spacecraft, disturbed from the equilibrium state of pure spin about the longitudinal axis by a disturbance, ω_{x0}, ω_{y0}, is oscillatory in the (oxy) plane (called the *lateral plane*), while the spin rate, $\omega_z = n$, remains unaffected. Such a motion is referred to as a *coning motion* of the disturbed body about the axis of symmetry, oz, because both the angular velocity vector, $\boldsymbol{\omega}$, and the spin axis, oz, describe separate cones about the constant angular momentum vector, $\mathbf{H} = \mathbf{J}\boldsymbol{\omega}$, with their apex at the origin, o, as depicted in Fig. 11.4.

This important characteristic of the solution given by Eq. (11.89) is the following constant amplitude of precession:

$$\omega_{xy}^2 = \omega_x^2 + \omega_y^2 = \omega_{x0}^2 + \omega_{y0}^2 = \text{const.}\,, \qquad (11.90)$$

which is responsible for the coning motion. Since precession is a constant amplitude oscillation, whose magnitude is the same as that of the applied disturbance, it follows that the motion of a rigid spacecraft about its axis of symmetry is unconditionally stable. Figure 11.4 shows the geometry of precession motion, wherein the angular velocity, $\boldsymbol{\omega}$, makes a constant angle, $\alpha = \tan^{-1}(\omega_{xy}/n)$ with the axis of symmetry, oz. Furthermore, the fixed angular momentum, $\mathbf{H} = J_{xx}(\omega_x\mathbf{i} + \omega_y\mathbf{j}) + J_{zz}n\mathbf{k}$, makes a constant angle, $\theta = \tan^{-1}(J_{xx}\omega_{xy}/J_{zz}n)$, with the axis of symmetry. As seen previously, the angle, θ, is termed the nutation angle in the general case of an asymmetric body, wherein it can vary with time in a nodding motion (or nutation) of the spin axis. However, in the present case of an axisymmetric body, the angle, θ, is constant. As shown in Fig. 11.4, the angular velocity vector describes a cone of semi-vertex angle α, called the *body cone*, about the axis of symmetry, and a cone of semi-vertex angle $\theta - \alpha$, called the *space cone*, about the angular momentum vector. In drawing Fig. 11.4, it is assumed that $J_{xx} > J_{zz}$; that is, $\theta > \alpha$, which implies that the spin stabilization takes place about the minor axis. While such a

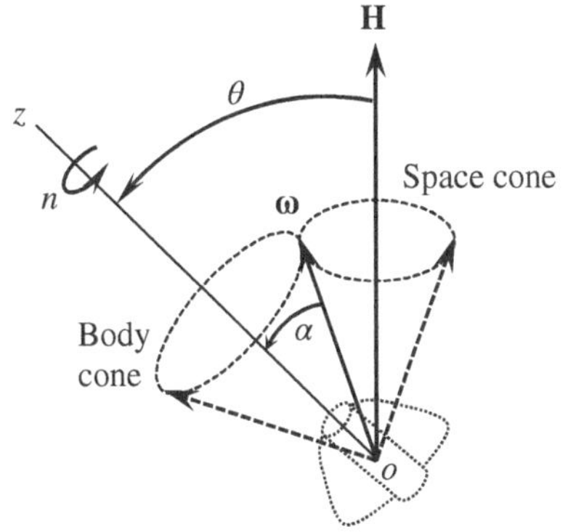

Figure 11.4 Precession of an axisymmetric spacecraft.

motion is stable for a rigid body, for an actual (semi-rigid) spacecraft, the only stable equilibrium is a spin about the major axis (i.e., for $J_{zz} > J_{xx}$; hence $\theta < \alpha$), which corresponds to the state of the smallest possible rotational kinetic energy, T_r, for a given angular momentum, H.

The attitude kinematics of an axisymmetric spacecraft, spin stabilized about the longitudinal axis, is described by the $(\psi)_3, (\theta)_1, (\phi)_3$ Euler angles relative to an inertial reference frame, $(\mathbf{I}_1, \mathbf{I}_2, \mathbf{I}_3)$, with the axis $\mathbf{I}_3$ pointed along $\mathbf{H}$. From Fig. 11.2 with $\epsilon = 0$, we have

$$\begin{aligned}\sin\theta &= \frac{J_{xx}\omega_{xy}}{H} = \frac{J_{xx}\omega_{xy}}{\sqrt{J_{xx}^2\omega_{xy}^2 + J_{zz}^2 n^2}} \\ \cos\theta &= \frac{J_{zz}n}{H} = \frac{J_{zz}n}{\sqrt{J_{xx}^2\omega_{xy}^2 + J_{zz}^2 n^2}} .\end{aligned} \tag{11.91}$$

The kinematical equations for the $(\psi)_3, (\theta)_1, (\phi)_3$ Euler angles are, in this case, the following:

$$\begin{Bmatrix} \dot\psi \\ \dot\theta \\ \dot\phi \end{Bmatrix} = \frac{1}{\sin\theta}\begin{pmatrix} \sin\phi & \cos\phi & 0 \\ \cos\phi\sin\theta & -\sin\phi\sin\theta & 0 \\ -\sin\phi\cos\theta & -\cos\phi\cos\theta & \sin\theta \end{pmatrix}\begin{Bmatrix} \omega_x \\ \omega_y \\ n \end{Bmatrix}, \tag{11.92}$$

where the fact of constant θ is substituted to yield

$$\omega_x \cos\phi = \omega_y \sin\phi , \tag{11.93}$$

or $\omega_{xy} = \omega_x / \sin\phi = \omega_y / \cos\phi$, resulting in

$$\begin{aligned}\dot\psi &= \frac{\omega_{xy}}{\sin\theta} \\ \dot\theta &= 0 \\ \dot\phi &= n - \frac{\omega_{xy}}{\tan\theta} .\end{aligned} \tag{11.94}$$

Since both θ and ω_{xy} are constants, the angular rates $\dot\psi$ and $\dot\phi$ are also constants, whose alternative expressions are the following:

$$\begin{aligned}\dot\psi &= \frac{H}{J_{xx}} = \frac{\sqrt{J_{xx}^2\omega_{xy}^2 + J_{zz}^2 n^2}}{J_{xx}} \\ \dot\theta &= 0 \\ \dot\phi &= n\left(1 - \frac{J_{zz}}{J_{xx}}\right) = -k .\end{aligned} \tag{11.95}$$

Hence, the angular rate $\dot\psi \sin\theta$ equals the constant precession amplitude, ω_{xy}, for the axisymmetric rigid body, $J_{xx} = J_{yy}$. The constant angular rate $\dot\psi = H/J_{xx}$ is termed the *precession rate*. Furthermore, we have $\mid \dot\phi \mid = \mid k \mid$, which implies that the magnitude $\mid \dot\phi \mid$ equals the natural frequency of precession, and is called the *inertial spin rate*.

If $J_{xx} > J_{zz}$, the axisymmetric body is a prolate body, and $\dot{\psi}$ has the same sign as that of $\dot{\phi}$. For the case of an oblate body ($J_{xx} < J_{zz}$), the angular rates $\dot{\psi}$ and $\dot{\phi}$ have opposite signs.

The solution for the Euler angles is obtained in a closed form by the integration of Eq. (11.92) – with the initial orientation at $t = 0$ specified as ψ_0, θ_0, ϕ_0 – to be the following:

$$\begin{aligned}
\psi &= \psi_0 + \frac{\sqrt{J_{xx}^2\omega_{xy}^2 + J_{zz}^2 n^2}}{J_{xx}}\, t \\
\theta &= \theta_0 \\
\phi &= \phi_0 - kt = \phi_0 - n\left(1 - \frac{J_{zz}}{J_{xx}}\right) t\,.
\end{aligned} \tag{11.96}$$

The angles ψ and ϕ thus vary linearly with time due to a constant precession rate, ω_{xy}.

11.8.2 Jacobian Elliptic Functions

Jacobi found a solution to the torque-free Euler's equations for an asymmetric rigid body in a closed-form as follows:

$$\begin{aligned}
\omega_x &= \omega_{xm}\ \mathrm{cn}(\Phi; m) \\
\omega_y &= -\omega_{ym}\ \mathrm{sn}(\Phi; m) \\
\omega_z &= \omega_{zm}\ \mathrm{dn}(\Phi; m)\,,
\end{aligned} \tag{11.97}$$

where

$$\omega_{xm} = \left[\frac{H^2 - 2I_{zz}T_r}{J_{xx}(J_{xx} - J_{zz})}\right]^{1/2}$$

$$\omega_{ym} = \left[\frac{H^2 - 2I_{zz}T_r}{J_{yy}(J_{yy} - J_{zz})}\right]^{1/2} \tag{11.98}$$

$$\omega_{zm} = \left[\frac{H^2 - 2I_{xx}T_r}{J_{zz}(J_{zz} - J_{xx})}\right]^{1/2} \tag{11.99}$$

are the maximum values of the respective angular velocity components, cn($\Phi; m$), sn($\Phi; m$), and dn($\Phi; m$) are the *Jacobian elliptic functions* with argument Φ and parameter m, with

$$\Phi = \omega_p(t - t_1)\,, \tag{11.100}$$

$$m = \frac{(J_{xx} - J_{yy})(H^2 - 2J_{zz}T_r)}{(J_{zz} - J_{yy})(H^2 - 2J_{xx}T_r)}\,, \tag{11.101}$$

and

$$\omega_p = -\left[\frac{(J_{zz} - J_{xx})(H^2 - 2J_{xx}T_r)}{J_{xx}J_{yy}J_{zz}}\right]^{1/2} = -\left[\frac{(J_{zz} - J_{xx})(J_{zz} - J_{yy})}{J_{xx}J_{yy}}\right]^{1/2} \omega_{zm} \tag{11.102}$$

with $J_{zz} > J_{yy} > J_{xx}$. Here the parameter, m, always falls between 0 and 1, and t_1 is the time when ω_x achieves its maximum value, $\omega_x(t_1) = \omega_{xm}$. The Jacobian elliptic functions and their properties are given in Appendix B. For $m = 0$, which is the case for an axisymmetric body ($J_{xx} = J_{yy}$), the Jacobian elliptic functions reduce to trigonometric functions.

A more convenient expression for the angular velocity components is derived in terms of the initial angular velocity, $\boldsymbol{\omega}(0) = (\omega_{x0}, \omega_{y0}, \omega_{z0})^T$, by substituting the addition laws of the Jacobi elliptic functions (Appendix B) into Eq. (11.97), resulting in the following:

$$\omega_x = \frac{1}{D}\left[\omega_{x0}\ \mathrm{cn}(\omega_p t; m) + \mathrm{sn}(\omega_p t; m)\mathrm{dn}(\omega_p t; m)\frac{\nu\omega_{y0}\omega_{z0}}{\omega_{zm}}\right], \tag{11.103}$$

$$\omega_y = \frac{1}{D}\left[\omega_{y0}\ \mathrm{cn}(\omega_p t; m)\mathrm{dn}(\omega_p t; m) - \mathrm{sn}(\omega_p t; m)\frac{\omega_{x0}\omega_{z0}}{\nu\omega_{zm}}\right], \tag{11.104}$$

$$\omega_z = \frac{1}{D}\left[\omega_{z0}\ \mathrm{dn}(\omega_p t; m) + \mu^2\ \mathrm{sn}(\omega_p t; m)\mathrm{cn}(\omega_p t; m)\frac{\omega_{x0}\omega_{y0}}{\nu\omega_{zm}}\right], \tag{11.105}$$

where

$$D = 1 - \left[\frac{\mu\omega_{y0}\ \mathrm{sn}(\omega_p t; m)}{\omega_{zm}}\right]^2, \tag{11.106}$$

$$\mu = \left[\frac{J_{yy}(J_{yy} - J_{xx})}{J_{zz}(J_{zz} - J_{xx})}\right]^{1/2}, \tag{11.107}$$

$$\nu = \left[\frac{J_{yy}(J_{yy} - J_{zz})}{J_{xx}(J_{xx} - J_{zz})}\right]^{1/2}, \tag{11.108}$$

$$\omega_{x0} = \omega_{xm}\ \mathrm{cn}(\omega_p t_1; m), \tag{11.109}$$

$$\omega_{y0} = \omega_{ym}\ \mathrm{sn}(\omega_p t_1; m), \tag{11.110}$$

$$\omega_{z0} = \omega_{zm}\ \mathrm{dn}(\omega_p t_1; m), \tag{11.111}$$

$$\omega_{zm} = \sqrt{\omega_{z0}^2 + \mu^2\omega_{y0}^2}. \tag{11.112}$$

11.8.3 Runge-Kutta Solution

An alternative to using the Jacobian elliptic functions for a closed-form solution to the torque-free Euler's equations is the direct numerical integration using a Runge-Kutta algorithm. Appendix A describes the iterative Runge-Kutta method for finding a solution to a set of first-order, ordinary differential equations (such as the Euler's equations). The following examples illustrate the use of such a procedure applied to solving Euler's equations.

Example 11.8.1 *Using the Jacobian elliptic functions, calculate the angular velocity at* $t = 30$ *s for the torque-free rotation of a rigid spacecraft with principal moments of inertia* $J_{xx} =$ 1000 kg.m^2, $J_{yy} = 2000$ kg.m^2, *and* $J_{zz} = 3500$ kg.m^2, *subject to the initial condition,* $t = 0$, $\omega_x(0) = 0.1$ rad./s, $\omega_y(0) = -0.1$ rad./s, *and* $\omega_x(0) = 0.2$ rad./s. *Compare the solution with that obtained by Runge-Kutta integration.*

The calculation steps are the following:

$$H = \sqrt{J_{xx}^2\omega_x(0)^2 + J_{yy}^2\omega_y(0)^2 + J_{zz}^2\omega_z(0)^2} = 734.846923 \text{ kg.m}^2/\text{s}$$

$$T_r = \frac{1}{2}(J_{xx}\omega_x(0)^2 + J_{yy}\omega_y(0)^2 + J_{zz}\omega_z(0)^2) = 85.0000 \text{ kg.m}^2/\text{s}^2$$

$$m = \frac{(J_{xx} - J_{yy})(H^2 - 2J_{zz}T_r)}{(J_{zz} - J_{yy})(H^2 - 2J_{xx}T_r)} = 0.0990991$$

$$\mu = \left[\frac{J_{yy}(J_{yy} - J_{xx})}{J_{zz}(J_{zz} - J_{xx})}\right]^{1/2} = 0.4780914$$

$$\nu = \left[\frac{J_{yy}(J_{yy} - J_{zz})}{J_{xx}(J_{xx} - J_{zz})}\right]^{1/2} = 1.0954451$$

$$\omega_p = -\left[\frac{(J_{zz} - J_{xx})(H^2 - 2J_{xx}T_r)}{J_{xx}J_{yy}J_{zz}}\right]^{1/2} = -0.2815772 \text{ rad./s}$$

$$\omega_{zm} = \sqrt{\omega_z(0)^2 + \mu^2\omega_y(0)^2} = 0.2056349 \text{ rad./s}$$

$$\omega_p t = -8.4473157 \text{ rad.}$$

$$cn(\omega_p t; m) = -0.3600334$$

$$sn(\omega_p t; m) = -0.9329394$$

$$dn(\omega_p t; m) = 0.9559009$$

$$D = 1 - \left[\frac{\mu\omega_y(0)sn(\omega_p t; m)}{\omega_{zm}}\right]^2 = 0.9529527$$

$$\omega_x = \frac{\omega_x(0)cn(\omega_p t; m) + \frac{\nu\omega_y(0)\omega_z(0)}{\omega_{zm}} sn(\omega_p t; m)dn(\omega_p t; m)}{D} = 0.0619246 \text{ rad./s}$$

$$\omega_y = \frac{\omega_y(0)cn(\omega_p t; m)dn(\omega_p t; m) - \frac{\omega_x(0)\omega_z(0)}{\nu\omega_{zm}} sn(\omega_p t; m)}{D} = 0.1230357 \text{ rad./s}$$

$$\omega_z = \frac{\omega_z(0)dn(\omega_p t; m) + \mu^2\frac{\omega_x(0)\omega_y(0)}{\nu\omega_{zm}} sn(\omega_p t; m)cn(\omega_p t; m)}{D} = 0.1970423 \text{ rad./s}.$$

Hence, the computed angular velocity vector at the given time is the following:

$$\boldsymbol{\omega} = \begin{Bmatrix} 0.0619246 \\ 0.1230357 \\ 0.1970423 \end{Bmatrix} \text{ rad./s.}$$

This solution is compared with a fourth-order Runge-Kutta solver (MATLAB function, ode45*) for a direct integration of the Euler's equations with a relative tolerance of* $\delta = 10^{-5}$ *in the range*

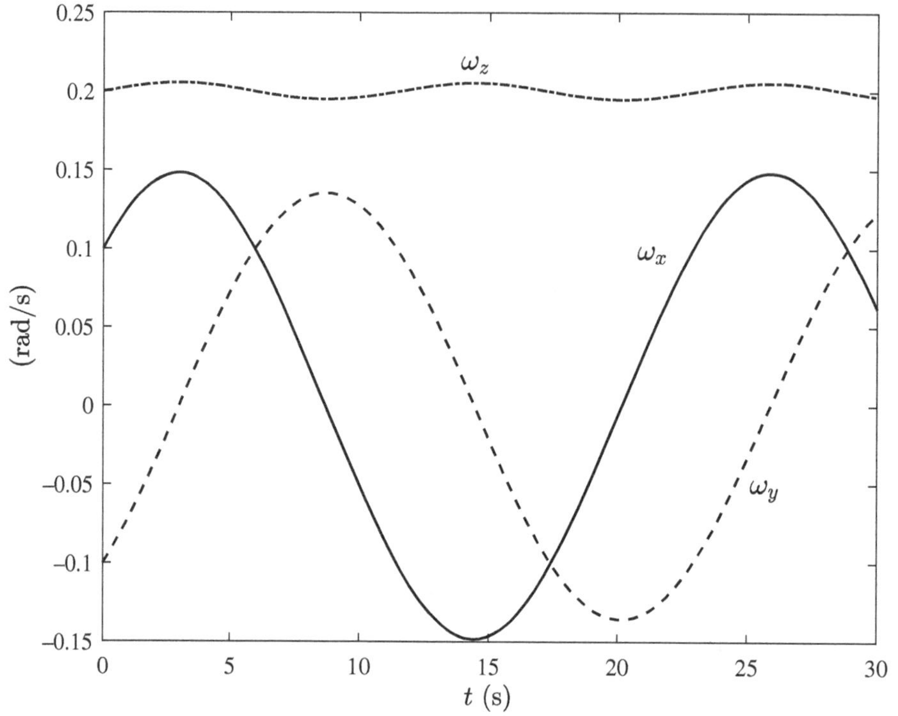

Figure 11.5 Solution of the torque-free Euler's equations of a rigid spacecraft by Runge-Kutta method.

$0 \le t \le 30$ s, and the result is plotted in Fig. 11.5. The final values of the angular velocity at $t = 30$ s are the following:

$$\boldsymbol{\omega} = \begin{Bmatrix} 0.0619234 \\ 0.1230353 \\ 0.1970423 \end{Bmatrix} \text{ rad./s},$$

which agree up to the fifth decimal place with those computed by the Jacobian elliptic functions.

Example 11.8.2 *Simulate the torque-free rotation of a rigid spacecraft with principal moments of inertia $J_{xx} = 1000$ kg.m^2, $J_{yy} = 2000$ kg.m^2, and $J_{zz} = 3500$ kg.m^2, to the following initial conditions at $t = 0$:*

(a) $\omega_x(0) = 0.001$ rad./s, $\omega_y(0) = -0.001$ rad./s, *and* $\omega_x(0) = 0.2$ rad./s.
(b) $\omega_x(0) = 0.001$ rad./s, $\omega_y(0) = 0.2$ rad./s, *and* $\omega_x(0) = -0.001$ rad./s.

The initial condition (a) corresponds to a small perturbation applied to the equilibrium state of pure spin about the major axis, oz, at the rate, $\omega_z = 0.2$ rad./s. This equilibrium state is found to be stable by the stability analysis conducted earlier; hence only a small deviation from the equilibrium is expected wherein the spacecraft's attitude undergoes precession and nutation.

The simulated initial response is computed by solving the torque-free Euler's equations with the use of a fourth-order Runge-Kutta algorithm (Appendix A), as encoded in the intrinsic MATLAB function, ode45.m. *Figures 11.6–11.8 plot the simulated response of the spacecraft to the given initial condition. Figure 11.6 shows the angular velocity components, ω_x, ω_y, on the plane normal to the spin axis, and describes the precession of the spacecraft. Neither of the two precessional angular velocity components exceed* 1.5×10^{-3} rad./s *in magnitude. The variation of the spin rate, ϵ, is shown in Fig. 11.7 to be bounded in magnitude by* $|\,\epsilon\,| \leq 6 \times 10^{-7}$ rad./s. *The nutation angle, θ, and the precession rate, ω_{xy}, are plotted in Fig. 11.8. While the nutation is seen to be bounded by* $|\,\theta(t) - 0.003\,| < 0.0083$ rad., *with a mean value of 0.003 rad., the precession rate has a mean value of 0.00142 rad./s, and its oscillation is bounded by* $|\,\omega_{xy}(t) - 0.00142\,| < 7 \times 10^{-5}$ rad./s, *thereby indicating a stable equilibrium. The nutation rate, $\dot{\theta}$, can be computed using Eq. (11.66) to be bounded by* $|\,\dot{\theta}\,| \leq 4 \times 10^{-4}$ rad./s, *which is about 3.75 times smaller in magnitude when compared to the precession rate, ω_{xy}. Instead of separately computing the angular rates, $\dot{\psi}, \dot{\theta}$, from the solution to the Euler's equations, one can directly solve the coupled set of differential equations, Eq. (11.37) and Eq. (11.51), by the Runge-Kutta method for the Euler angles, ψ, θ, ϕ, as well as the angular velocity components,* $(\omega_x, \omega_y, \omega_z)$.

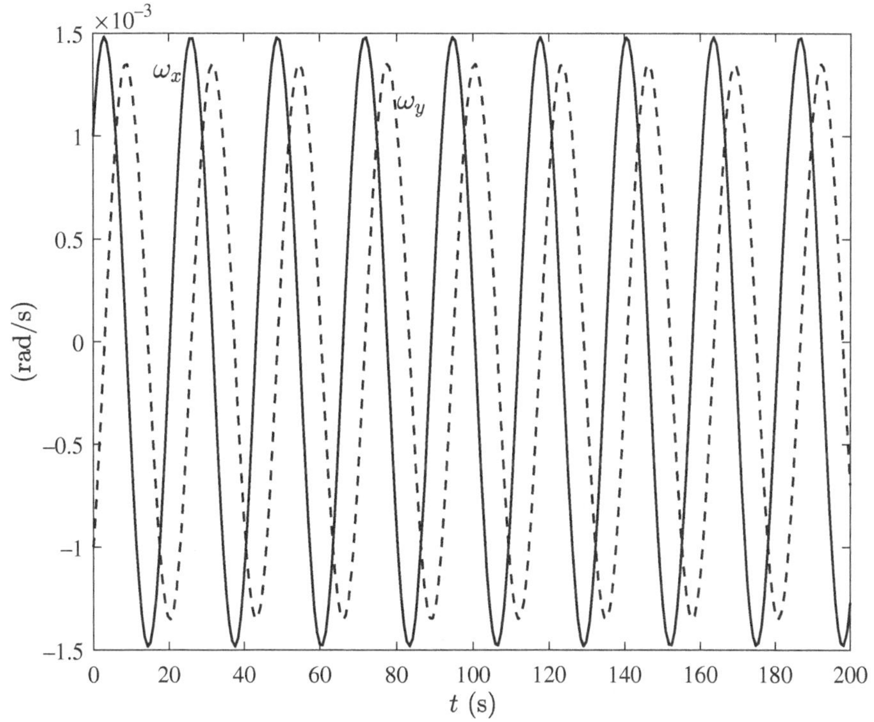

Figure 11.6 The angular-velocity components, ω_x, ω_y, in a perturbed spin about the major axis, oz, computed by Runge-Kutta method.

The initial condition (b) corresponds to a small perturbation applied to the equilibrium state of pure spin about the intermediate axis, oy, at the rate, $\omega_y = 0.2$ rad./s. *Figure 11.9 plots the*

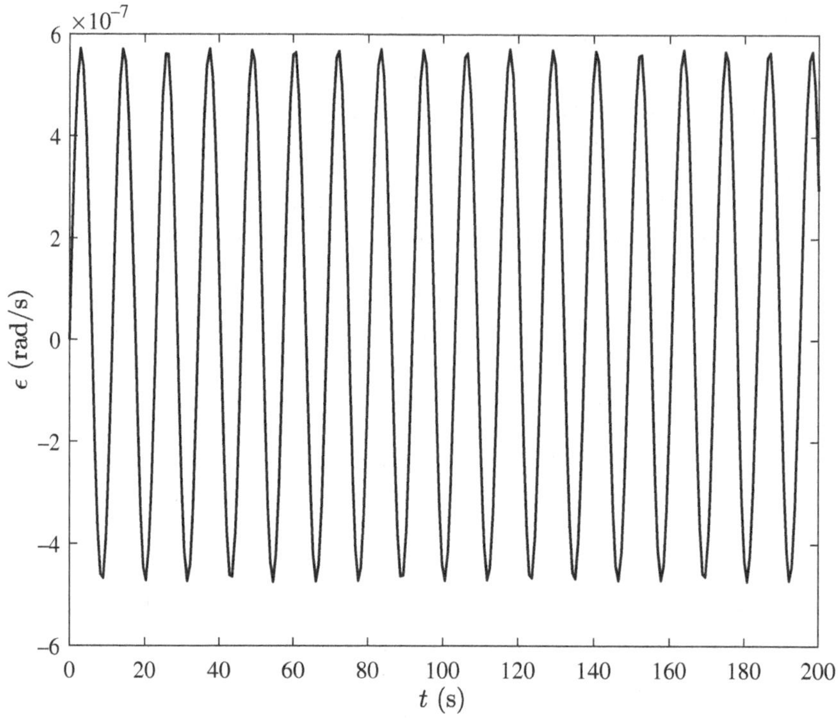

Figure 11.7 The perturbation in the spin rate, ϵ, about the major axis, oz, computed by Runge-Kutta method.

simulated angular velocity components, $\omega_x, \omega_y, \omega_z$, computed by the fourth-order Runge-Kutta method, and describe a departure from the equilibrium state. The component, ω_y, is seen to oscillate between ± 0.2 rad./s, *which indicates a* tumbling motion *of the original spin axis. The other two angular velocity components, ω_x, ω_z, which were initially very small, increase by up to 225 times, and oscillate wildly between two extreme values. There is a periodic transfer of the angular momentum from the original spin axis to the other two axes, and back. Such a motion indicates an instability of the equilibrium state, $\omega_x = \omega_z = 0$, which is expected from the stability analysis presented earlier.*

11.9 Gravity-Gradient Stabilization

A spacecraft in a low orbit around a central body experiences an appreciable torque due to the variation of gravity along the spacecraft's body-fixed axes. Such a torque is called the *gravity-gradient* torque. While small in magnitude and therefore considered to be negligible in atmospheric flight due to the much larger aerodynamic moments, the gravity-gradient torque is sufficiently large to exert a stabilizing (or destabilizing) influence over a spacecraft. The magnitude of gravity-gradient torque can be increased by increasing the length in the desired direction, which creates a longer arm for the moments due to gravity. For a large spacecraft in

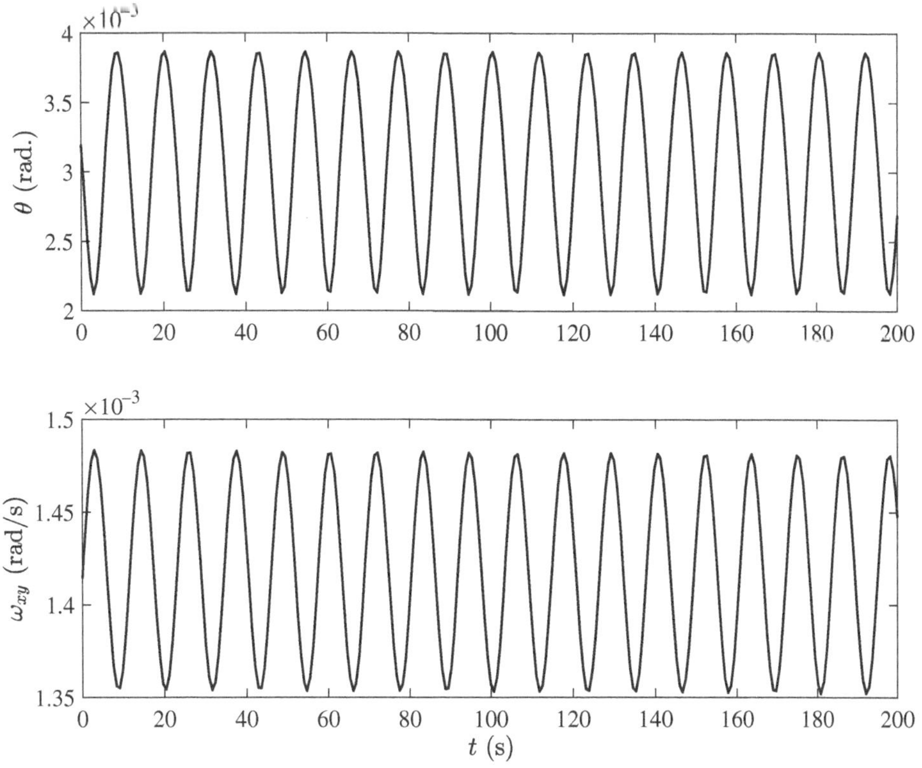

Figure 11.8 The nutation angle, θ, and the precession rate, ω_{xy}, in a perturbed spin about the major axis, oz, computed by Runge-Kutta method.

a low orbit (such as the *International Space Station*), the gravity-gradient torque is capable of overwhelming the attitude control system over a period of time if not properly compensated for. This was an important reason why the *Skylab* mission came to a premature end in the 1970s. Most natural satellites orbiting close to the central body (such as a planet or the sun) experience the phenomenon of *tidal lock*, wherein the gravity-gradient torque makes the principal frame of the orbiting body point in specific stable orientations. The Earth's moon is tidally locked to Earth in its nearly circular orbit, such that it always presents the same face towards Earth. The planet Mercury experiences a more complicated 3:2 spin-orbit resonance caused by the sun's gravity, which means that it rotates three times on its own axis for every two revolutions around the sun. The high eccentricity of Mercury's orbit makes this resonance stable at the perihelion, where the solar gravity is the strongest. This section models a spacecraft's dynamics in the presence of a gravity-gradient torque, and analyzes the stabilizing influence of the coupling between the orbital and attitude dynamics due to such a torque.

Consider a spacecraft in a low circular orbit of radius, $r =$ const., around a spherical body of gravitational constant, $\mu = GM$. Such a model is adequate for understanding gravity-gradient dynamics, because the oblateness effects have a negligible influence on the gravity-gradient

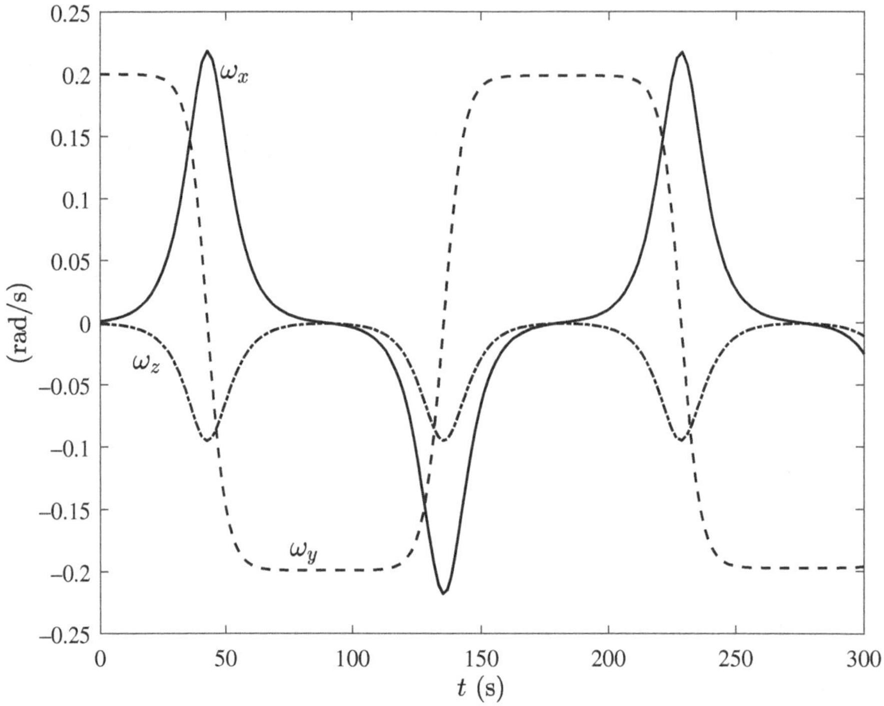

Figure 11.9 The angular-velocity components, $\omega_x, \omega_y, \omega_y$, in a perturbed spin about the intermediate axis, *oy*, computed by Runge-Kutta method.

torque. The gravity-gradient torque experienced by the craft is expressed as follows:

$$\mathbf{M}_\mathrm{g} = \int \boldsymbol{\rho} \times \mathbf{g} \mathrm{d}m\,, \tag{11.113}$$

where $\boldsymbol{\rho}$ locates an elemental mass, dm, relative to the spacecraft's centre of mass, o (Fig. 11.10). The acceleration due to gravity, $\mathbf{g}$, is given by Newton's law of gravitation for the spherical central body, and expanded using the binomial theorem as follows:

$$\begin{aligned}\mathbf{g} = -\mu \frac{\mathbf{r} + \boldsymbol{\rho}}{|\,\mathbf{r} + \boldsymbol{\rho}|^3} &= \frac{-\mu(\mathbf{r} + \boldsymbol{\rho})}{r^3\left(1 + \frac{\rho^2}{r^2}\right)^{3/2}} \\ &= -\frac{\mu(\mathbf{r} + \boldsymbol{\rho})}{r^3}\left[1 - \frac{3}{2}\left(\frac{\rho}{r}\right)^2 + \cdots\right]\,. \end{aligned} \tag{11.114}$$

As the spacecraft dimensions are quite small in comparison with the orbital radius, we have $\rho = |\,\boldsymbol{\rho}\,| \ll r$; hence the second- and higher-order terms involving the ratio $(\rho/r)^2$ in Eq. (11.114) can be neglected, resulting in the following:

$$\mathbf{g} \simeq -\frac{\mu(\mathbf{r}+\rho)}{r^3}\left[1-\frac{3}{2}\left(\frac{\rho}{r}\right)^2\right]$$
$$= -n^2\left(1-\frac{3}{2}\frac{\rho^2}{r^2}\right)(\mathbf{r}+\rho)\,, \tag{11.115}$$

where $n = \sqrt{\mu/r^3}$ is the constant frequency of the circular orbit. A substitution of Eq. (11.115) into Eq. (11.113) yields the following:

$$\mathbf{M_g} = -n^2\int\left(1-\frac{3}{2}\frac{\rho^2}{r^2}\right)\rho\times\mathbf{r}\mathrm{d}m$$
$$= -n^2\left(\int\rho\mathrm{d}m\right)\times\mathbf{r}+\frac{3n^2}{2}\int\frac{\rho^2}{r^2}(\rho\times\mathbf{r})\mathrm{d}m \tag{11.116}$$
$$= \frac{3n^2}{2r^2}\int\rho^2(\rho\times\mathbf{r})\mathrm{d}m\,, \tag{11.117}$$

where $\int\rho\mathrm{d}m = \mathbf{0}$ by virtue of o being the centre of mass of the spacecraft.

To carry out the integral in Eq. (11.116), the position vectors are resolved in the principal body-fixed frame, $(\mathbf{i},\mathbf{j},\mathbf{k})$, of the spacecraft with origin at o (Fig. 11.10) as follows:

$$\rho = x\mathbf{i}+y\mathbf{j}+z\mathbf{k}$$
$$\mathbf{r} = X\mathbf{i}+Y\mathbf{j}+Z\mathbf{k} \tag{11.118}$$
$$\rho\times\mathbf{r} = (yZ-zY)\mathbf{i}-(xZ-zX)\mathbf{j}+(xY-yX)\mathbf{k}\,.$$

The integral in Eq. (11.116) is thus the following:

$$\int\rho^2(\rho\times\mathbf{r})\mathrm{d}m = \int(x^2+y^2+z^2)\begin{Bmatrix} yZ-zY \\ zX-xZ \\ xY-yX \end{Bmatrix}\mathrm{d}m$$
$$= 2\int\begin{Bmatrix} YZ[(x^2+y^2)-(x^2+z^2)] \\ XZ[(y^2+z^2)-(x^2+y^2)] \\ XY[(x^2+z^2)-(y^2+z^2)] \end{Bmatrix}\mathrm{d}m$$
$$= 2\begin{Bmatrix} YZ(J_{zz}-J_{yy}) \\ XZ(J_{xx}-J_{zz}) \\ XY(J_{yy}-J_{xx}) \end{Bmatrix}, \tag{11.119}$$

where the following have been substituted due to the facts that o is the centre of mass and $oxyz$ is the principal body-fixed frame:

$$\int x\mathrm{d}m = \int y\mathrm{d}m = \int z\mathrm{d}m = 0$$

$$\int xy\mathrm{d}m = \int yz\mathrm{d}m = \int xz\mathrm{d}m = 0$$

$$\int (y^2 + z^2)\mathrm{d}m = J_{xx}\ ; \quad \int (x^2 + z^2)\mathrm{d}m = J_{yy}\ ; \int (x^2 + y^2)\mathrm{d}m = J_{zz}\,.$$

Therefore, we have the following expression for the gravity-gradient torque:

$$\mathbf{M_g} = M_{gx}\mathbf{i} + M_{gy}\mathbf{j} + M_{gz}\mathbf{k}\,, \tag{11.120}$$

where

$$\begin{aligned} M_{gx} &= \frac{3n^2}{r^2} YZ(J_{zz} - J_{yy}) \\ M_{gy} &= \frac{3n^2}{r^2} XZ(J_{xx} - J_{zz}) \\ M_{gz} &= \frac{3n^2}{r^2} XY(J_{yy} - J_{xx})\,. \end{aligned} \tag{11.121}$$

Substituting the gravity gradient torque components into the Euler's equations, Eq. (11.16), we have

$$\begin{aligned} J_{xx}\dot{\omega}_x + \omega_y\omega_z(J_{zz} - J_{yy}) &= \frac{3n^2}{r^2} YZ(J_{zz} - J_{yy}) \\ J_{yy}\dot{\omega}_y + \omega_x\omega_z(J_{xx} - J_{zz}) &= \frac{3n^2}{r^2} XZ(J_{xx} - J_{zz}) \\ J_{zz}\dot{\omega}_z + \omega_x\omega_y(J_{yy} - J_{xx}) &= \frac{3n^2}{r^2} XY(J_{yy} - J_{xx})\,. \end{aligned} \tag{11.122}$$

The equations of motion, Eq. (11.122), possess three distinct equilibrium attitudes (and their mirror images) for which any two of the angular velocity components vanish, and the third equals the orbital frequency, n. This requires that one of the principal axes of the spacecraft must be normal to the orbital plane in an equilibrium attitude. The relative magnitudes of the principal moments of inertia, J_{xx}, J_{yy}, J_{zz}, determine the stability of the equilibrium points.

Consider a perturbation from the general equilibrium attitude represented by the 3-2-1 Euler angles, $(\psi_3, \theta_2, \phi_1)$, where ψ is termed the *yaw angle*, θ the *pitch angle*, and ϕ the *roll angle*. If the equilibrium attitude of the spacecraft's principal axes is represented by $(\mathbf{i_e}, \mathbf{j_e}, \mathbf{k_e})$, and their perturbed attitude by $(\mathbf{i}, \mathbf{j}, \mathbf{k})$, then the transformation between the two frames in terms of the $(\psi_3, \theta_2, \phi_1)$ Euler angle sequence is given by:

$$\begin{Bmatrix} \mathbf{i} \\ \mathbf{j} \\ \mathbf{k} \end{Bmatrix} = \mathbf{C}_1(\phi)\mathbf{C}_2(\theta)\mathbf{C}_3(\psi) \begin{Bmatrix} \mathbf{i_e} \\ \mathbf{j_e} \\ \mathbf{k_e} \end{Bmatrix}. \tag{11.123}$$

The rotation matrix corresponding to the Euler-angle representation, $(\psi_3, \theta_2, \phi_1)$, is the following:

$$\mathbf{C} = \mathbf{C}_1(\phi)\mathbf{C}_2(\theta)\mathbf{C}_3(\psi)$$

$$= \begin{pmatrix} c_{11} & c_{12} & c_{13} \\ c_{21} & c_{22} & c_{23} \\ c_{31} & c_{32} & c_{33} \end{pmatrix}, \tag{11.124}$$

where the elements, c_{ij}, are the following:

$$c_{11} = \cos\theta\cos\psi \tag{11.125}$$

$$c_{12} = \cos\theta\sin\psi \tag{11.126}$$

$$c_{13} = -\sin\theta \tag{11.127}$$

$$c_{21} = \sin\phi\sin\theta\cos\psi - \cos\phi\sin\psi \tag{11.128}$$

$$c_{22} = \sin\phi\sin\theta\sin\psi + \cos\phi\cos\psi \tag{11.129}$$

$$c_{23} = \sin\phi\cos\theta \tag{11.130}$$

$$c_{31} = \cos\phi\sin\theta\cos\psi + \sin\phi\sin\psi \tag{11.131}$$

$$c_{32} = \cos\phi\sin\theta\sin\psi - \sin\phi\cos\psi \tag{11.132}$$

$$c_{33} = \cos\phi\cos\theta\,. \tag{11.133}$$

The attitude kinematics equations for the $(\psi_3, \theta_2, \phi_1)$ representation are the following:

$$\begin{aligned} \boldsymbol{\omega} &= \dot{\psi}\mathbf{k_e} + \dot{\theta}\mathbf{j_e}' + \dot{\phi}\mathbf{i} \\ &= \omega_x\mathbf{i} + \omega_y\mathbf{j} + \omega_z\mathbf{k}, \end{aligned} \tag{11.134}$$

where $\mathbf{j_e}' = \mathbf{j_e}''$ is the second axis of the reference frame following the first and second elementary rotations, and is given by

$$\mathbf{j_e}' = \mathbf{j_e}'' = \mathbf{j}\cos\phi - \mathbf{k}\sin\phi\,. \tag{11.135}$$

The third axis of the reference frame resolved in the body-fixed frame is the following:

$$\mathbf{k_e} = -\mathbf{i}\sin\theta + \mathbf{j}\sin\phi\cos\theta + \mathbf{k}\cos\phi\cos\theta\,. \tag{11.136}$$

Substituting Eqs. (11.135) and (11.136) into Eq. (11.134), and comparing the components, yields the following angular velocity components:

$$\begin{Bmatrix} \omega_x \\ \omega_y \\ \omega_z \end{Bmatrix} = \begin{pmatrix} -\sin\phi & 0 & 1 \\ \sin\phi\cos\theta & \cos\phi & 0 \\ \cos\phi\cos\theta & -\sin\phi & 0 \end{pmatrix} \begin{Bmatrix} \dot{\psi} \\ \dot{\theta} \\ \dot{\phi} \end{Bmatrix}. \tag{11.137}$$

For conducting an attitude stability analysis, only small perturbations from the equilibrium attitude need be considered. The assumption of small Euler angle magnitudes, ψ, θ, ϕ, corresponding to a small deviation from the equilibrium attitude, results in the cosines of the angles being approximated by unity (e.g., $\cos\phi \simeq 1$), and sines by the angles themselves (e.g., $\sin\phi \simeq \phi$). Furthermore, the angular rates are also assumed to be small, i.e., $\dot{\psi} \ll n$, $\dot{\theta} \ll n$, and $\dot{\phi} \ll n$. Consequently, the products of any two angles (and their rates) are approximated to be zero. Making these approximations in Eqs. (11.123) and (11.137) yields the following approximate relationships for the perturbed attitude, $(\mathbf{i}, \mathbf{j}, \mathbf{k})$, and the perturbation in the angular velocity components, $\delta\boldsymbol{\omega} = (\delta\omega_x, \delta\omega_y, \delta\omega_z)^T$:

$$\begin{Bmatrix} \mathbf{i} \\ \mathbf{j} \\ \mathbf{k} \end{Bmatrix} \simeq \begin{pmatrix} 1 & \psi & -\theta \\ -\psi & 1 & \phi \\ \theta & -\phi & 1 \end{pmatrix} \begin{Bmatrix} \mathbf{i_e} \\ \mathbf{j_e} \\ \mathbf{k_e} \end{Bmatrix} \tag{11.138}$$

and

$$\begin{Bmatrix} \delta\omega_x \\ \delta\omega_y \\ \delta\omega_z \end{Bmatrix} \simeq \begin{pmatrix} -\phi & 0 & 1 \\ \phi & 1 & 0 \\ 1 & -\phi & 0 \end{pmatrix} \begin{Bmatrix} \dot{\psi} \\ \dot{\theta} \\ \dot{\phi} \end{Bmatrix}. \tag{11.139}$$

Now, consider the equilibrium attitude of the spacecraft, $(\mathbf{i_e}, \mathbf{j_e}, \mathbf{k_e})$, to be defined by the axis $\mathbf{k_e}$ in the radial direction, i.e., $\mathbf{k_e} = \mathbf{r}/r$, and the axis, $\mathbf{j_e} = \mathbf{h}/h$, along the orbital angular momentum vector normal to the orbital plane. Then the axis $\mathbf{i_e} = \mathbf{j_e} \times \mathbf{k_e}$ points along the orbit direction. This attitude of the equilibrium axes is shown in Fig. 11.10. The net angular velocity is the sum of the angular velocity of the equilibrium axes, $\boldsymbol{\omega}_e = n\mathbf{j_e}$, and the small perturbation, $\delta\boldsymbol{\omega}$, which are resolved in the current body-fixed principal frame, $(\mathbf{i}, \mathbf{j}, \mathbf{k})$, with the use of Eqs. (11.138) and (11.139) as follows:

$$\begin{aligned} \boldsymbol{\omega}_e &= n\mathbf{j_e} = n(\psi\mathbf{i} + \mathbf{j} - \psi\mathbf{k}) \\ \delta\boldsymbol{\omega} &= (\dot{\phi} - \phi p\dot{s}i)\mathbf{i} + (\dot{\theta} + \phi\dot{\psi})\mathbf{j} + (\dot{\psi} - \phi\dot{\theta})\mathbf{k} \\ &\simeq \dot{\phi}\mathbf{i} + \dot{\theta}\mathbf{j} + \dot{\psi}\mathbf{k} \\ \boldsymbol{\omega} &= \boldsymbol{\omega}_e + \delta\boldsymbol{\omega} = (\dot{\phi} + n\psi)\mathbf{i} + (\dot{\theta} + n)\mathbf{j} + (\dot{\psi} - n\psi)\mathbf{k}\,, \end{aligned} \tag{11.140}$$

where the products of the small perturbations have been neglected. The resolution of the position vector, $\mathbf{r}$, in the perturbed body-fixed principal frame, $(\mathbf{i}, \mathbf{j}, \mathbf{k})$, yields

$$\begin{aligned} \mathbf{r} &= r(-\sin\theta\mathbf{i} + \sin\phi\cos\theta\mathbf{j} + \cos\phi\cos\theta\mathbf{k}) \\ &\simeq r(-\theta\mathbf{i} + \phi\mathbf{j} + \mathbf{k})\,, \end{aligned} \tag{11.141}$$

which leads to $X \simeq -r\theta$, $Y \simeq r\phi$, and $Z \simeq r$ for the small perturbation.

When Eqs. (11.140) and (11.141) are substituted into (11.122), we have the following linearized equations of perturbed motion:

$$\ddot{\phi} = \frac{(J_{xx} - J_{yy} + J_{zz})n}{J_{xx}}\dot{\psi} - \frac{4n^2(J_{yy} - J_{zz})}{J_{xx}}\phi\,. \tag{11.142}$$

$$\ddot{\theta} = -\frac{3n^2(J_{xx} - J_{zz})}{J_{yy}}\theta\,. \tag{11.143}$$

$$\ddot{\psi} = -\frac{(J_{xx} - J_{yy} + J_{zz})n}{J_{zz}}\dot{\phi} - \frac{n^2(J_{yy} - J_{xx})}{J_{zz}}\psi\,. \tag{11.144}$$

Clearly, the small-disturbance, linear pitching motion is decoupled from the roll-yaw dynamics, and can be solved in a closed form. If $J_{xx} > J_{zz}$, the pitching motion is a stable oscillation of a constant amplitude, and for an initial condition, $\theta(0) = \theta_0$, $\dot{\theta}(0) = 0$, has the solution

$$\theta(t) = \theta_0 \cos \lambda_p t \,, \tag{11.145}$$

where

$$\lambda_p = n\sqrt{\frac{3(J_{xx} - J_{zz})}{J_{yy}}} \tag{11.146}$$

is the *pitch natural frequency*. This undamped pitching oscillation is called *libration*, and requires an active damping mechanism, such as through a reaction wheel (Chapter 12). The coupled roll-yaw dynamics, Eqs. (11.143) and (11.144) – termed *nutation* – has the following characteristic equation:

$$s^4 + n^2(1 + 3j_x + j_x j_z)s^2 + 4n^4 j_x j_z = 0 \,, \tag{11.147}$$

where

$$\begin{aligned} j_x &= \frac{J_{yy} - J_{zz}}{J_{xx}} \\ j_z &= \frac{J_{yy} - J_{xx}}{J_{zz}} \,. \end{aligned} \tag{11.148}$$

For stability, all roots, s, of the characteristic equation should have non-positive real parts, indicating either a decaying, or a constant amplitude oscillation. However, since there is no damping provided by the gravity-gradient torque (gravity being a conservative force), one can only expect real and negative values of both the quadratic roots, $s_1^2 = -\lambda_1^2$ and $s_2^2 = -\lambda_2^2$, which leads to imaginary values of $s_{1,3} = \pm i\lambda_1$ and $s_{2,4} = \pm i\lambda_2$, $i = \sqrt{-1}$, corresponding to zero real parts, and a constant amplitude pitch-yaw oscillation, where λ_1, λ_2 are the *roll-yaw natural frequencies* of the stable oscillation. The requirement of real and negative roots, s^2, of Eq. (11.147) results in the following necessary and sufficient stability conditions:

$$\begin{aligned} 1 + 3j_x + j_x j_z &\geq 4\sqrt{j_x j_z} \\ j_x j_z &> 0 \,. \end{aligned} \tag{11.149}$$

As discussed earlier, an actual spacecraft is only semi-rigid, and has an internal energy dissipation until the lowest rotational kinetic energy is reached. Hence the only stable gravity-gradient attitude is the one with $J_{yy} > J_{xx} > J_{zz}$, because it yields the lowest rotational kinetic energy, $T_r = H^2/2J_{yy}$, apart from satisfying the stability criteria, Eq. (11.149). Thus, the minor axis, $\mathbf{k_e}$, should point towards (or away from) the central body, while the major axis, $\mathbf{j_e}$, should be normal to the orbital plane. Such an attitude is adopted for most asymmetric spacecraft in low orbits. For small – or nearly axisymmetric – satellites a long boom with an end-mass can provide an effective gravity gradient stabilization by satisfying the conditions of Eq. (11.149).

Example 11.9.1 *Consider the* International Space Station *with the following inertia tensor expressed in a body-fixed frame, $(ox'y'x')$:*

$$\mathbf{J} = \begin{pmatrix} 127908568 & 3141229 & 7709108 \\ 3141229 & 107362480 & 1345279 \\ 7709108 & 1345279 & 200432320 \end{pmatrix} \text{kg.m}^2 .$$

Determine a stable gravity-gradient orientation of the spacecraft in a circular orbit around Earth of 93 min. period, and find the natural frequencies of the linearized pitch and roll-yaw dynamics.

The principal inertia tensor is first determined as follows:

$$\mathbf{J_p} = \mathbf{C_p}^T\mathbf{J}\mathbf{C_p} = diag(J_1, J_2, J_3)$$
$$= \begin{pmatrix} 106892554.98 & 0 & 0 \\ 0 & 127538483.85 & 0 \\ 0 & 0 & 201272329.17 \end{pmatrix} \text{kg.m}^2 ,$$

where

$$\mathbf{C_p} = \begin{pmatrix} 0.1471 & 0.9835 & 0.1052 \\ -0.9891 & 0.1460 & 0.0178 \\ 0.0021 & -0.1067 & 0.9943 \end{pmatrix} .$$

For a stable gravity gradient attitude, we require $J_{yy} > J_{xx} > J_{zz}$. In terms of the principal frame, this requires that

$$J_{zz} = J_1 = 106892554.98 \text{ kg.m}^2$$
$$J_{xx} = J_2 = 127538483.85 \text{ kg.m}^2$$
$$J_{yy} = J_3 = 201272329.17 \text{ kg.m}^2 ,$$

which implies that the first principal axis is the minor axis, oz, the second principal axis is the intermediate axis, ox, and the third principal axis is the major axis, oy. Hence, the stable attitude, $(\mathbf{i_e}, \mathbf{j_e}, \mathbf{k_e})$, is given by the minor axis, oz, $(\mathbf{k_e})$, pointing either directly towards or away from the central body, with the major axis, oy, $(\mathbf{j_e})$ being normal to the orbital plane. The axis, ox, $(\mathbf{i_e})$ is tangential to the orbit direction.

In terms of the given body axes, $(ox'y'x')$, represented by the unit vectors, $(\mathbf{i}', \mathbf{j}', \mathbf{k}')$, in which the inertia tensor, $\mathbf{J}$, is prescribed, the transformation to the stable orientation, $(\mathbf{i_e}, \mathbf{j_e}, \mathbf{k_e})$, is given by the following:

$$\begin{Bmatrix} \mathbf{k}' \\ \mathbf{i}' \\ \mathbf{j}' \end{Bmatrix} = \mathbf{C_p} \begin{Bmatrix} \mathbf{i_e} \\ \mathbf{j_e} \\ \mathbf{k_e} \end{Bmatrix} ,$$

or

$$\begin{Bmatrix} \mathbf{i_e} \\ \mathbf{j_e} \\ \mathbf{k_e} \end{Bmatrix} = \mathbf{C_p} \begin{Bmatrix} \mathbf{k}' \\ \mathbf{i}' \\ \mathbf{j}' \end{Bmatrix} .$$

The linearized pitch dynamics is described by Eq. (11.145), with the pitch natural frequency calculated by Eq. (11.146) to be the following:

$$\lambda_p = n\sqrt{\frac{3(J_{xx} - J_{zz})}{J_{yy}}} = 0.000625 \text{ rad./s} .$$

The linearized roll-yaw dynamics is represented by Eqs. (11.143) and (11.144), with the roll-yaw natural frequencies, λ_1, λ_2, obtained from the real and negative quadratic roots, $s_1^2 = -\lambda_1^2$ and $s_2^2 = -\lambda_2^2$ of the characteristic equation, Eq. (11.147), given by:

$$s^4 + 3.73n^2s^2 + 2.0418n^4 = 0 ,$$

which yields

$$s_1^2 = -0.6664n^2 ; \qquad s_2^2 = -3.0641n^2 .$$

For $n = 2\pi/(93 \times 60) = 0.001126$ rad./s, *these result in the following roll-yaw natural frequencies:*

$$\lambda_1 = 0.000919 \text{ rad./s} ; \qquad \lambda_2 = 0.00197 \text{ rad./s}$$

Exercises

1. Express the rotational kinetics equations in a *space-fixed frame*, which is a non-rotating frame with the origin at the centre of mass of the spacecraft.

2. Calculate the principal inertia tensor, $\mathbf{J_p}$, and the principal rotation matrix, $\mathbf{C_p}$, for a spacecraft with the following inertia tensor expressed in a body-fixed frame, $(ox'y'x')$:

$$\mathbf{J} = \begin{pmatrix} 10000 & -500 & -800 \\ -500 & 50500 & -500 \\ -800 & -500 & 20000 \end{pmatrix} \text{kg.m}^2 .$$

   Use the result to find the angular velocity in the principal frame, if the angular velocity in the given body axes is $\boldsymbol{\omega} = (-0.1,\ 0.2,\ 0.5)^T$ rad./s.

3. An axisymmetric spacecraft with the principal moments of inertia $J_{xx} = J_{yy} = 25000$ kg.m^2 and $J_{zz} = 5000$ kg.m^2 is initially in a state of pure spin about the axis of symmetry, oz, with a spin rate $n = 0.001$ rad./s. Calculate the angular velocity, $\boldsymbol{\omega} = (\omega_x,\ \omega_y,\ \omega_z)^T$, as a function of time if an initial perturbation, $\omega_x(0) = 0.1n$, is applied at $t = 0$.

4. A spacecraft with the following inertia tensor in a body-fixed frame has the angular velocity $\boldsymbol{\omega} = \frac{1}{10\sqrt{2}}(-1,\ 0,\ 1)^T$ rad/s at $t = 0$:

$$\mathbf{J} = \begin{pmatrix} 200 & 0 & -10 \\ 0 & 100 & 0 \\ -10 & 0 & 200 \end{pmatrix} \text{kg.m}^2 .$$

   Determine the angular velocity vector at $t = 10$ s.

5. Use the method of Jacobian elliptic functions to solve the problem given in Exercise 4 if the inertia tensor is changed to the following:

$$\mathbf{J} = \begin{pmatrix} 200 & -5 & -10 \\ -5 & 100 & -2 \\ -10 & -2 & 200 \end{pmatrix} \text{kg.m}^2 .$$

6. Repeat Exercise 5 by the Runge-Kutta method (Appendix A).

7. A semi-rigid spacecraft with the principal moments of inertia, $J_{xx} > J_{yy} > J_{zz}$, is initially spinning about the axis, oy, with a rate n. What is the angular velocity of the spacecraft in the stable equilibrium state?

8. Write a computer programme to simulate the response of an axisymmetric spacecraft platform with an oblate rotor in a dual-spin configuration, using the Runge-Kutta method (Appendix A). The spacecraft has a moment of inertia of 1500 kg.m^2 about its spin axis, and 4000 kg.m^2 about a lateral principal axis. The rotor's moment of inertia about its spin axis is 300 kg.m^2, and 100 kg.m^2 about a lateral principal axis. The centres of mass of the platform and the rotor are offset from the centre of mass of the dual-spin configuration by 1.0 m and 2.0 m, respectively. Initially, both the platform and the rotor are spinning in the same direction, with angular speeds of 7×10^{-5} rad./s and 1 rad./s, respectively, when a lateral angular velocity disturbance of $\omega_{xy} = 0.001$ rad./s is encountered. Neglect the bearing friction between the rotor and the platform.

9. Determine a stable gravity-gradient orientation of the body-fixed frame, $(ox'y'x')$, in Exercise 2, assuming that the spacecraft is in a circular orbit of 90 min. period. Also find the natural frequencies of the linearized pitch and roll-yaw dynamics in the stable orientation.

10. Estimate the natural frequencies of a stable gravity-gradient oscillation of a spacecraft with the following characteristics:

$$J_{xx} = J_{yy} = 25000 \text{ kg.m}^2$$

$$J_{zz} = 3000 \text{ kg.m}^2$$

$$n = 0.00105 \text{ rad./s} .$$

12

Attitude Manoeuvres

A spacecraft must vary its attitude for pointing its body-fixed axes in specific directions as part of its mission. An example is a *nadir-pointing* satellite, one of whose axes must be always pointed down towards the central body in its circular orbit of radius r for remote sensing, imaging, and communications purposes. If such a satellite is not spinning at a constant rate, $n = \sqrt{\mu/r^3}$, about an axis passing through its own centre of mass, o, and normal to the plane of the orbit, such a spacecraft would very soon fail to perform its mission. While the methods of Chapter 11 can be used to achieve a stable equilibrium attitude, such an attitude may require a modification as the spacecraft travels in its orbit. It is thus necessary to perform attitude manoeuvres to rotate the spacecraft, either to a new equilibrium attitude, or to correct the deviations resulting from environmental torques and/or variations in the spacecraft's mass distribution. Such manoeuvres are performed by applying an external torque by rocket thrusters or magnetic torquers, or through the internal torques generated by rotors mounted on the spacecraft. This chapter presents several introductory concepts of attitude manoeuvres commonly used by spacecraft. In this chapter, the attitude of a spacecraft is described by the orientation of the body-fixed principal frame ($oxyz$) with the axes $\mathbf{i}_1, \mathbf{i}_2, \mathbf{i}_3$, respectively.

12.1 Impulsive Manoeuvres with Attitude Thrusters

Spin stabilization of torque-free spacecraft is a propellant-free (hence economical) way of maintaining a stable equilibrium attitude. However, controlling the motion of a spinning body for carrying out the necessary attitude manoeuvres is much more demanding than controlling the motion of a non-spinning body. Most spacecraft have a *reaction control system* (RCS) consisting of a pair of rocket thrusters mounted about each principal axis. These are termed *attitude thrusters* because they apply a torque about the given principal axis to perform attitude manoeuvres. Chapter 5 showed how a general orientation of a coordinate frame can be achieved by a set of elementary rotations. In the case of a spacecraft equipped with RCS thrusters, such an elementary rotation is produced by each pair of thrusters. For example, a 3-2-1 sequence of elementary rotations can be used to achieve an arbitrary orientation, which requires that torques be applied in turn about each of the principal axes. A spacecraft having the ability to individually

Foundations of Space Dynamics, First Edition. Ashish Tewari.

rotate about all the principal axes is said to be *three-axis stabilized*. It is not necessary for a general attitude manoeuvre to apply a three-axis torque. As an example, a 3-1-3 sequence of elementary rotations can produce the same orientation as that achieved by a 3-2-1 sequence via torques applied about only the first and the third principal axes. However, consideration must be given to the singularities of the Euler angles, e.g., a $(\psi_3, \theta_1, \phi_3)$ rotation becomes useless for $\theta = 0, \pm\pi$, and a $(\psi_3, \theta_2, \phi_1)$ fails for $\theta = \pm\pi/2$.

The attitude thrusters of an RCS are operated in pairs with equal and opposite thrust vectors, such that the net external force remains unaffected and the vehicle is undisturbed in its orbit. The firing of the thrusters is limited to short bursts, which can be approximated by *torque impulses*. A torque impulse is defined as a torque of infinite magnitude acting for an infinitesimal duration, which causes an instantaneous change in the angular momentum of the spacecraft about the axis of application. The concept of the torque impulse is very useful in analyzing the single-axis (or elementary) rotations, as it allows us to utilize the concept of the *unit-impulse function*, $\delta(t)$. The unit-impulse function (also known as the *Dirac delta function*), $\delta(t - t_0)$, denotes an impulse of unit magnitude applied at time $t = t_0$, and is defined by

$$\delta(t - t_0) = \begin{cases} \infty, & t = t_0 \\ 0, & t \neq t_0 \end{cases}, \tag{12.1}$$

such that

$$\int_{-\infty}^{\infty} \delta(\tau) \mathrm{d}\tau = 1 . \tag{12.2}$$

It has the following useful property:

$$\int_{-\infty}^{\infty} f(t)\delta(t - t_0) \mathrm{d}t = f(t_0) , \tag{12.3}$$

where $f(t)$ is a single-valued function.

The change in angular momentum, $\Delta\mathbf{H}$, caused by an impulsive torque applied at $t = 0$, $\mathbf{M}(t) = \mathbf{M}\delta(t)$, is seen to be the total area under the torque *vs*. time graph:

$$\Delta\mathbf{H} = \int_{-\infty}^{\infty} \mathbf{M}(t) \mathrm{d}t = \int_{-\infty}^{\infty} \mathbf{M}\delta(t) \mathrm{d}t = \mathbf{M} . \tag{12.4}$$

Thus, the torque impulse causes an instantaneous change in the angular momentum, equal to the value of the torque at the instant of impulse application, $t = 0$, and is therefore said to produce an *impulsive manoeuvre*.

12.1.1 Single-Axis Rotation

A general attitude manoeuvre can be achieved after a sequence of single-axis rotations. When the rotations are performed using attitude thrusters, the concept of torque impulse offers simplicity in designing such manoeuvres. Since only the maximum possible torque impulse is applied at any given time, the resulting rotation fulfills the requirements of *time-optimal control theory* (Tewari, 2011), wherein the inputs of the maximum possible magnitude applied in either the positive or the negative direction minimize the total time of rotation by a given angle.

Consider a rigid spacecraft with moment of inertia, J_{zz}, about the axis of desired rotation, oz, equipped with a pair of attitude thrusters capable of exerting a maximum torque, M_z, for an infinitesimal duration, $\Delta t \to 0$, which causes an instantaneous change in the angular momentum by $\Delta H_z = M_z$. The torque as a function of time, $M_z(t) = M_z\delta(t)$, is substituted into the Euler's equations (Chapter 11) and results in the following:

$$\begin{aligned}\dot{\omega}_x &= 0\\ \dot{\omega}_y &= 0\\ J_{zz}\dot{\omega}_z &= M_z\delta(t)\,.\end{aligned} \tag{12.5}$$

The last of Eq. (12.5) is expressed in terms of the angular displacement, $\theta(t)$, about oz as follows:

$$\ddot{\theta} = \frac{M_z}{J_{zz}}\delta(t)\,, \tag{12.6}$$

whose solution is obtained by successive integrations in time as follows:

$$\begin{aligned}\omega_z(t) = \dot{\theta} &= \omega_z(0) + \frac{M_z}{J_{zz}}u_s(t)\\ \theta(t) &= \theta(0) + \omega_z(0)t + \frac{M_z}{J_{zz}}r(t)\,,\end{aligned} \tag{12.7}$$

where $\theta(0)$, $\omega_z(0)$ refer to the initial condition immediately before the torque application, $u_s(t) = \int \delta(t)\mathrm{d}t$ is the *unit-step function* applied at $t = 0$, which is defined by

$$u_s(t - t_0) = \begin{cases} 0\,, & t < t_0\\ 1\,, & t \geq t_0 \end{cases}\,, \tag{12.8}$$

and $r(t) = \int u_s(t)\mathrm{d}t$ is the *unit-ramp function* applied at $t = 0$, defined by

$$r(t - t_0) = \begin{cases} 0\,, & t < t_0\\ t - t_0\,, & t \geq t_0 \end{cases}\,. \tag{12.9}$$

Equation (12.7) implies that the response to a single impulse is a linearly increasing angular displacement, $\theta(t)$, and a step change in the angular velocity, $\omega_z(t)$. In a *spin-up manoeuvre*, defined as a step change in the angular velocity, ω_z, by a given amount, only a single impulse is needs to be applied. However, if a given angular displacement is desired – called a *rest-to-rest manoeuvre* – at least two impulses need to applied: the first to begin the rotation, and the other in an opposite direction to stop the rotation at the desired angle.

In terms of a time-optimal rest-to-rest manoeuvre, the first impulse applied at $t = 0$ is given by $M_z\delta(t)$, while the second impulse applied at the final time $t = \tau$ is given by $-M_z\delta(t - \tau)$, where M_z is the maximum possible torque generated by the thrusters. This situation is depicted in Fig. 12.1. Since the governing differential equation, Eq. (12.6), is linear, its solution obeys the *principle of linear superposition*, which allows a weighted addition of the responses to individual impulses, to yield the total displacement caused by multiple impulses. Therefore, the net response to two equal and opposite impulses applied at after an interval of $t = \tau$, is given by

$$\begin{aligned}\omega_z(t) &= \frac{M_z}{J_{zz}}[u_s(t) - u_s(t - \tau)]\\ \theta(t) &= \frac{M_z}{J_{zz}}[r(t) - r(t - \tau)] + \omega_z\tau = \theta_d\,.\end{aligned} \tag{12.10}$$

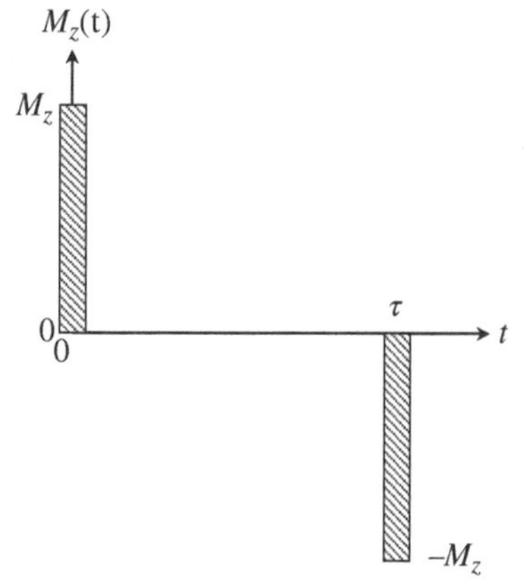

Figure 12.1 Two equal and opposite torque impulses separated by time interval, τ, comprising a time-optimal, rest-to-rest manoeuvre about oz.

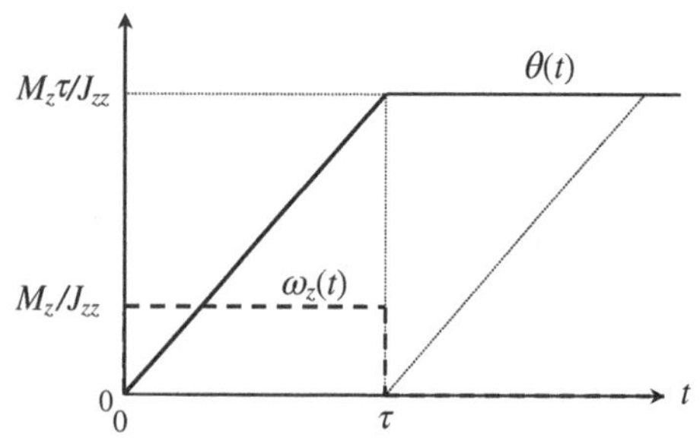

Figure 12.2 Angular displacement, $\theta(t)$, and angular velocity, $\omega_z(t)$, produced by the torque impulses of Fig. 12.1 comprising a time-optimal, rest-to-rest manoeuvre about oz.

Hence, the angular velocity becomes zero, and a desired constant displacement, $\theta(t) = \theta_d$, is reached at $t = \tau$, as shown in Fig. 12.2. The magnitude of θ_d can be controlled by varying the time τ at which the second impulse is applied. The application of two equal and opposite impulses of maximum magnitude for achieving a time-optimal displacement is called *bang-bang* control. The bang-bang (time-optimal) control is exactly applicable to any linear system without resistive and dissipative external forces. However, even when a small damping force is present, one can approximately apply this approach to control linear systems.

12.1.2 Rigid Axisymmetric Spin-Stabilized Spacecraft

Attitude thrusters can be used for varying the attitude of a spin-stabilized, rigid spacecraft via precession and nutation of the spin axis. As discussed in Chapter 11, an axisymmetric spacecraft spinning about oz has a constant spin rate, $\omega_z = n$, a constant nutation angle, θ, and the governing differential equations describing the precession, ω_{xy}, are linear. Hence the time-optimal,

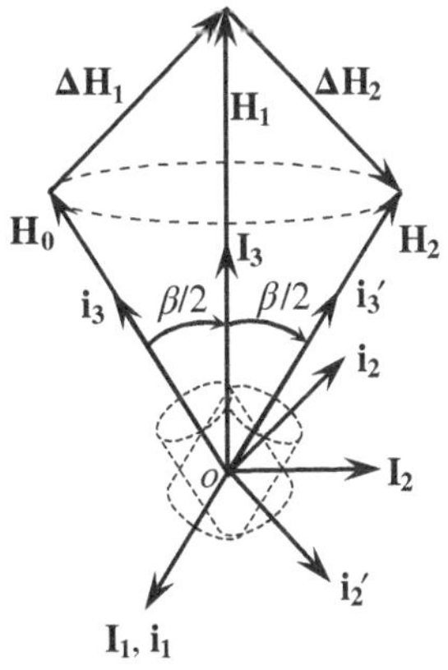

Figure 12.3 Two-impulse, time-optimal attitude manoeuvre of a rigid, axisymmetric spacecraft for rotating the spin axis by angle β in time $t = t_{1/2}$.

bang-bang manoeuvring strategy can be employed in a similar manner as for the single-axis rotation. To do so, the precession is excited by applying a torque impulse normal to the spin axis, and then exerting another equal and opposite impulse to stop the precession when the desired spin axis orientation has been reached. Since the principal axes of a precessing body are not fixed in space, the directions of the two torque impulses are referred to an inertial frame ($oXYZ$) with axes $(\mathbf{I}_1, \mathbf{I}_2, \mathbf{I}_3)$, respectively.

Let the spin axis be varied by a desired angle, β, by the application of the two torque impulses, as shown in Fig. 12.3. After the application of the first impulse, $\Delta\mathbf{H}_1$, at the time $t = 0$, the angular momentum changes instantaneously from $\mathbf{H}_0 = J_{zz}n\mathbf{k}$, to its new value, $\mathbf{H}_1 = \mathbf{H}_0 + \Delta\mathbf{H}_1$, such that a nutation angle of $\theta = \beta/2$ is obtained. The orientation of the inertial frame is selected such that oZ is along the intermediate angular momentum vector, $\mathbf{H}_1$, and oX coincides with the principal axis ox at time $t = 0$, i.e., $\mathbf{i}_1 = \mathbf{I}_1$. Thus we have $\psi(0) = 0, \theta(t) = \beta/2,\ \phi(0) = 0$ in terms of the 3-1-3 Euler angle kinematics (Chapter 11). Figure 12.3 shows that the first torque impulse applied normal to the spin axis at $t = 0$ is given by

$$\Delta\mathbf{H}_1 = J_{zz}n \tan\frac{\beta}{2}\left(\cos\frac{\beta}{2}\mathbf{I}_2 + \sin\frac{\beta}{2}\mathbf{I}_3\right) = J_{zz}n\tan\frac{\beta}{2}\mathbf{i}_2\,, \tag{12.11}$$

and causes a positive rotation of the angular momentum vector about $-\mathbf{I}_1$. Since the angular momentum has been deflected from the spin axis, the precession is excited, and is allowed to continue for half the inertial rotation ($\phi = \pi$), until the first axis has rotated to the direction opposite of its original orientation, $\mathbf{i}_1' = -\mathbf{I}_1$. At that precise instant, $t = t_{1/2}$, the second impulse is applied, given by

$$\Delta\mathbf{H}_2 = J_{zz}n \tan\frac{\beta}{2}\left(\cos\frac{\beta}{2}\mathbf{I}_2 - \sin\frac{\beta}{2}\mathbf{I}_3\right) = J_{zz}n\tan\frac{\beta}{2}\mathbf{i}_2'\,, \tag{12.12}$$

in order to stop the precession by causing a positive rotation of the angular momentum vector about $\mathbf{I}_1$.

The angular momenta, $\mathbf{H}_1$ and $\mathbf{H}_2$, at the beginning and the end of the precession, respectively, are resolved in the instantaneous principal axes as follows:

$$\mathbf{H}_1 = J_{zz}n\mathbf{i}_3 + J_{zz}n \tan\frac{\beta}{2}\mathbf{i}_2$$
$$\mathbf{H}_2 = J_{zz}n\mathbf{i}_3' . \tag{12.13}$$

The frames, $(\mathbf{i}_1, \mathbf{i}_2, \mathbf{i}_3)$ and $(\mathbf{i}_1', \mathbf{i}_2', \mathbf{i}_3')$, correspond to the instantaneous orientations of the body-fixed principal axes at the time instants, $t = 0$, and $t = t_{1/2}$, respectively, which are separated in time by half the inertial spin time period, $t_{1/2}$. Recall from Chapter 11 that the inertial spin rate (the natural frequency of precession), $|\, k\,|$, is given by

$$|\, k\,| = |\, \dot{\phi}\,| = n\left|1 - \frac{J_{zz}}{J_{xx}}\right| . \tag{12.14}$$

Hence the time taken to undergo half the precession is given by

$$t_{1/2} = \frac{\pi}{|\,\dot{\phi}\,|} = \frac{J_{xx}\pi}{n\,|\, J_{xx} - J_{zz}\,|} . \tag{12.15}$$

It follows from Eq. (12.15) that the time required for the manoeuvre is large if the spin rate, n, is small, or if the two moments of inertia, J_{xx} and J_{zz}, are nearly equal.

Although the two torque impulses are opposite in direction relative to the inertial frame ($oXYZ$), they have the same orientation in the body-fixed principal frame ($oxyz$) (i.e., along the instantaneous second axis of the frame). Hence the same pair of attitude thrusters can be used to both begin and end the precession after multiples of the half inertial spin ($\phi = \pm\pi, 2\pi, \dots$). However, in order to achieve the largest possible deflection, β, of the spin axis – which happens when $\mathbf{H}_0$, $\mathbf{H}_1$, and $\mathbf{H}_2$ all lie in the same plane – the precession angle, ψ, must have changed exactly by $\pm 180°$ at the end of the manoeuvre. This requires that $\dot{\psi} = |\,\dot{\phi}\,|$. From Eq. (11.94), it follows that the matching of the precession rate with the inertial-spin rate is possible if and only if

$$J_{xx}^2\omega_{xy}^2 = J_{xx}^2 n^2 - 2J_{xx}J_{zz}n^2 , \tag{12.16}$$

which, substituted into Eq. (11.91), yields

$$\cos\frac{\beta}{2} = \frac{J_{zz}n}{\sqrt{J_{xx}^2\omega_{xy}^2 + J_{zz}^2 n^2}}$$
$$= \frac{J_{zz}n}{\sqrt{J_{xx}^2 n^2 + J_{zz}^2 n^2 - 2J_{xx}J_{zz}n^2}} = \frac{J_{zz}}{\sqrt{(J_{xx} - J_{zz})^2}}$$
$$= \frac{J_{zz}}{|\, J_{xx} - J_{zz}\,|} . \tag{12.17}$$

Since the cosine cannot exceed unity, this implies that precession and inertial spin can be synchronized only for prolate bodies with $J_{xx} > 2J_{zz}$.

Equation (12.17) gives the largest possible angular deflection of the spin axis (β) that can be achieved with a given pair of attitude thrusters, and is obtained when $\psi = |\,\phi\,| = \pi$. Since the nutation angle, $\theta = \frac{\beta}{2}$, is determined purely by the impulse magnitude, its value can be different from that given by Eq. (12.17), in which case the total angular deviation of the spin axis is less than β. A greater control on the attitude of the spin axis is provided by using more than two torque impulses.

The torque magnitude required for each impulse, M_y, is proportional to $\tan(\beta/2)$. Hence changing the spin axis of a spin-stabilized spacecraft is quite expensive even by a moderate angle β, just as it is expensive to change an orbital plane. For $\beta = \pi$, an infinitely large torque is required. However, the impulsive manoeuvre is an idealization of the actual case where the thruster firing takes place in a non-zero duration, Δt, and the average thruster torque, $\mathbf{M_y} = \Delta \mathbf{H}/\Delta t$, does not result in an instantaneous momentum change. Therefore, an actual (non-impulsive) manoeuvre is more expensive than the ideal impulsive manoeuvre. This can be seen by the following expression for the initial and final angular momentum vectors:

$$\mathbf{H}(t) = \mathbf{H}(0) + \int_0^t \mathbf{M_y}(\tau)\mathrm{d}\tau , \tag{12.18}$$

where the integral represents the net cost to be the total area of the torque vs. time graph.

Example 12.1.1 *Consider an axisymmetric, spin-stabilized, rigid spacecraft with principal moments of inertia* $J_{xx} = J_{yy} = 5000$ kg.m^2 *and* $J_{zz} = 2000$ kg.m^2, *and spin rate* $\omega_z = 0.5$ rad./s. *A pair of attitude thrusters mounted normal to the spin axis produces a constant torque at each one-hundredth-second firing. Calculate the maximum possible spin axis deflection, the required time, and the corresponding magnitude of the two bang-bang torque impulses separated by half the precession period.*

The maximum spin axis deviation is achieved with a synchronization of the inertial spin with precession, such that $\psi = -\phi = \pi$ *at the end of the second impulse. Thus we have*

$$\beta = 2\cos^{-1}\left(\frac{J_{zz}}{\mid J_{xx} - J_{zz} \mid}\right) = 1.6821 \text{ rad. } (96.3794°) .$$

The semi-period of precession is computed as follows:

$$t_{1/2} = \frac{J_{xx}\pi}{n \mid J_{xx} - J_{zz} \mid} = \frac{J_{xx}\pi \cos\frac{\beta}{2}}{nJ_{zz}} = 10.472 \text{ s} .$$

The magnitude of each torque impulse for an ideal manoeuvre is the following:

$$\Delta H_y = J_{zz} n \tan\frac{\beta}{2} = 1118.04 \text{ N.m.s} .$$

The thruster torque magnitude for this manoeuvre is calculated as follows:

$$M_y = \frac{J_{zz} n \tan\frac{\beta}{2}}{\Delta t} = 111803.4 \text{ N.m} ,$$

which is expectedly large, thus agreeing with the assumption of an impulsive torque.

12.1.3 Spin-Stabilized Asymmetric Spacecraft

The case of spin-stabilized asymmetric spacecraft is not amenable to a treatment similar to that of the axisymmetric spacecraft, mainly due to the non-linear nature of the former's governing Euler's equations, wherein the principle of linear superposition of impulses becomes invalid. The manoeuvre can then be designed and analyzed by the numerical solution to the Euler's equations, either by the Jacobian elliptic functions, or via direct integration using a Runge-Kutta – type method, as described in Chapter 11. However, if the elementary rotations constituting a manoeuvre are of small angular rates, then the Euler's equations are rendered linear by approximation, and the bang-bang approach could be approximately applied. Hence an arbitrarily large, multi-axis, rest-to-rest manoeuvre can be broken up into a series of small elementary rotations. Since the single-axis rotations of a rigid body have already been covered earlier, the approximation of sequential multiple elementary rotations does not require a further discussion.

12.2 Attitude Manoeuvres with Rotors

The repeated use of an attitude thruster-based reaction-control system (RCS) for manoeuvring a spacecraft entails a large propellant expenditure. Since the cost of a space mission greatly depends upon the propellant mass to be carried by the spacecraft, a saving of the propellant by employing other means of attitude manoeuvring is much sought after. One such method is the use of rotors mounted on the spacecraft. The rotors are axisymmetric rigid bodies rotating relative to the spacecraft which itself is assumed to be a separate rigid body. The rotor axes are either fixed or variable relative to the spacecraft's body-fixed principal frame. Since the rotors exchange their angular momentum with that of the spacecraft in order to rotate the latter, they are called *momentum-exchange devices* (MED). In a three-axis stabilized spacecraft, there is at least one rotor capable of rotating the spacecraft about each principal axis. As the MED are driven by electric motors which derive their power from the solar arrays of the spacecraft, they provide a propellant-free means of attitude manoeuvring.

Consider a spacecraft with the principal inertia tensor, $\mathbf{J}$, and angular velocity resolved in the principal axes, $\boldsymbol{\omega} = (\omega_x, \omega_y, \omega_z)^T$. A rotor with the inertia tensor, $\mathbf{J_r}$, resolved in the spacecraft's principal axes rotates relative to the spacecraft with the angular velocity, $\boldsymbol{\omega}_r = (\omega_{rx}, \omega_{ry}, \omega_{rz})^T$. A transformation of the rotor's inertia tensor, $\mathbf{J_r}$, from the rotor's principal frame to that in the spacecraft's principal frame can be performed via the *parallel axes theorem* and a rotation matrix. The parallel axes theorem states that the inertia tensor, $\mathbf{J_r''}$, of a mass, m, about a parallely displaced body-fixed frame is derived from that in the original body frame, $\mathbf{J_r'}$, as follows:

$$\mathbf{J_r''} = \mathbf{J_r'} + m\begin{pmatrix} \Delta y^2 + \Delta z^2 & -\Delta x\Delta y & -\Delta x\Delta z \\ -\Delta x\Delta y & \Delta x^2 + \Delta z^2 & -\Delta y\Delta z \\ -\Delta x\Delta z & -\Delta y\Delta z & \Delta x^2 + \Delta y^2 \end{pmatrix}, \tag{12.19}$$

where $\Delta x, \Delta y, \Delta z$ are the components of the parallel displacement of the body frame. The parallel axis theorem is useful in deriving the inertia tensor of a complex-shaped body composed of several simpler bodies with known inertia tensors. After translating the principal frame of the rotor to the spacecraft's centre of mass, o, by parallel displacement, a rotation is performed about o to align the rotor's principal axes with the principal axis of the spacecraft. Let this rotation be represented by the following coordinate transformation:

$$\begin{Bmatrix} \mathbf{i}_{1r} \\ \mathbf{i}_{2r} \\ \mathbf{i}_{3r} \end{Bmatrix} = \mathbf{C}_p \begin{Bmatrix} \mathbf{i}_{1p} \\ \mathbf{i}_{2p} \\ \mathbf{i}_{3p} \end{Bmatrix}, \tag{12.20}$$

where $(\mathbf{i}_{1r}, \mathbf{i}_{2r}, \mathbf{i}_{3r})$ are the rotor's principal axes and $(\mathbf{i}_{1p}, \mathbf{i}_{2p}, \mathbf{i}_{3p})$ are the spacecraft's principal axes. Then the rotor's inertia tensor transformed by the rotation is given by

$$\mathbf{J}_r = \mathbf{C}_p{}^T \mathbf{J}_r'' \mathbf{C}_p . \tag{12.21}$$

The angular momentum of the spacecraft is $\mathbf{H}_s = \mathbf{J}\boldsymbol{\omega}$ while that of the rotor is given by $\mathbf{H}_r = \mathbf{J}_r(\boldsymbol{\omega} + \boldsymbol{\omega}_r)$. The net angular momentum of the system (spacecraft and rotor) is the following:

$$\mathbf{H} = \mathbf{H}_s + \mathbf{H}_r = \mathbf{J}\boldsymbol{\omega} + \mathbf{J}_r(\boldsymbol{\omega} + \boldsymbol{\omega}_r), \tag{12.22}$$

the time derivative of which is zero (because no external torque acts on the system), and is expressed as follows:

$$\frac{d\mathbf{H}}{dt} = \frac{d\mathbf{H}_s}{dt} + \frac{d\mathbf{H}_r}{dt} = 0, \tag{12.23}$$

or

$$\frac{d\mathbf{H}_s}{dt} = -\frac{d\mathbf{H}_r}{dt}. \tag{12.24}$$

Hence the rate of change of the spacecraft's angular momentum is equal and opposite to that of the rotor. On comparison with the Euler's equations for a rigid body (Chapter 11), it is evident from Eq. (12.24) that the spacecraft can be regarded as a separate body on which the rotor applies a torque given by the terms on the right-hand side of Eq. (12.24). Equation (12.24) can be expressed as follows:

$$\begin{aligned} \mathbf{J}\frac{\partial \boldsymbol{\omega}}{\partial t} + \boldsymbol{\omega} \times \mathbf{J}\boldsymbol{\omega} = -\mathbf{J}_r \left[\frac{\partial(\boldsymbol{\omega} + \boldsymbol{\omega}_r)}{\partial t} + \boldsymbol{\omega} \times \boldsymbol{\omega}_r \right] - \frac{\partial \mathbf{J}_r}{\partial t}(\boldsymbol{\omega} + \boldsymbol{\omega}_r) \\ -(\boldsymbol{\omega} + \boldsymbol{\omega}_r) \times \mathbf{J}_r(\boldsymbol{\omega} + \boldsymbol{\omega}_r), \end{aligned} \tag{12.25}$$

where the partial derivative, $\partial(.)/\partial t$, denotes the variation relative to the spacecraft's principal frame. Reorganizing the terms in Eq. (12.25) results in the following:

$$\begin{aligned} (\mathbf{J} + \mathbf{J}_r)\frac{\partial \boldsymbol{\omega}}{\partial t} + \boldsymbol{\omega} \times (\mathbf{J} + \mathbf{J}_r)\boldsymbol{\omega} = -\mathbf{J}_r \frac{\partial \boldsymbol{\omega}_r}{\partial t} - \frac{\partial \mathbf{J}_r}{\partial t}(\boldsymbol{\omega} + \boldsymbol{\omega}_r) - \boldsymbol{\omega} \times \mathbf{J}_r \boldsymbol{\omega}_r \\ -\boldsymbol{\omega}_r \times \mathbf{J}_r(\boldsymbol{\omega} + \boldsymbol{\omega}_r). \end{aligned} \tag{12.26}$$

The torque generated by the rotor is caused by the variation of the angular velocity of the rotor relative to the spacecraft, $\boldsymbol{\omega}_r$, as well as by the variation of the rotor's inertia tensor, $\mathbf{J}_r$, which is resolved in the spacecraft's principal axes. Since a rotor used for attitude control has much smaller moments of inertia when compared to those of the spacecraft, one can use the approximation, $\mathbf{J} + \mathbf{J}_r \simeq \mathbf{J}$. Furthermore, in order to exert an appreciable torque on the spacecraft, the rotor's spin rate, ω_r, must be much larger than the angular speed, ω, of the spacecraft. Thus the approximation, $\omega_r \gg \omega$, can be applied for all practical considerations. These approximations result in the following:

$$\mathbf{J}\frac{\partial \boldsymbol{\omega}}{\partial t} + \boldsymbol{\omega} \times \mathbf{J}\boldsymbol{\omega} = -\mathbf{J_r}\frac{\partial \boldsymbol{\omega}_r}{\partial t} - \frac{\partial \mathbf{J_r}}{\partial t}\boldsymbol{\omega}_r - \boldsymbol{\omega}_r \times \mathbf{J_r}\boldsymbol{\omega}_r \,. \tag{12.27}$$

If there are N rotors mounted on the spacecraft, the right-hand side of Eq. (12.27) is replaced by a summation of the corresponding terms of all the rotors as follows:

$$\mathbf{J}\frac{\partial \boldsymbol{\omega}}{\partial t} + \boldsymbol{\omega} \times \mathbf{J}\boldsymbol{\omega} = -\sum_{i=1}^{N}\left(\mathbf{J_{r}}_i\frac{\partial \boldsymbol{\omega}_{r_i}}{\partial t} + \frac{\partial \mathbf{J_{r}}_i}{\partial t}\boldsymbol{\omega}_{r_i} + \boldsymbol{\omega}_{r_i} \times \mathbf{J_{r}}_i\boldsymbol{\omega}_{r_i}\right) . \tag{12.28}$$

Equation (12.28) is a general equation for the rotation of a spacecraft with rotors, whose angular velocity can be changing in time due to a varying spin rate, as well as a varying spin axis relative to the spacecraft. A rotor whose spin axis is fixed and also has a constant spin rate, ω_r, relative to the spacecraft's principal frame is termed a *momentum wheel.* A momentum wheel mounted about one of the spacecraft's principal axes provides a coupling of the rotational motion between the remaining two axes. A rotor with a fixed spin axis but a variable spin rate relative to the spacecraft is called a *reaction wheel*, and applies a torque about the spin axis by varying the wheel's speed, ω_r. If mounted parallel to a principal axis, a reaction wheel's angular momentum is directly exchanged with that of the spacecraft about the given principal axis. The rotor with a fixed spin rate, ω_r, but a variable spin axis relative to the spacecraft is called a *control-moment gyro* (CMG). A CMG applies a *gyroscopic* torque arising out of the last term on right-hand side of Eq. (12.28), and such a torque is produced normal to the spin axis. The rotor with a variable spin rate as well a variable axis is called a *variable-speed control-moment gyro*, and can be regarded to be the most general momentum exchange device because it can control all the three principal axes of the spacecraft.

12.2.1 Reaction Wheel

A reaction wheel spinning about an axis fixed relative to the spacecraft always has the rotor's angular momentum vector, $\mathbf{H_r} = \mathbf{J_r}\boldsymbol{\omega}_r$, along the spin axis, but can change the spacecraft's angular momentum by varying the spin rate, ω_r. This results in the following equation of motion by substituting $\mathbf{J_r} =$ const. and $\boldsymbol{\omega}_r \times \mathbf{H_r} = \mathbf{0}$ into Eq. (12.27):

$$\mathbf{J}\frac{\partial \boldsymbol{\omega}}{\partial t} + \boldsymbol{\omega} \times \mathbf{J}\boldsymbol{\omega} = -\mathbf{J_r}\frac{\partial \boldsymbol{\omega}_r}{\partial t} . \tag{12.29}$$

Since the spacecraft's rate of rotation, ω, is usually thousands of times slower than the spin rate, ω_r, it is a reasonable assumption that the spacecraft's angular momentum undergoes a step change due to the reaction wheel. For example, a reaction wheel mounted parallel to the principal axis, oz, creates a torque of the following magnitude:

$$M_z(t) = \dot{H}_z u_s(t) = -J_{rz}\dot{\omega}_r u_s(t) , \tag{12.30}$$

where J_{rz} is the moment of inertia of the wheel about the spacecraft's principal axis, oz, and $u_s(t)$ is the unit step function given by Eq. (12.8). Since a reaction wheel is very small compared to the spacecraft, we have $J_{rz} \ll J_{zz}$; hence to generate a torque of an appreciable magnitude,

M_z, requires a large wheel acceleration, ω_r. When a manoeuvre requires a continuous variation of the torque, it can be produced by a smooth variation of the wheel speed, $\omega_r(t)$.

A combination of momentum wheels and reaction wheels can provide three-axis control of the spacecraft. For example, a spacecraft equipped with a pair of reaction wheels about its two principal axes, and a momentum wheel about any one of the two given axes, can be manoeuvred about all the three axes.

12.2.2 Control-Moment Gyro

A control-moment gyro (CMG) spinning at a constant rate produces a torque normal to its spin axis by deflecting it relative to the spacecraft.

A CMG mounted in such a way that its spin axis is free to rotate about both the normal axes is termed a *fully-gimballed gyroscope*, and the arrangement that allows it to rotate freely about the spacecraft is called *gimballing*. Gimballing can be carried out either using mechanical rotor supports which are hinged about the three principal axes of the spacecraft (called *gimbals*), or by a magnetic suspension. Of these, the former is more commonly employed. Motors are used to apply the necessary torques normal to the CMG rotor's spin axis relative to the spacecraft, in order to tilt the rotor's axis in a desired manner, thereby controlling the motion of the spacecraft.

Consider a CMG which can tilt its spin axis about only one *gimbal axis*, $\mathbf{k}$, which is inclined at a constant angle, σ, with respect to the spacecraft principal axis, $\mathbf{i}_3$, as shown in Fig. 12.4. If the angle $\alpha(t)$ is the deflection angle of the rotor axis measured from the spacecraft axis, $\mathbf{i}_1$ (Fig. 12.4), then the orientation of the gimbal axis is given by the Euler-angle sequence, $(\sigma)_1, (\alpha)_3$ as follows:

$$\mathbf{k} = -\sin\sigma\mathbf{i}_2 + \cos\sigma\mathbf{i}_3 \,. \tag{12.31}$$

The CMG can be gimballed at a rate, $\dot{\alpha}$, relative to the spacecraft, thereby generating a control torque given by

$$\mathbf{M} = \dot{\alpha}\mathbf{k} \times H_r(\mathbf{i}_1 \cos\alpha + \mathbf{i}_2' \sin\alpha) \,, \tag{12.32}$$

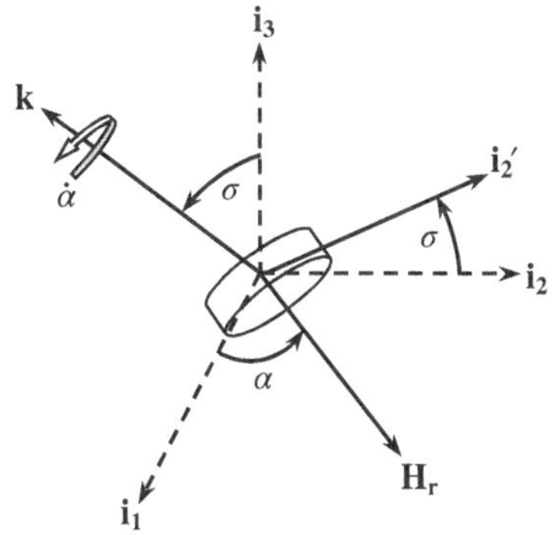

Figure 12.4 A control-moment gyro with spin angular momentum, $\mathbf{H}_r$, and gimbal axis, $\mathbf{k}$, inclined at a constant angle, σ, with respect to the spacecraft's body axis, $\mathbf{i}_3$.

where

$$\mathbf{i}_2' = \cos\sigma\mathbf{i}_2 + \sin\sigma\mathbf{i}_3 . \tag{12.33}$$

The spin rate (hence H_r) of the CMG is usually chosen to be large enough to require only small deflections for a given control torque. Therefore, with the small α assumption, we have

$$\mathbf{M} \simeq \dot{\alpha}\mathbf{k} \times H_r(\mathbf{i}_1 + \alpha\cos\sigma\mathbf{i}_2 + \alpha\sin\sigma\mathbf{i}_3) = H_r\dot{\alpha}(-\alpha\mathbf{i}_1 + \cos\sigma\mathbf{i}_2 + \sin\sigma\mathbf{i}_3) . \tag{12.34}$$

Hence, a single CMG is capable of affecting rotations about all spacecraft axes. By appropriately choosing the constant angle σ, one can select the control effectiveness of the CMG rotor about the various axes. The gimbal speed, $\dot{\alpha}$, and the deflection angle, α, are the variables of the CMG actuator. A desired magnitude of roll, pitch, and yaw control moments can thus be generated by the following non-linear relationships:

$$\begin{aligned} \dot{H}_x &= -H_r\alpha\dot{\alpha} \\ \dot{H}_y &= H_r\dot{\alpha}\cos\sigma \\ \dot{H}_z &= H_r\dot{\alpha}\sin\sigma . \end{aligned} \tag{12.35}$$

For a linear servomotor driving the CMG gimbal, the gimbal rate can be treated as a step input, $\dot{\alpha} = a\, u_s(t)$, which results in a ramp deflection $\alpha(t) = at\, u_s(t)$. Since the gimbal rate is an order of magnitude larger than the spacecraft's angular velocity, such an approximation is usually valid.

12.2.3 Variable-Speed Control-Moment Gyro

A fully gimballed CMG rotor with a variable spin rate is called a *variable-speed control-moment gyro* (VSCMG). This is a rotor in which both the angles, α and σ, as well as the spin rate (thus the angular momentum magnitude, H_r) (Fig. 12.4) can be varied as desired. Due to its variable speed, a single VSCMG offers a greater controllability of all the spacecraft axes when compared to a fully gimballed CMG. Let $\mathbf{M_r}$ be the motor torque applied on the VSCMG rotor. Then the equations of motion of the rotor relative to the spacecraft are expressed as follows:

$$\mathbf{M_r} = \mathbf{J_r}\frac{\partial\boldsymbol{\omega}_r}{\partial t} + \mathbf{S}(\boldsymbol{\omega}_r)\mathbf{J_r}\boldsymbol{\omega}_r , \tag{12.36}$$

where $\mathbf{S}(\boldsymbol{\omega}_r)$ is the skew-symmetric matrix multiplication form of the vector product, $\boldsymbol{\omega}_r \times (.)$, given by Eq. (11.19).

The motion of the spacecraft is described by the kinetic equations, Eq. (12.27), and the kinematic equations representing the attitude. Since the instantaneous attitude of the spacecraft's principal axes can be arbitrary, the non-singular quaternion representation, $(\mathbf{q}, q_4)$ (Chapter 5) is more appropriate in this case than the Euler angles. The attitude kinematics of the spacecraft in terms of the quaternion are given by Tewari (2006):

$$\frac{\mathrm{d}\{\mathbf{q}, q_4\}^T}{\mathrm{d}t} = \frac{1}{2}\boldsymbol{\Omega}\{\mathbf{q}(t), q_4(t)\}^T , \tag{12.37}$$

where $\boldsymbol{\Omega}$ is the following skew-symmetric matrix of the spacecraft's angular velocity components:

$$\boldsymbol{\Omega} = \begin{pmatrix} 0 & \omega_z & -\omega_y & \omega_x \\ -\omega_z & 0 & \omega_x & \omega_y \\ \omega_y & -\omega_x & 0 & \omega_z \\ -\omega_x & -\omega_y & -\omega_z & 0 \end{pmatrix}. \quad (12.38)$$

For the general simulation of an attitude manoeuvre, Eqs. (12.27), (12.36), and (12.37) must be integrated in time, with the given initial conditions, $\boldsymbol{\omega}(0), \omega_r(0)$, and $\mathbf{q}(0), q_4(0)$, and a prescribed motor torque profile, $\mathbf{M}_\mathrm{r}(t)$. In addition, the rotor's inertia tensor, $\mathbf{J}_\mathrm{r}$, which depends on the orientation of the rotor relative to the spacecraft, must be known at the beginning of the manoeuvre.

Exercises

1. A non-spinning, rigid spacecraft with principal moments of inertia, $J_{xx} = J_{yy} = 1000$ kg.m^2 and $J_{zz} = 250$ kg.m^2, needs to be re-oriented by rotating it about the principal axis, ox, by angle 180°. This is achieved by firing a pair of rocket thrusters capable of generating an impulsive torque, $M_x = 10\delta t$ N.m. Calculate the shortest time required for the manoeuvre. (*Ans*. 314.15 s.)

2. A spacecraft with the principal moments of inertia J_{xx}, J_{yy}, J_{zz} is equipped with a pair of rocket thrusters, which can apply a torque impulse of $\pm\Delta H$ about the principal axis, ox, and another pair which can apply the same impulse about the principal axis, oz. Devise a multi-impulse manoeuvre to achieve a final orientation of the spacecraft's principal frame, which is represented by the following rotation matrix relative to its initial orientation:

$$\mathbf{C} = \begin{pmatrix} \cos\psi & \sin\psi & 0 \\ -\sin\psi\cos\theta & \cos\psi\cos\theta & \sin\theta \\ \sin\psi\sin\theta & -\cos\psi\sin\theta & \cos\theta \end{pmatrix}.$$

   Find the expression for the total time for the manoeuvre in terms of ψ, θ, ΔH, J_{xx}, J_{yy}, J_{zz}.

3. A GEO satellite with principal moments of inertia $J_{xx} = J_{yy} = 10$ kg.m^2 and $J_{zz} = 2$ kg.m^2 is spinning with an angular velocity, $\omega_z = 1$ r.p.m. It is equipped with a pair of rocket thrusters mounted normal to the spin axis, which are capable of exerting a moment of 0.1 N.m. about the centre of mass. Estimate the total time required (including the duration for rocket firings) to change the spin axis of the spacecraft by 30°.

4. An axisymmetric spacecraft ($J_{xx} = J_{yy}$) spinning at a constant rate, $\omega_z = n$, about the principal axis, oz, is equipped with a pair of rocket thrusters, which can apply a torque impulse of $\Delta H = J_{zz}n/100$ about the principal axis, ox. What is the maximum possible deviation, β, of the spacecraft's spin axis in a two-impulse manoeuvre, and what is the total time required for such a change?

5. An axisymmetric, non-spinning spacecraft ($J_{xx} = J_{yy}$) has a small rotor of moment of inertia, J_r, about its spin axis, which is spinning about the spacecraft's axis of symmetry at a rate, ω_r. Determine the angular velocity achieved by the spacecraft if:

(a) The rotor's axis is turned by a 45° angle relative to the axis of symmetry.
(b) The rotor is spun up to twice the original spin rate.

6. For a satellite in a circular orbit, show that the orientation of the principal body axes, $(\mathbf{i}, \mathbf{j}, \mathbf{k})$ – such that the axis, $\mathbf{i}$, points towards the central body, and the axis, $\mathbf{j}$, is tangential to the orbit – is an equilibrium condition for the satellite's rotational kinetics.

7. Derive the governing equations of motion for a rigid, asymmetric spacecraft equipped with two identical reaction wheels, having their spin axes along the major and minor axes of the spacecraft, respectively.

8. Can a rigid, symmetric ($J_{xx} = J_{yy} = J_{zz}$) spacecraft, initially spinning about the principal axis, oy, at a constant rate, n, be controlled with separate reaction wheels about the principal axes, oy and oz, respectively?

9. A rigid, axisymmetric spacecraft with $J_{xx} = J_{yy}$ and $J_{zz} = 2J_{yy}$, initially spinning about the pitch principal axis, oy, at a constant rate, n, has a momentum wheel mounted about the axis, oy, with $H^e_{Ry} = 0.01J_{yy}n$. In addition, the spacecraft has reaction wheels mounted about the axes, ox and oy. Derive the equations of rotational kinetics of the spacecraft, and determine whether the angular velocity component, ω_z, can be controlled with the reaction wheels.

10. Consider the single-axis, gyro-stabilized platform shown in Fig. 12.5 with the spacecraft principal body-fixed frame, $(\mathbf{i}, \mathbf{j}, \mathbf{k})$. The bearing frictional torque in the gimbal and platform axes is modeled by

$$M_g = -c_g\dot{\theta}\,; \qquad M_p = -c_p\dot{\theta}\,,$$

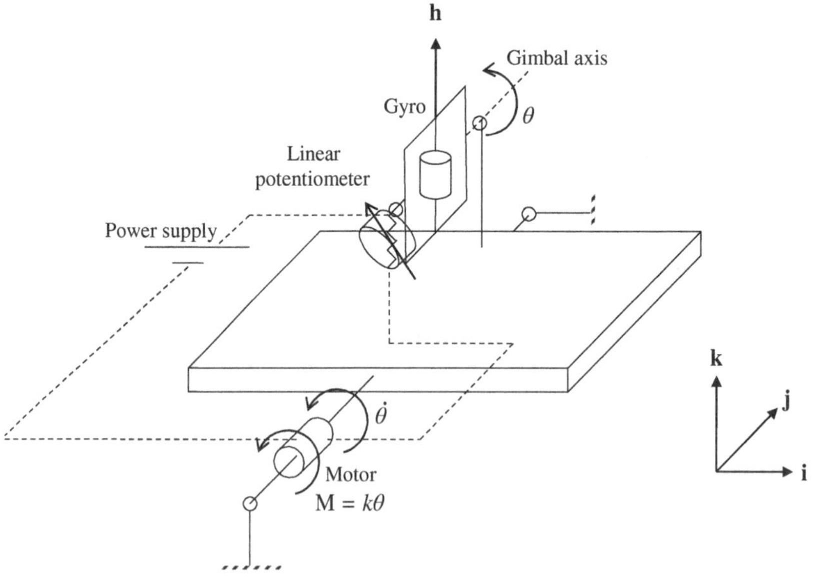

Figure 12.5 A single-axis, gyro-stabilized platform.

respectively, where c_g and c_p are linear damping constants. The moments of inertia of the gimbal and the platform about their respective axes are J and J_p, respectively. A linear potentiometer mounted on the gimbal axis is used to drive a DC motor via the electrical circuit shown in the figure, such that the torque, M, applied by the motor is directly proportional to the gimbal displacement, θ:

$$M = k\theta\,,$$

with k being a constant amplifier gain. Derive the equations of motion of the gimbal and platform.

References

Tewari A 2006. *Atmospheric and Space Flight Dynamics*. Birkhäuser, Boston.
Tewari A 2011. *Advanced Control of Aircraft, Spacecraft and Rockets*. Wiley, Chichester.

A

Numerical Solution of Ordinary Differential Equations

A general set of n first-order, ordinary differential equations is expressed in terms of an n-dimensional real vector, $\mathbf{y}(t)$, which varies with time, t, according to

$$\frac{d\mathbf{y}}{dt} = \mathbf{f}(t, \mathbf{y}(t)), \tag{A.1}$$

with the initial condition at $t = 0$ given by

$$\mathbf{y}(0) = \mathbf{y_0}. \tag{A.2}$$

The non-linear function, $\mathbf{f}(.)$, is assumed to possesses continuous partial derivatives with time up to an indefinite order, and satisfies the following *Lipschitz condition*:

$$\mid \mathbf{f}(t, \mathbf{y_1}) - \mathbf{f}(t, \mathbf{y_2}) \mid \leq c \mid \mathbf{y_1} - \mathbf{y_2} \mid, \tag{A.3}$$

for a real constant c, where $\mid \mathbf{y} \mid = \sqrt{\sum_{i=1}^{n} y_i^2}$, $i = 1, 2, \ldots, n$, is the Euclidean norm of $\mathbf{y}$. Then the system described by Eq. (A.1) has a unique solution called the system's trajectory, $\mathbf{y}(t)$, $t \geq 0$, beginning from the initial condition, Eq. (A.2). Finding a system's trajectory often requires numerical approximation. The approximate numerical solution to Eq. (A.1) involves a series expansion of the solution around the given initial condition, and can be divided into various categories, depending upon the determination of the coefficients of the series.

A.1 Fixed-Step Runge-Kutta Algorithms

A Runge-Kutta algorithm of a fixed integration time step h, order p, and s stages approximates the solution by the following truncated series (Atkinson, 2001):

$$\mathbf{y}(h) = \mathbf{y_0} + h \sum_{k=0}^{s-1} a_k \mathbf{f_k} + \mathcal{O}(h^{p+1}), \tag{A.4}$$

Foundations of Space Dynamics, First Edition. Ashish Tewari.

where the neglected part of the series, $\mathcal{O}(h^{p+1})$, is called the *truncation error*, Δ_{TE}. The term "stages" refers to the number of evaluations of the function, $\mathbf{f}(.)$, required at each time step, apart from that at the initial time ($t = 0$):

$$\mathbf{f_0} = \mathbf{f}(0, \mathbf{y_0})$$
$$\mathbf{f_k} = \mathbf{f}\left(b_k h, \mathbf{y_0} + h \sum_{i=0}^{k-1} c_{ki}\mathbf{f_i}\right) \quad (k = 1, 2, \dots, s-1). \tag{A.5}$$

The coefficients, a_k, b_k, c_{ki}, are chosen such that the solution is identical to that of a Taylor series approximation of the same order, p, given by:

$$\mathbf{y}(h) = \mathbf{y_0} + h \sum_{k=1}^{p} \frac{h^k}{k!} \frac{\partial^{k-1}\mathbf{f}}{\partial t^{k-1}}\bigg|_{t_0,\mathbf{y_0}} . \tag{A.6}$$

The magnitude of truncation error is estimated by the Euclidean norm of the largest neglected term in the series, and must be less than a given constant, $\delta > 0$, called the *tolerance*:

$$| \mathcal{O}(h^{p+1}) | = \left| \frac{h^{(p+1)}}{(p+1)!} \frac{\partial^p \mathbf{f}}{\partial t^p} \right|_{t_0,\mathbf{y_0}} \leq \delta . \tag{A.7}$$

The determination of the unknown coefficients, a_k, b_k, c_{ki}, by comparison with the equivalent Taylor series, involves *constraint equations*, which are typically smaller in number than the number of unknowns. Therefore, some of the unknown coefficients are chosen arbitrarily, and hence are termed the *free parameters*. As the order of the method increases, the number of free parameters also increases. The fixed-step algorithms suffer from a large order, p, required for a given tolerance, δ, which results in a large number of constraint equations per time step.

A.2 Variable-Step Runge-Kutta Algorithms

To improve the efficiency of Runge-Kutta algorithms, the time step size, h, is made variable in Eq. (A.4), such that the accuracy of the next higher-order algorithm is achieved at each step. The higher-order solution, $\hat{\mathbf{y}}$, of stage r is given by

$$\hat{\mathbf{y}}(h) = \mathbf{y_0} + h \sum_{k=0}^{r-1} \hat{a}_k \mathbf{f_k} + \mathcal{O}(h^{p+2}), \tag{A.8}$$

where the functional evaluations are now carried out as follows:

$$\mathbf{f_0} = \mathbf{f}(0, \mathbf{y_0})$$
$$\mathbf{f_k} = \mathbf{f}\left(b_k h, \mathbf{y_0} + h \sum_{i=0}^{k-1} c_{ki}\mathbf{f_i}\right) ; \qquad (k = 1, 2, \dots, m-1). \tag{A.9}$$

The number of stages m in the solution is the higher of s and r ($m = \max(s, r)$). Note that the coefficients b_k, c_{ki} remain the same for the two solutions of adjacent order. The time step is chosen such that the truncation error, which is defined to be the difference between the adjacent order solutions,

$$\Delta_{TE} = \mathbf{y}(h) - \hat{\mathbf{y}}(h) = h \sum_{k=0}^{r-1} (a_k - \hat{a}_k)\mathbf{f_k} + \mathcal{O}(h^{p+2}), \tag{A.10}$$

remains bounded by a specified tolerance, $|\Delta_{TE}| \leq \delta$. By expressing Eq. (A.10) as follows:

$$\Delta_{TE} = Kh^{p+1}, \tag{A.11}$$

one can select the time step size of the next step, h', from the specified tolerance as follows:

$$h' = h\left(\frac{\delta}{\Delta_{TE}}\right)^{\frac{1}{p+1}}; \qquad \Delta_{TE} \leq \delta, \tag{A.12}$$

assuming K remains constant over the next step. In this manner, an accuracy of order $p+1$ is achieved with a method of order p, albeit with an increased number of stages. Therefore, the variable-step Runge-Kutta algorithms are also referred to as Runge-Kutta methods of order $p(p+1)$.

As an example of the Runge-Kutta method, consider the Runge-Kutta 4(5) algorithm (Fehlberg, 1969), implemented in the MATLAB® intrinsic function *ode45.m*, which has the coefficients tabulated in Table A.1, and leads to the following expression for the truncation error:

$$\Delta_{TE} = h\left(-\frac{1}{360}\mathbf{f_0} + \frac{128}{4275}\mathbf{f_2} + \frac{2197}{75240}\mathbf{f_3} - \frac{1}{50}\mathbf{f_4} - \frac{2}{55}\mathbf{f_5}\right). \tag{A.13}$$

The coefficients in Table A.1 are by no means unique, since they depend upon the choice of the free parameter values. Fehlberg (1969) gives two sets of values for these coefficients. Expressing the coefficients in fractional form makes them independent of the machine round-off errors.

Apart from the Runge-Kutta methods, there are other choices available for low-order integration algorithms:

(a) The *finite-difference* (or Euler's) methods where approximate values are prescribed for the solution at a number of grid points, at which the solution is propagated in time using the values at previous times. These methods, while simple to implement, are computationally inefficient.
(b) The *multi-step* explicit and implicit algorithms, such as *Adams*, *Adams-Bashforth*, and *Adams-Moulton* algorithms (Atkinson, 2001).
(c) The *predictor-corrector* methods, such as those by *Milne* and Shampine-Gordon (Atkinson, 2001).

The multi-step and predictor-corrector methods require sophisticated programming, as their dependence on starting estimates and step sizes may cause convergence and numerical stability problems. In comparison, Runge-Kutta methods are much simpler to programme, mainly because their solution begins from a known initial condition, and their truncation error is easily controlled in a straightforward manner by a variable step size. For solving a set of differential equations with a large difference in the time scales (called *stiff equations*) certain implicit multi-step algorithms have been especially adapted, such as the *ode23tb* and *ode23s* algorithms of MATLAB®.

Table A.1 Coefficients of the Runge-Kutta 4(5) algorithm

k:	0	1	2	3	4	5
a_k:	$\frac{25}{216}$	0	$\frac{1408}{2565}$	$\frac{2197}{4104}$	$-\frac{1}{5}$	—
$\hat{a}_k$:	$\frac{16}{135}$	0	$\frac{6656}{12825}$	$\frac{28561}{56430}$	$-\frac{9}{50}$	$\frac{2}{55}$
b_k:	0	$\frac{1}{4}$	$\frac{3}{8}$	$\frac{12}{13}$	1	$\frac{1}{2}$
c_{k0}:	0	$\frac{1}{4}$	$\frac{3}{32}$	$\frac{1932}{2197}$	$\frac{439}{216}$	$-\frac{8}{27}$
c_{k1}:	—	—	$\frac{9}{32}$	$-\frac{7200}{2197}$	−8	2
c_{k2}:	—	—	—	$\frac{7296}{2197}$	$\frac{3680}{513}$	$-\frac{3544}{2565}$
c_{k3}:	—	—	—	—	$-\frac{845}{4104}$	$\frac{1859}{4104}$
c_{k4}:	—	—	—	—	—	$-\frac{11}{40}$

A.3 Runge-Kutta-Nyström Algorithms

When solving a certain class of astronautical problems, such as those involving Cowell's and Encke's formulations for lunar and interplanetary travel, the low-order time-integration methods given in Section A.2 prove unsuitable, as they result in an accumulation of truncation errors over the long times of flight. For such problems, the equations of motion can be written in the following form:

$$\frac{\mathrm{d}\mathbf{x}}{\mathrm{d}t} = \mathbf{y}(t)$$
$$\frac{\mathrm{d}\mathbf{y}}{\mathrm{d}t} = \mathbf{f}(t, \mathbf{x}), \tag{A.14}$$

with the initial condition

$$\mathbf{x}(0) = \mathbf{x_0};\ \ \mathbf{y}(0) = \mathbf{y_0}. \tag{A.15}$$

The *Runge-Kutta-Nyström* (RKN) method is suitable for integrating the set of implicit differential equations in Eq. (A.14) with a high order (thus small truncation error), and a relatively smaller number of stages compared with the traditional Runge-Kutta algorithm of the same order. The RKN solution of order p and stages s is expressed as follows:

$$\mathbf{x}(h) = \mathbf{x_0} + h\mathbf{y_0} + h^2 \sum_{k=0}^{s-1} a_k \mathbf{z_k} + \mathcal{O}(h^{p+1})$$
$$\mathbf{y}(h) = \mathbf{y_0} + h \sum_{k=0}^{s-1} b_k \mathbf{z_k} + \mathcal{O}(h^{p+1}), \tag{A.16}$$

where $\mathbf{z_k}$ refers to the following functional evaluations:

$$\mathbf{z_k} = \mathbf{f}\left(\alpha_k h, \mathbf{x_0} + \alpha_k h \mathbf{y_0} + h^2 \sum_{i=0}^{k-1} c_{ki}\mathbf{z_i}\right) ; \qquad (k = 1, 2, \ldots, m-1). \tag{A.17}$$

Although the number of stages for a given order is reduced in the RKN algorithms, the determination of coefficients, $a_k, b_k, \alpha_k, c_{ki}$, requires the solution of non-linear constraint equations.

Battin (1999) presents the procedure for solving the constraint equations for RKN algorithms of up to eighth order. For $p = 8$ we have $s = 8$, but 36 constraint equations and 5 free parameters. Consequently, a significant effort is necessary for evaluating the coefficients, which are then stored and utilized in the solution.

It is possible to achieve a still higher accuracy with the use of time-step control in a high-order RKN algorithm. In such a case, time-step control may consider minimizing the truncation error in either $\mathbf{x}$ only (Fehlberg, 1972), or both $\mathbf{x}, \mathbf{y}$ simultaneously (Bettis, 1973). In such applications with adjacent order solutions, the higher-order solution is used to control the step size, h, while the free parameters of the lower-order solution are retained for efficiency.

As an example of the RKN method, consider the following formulation for a sixth-order RKN algorithm with only five stages (Battin, 1999):

$$\begin{aligned}
\mathbf{x}(h) &= \mathbf{x_0} + h\mathbf{y_0} + \frac{1}{24}h^2[2\mathbf{z_0} + (5+\sqrt{5})\mathbf{z_2} + (5-\sqrt{5})\mathbf{z_3}] + \mathcal{O}(h^7) \\
\mathbf{y}(h) &= \mathbf{y_0} + \frac{1}{12}h(\mathbf{z_0} + 5\mathbf{z_2} + 5\mathbf{z_3} + \mathbf{z_4}) + \mathcal{O}(h^7),
\end{aligned} \tag{A.18}$$

where

$$\begin{aligned}
\mathbf{z_0} &= \mathbf{f}(0, \mathbf{x_0}) \\
\mathbf{z_1} &= \mathbf{f}\left[\frac{1}{20}(5-\sqrt{5})h, \mathbf{x_0} + \frac{1}{20}(5-\sqrt{5})h\mathbf{y_0} + \frac{1}{80}(3-\sqrt{5})h^2\mathbf{z_0}\right] \\
\mathbf{z_2} &= \mathbf{f}\left\{\frac{1}{10}(5-\sqrt{5})h, \mathbf{x_0} + \frac{1}{10}(5-\sqrt{5})h\mathbf{y_0}\right. \\
&\qquad \left. + \frac{1}{60}h^2[(3-\sqrt{5})\mathbf{z_0} + (6-2\sqrt{5})\mathbf{z_1}]\right\} \\
\mathbf{z_3} &= \mathbf{f}\left\{\frac{1}{10}(5+\sqrt{5})h, \mathbf{x_0} + \frac{1}{10}(5+\sqrt{5})h\mathbf{y_0}\right. \\
&\qquad \left. + \frac{1}{60}h^2[(6+2\sqrt{5})\mathbf{z_0} - (8+4\sqrt{5})\mathbf{z_1} + (11+5\sqrt{5})\mathbf{z_2}]\right\} \\
\mathbf{z_4} &= \mathbf{f}\left\{h, \mathbf{x_0} + h\mathbf{y_0} - \frac{1}{12}h^2[(3+\sqrt{5})\mathbf{z_0}\right. \\
&\qquad \left. - (2+6\sqrt{5})\mathbf{z_1} + (2+2\sqrt{5})\mathbf{z_2} - (9-3\sqrt{5})\mathbf{z_3}]\right\}.
\end{aligned} \tag{A.19}$$

This algorithm involves only four free parameters.

References

Atkinson KE 2001. *An Introduction to Numerical Analysis*. Wiley, New York.

Battin RH 1999. *An Introduction to the Mathematics and Methods of Astrodynamics*. American Institute of Aeronautics and Astronautics (AIAA) Education Series, Reston, VA.

Bettis D 1973. Runge-Kutta-Nyström algorithms. *Celestial Mechanics* **8**, 229–233.

Fehlberg E 1969. *Low Order Classical Runge-Kutta Formulas with Stepsize Control and Their Application to Some Heat Transfer Problems*. NASA TR R-315.

Fehlberg E 1972. *Classical Eighth- and Lower-Order Runge-Kutta-Nyström Formulas with Stepsize Control for Special Second-Order Differential Equations*. NASA TR R-381.

B

Jacobian Elliptic Functions

The solution to the torque-free Euler's equations (Chapter 11) are expressed in a closed form by the use of elliptic functions of Jacobi, which arise out of the inversion of the following incomplete elliptic integral of the first kind (Abramowitz and Stegun, 1974):

$$u = \int_0^{\Phi} \frac{\mathrm{d}\beta}{\sqrt{1 - m \sin^2\beta}} . \tag{B.1}$$

The angle Φ is termed the *amplitude* and the constant $0 \leq m \leq 1$ is the *parameter* of the elliptic functions. The *elliptic sine*, $\mathrm{sn}(u)$, is denoted in terms of the amplitude and the parameter by $\mathrm{sn}(\Phi; m)$, whereas the *elliptic cosine*, $\mathrm{cn}(u)$, is denoted $\mathrm{cn}(\Phi; m)$. The *delta amplitude*, $\mathrm{dn}(u)$, is denoted $\mathrm{dn}(\Phi; m)$. These functions are defined by the following:

$$\begin{aligned} \mathrm{cn}(\Phi; m) &= \cos\Phi \\ \mathrm{sn}(\Phi; m) &= \sin\Phi \end{aligned} \tag{B.2}$$

$$\mathrm{dn}(\Phi; m) = \sqrt{1 - m \sin^2\Phi} . \tag{B.3}$$

These definitions immediately give rise to the following properties:

$$\mathrm{cn}^2(\Phi; m) + \mathrm{sn}^2(\Phi; m) = 1 \tag{B.4}$$

$$\mathrm{dn}^2(\Phi; m) + m\ \mathrm{sn}^2(\Phi; m) = 1 . \tag{B.5}$$

The Jacobian elliptic functions are periodic with period $4K$, where $\Phi = \pi/2$ corresponds to $u = K$, the quarter period. In this sense, they can be considered to be a generalization of trigonometric functions. For $m = 0$, the Jacobian elliptic functions reduce to the trigonometric functions. The derivatives with respect to the amplitude, Φ, are the following:

$$\frac{\mathrm{d}}{\mathrm{d}\Phi} \mathrm{cn}(\Phi; m) = -\mathrm{sn}(\Phi; m)\mathrm{dn}(\Phi; m) \tag{B.6}$$

$$\frac{\mathrm{d}}{\mathrm{d}\Phi} \mathrm{sn}(\Phi; m) = \mathrm{cn}(\Phi; m)\mathrm{dn}(\Phi; m) , \tag{B.7}$$

Foundations of Space Dynamics, First Edition. Ashish Tewari.

$$\frac{\mathrm{d}}{\mathrm{d}\Phi}\,\mathrm{dn}(\Phi; m) = -m\,\mathrm{sn}(\Phi; m)\mathrm{cn}(\Phi; m)\,. \tag{B.8}$$

The reversal of the sign of the amplitude yields the following relationships:

$$\mathrm{cn}(-\Phi; m) = \mathrm{cn}(\Phi; m)\,, \tag{B.9}$$

$$\mathrm{sn}(-\Phi; m) = -\mathrm{sn}(\Phi; m) \tag{B.10}$$

$$\mathrm{dn}(-\Phi; m) = \mathrm{dn}(\Phi; m)\,. \tag{B.11}$$

The addition theorems satisfied by the elliptic functions are the following:

$$\mathrm{sn}(u+v; m) = \frac{\mathrm{sn}(u; m)\mathrm{cn}(v; m)\mathrm{dn}(v; m) + \mathrm{sn}(v; m)\mathrm{cn}(u; m)\mathrm{dn}(u; m)}{1 - m\,\mathrm{sn}^2(u; m)\mathrm{sn}^2(v; m)} \tag{B.12}$$

$$\mathrm{cn}(u+v; m) = \frac{\mathrm{cn}(u; m)\mathrm{cn}(v; m) - \mathrm{sn}(u; m)\mathrm{dn}(u; m)\mathrm{sn}(v; m)\mathrm{dn}(v; m)}{1 - m\,\mathrm{sn}^2(u; m)\mathrm{sn}^2(v; m)} \tag{B.13}$$

$$\mathrm{dn}(u+v; m) = \frac{\mathrm{dn}(u; m)\mathrm{dn}(v; m) - m\,\mathrm{sn}(u; m)\mathrm{cn}(u; m)\mathrm{sn}(v; m)\mathrm{cn}(v; m)}{1 - m\,\mathrm{sn}^2(u; m)\mathrm{sn}^2(v; m)}\,. \tag{B.14}$$

The following approximations can be applied when the parameter is very small compared to unity, $m \ll 1$, which allows an evaluation of the elliptic functions in terms of the trigonometric functions:

$$\mathrm{sn}(\Phi; m) \simeq \sin\Phi - \frac{1}{4}m(\Phi - \sin\Phi\cos\Phi)\cos\Phi \tag{B.15}$$

$$\mathrm{cn}(\Phi; m) \simeq \cos\Phi + \frac{1}{4}m(\Phi - \sin\Phi\cos\Phi)\sin\Phi \tag{B.16}$$

$$\mathrm{dn}(\Phi; m) \simeq 1 - \frac{1}{2}m\sin^2\Phi\,. \tag{B.17}$$

Several numerical algorithms are available for the computation of Jacobian elliptic functions, such as the functions *jacobiSN*, *jacobiCN*, and *jacobiDN* of MATLAB®.

Reference

Abramowitz M and Stegun IA 1974. *Handbook of Mathematical Functions*. Dover, New York.

Index

Foundations of Space Dynamics, First Edition. Ashish Tewari.